Vol. XXVII: **Condensation Monomers**
Edited by John K. Stille and Tod W. Campbell

Vol. XXVIII: **Allyl Compounds and Their Polymers** (Including Polyolefins)
By Calvin E. Schildknecht

MACROMOLECULES IN SOLUTION

HIGH POLYMERS

A SERIES OF MONOGRAPHS ON THE CHEMISTRY, PHYSICS, AND TECHNOLOGY OF HIGH POLYMERIC SUBSTANCES

VOLUME XXI

MACROMOLECULES IN SOLUTION

Second Edition

HERBERT MORAWETZ

Polytechnic Institute of New York
Brooklyn, New York

A WILEY-INTERSCIENCE PUBLICATION

JOHN WILEY & SONS, INC.

New York • London • Sydney • Toronto

Library of Congress Cataloging in Publication Data:

Morawetz, Herbert.
Macromolecules in solution.

(High polymers: a series of monographs on the chemistry, physics and technology of high polymeric substances; v. 21)
"A Wiley-Interscience publication."
Bibliography: p.
Includes index.
1. Macromolecules. 2. Solution (Chemistry)
I. Title. II. Series: High polymers; v. 21.

QD381.M66 1975 547′.1′2254 74–26655
ISBN 0–471–39321–5

Printed in the United States of America

10 9 8 7 6 5 4 3 2 1

To the memory of my father
Richard Morawetz
who first introduced me to the enjoyment of science

PREFACE

In submitting this book to the public, I should like to comment briefly on the motivations that guided me in the selection of the material and the manner of presentation. I have tried to impress the reader with the variety of phenomena which can be observed in solutions of macromolecules. This richness of the field has been my main inspiration. Not all the phenomena observed are clearly understood, and I have made a conscious effort to avoid the common tendency to ignore observations for which there is currently no satisfactory interpretation. As a result, it is hoped, the reader will not be left with the impression that all that remains to be done are minor additions to the work of the great masters, but rather will be stimulated to search for exciting new discoveries yet to be made.

Every author is faced with the difficult problem of defining for himself the audience for which he is writing. The first edition of this book developed from a set of lecture notes for a course in "Solution Properties of High Polymers" given at the Polytechnic Institute of Brooklyn (now the Polytechnic Institute of New York), and the graduate students attending this course influenced, without doubt, my image of the potential reader. It was also my ambition to break down the barriers that separate those working on synthetic polymers from those who investigate biological macromolecules and to present the study of solutions of macromolecules—whatever their origin—as a unified field. Since I aimed at a coverage of a broad range of phenomena within a book of modest size, I did not try to describe experimental methods and I restricted myself to an attempt to convey a qualitative comprehension of the manner in which theoretical results were obtained. However, the extensive bibliography should make it easy for the reader to find desired sources for a detailed description of experimental procedures or theoretical treatments. Something else has also been sacrificed, very reluctantly, to reduce the size of the book: on many occasions I felt tempted to follow up the development of a concept from its early beginnings, but it seemed wiser to concentrate on a description of the current "state of the art." Finally, practical considerations led me to the decision to emphasize literature references in English, although I fully realize that this has resulted in an inadequate acknowledgment of distinguished contributions published in other languages.

When I embarked on the preparation of a revised edition of a book published only 10 years ago, I had no idea how different the new manuscript would turn out to be in comparison to the original text. If almost half of the book is new, the changes are due to a variety of causes. In some cases, errors were corrected; in others, my understanding of phenomena has, it is hoped, improved over the years. Some subjects that now seem important were inexplicably omitted 10 years ago. Some areas have been almost miraculously transformed by recent research.

I was fortunate in having the complete text read by Professor James A. Moore of Rensselaer Polytechnic Institute, Professor Nan-Loh Yang of Richmond College of the City University of New York, and Professor Nancy Tooney of the Polytechnic Institute of New York. In addition, Professor G. Allegra of the Politecnico in Milan read Chapter III, Professor Sonja Krause of Rensselaer Polytechnic Institute read the first three chapters and Dr. Robert Ullman of the Ford Motor Company Research Laboratories read Chapter VI. To all these friends I feel greatly indebted—they drew my attention to a number of errors and made many helpful suggestions. Finally, I want to acknowledge my gratitude to Mrs. Dorothy Luyster, who again helped me, with infinite patience, by typing the revised text.

I have derived, over the years, an enormous amount of pleasure from my career as a teacher and research scientist. In undertaking the rather exacting task of writing this book, I was motivated in large part by a desire to repay a debt to the happy fortune that has allowed me to spend so many years in this way. I can only hope that the book reflects in some measure this guiding sentiment.

HERBERT MORAWETZ

Brooklyn, New York
August 1974

CONTENTS

GLOSSARY OF SYMBOLS

a	activity (persistence length in chap. III. and V)
a_1	semiaxis of revolution of ellipsoidal particle
a_2	equatorial radius of ellipsoidal particle
A_2, A_3	second, third virial coefficient
A_S	surface area
b	length of chain link
b_0	optical rotatory dispersion parameter
b_s	length of statistical chain element
c	concentration (g/cm^3)
d	diameter of rod-shaped particle
D	linear diffusion constant
D_r	rotatory diffusion constant
$\mathscr{D}$	dielectric constant
$\boldsymbol{e}$	charge of electron
E	energy
$\mathscr{E}$	electromotive force
f	frictional coefficient
f^*	functionality of branch point
F	Force
F	Helmholtz free energy
$\mathfrak{F}$	Faraday constant
g	gauche
$\mathbf{g}$	terrestrial gravitational constant
g^*	ratio of the radius of gyration in branched and linear chains
G	Gibbs free energy
h	elevation
h	distance between chain ends of flexible chain molecules
H	enthalpy
k	$4\pi \sin(\theta/2)/\lambda'$
k	reaction rate constant
$\boldsymbol{k}$	Boltzmann constant
l	optical path length
L	length of rod-shaped particlc, contour lcngth of chain molecule
L_f	hydrodynamic shielding length
m	mass of particle
m_s m_s'	molarity of uni-univalent salt

M	molecular weight
n	refractive index
n	number of moles
N	number of molecules
$\boldsymbol{N}$	Avogadro's number
p	axial ratio of ellipsoidal particles
p	vapor pressure
P	degree of polymerization
P	pressure
$\mathscr{P}$	factor describing the attenuation of scattered light owing to destructive interference
P_s	surface pressure
q	shear gradient
q_i	charge on the *i*th polymer chain segment
Q	charge of polyion
r	distance from center of molecular coil of distance from the axis of an ultracentrifuge
r_{ij}	distance between light scattering elements
R_s	radius of sphere
R_t, R_r, R_η	radius of hydrodynamically equivalent sphere
$\boldsymbol{R}$	gas constant
$\mathscr{R}$	Rayleigh ratio
s	radius of gyration
$\mathbf{s}$	sedimentation constant
S	entropy
t	trans
t	time
T	temperature (°K)
u	velocity
$\bar{u}$	electrophoretic mobility
U	molar excluded volume
v	volume of particle
$\bar{\mathrm{v}}$	specific volume
$\mathbf{v}$	reaction velocity
V_i	molar volume of species *i*
V_e	molar volume of hydrodynamically equivalent spheres
$\mathscr{W}$	probability distribution function
x	coordinate
x_i	mole fraction of species *i*
X	electrical field
y	coordinate
z	coordinate

$\mathbf{z}$	parameter describing the permeability of molecular coils
z^*	dissymmetry of light scattering
Z	number of chain segments
Z_s	number of statistical chain elements
α	polarizability
α_i	degree of ionization
α_e	linear expansion coefficient
$\boldsymbol{\alpha}$	ratio of shear gradient to rotary diffusion constant
η	viscosity
η_0	viscosity of solvent
$[\eta]$	intrinsic viscosity (cm^3/g)
θ	bond angle (Chapter III); scattering angle (Chapter V)
Θ	characteristic temperature for polymer-solvent system
κ	reciprocal of Debye-Hückel shielding length
λ	wavelength
λ'	λ/n
ν	number of charges
$\boldsymbol{\nu}$	frequency
ξ	extinction angle
Ξ	hydrodynamic permeability parameter
II	osmotic pressure
ρ	density
$\boldsymbol{\rho}$	charge density
σ^*	stereoregularity parameter
σ	standard deviation
τ	relaxation time
ϕ	volume fraction (Chapter II); internal angle of rotation (Chapter III)
φ	orientation angle
Φ'	constant relating the intrinsic viscosity to the molecular weight and the radius of gyration of chain molecules
χ	Flory-Huggins interaction parameter
ψ	internal angle of rotation (Chapter III); electrostatic potential (Chapter VII)
Ψ	Flory entropy parameter
Ω	element of solid angle
ω	angular velocity or frequency (radians/see)

MACROMOLECULES IN SOLUTION

Chapter I

GENERAL CONSIDERATIONS

About 50 years have now elapsed since it was clearly recognized that some important materials occurring in nature or synthesized in the. laboratory consist of very large molecules. Since then, the field of macromolecular chemistry has grown at a rate which seems spectacular even in our age of scientific and technological revolution. The driving force for this advance has been twofold. On the one hand, scientists learned to understand the relationship between the useful properties of such materials as cotton and Hevea rubber and what they came to call their "molecular architecture." This understanding proved invaluable in "designing" a wide variety of new materials such as synthetic fibers, synthetic rubbers, and plastics which duplicated and frequently surpassed the technically useful properties of materials available in nature. On the other hand, it was recognized that some of the most important constituents of living organisms are macromolecules, most notably perhaps, the proteins and nucleic acids. It also became increasingly apparent that an understanding of the relation between the molecular properties of these materials and their functions in the living cell is one of the most hopeful pathways of advance in biology. The discipline dealing with such relationships came to be called "molecular biology."

Unfortunately, the difference in the aims pursued by the "molecular architects" and the "molecular biologists" makes it inevitable that they should often go their separate ways, each group knowing relatively little about the activities of the other. This is but another example of the specialization brought about inevitably by the vastness of scientific literature. Yet, when an investigator attempts to answer questions concerning the size and shape of macromolecules, it may make little difference as to the methods he will employ whether the material to be investigated is a nucleic acid whose function is to transmit the genetic message in successive cell divisions, or a synthetic fiber to be used for making socks. Until a few years ago a dividing line appeared to exist, becausc of thc fact that macromolecules important in the life process usually have rather rigorously defined shapes, whereas synthetic polymers consist generally of flexible chain molecules whose shape, highly dependent on conditions, could be discussed only in statistical terms. However, the demonstration since then that the rigid structures characteristic of proteins and nucleic acids can be simulated by synthetic polypeptides and polynucleotides has led to a strengthening of the conceptual unity of the polymer field.

A. COMPARISON OF METHODS USED FOR THE STUDY OF SMALL AND LARGE MOLECULES

In classical physical chemistry, dealing with small molecules, the interpretation of measured properties in terms of the properties of the molecular constituents of a sample is accomplished most easily for very dilute gases and crystals. Investigations of the gaseous state have the advantage that samples can be easily dealt with under conditions where the effects of molecular interactions are small and can be taken care of by a suitable extrapolation procedure. The macroscopic properties of the gas may then be considered as made up additively from the properties of the individual molecules. At the other extreme, in the crystalline state the constituent molecules of a sample are arranged in an almost perfect three-dimensional order, and x-ray diffraction analysis methods may be used to define, except for thermal vibrations, the position of every atom in the crystal lattice. The theory of the liquid state is incomparably more difficult. Here the distances of molecules from each other are small, so that effects due to molecular interactions are dominant but are complicated by the absence of a long-range order. Nevertheless, the study of the liquid state, particularly of dilute solutions, is an important branch of classical physical chemistry. If the solution is so dilute that the solute molecules may be considered to be effectively separated from one another, the property of the individual solute molecule is independent of the number of such molecules in the system, and we may then study the molecular properties of the solute by comparing the macroscopic behavior of the solution and the pure solvent. Thus there is an analogy between the study of dilute gases and the study of solutes in dilute solutions. The properties of the solute molecule may, of course, be strongly affected by the nature of the solvent medium.

The basic principles used by physical chemists are equally applicable to large and small molecules. Nevertheless, several considerations make it convenient to deal with the physical chemistry of high polymers as a separate subject. We may subdivide these considerations into three categories. On the one hand, we find that many important techniques widely used in classical physical chemistry are not practical with macromolecules. Conversely, we shall see that a number of effects are too small to be conveniently measurable with small molecules, but become prominent and most useful when molecules are large. Finally, there is a significant qualitative difference in the questions to be answered by the physical chemist, depending on whether he concerns himself with substances composed of small molecules or with polymers.

1. Methods More Applicable to Small than to Large Molecules

Many of the methods most useful with small molecules are either experimentally inaccessible with high polymers or else have to be modified before they can be utilized. This limitation particularly concerns techniques used in the determination of molecular weights.

It is obvious that high polymers cannot be studied in the gaseous state. In Fig. I.1 the equilibrium vapor pressures at 100 °C of liquid normal paraffin hydrocarbons are plotted as a function of the number n of carbon atoms in

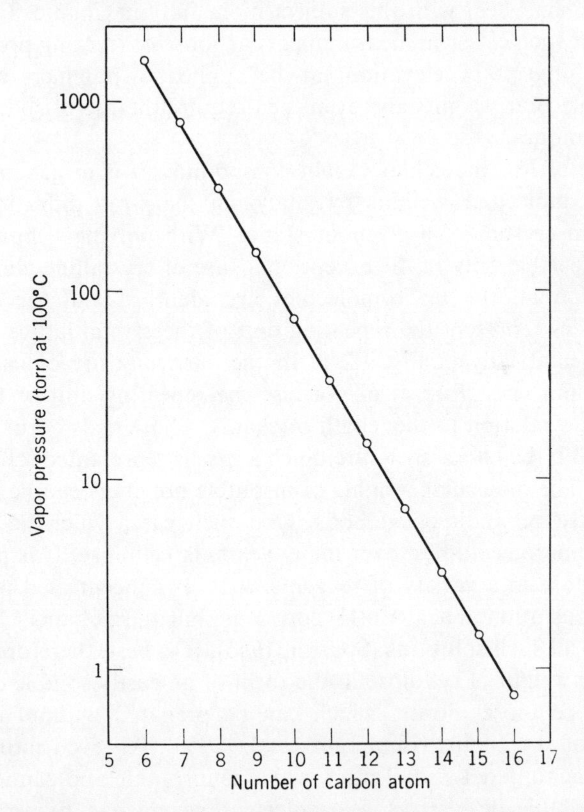

Fig. I.1. Dependence of equilibrium vapor pressure at 100°C on the chain length of normal alkanes.

the chain from hexane to hexadecane. The data correspond closely to the relation $\log p = 5.55 - 0.33n$, where the presure p is given in torr. On this basis, a normal paraffin with 100 carbon atoms would have a vapor pressure at 100°C of only 3.4×10^{-28} torr. This pressure corresponds to a single molecule contained in a volume of more than 10^8 liters. It is apparent that no conceivable technique could demonstrate the presence of a substance at such a dilution. For more polar compounds, the energy to separate a molecule of any given size from its neighbors would be larger and such minute vapor pressures would be attained at correspondingly shorter chain lengths.

When a substance is studied in solution, some characteristic effects, the so-called colligative properties, are proportional in the limit of high dilution to the number of molecules contained in the solute. As the size of the molecules increases, the number of solute molecules present at any solution concentration is reduced and the effects due to the colligative properties decrease. We shall deal with this subject more fully in Chapter IV and note here only that such classical measurements as those of freezing point depression and boiling point elevation can be applied to polymers up to only moderate molecular weights and even then require the use of special, highly sensitive techniques.

In the case of low molecular weight compounds, it is in general possible to determine molecular weights by analyzing the x-ray diffration pattern obtained from crystals of a given substance. With polymers, however, this approach is feasible only in the exceptional case of crystalline globular proteins, in which all the macromolecules are identical and the individual macromolecules represent the repeating unit of the crystal lattice. This case is shown schematically in Fig. I.2 *a*. In the vast majority of cases, either polymers do not crystallize at all, or else the repeating unit of the crystal lattice bears no relation to the length of chains, whose ends occur randomly dispersed within the lattice structure. Such a case is represented schematically by Fig. 12 *b*. The molecular weights of insoluble polymers can be estimated, therefore, only be indirect methods. One such case, which has been the subject of numerous studies over many years, is cellulose. It is possible to dissolve cellulose in a variety of reagents, such as concentrated phosphoric acid or aqueous ammoniacal Cu(II) solutions, but the polymer chains tend to degrade in all such solutions (Spurlin, 1955), It is best, therefore, to study the molecular weight of cellulose in the form of an easily soluble derivative, for instance, cellulose nitrate, which can be prepared without significant degradation of the chains (Immergut et al., 1953), An even more difficult problem is encountered with Teflon, which is insoluble and cannot be converted into soluble derivatives. Its molecular weight can be estimated by comparing the physical properties of a sample with those of calibration standards prepared with a radioactive chain initiator, so that the number of chain ends may be obtained from a determination of radioactivity (Berry and Peterson, 1951). However, the use of such indirect methods is generally undesirable and should be minimized as much as possible.

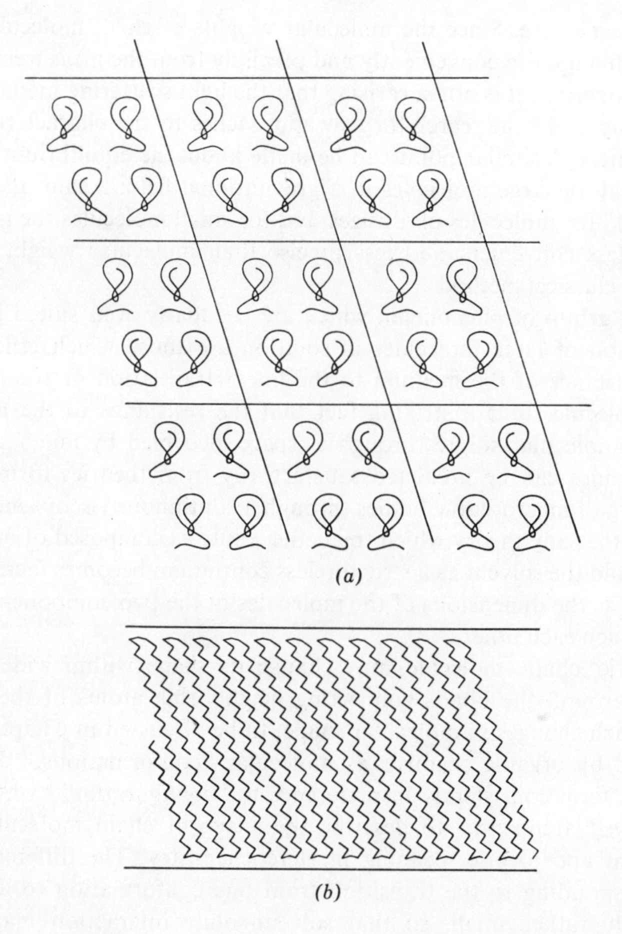

Fig. I.2. The nature of macromolecular crystals. (*a*) Schematic representation of a crystal of a typical globular protein. The unit cell contains several protein molecules. (*b*) Schematic representation of a crystal of a typical chain polymer. Chain ends are distributed at random within the crystal, and the repeat pattern bears no relation to the size of the molecules.

2. Methods Specifically Suited to Macromolecules

Just as some experimental techniques become inaccessible with increasing molecular size, other effects tend to become more prominent. For instance, the relation between the light scattering of a solution and the molecular weight of the solute, which we shall discuss in Chapter. V, is equally applicable to large and small solute molecules, but the effects to be expected in solution of low molecular weight solutes are small and can be studied only by using extreme care. Since the molecular weights of small molecules can be obtained much more conveniently and precisely from the measurement of colligative properties, it is not surprising that the light scattering method was developed only in the the search for new approaches to the characterization of high polymers. A similar point can be made about the equilibrium distribution of small or large molecules in a gravitational field. Again, the same principles hold for molecules of all sizes, but for small molecules the method is generally less convenient and less precise than molecular weight determinations by classical methods.

A different group of phenomena which are peculiarly well suited for the characterization of large molecules in solution are those which reflect the frictional resistance of the medium to the linear translation or rotation of the macromolecule. It is a striking fact that the resistance to the motion of the macromolecular solute through a space occupied by much smaller solvent molecules can be predicted satisfactorily from theories formulated for the motion of macroscopic bodies through a continuous viscous medium. It is obvious that any theory which treats the solute as composed of molecular particles and the solvent as a structureless continuum becomes increasingly unrealistic as the dimensions of the molecules of the two components of a system approach each other.

A polymeric chain molecule may change its shape within wide limits by rotation around the bonds connecting neighboring atoms of the chain backbone. Such changes in molecular shape, to be discussed in Chapter. III, are described by organic chemists as molecular "conformations," and we shall use this term consistently in preference to "configuration," which has been employed frequently to describe the shape of chain molecules but has a different and precise meaning in stereochemistry. The differences in energy corresponding to the transition from one conformation to another are frequently rather small, so that solvent-solute interaction may alter profoundly the probability distribution of conformations in which the solute can exist. Such effects have been observed in small molecules by techniques such as Raman and infrared spectroscopy (Mizushima, 1954; Oi and Coetzee, 1969). With a solute consisting of long-chain molecules we are able, in addition, to observe the overall expansion or contraction of the chains by following changes in the frictional properties of the solution. We shall also see that the characteristic dimensions of large molecules may be estimated from the angular distribution of scattered light.

The measurement of osmotic pressure occupies a peculiar intermediate position between the phenomena discussed in this section and those considered in section I.A1. On the one hand, as one of the colligative properties, osmotic pressure decreases, at constant solution concentration, with the size of the solute molecule, and the measurement of osmotic pressure for molecular weight determination becomes impracticable if the solute molecule is too large. On the other hand in classical physical chemistry osmotic pressure was "much talked about but little used"—and for good reason. The phenomenon depends on the existence of a semipermeable membrane which will transmit solvent but not solute, and such selectivity is hard to achieve without a considerable difference in the molecular dimensions of the two components of the system.

The very large deviations from ideal solution behavior observed in osmotic pressure measurements on polymer solutions historically constituted one of the serious obstacles to the acceptance of the existence of macromolecules. Later, when Staudinger proved that very large molecules did in fact exist, a need arose to explain on theoretical grounds why their solutions should be so nonideal. Qualitatively, this effect arises because bulk specimens of high polymers composed of flexible chain molecules must have the shapes of neighboring molecules correlated with one another, while this restriction is gradually eliminated when the system is diluted. It is interesting, in retrospect, that an analogous effect was never considered as a possible contributing cause of deviations from solution ideality in systems composed only of small molecules until data on macromolecular solutions became available. This is not surprising, however, since the effect remains rather small unless the chain length of the solute molecules becomes quite large.† Before the advent of polymer chemistry the theory of the excluded volume had been developed specifically for nonideal gases, in which the molecules may be treated as compact spheres or ellipsoids and there is no need to consider the behavior of flexible chains.

3. Motivation Characteristic of the Study of Macromolecules

It is, perhaps, not sufficiently emphasized that the questions which a physical chemist dealing with high polymers tries to answer are in general *qualitatively* different from the questions that concern a physical chemist who deals with substances composed of small molecules. It would be pointless, for instance, to study the properties of a solution containing a mixture of alcohols varying in chainlength, degree of branching, and the number of hydroxyls carried by a molecule. Since substances of low molecular weight can be obtained in a state of high purity, the study of their solutions is usually restricted to systems of only two components to facilitate interpretation of the experimental data. With high polymers, on the other hand, pure molecular species may be isolated only in a restricted group of naturally occurring proteins and nucleic acids. In the case of synthetic macromolecules and many macromolecular preparations derived from living organisms we

†Particularly if the larger species is the dilute component of the solution, as would normally be the case in the study of colligative properties.

have to deal with complex mixtures. It is true that such mixtures may be subdivided by fractional precipitation, fractional extraction, or other techniques, but even the sharpest fraction thus obtained is far from the ideal of "chemical purity" in the classical sense of consisting of a single molecular species. The task of the physical chemist is then twofold: not only does he want to *describe* and *explain* the properties of solutions of his complex solute, but also he is called upon to *characterize* the solute both in terms of the properties of its "average" molecule and in terms of the variability of these properties.

To obtain a feeling for the magnitude of the problem, let us consider a "simple" polymer, such as polystyrene, whose formula is written schematically as follows:

$$R-(-CH_2-\underset{\displaystyle C_6H_5}{\underset{|}{CH}}-)_P-R'$$

where R and R′ may be various end groups which will make little difference to the properties of the substance if the chain is very long. No means are as yet known by which high molecular weight polystyrene with a uniquely defined degree of polymerization P can be prepared, and precise characterization of a sample requires, therefore, specification of the distribution function of the molecular weight M. We define the distribution $N(M)$ so that

$$\int_{M_1}^{M_2} N(M)\,dM$$

is the fraction of chains with molecular weights in the interval $M_1 \leqslant M \leqslant M_2$. For unfractionated polystyrene samples, the form of $N(M)$ will depend on the polymerization conditions in a manner to be considered in the following section.

If branched chains can be formed during the polymerization, specification of the probability distribution of the various types of polymer molecules becomes quite unwieldy, unless we introduce certain assumptions, which may be justified by mechanistic and kinetic considerations of the polymerization reaction. For instance, we may assume that the number of branches is proportional to the length of the chain backbone and that the probability of chain termination at every stage of monomer addition is the same in the growing chain backbone and the growing branch. It is important to emphasize that such assumptions do not have general validity but depend on the precise manner in which the branched macromolecules were formed.

Yet another variable is of great importance in studies of synthetic polymers. In chain molecules produced by polymerizing any vinyl or unsymmetrically substituted vinylidene monomer $CH_2{=}CXY$, two stereoisomeric configurations are possible at every second carbon atom. Depending on the conditions of the polymerization, we may have a completely random sequence of the steric configurations of these centers, or we may produce chains in which long sequences of these centers have either the same or regularly alternating configurations. Such polymers, represented schematically below, are called atactic, isotactic, and syndiotactic.

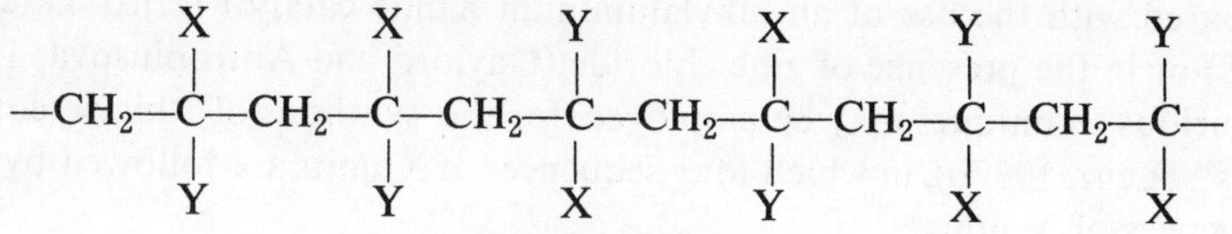

$$\begin{array}{ccccccccccccc}
 & X & & X & & Y & & X & & Y & & Y & \\
 & | & & | & & | & & | & & | & & | & \\
-CH_2- & C & -CH_2- & C & -CH_2- & C & -CH_2- & C & -CH_2- & C & -CH_2- & C & - \\
 & | & & | & & | & & | & & | & & | & \\
 & Y & & Y & & X & & Y & & X & & X &
\end{array}$$

Atactic chain

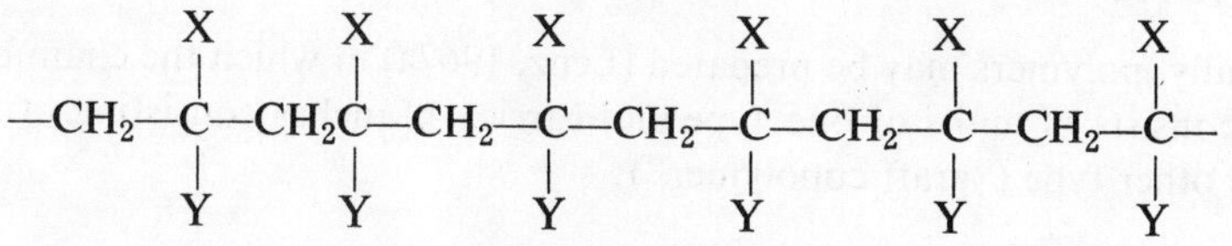

$$\begin{array}{ccccccccccccc}
 & X & & X & & X & & X & & X & & X & \\
 & | & & | & & | & & | & & | & & | & \\
-CH_2- & C & -CH_2 & C & -CH_2- & C & -CH_2- & C & -CH_2- & C & -CH_2- & C & - \\
 & | & & | & & | & & | & & | & & | & \\
 & Y & & Y & & Y & & Y & & Y & & Y &
\end{array}$$

Isotactic chain

$$\begin{array}{ccccccccccccc}
 & X & & Y & & X & & Y & & X & & Y & \\
 & | & & | & & | & & | & & | & & | & \\
-CH_2- & C & -CH_2- & C & -CH_2- & C & -CH_2- & C & -CH_2- & C & -CH_2- & C & - \\
 & | & & | & & | & & | & & | & & | & \\
 & Y & & X & & Y & & X & & Y & & X &
\end{array}$$

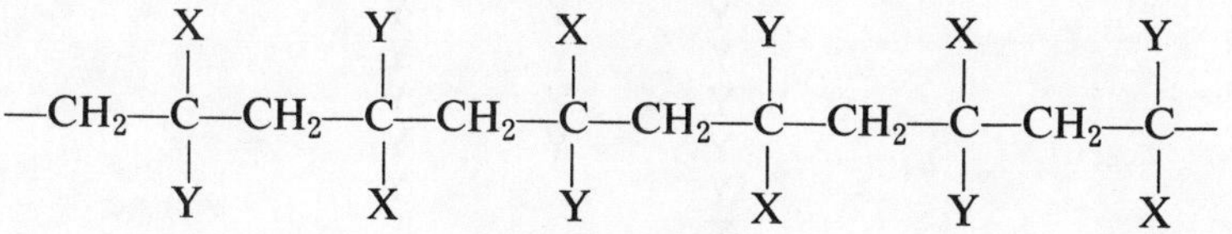

Syndiotactic chain

Complete randomness or complete order in the sequence of the asymmetric centers represents an extreme case, and polymers prepared under a given set of conditions may show a systematic bias toward the isotactic or syndiotactic structure without the appearance of long, ordered sequences. Special methods have been developed, therefore, for characterization of the "tacticity" of polymer chains.

The task of defining the nature of a sample becomes even more involved when we deal with copolymers. In that case, complete characterization requires, in addition to the factors discussed above, a description of the relative frequency of the monomeric units as well as their distribution in the macromolecule. For instance, a copolymer containing an equal number of two kinds of monomer residues, X and Y, may have them distributed at random:

X—Y—X—X—Y—X—Y—Y—Y—X—Y—X—X—Y—X—Y—X—Y—Y

or there may be a bias toward alternation of dissimilar monomer residues, leading in the extreme case to the regularly alternating structure:

X—Y—X—Y—X—Y—X—Y—X—Y—X—Y—X—Y—X—Y—X—Y

In some cases, it is possible to vary the tendency toward such alternation by the selection of the polymerization catalyst. For instance, copolymers of styrene with methyl acrylate, methyl methacrylate, or acrylonitrile containing 50 mole % of styrene residues have a much less ordered sequence of the comonomer residues when the copolymer is prepared by free-radical polymerization, compared to the regularly alternating structure if the polymer is prepared with the use of an alkylaluminum halide catalyst (Hirooka et al., 1968) or in the presence of zinc chloride (Gaylord and Antropiusova, 1969).

Various techniques may be employed for the synthesis of "block copolymers" (Lenz, 1967b), in which long sequences of X units are followed by long sequences of Y units:

X—X—X—X—Y—Y—Y—Y—Y—Y—Y—Y—Y—X—X—X—X—X

Finally, polymers may be prepared (Lenz, 1967a) in which the chain backbone consists of units of one type and carries branches consisting of units of the other type ("graft copolymer"):

```
XXXXXXXXXXXXXXXXX
   Y     Y     Y
   Y     Y     Y
   Y     Y     Y
   Y     Y     Y
   Y           Y
   Y
   Y
```

Moreover, just as a sample of homopolymer does not consist of identical molecules but should be described by distribution functions of chain lengths, degrees of branching, and stereoregularity, so the compositional variables of a copolymer will also cover a range described by probability distribution functions. This presents the polymer chemist with a particularly difficult problem, since fractionation procedures are generally more or less sensitive to all these variables. For instance, the solubility of polyethylene tends to increase with both increasing branching and decreasing molecular weight, or the solubility of cellulose acetate tends to increase with both increasing degree of acetylation and decreasing chain length. Methods which can be used to study compositional variation in isolation from chain length variation are particularly valuable. We shall see in Chapters IV and V how the study of ultracentrifugation and light scattering can contribute to the solution of these problems.

We have already touched on another class of molecular properties of special interest to the student of macromolecules—namely, the study of molecular shape. In a typical synthetic chain molecule whose "backbone" contains a sequence of a large number of single bonds, the possibility of rotation around these bonds enables the molecule to assume a large number of conformations. It is then obvious that we cannot deal with "a shape" of the molecular chain, but must consider a distribution function of such shapes or some particular average values. The average extension of a chain molecule in solution gives us an important insight into the relative energy of the available conformations and into interactions of the polymer with the solvent medium.

The aims of the physical chemist dealing with macromolecular solutions are somewhat different if he works with materials of biological origin, which may contain a single kind of macromolecule or a relatively small number of chemical species. Such materials may be highly complex chemically; for instance, proteins may consist of a sequence of up to 20 different amino acids constituting the building blocks of the chain. However, the molecules of a given species are uniquely defined, not only with respect to the exact sequence of these amino acid units along the chain, but even with respect to the conformation of the chain backbone. This precise definition of the molecular shape may be dealt with by the experimentalist on two levels. The historically older approach deals with the "overall shape" of the molecule, in which its intricate form is approximated by an "equivalent ellipsoid of revolution." This is the type of information obtainable, in principle, from a study of the frictional properties of the solutions of rigid macromolecules. On the other hand, methods are available for studying the formation of an ordered sequence of certain preferred conformations in the chain backbone. This is the subject discussed in Chapters III and V.

If the solution contains several well-defined macromolecular species we may wish to analyze the system without the need of separating the macromolecular components. This may be accomplished by the use of ultracentrifugation, electrophoresis, or gel permeation chromatography, techniques that we shall have occasion to discuss in later chapters of this book.

B. DISTRIBUTION FUNCTIONS OF CHAIN LENGTHS AND COMPOSITION

We saw in the preceding section that the characterization of a polymer sample may involve not only the evaluation of a number of parameters but also distribution functions for the values of these parameters. A rigorous solution of this problem for a given sample is extremely difficult if not impossible. On the other hand, the interpretation of data obtained from measurements of the solution properties of the sample may be greatly facilitated by knowledge of the manner in which the sample was prepared. With this additional information, we may use kinetic arguments to assume that we are dealing with certain *types of distributions* and our task is then simplified to the evaluation of a small number of parameters by which these distributions are characterized. We shall discuss first the nature and the origin of some simple distributions and consider later the causes for more complex distribution functions.

1. The "Most Probable Distribution" of Chain Lengths

A very frequently assumed distribution of chain lengths in a sample of linear macromolecules is commonly referred to as "the most probable distribution" (alternatively, as a "normal distribution"). It would be obtained if we joined all the monomer units in the system into a single linear chain and cut subsequently at random a fraction ϵ of the links. The fraction of chains containing P units will then be proportional to the probability that in proceeding from one cut to the next we shall find P monomer units linked together, while the next link is broken. This probability is

$$N(\mathrm{P}) = A\,(1 - \epsilon)^{\mathrm{P}-1}\epsilon \qquad \text{(I.1)}$$

For long average lengths of the chain, $\epsilon \ll 1$ and we may approximate (I.1) by a continuous distribution function:

$$N(\mathrm{P})\,d\mathrm{P} = A\epsilon e^{-\epsilon \mathrm{P}}\,d\mathrm{P} \qquad \text{(I.2)}$$

Since the normalization constant A is equal to unity,

$$N(\mathrm{P})d\mathrm{P} = \epsilon e^{-\epsilon \mathrm{P}}d\mathrm{P} \qquad \text{(I.3)}$$

This distribution function is applicable to several cases of practical importance. Consider, for instance, a typical polycondensation reaction of two bifunctional reagents, such as a polyesterification:

$$\mathrm{HOC(=O){-}(CH_2)_n{-}C(=O)OH + HO{-}(CH_2)_m{-}OH \rightarrow}$$
$$\mathrm{HOC(=O){-}(CH_2)_n{-}C(=O)O{-}(CH_2)_m{-}OH \rightarrow}$$
$$\mathrm{HO{-}\left[{-}C(=O){-}(CH_2)_n{-}C(=O)O{-}(CH_2)_m{-}O{-}\right]_{P/2}{-}OH} \qquad \text{(I.4)}$$

Since it has been shown that the reactivity of the hydroxyl or carboxyl groups does not depend on the length of the chain to which they are attached (Flory, 1939, 1940b), the reasoning leading to distribution function (I.3) will describe the product of a polyesterification. The parameter ϵ stands for the fraction of the total number of carboxyl and hydroxyl groups which have remained unaltered at a given stage of the reaction if the two reactive groups were originally present in equivalent amounts.

The same type of distribution may be obtained in polymerization processes in which a single monomer at a time is added to the growing chain, provided that the length of the chain is determined by a termination step, that this termination does not involve the coupling of two chains, and that the polymerization conditions remain constant during the preparation of the sample. Among numerous examples of such processes we may cite the following:

(a) The free-radical polymerization of a vinyl monomer such as styrene in the presence of sufficient concentrations of chain transfer agents (e.g., mercaptans) to ensure that chain transfer controls the length of the macromolecules:

$$R-(CH_2CH(C_6H_5))_n-CH_2\dot{C}H(C_6H_5) + CH(C_6H_5)=CH_2 \xrightarrow{k_p} R-(CH_2CH(C_6H_5))_{n+1}-CH_2\dot{C}H(C_6H_5) \quad (I.5a)$$

$$R-(CH_2CH(C_6H_5))_n-CH_2\dot{C}H(C_6H_5) + R'SH \xrightarrow{k_{tr}} R-(CH_2CH(C_6H_5)-)_n-CH_2CH_2(C_6H_5) + R'S^{\cdot} \quad (I.5b)$$

$$R'S^{\cdot} + CH_2=CH(C_6H_5) \xrightarrow{k_p} R'SCH_2\dot{C}H(C_6H_5) \quad (I.5c)$$

(b) The ionic polymerization of formaldehyde with water acting as chain transfer agent:

$$HO-(CH_2O)_n-CH_2O^- + CH_2O \xrightarrow{k_p} HO-(CH_2O)_{n+1}-CH_2O^- \quad (I.6a)$$

$$HO-(CH_2O)_n-CH_2O^- + H_2O \xrightarrow{k_{tr}} HO-(CH_2O)_n-CH_2OH + OH^- \quad (I.6b)$$

$$OH^- + CH_2O \xrightarrow{k_i} HOCH_2O^- \quad (I.6c)$$

For both processes (I.5) and (I.6), the parameter ϵ governing the chain length distribution function is given by

$$\epsilon = \frac{k_{tr}\,(\mathrm{S})}{k_{tr}\,(\mathrm{S}) + k_p\,(\mathrm{M})} \tag{I.7}$$

where (S) and (M) are concentrations of chain transfer agent and monomer, respectively.

(c) Ionic polymerizations in which ion recombination or proton elimination terminates the chain:

$$\underset{\underset{C_6H_5}{|}}{\mathrm{H{-}(CH_2CH)_n}}\mathrm{{-}CH_2}\underset{|\,C_6H_5}{\overset{+}{\mathrm{C}}}\mathrm{H\ldots X^-} + \mathrm{CH_2{=}}\underset{|\,C_6H_5}{\mathrm{CH}} \xrightarrow{k_p}$$

$$\mathrm{H{-}(CH_2\underset{|\,C_6H_5}{C}H)_{n+1}{-}CH_2\underset{|\,C_6H_5}{\overset{+}{C}}H\ldots X^-} \tag{I.8a}$$

$$\mathrm{H{-}(CH_2\underset{|\,C_6H_5}{C}H)_n{-}CH_2\underset{|\,C_6H_5}{\overset{+}{C}}H\ldots X^-} \begin{cases} \xrightarrow{k_t} \mathrm{H{-}(CH_2\underset{|\,C_6H_5}{C}H)_n{-}CH_2\underset{|\,C_6H_5}{C}HX} & (1.8b) \\ \xrightarrow{k'} \mathrm{H{-}(CH_2\underset{|\,C_6H_5}{C}H)_n{-}CH{=}\underset{|\,C_6H_5}{C}II + IIX} & (I.8c) \end{cases}$$

In this case ϵ is given by

$$\epsilon = \frac{k_t + k_t'}{(k_t + k_t') + k_p\,(\mathrm{M})} \tag{I.9}$$

(d) Free radical polymerization terminated by radical disproportionation:

$$\mathrm{R(CH_2\underset{|\,X}{C}H)_n{-}CH_2\underset{|\,X}{\dot{C}}H + R(CH_2\underset{|\,X}{C}H)_m{-}CH_2\underset{|\,X}{\dot{C}}H} \xrightarrow{k_t}$$

$$\mathrm{R(CH_2\underset{|\,X}{C}H)_nCH_2\underset{|\,X}{C}H_2 + R(CH_2\underset{|\,X}{C}H)_mCH{=}\underset{|\,X}{C}H} \tag{I.10}$$

For this case the characteristic parameter of the distribution function is

$$\epsilon = \frac{k_t(\mathrm{R}\cdot)}{k_p(\mathrm{M}) + k_t(\mathrm{R}\cdot)} \tag{I.11}$$

when (R·) is the radical concentration. If we assume steady-state conditions with the rate of chain initiation I equal to the chain termination rate, we can express (I.11) as

$$\epsilon = \frac{(Ik_t)^{1/2}}{k_p(\mathrm{M}) + (Ik_t)^{1/2}} \tag{I.12}$$

The most probable distribution is also obtained if a very long chain molecule is degraded under conditions such that all the links between the monomer residues have an equal probability of being broken. This condition is most likely to be met in various solvolytic degradations, although complications may arise from a slight difference between the reactivities of linkages close to the chain end and those in the interior of the chain (Freudenberg and Blomquist, 1935).

Finally, it should be pointed out that this type of chain length distribution corresponds to a state of maximum entropy with respect to the distribution of monomer units among a fixed number of macromolecules. Since the transfer of a monomer residue from one chain to a chain of different length does not involve any energy change, the "most probable distribution" of chain lengths is characteristic of a state of chemical equilibrium. This will be attained, for instance, if a polyester sample is heated with a transesterification catalyst or if a vinyl polymer is heated with a source of free radicals which will catalyze both polymerization and depolymerization.

2. Chain Length Distribution in Vinyl Polymerization Terminated by Radical Recombination

If free-radical-catalyzed vinyl polymerizations are carried out under conditions such that chain transfer is negligible, the chain termination may involve a combination of two chain radicals:

$$\mathrm{R(CH_2\underset{|}{\overset{}{C}}HX)_{P'}{-}CH_2\dot{C}HX + R(CH_2CHX)_{P-P'}{-}CH_2\dot{C}HX} \xrightarrow{k_t}$$

$$\mathrm{R(CH_2CHX)_{P'}CH_2CHXCHXCH_2(CHXCH_2)_{P-P'}R} \tag{I.13}$$

(In each formula, X is attached by a vertical bond beneath each CH.)

In this case a chain with a degree of polymerization P will form whenever two chains combine which contain P′ and P − P′ monomer units, respectively. The fraction of such chains will be, therefore,

$$N(\mathrm{P}) = A' \sum_{\mathrm{P'}=1}^{\mathrm{P'}=\mathrm{P}-1} (1-\epsilon)^{\mathrm{P'}}\,\epsilon\,(1-\epsilon)^{\mathrm{P}-\mathrm{P'}}\,\epsilon$$

$$= A' \sum_{\mathrm{P'}=1}^{\mathrm{P'}=\mathrm{P}-1} (1-\epsilon)^{\mathrm{P}}\epsilon^2 \tag{I.14}$$

where ϵ has the significance given in (I.12). Since all values of P′ are equally probable and the normalization constant $A' = 1$, the continuous distribution function becomes

$$N(\mathrm{P})\,d\mathrm{P} = \mathrm{P}\epsilon^2 e^{-\epsilon \mathrm{P}}\,d\mathrm{P} \tag{I.15}$$

It is qualitatively obvious that this distribution is narrower than the "most probable distribution." This may be seen most easily by comparing a free-radical polymerization in which chain termination takes place by recombination with one that occurs by the disproportionation of two chain radicals. If the recombination always involved two chains of equal length, the chain length would merely be doubled and the two distributions would be similar. Since, however, the chain combination may involve chains of different length, an averaging process is superimposed on the chain doubling and the chain length distribution is narrowed.

3. The Poisson Distribution

Much narrower chain length distributions are obtained if a fixed number of chains are allowed to add for the same period one monomer unit at a time in the absence of any chain termination process.† This case was first analyzed by Flory (1940a) for the polymerization of ethylene oxide. The methoxide-catalyzed reaction would now be represented by

$$CH_3O^- + \overset{O}{\overbrace{CH_2\text{—}CH_2}} \rightarrow CH_3OCH_2CH_2O^- \tag{I.16a}$$

$$CH_3(OCH_2CH_2)_n\text{—}O^- + \overset{O}{\overbrace{CH_2\text{—}CH_2}} \rightarrow CH_3(OCH_2CH_2)_{n+1}\text{—}O^- \tag{I.16b}$$

and it would be assumed that (I.16a) is at least as fast as the subsequent steps. This type of kinetics leads to a Poisson distribution of chain lengths:

$$N(\mathrm{P}) = \frac{e^{-\mathrm{r}}\mathrm{r}^{\mathrm{P}-1}}{(\mathrm{P}-1)!} \tag{I.17}$$

where r is the ratio of the number of molecules of monomer consumed to the number of growing polymer chains. For large values of r this distribution is extremely narrow, and a high polymer sample with this distribution would be indistinguishable, with the available experimental precision, from a unique

†The nature of this distribution may be visualized as follows. Let us select a time τ such that there is a 50% chance that a growing chain will add a unit. The fractional probability $N(\mathrm{P})$ that P units are added at time t is then equivalent to the probability of obtaining P "heads" in t/τ tosses of a true coin.

chemical species. Careful studies of the methoxide-initiated polymerization of ethylene oxide have confirmed the absence of a chain termination or chain transfer process (Gee et al., 1959) and the production of a polymer which seems to correspond closely to the narrow Poisson distribution of chain lengths (Weibull and Nycander, 1954; Wojtech, 1963).

Another case in which a stepwise addition of monomer to growing polymer chains can take place without chain termination is the polymerization of *N*-carboxy-α-amino acid anhydrides to polypeptides, for example:

$$RNH_2 + \underset{\substack{|\\ HN\diagdown_{\underset{\underset{O}{\|}}{C}}\diagup O}}{\overset{\overset{R'}{|}}{HC}}\!-\!\!-\!\!-CO \rightarrow RNH\underset{\underset{O}{\|}}{C}\!-\!\overset{\overset{R'}{|}}{C}HNH_2 + CO_2 \qquad (I.18a)$$

$$RNH(-\underset{\underset{O}{\|}}{C}\overset{\overset{R'}{|}}{C}H\underset{\underset{H}{|}}{N}-)_n-H + HN\!\!\!\!\overset{\overset{\overset{R'}{|}}{HC}\!-\!\!-\!\!-CO}{}\diagdown_{\underset{\underset{C}{\|}}{O}}\diagup O \rightarrow$$

$$RNH(-\underset{\underset{O}{\|}}{C}\overset{\overset{R'}{|}}{C}H\underset{\underset{H}{|}}{N}-)_{n+1}-H + CO_2 \qquad (I.18b)$$

Finally, anion-catalyzed vinyl polymerization can be carried out under conditions in which the chains do not terminate, and such processes may lead to a product with a narrow Poisson distribution (Waack et al., 1957). These processes are initiated by the product of an electron transfer from metallic alkali to an aromatic hydrocarbon D (such as naphthalene) in a suitable solvent medium:

$$Na + D \rightarrow Na^+ + D^{\overset{.}{-}} \qquad (I.19a)$$

On addition of a monomer such as styrene, the subsequent reaction steps are presumably

$$D^{\overset{.}{-}} + CH_2{=}\underset{\underset{C_6H_5}{|}}{CH} \rightleftharpoons D + \dot{C}H_2-\underset{\underset{C_6H_5}{|}}{\bar{C}H} \qquad (I.19b)$$

$$2\dot{C}H_2{-}\bar{C}H(C_6H_5) \rightarrow \bar{C}H(C_6H_5)CH_2CH_2\bar{C}H(C_6H_5) \qquad (I.19c)$$

followed by monomer addition to the dianion. In practice, a number of complications may arise in the kinetic scheme which will lead to a pronounced broadening of the molecular weight distribution as compared to the Poisson formula (I.17) .The factors which have to be controlled if such a broadening is to be avoided will be discussed in Section I.B.5.

The narrow Poisson distributions of molecular weights can also be obtained in a completely different type of polymerization. The enzyme phosphorylase catalyzes the reaction of glucose-1-phosphate with amylose, which results in the addition of an anhydroglucose unit to the end of the polymer chain:

$$\text{(glucose-1-)}OPO_3^{2-} + H{-}[O{-}C_6H_{10}O_4]_n{-}OH \longrightarrow HOPO_3^{2-} + \text{(glucose)}{-}[O{-}C_6H_{10}O_4]_{n+1}{-}OH$$

A systematic study of this reaction was first undertaken by Husemann et al. (1958). The polymerization requires initially the presence of an oligosaccharide to which the enzyme can attach itself, and Pfannemüller and Burchard (1969) have demonstrated that the use of the proper "primer" leads to the synthesis of amyloses of extremely narrow molecular weight distribution. The reaction is completely analogous to the anionic polymerizations discussed above in that the number of growing chains is fixed by the number of catalyst molecules, so that the chain length of the polymer produced in any given experiment is proportional to the polymer yield.

A very different mechanism for the production of polymers with an extremely narrow distribution of chain lengths is implied in a study of cellulose biosynthesis by Marx-Figini and Schulz (1966). Following very carefully the time dependence of the amount and the molecular weight of cellulose during the maturation of cotton fibers, they found that 90% of the cellulose is material with an extremely narrow distribution of the degree of polymerization ($\bar{P} = 14{,}000$) and that the chain length of this material remains constant during most of the time in which it is formed. This result suggests that the polymer is formed on some matrix which controls its molecular weight.

4. Molecular Weight Averages

When a procedure suitable for the determination of molecular weights is applied to a sample containing molecules of different sizes, it yields a value which we may call the "average molecular weight." The nature of the averaging process depends on the property being measured, and the magnitude of the average molecular weight will tend to increase with increasing sensitivity of the measured effects to the weight of a molecule.

In the determination of molecular weights by the colligative properties of solutions, each molecule, large or small, makes the same contribution to the observed effect. This effect then remains unchanged if the total weight is shared equally among the molecules of the system, a procedure leading to the *number-average* molecular weight $\bar{M}_n$ given by

$$\bar{M}_n = \int_0^\infty MN(M)\, dM \tag{I.20}$$

On the other hand, we may be interested in an effect which, for a given solution concentration, is proportional to the molecular weight of the solute. The effect obtained with a polydisperse solute will then depend on its *weight-average* molecular weight $\bar{M}_w$, defined by

$$\bar{M}_w = \int_0^\infty MW(M)\, dM \tag{I.21}$$

where the distribution function $W(M)$ is defined so that

$$\int_{M_1}^{M_2} W(M)\, dM$$

is the weight fraction of material with molecular weights in the range $M_1 \leqslant M \leqslant M_2$. Since $W(M)$ is related to $N(M)$ by

$$W(M)dM = \frac{MN(M)dM}{\int_0^\infty MN(M)dM} \tag{I.22}$$

we may express $\bar{M}_w$ as

$$\bar{M}_w = \frac{\int_0^\infty M^2 N(M)\, dM}{\int_0^\infty M N(M)\, dM} \tag{I.23}$$

A comparison of (I.23) with (I.20) shows that the large species are weighted more heavily in the weight average than in the number average, so that for any polydisperse sample $\bar{M}_w > \bar{M}_n$. In the absence of precise information on the molecular weight distribution function, the ratio $\bar{M}_w/\bar{M}_n$ is frequently used to characterize the "polydispersity" of a sample. In Germany it has become customary to use the parameter U (*Uneinheitlichkeit*), defined by $U = (\bar{M}_w/\bar{M}_n) - 1$.†

Another average which plays a role in the interpretation of the behavior of polymer solutions is the so-called "*z*-average," defined by

$$\bar{M}_z = \frac{\int_0^\infty M^3 N(M)\, dM}{\int_0^\infty M^2 N(M)\, dM} \tag{I.24}$$

It is influenced by the high molecular weight species even more than the weight average.

Whenever we are justified in assuming that the molecular weight distribution has a form derived from the simple kinetic patterns discussed in sections I.B.1, B.2, and B.3, a single molecular weight average will define the complete molecular weight distribution function. For the "most probable distribution" the molecular weight averages are related to the parameter ϵ and to M_0, the molecular weight of the monomer, by

$$\epsilon = \frac{M_0}{\bar{M}_n} = \frac{2M_0}{\bar{M}_w} = \frac{3M_0}{\bar{M}_z} \tag{I.25}$$

(most probable distribution)

For radical polymerization with termination by recombination of chain radicals we have

$$\epsilon = \frac{2M_0}{\bar{M}_n} = \frac{3M_0}{\bar{M}_w} = \frac{4M_0}{\bar{M}_z} \tag{I.26}$$

(termination by recombination)

†As will be shown in Chapters IV and V, $\bar{M}_w$ and $\bar{M}_n$ are usually obtained from light scattering and osmometry, respectively. However, since synthetic polymers usually contain a low molecular weight fraction which may pass through the osmotic membrane, the measured osmotic pressure is too low and the estimate of $\bar{M}_n$ too high. As a result, it is probable that polymer polydispersities are frequently underestimated when based on measured $\bar{M}_w/\bar{M}_n$ ratios.

For the Poisson distribution, the characteristic parameter is r, the ratio of monomer moles consumed to moles of chains initiated. It is obvious that $r = M_n/M_0$. The ratio of M_w to M_n is given by

$$\frac{\bar{M}_w}{\bar{M}_n} = 1 + \frac{r}{(r+1)^2} \tag{I.27}$$

Thus, even for chains of no more than 100 units, $\bar{M}_w$ is only 1% larger than $\bar{M}_n$, well within the experimental error of molecular weight determinations.

5. Real Distributions and Generalized Molecular Weight Distribution Functions

In the preceding sections we discussed chain length distributions on the basis of kinetic considerations that apply under idealized conditions. In practice, however, these distributions will be modified by a number of complicating factors. If the polymerization of a vinyl monomer is carried to an appreciable conversion, the ratio of the chain propagation and chain termination rates will gradually change and produce a drift in the value of the parameter ϵ. As a result, the polymer will be characterized by a superposition of the distributions of chain lengths obtained at any given time, and the actual distribution will be broader than predicted from (I.3) or (I.15). A similar broadening may occur even for the polydispersity of the polymer obtained at a given time, if reaction conditions vary from point to point within the polymerizing system. This happens if the polymerization is carried out in a viscous medium, so that temperature is subject to significant fluctuations, a condition not uncommon in industrial processes. Very often, several chain terminating mechanisms, for which different forms of molecular weight distribution functions are predicted, occur in the same system. For instance, the termination of methyl methacrylate chains occurs both by chain disproportionation and by chain combination (Bevington et al., 1954). Such superposition of several chain terminating mechanisms has a particularly drastic effect on polymerizations which should, under ideal conditions, yield the very narrow Poisson distribution of molecular weights. For instance, Sela and Berger (1953) have shown that a *N*-carboxy-α-amino acid anhydride unit need not always add to a growing polypeptide chain as indicated in (I.18b), but may also add so as to produce a species

$$\begin{array}{l} \mathrm{R(COCHNH)_{\mathit{n}}CONHCHCOOH} \\ \qquad\quad | \qquad\qquad\qquad\quad | \\ \qquad\;\; \mathrm{R'} \qquad\qquad\qquad \mathrm{R} \end{array}$$

which will not add any further monomer. Such a monomer addition represents, therefore, a chain termination step, and if it occurs with sufficient frequency the polymer will be characterized by a normal, rather than a Poisson, distribution of chain lengths. Another factor leading to a broadening of the polydispersity in polymers for which a Poisson distribution is expected is a relatively low rate of the chain initiation process. For instance, Becker and Stahmann (1952) have shown that the initiation of *N*-carboxy-α-amino acid anhydride polymerizations by water is slow compared to the subsequent rate of chain growth. In the electron-transfer-initiated polymerizations of vinyl monomers represented in (I.19) complications may arise not only from chain termination by impurities and an insufficiently rapid initiation rate, but also from the reversibility of the addition of monomer to the growing chain (Figini and Schulz, 1960; Wenger, 1960a, b). In addition, a broadening of the molecular weight distribution may result from unequal rates of propagation of chains in which the terminal anions either are in contact with their counterions or are separated from them by molecules of the solvent (Lohr and Schulz, 1964; Bhattacharya et al., 1964, 1965; Figini, 1967; Szwarc, 1966, 1970). An approach to the ideal of a Poisson distribution can be realized, therefore, only if the experimental conditions are very carefully chosen and controlled.

Frequently we are faced with the need to take account of the polydispersity of a polymer in the theoretical interpretation of its solution properties. Because of the various complications in the reaction mechanism alluded to above, the distribution functions (I.3), (I.15), and (I.17), based on simple kinetic schemes, will rarely describe a sample under investigation. It will then be advantageous to use *generalized* molecular weight distribution functions in which the breadth of the distribution is characterized by an adjustable parameter. Such a function was proposed by Schulz (1939) and applied by Zimm (1948) to the treatment of light scattering from a solution of a polydisperse polymer. It has the form

$$N(\mathrm{P}) = \frac{\mathrm{y}^{\mathrm{z}}}{\Gamma(\mathrm{z})} \mathrm{P}^{\mathrm{z}-1} \exp(-\mathrm{yP}) \tag{I.28}$$

where Γ is the gamma function and the number-, weight-, and z-average molecular weights are related to the parameters y and z by

$$\begin{aligned} \bar{M}_n &= M_0 \frac{\mathrm{z}}{\mathrm{y}} \\ \bar{M}_w &= \frac{M_0(\mathrm{z}+1)}{\mathrm{y}} \\ \bar{M}_z &= \frac{M_0(\mathrm{z}+2)}{\mathrm{y}} \end{aligned} \tag{I.29}$$

We may note that $\mathrm{z} = 1$ corresponds to the "most probable" distribution, while $\mathrm{z} = 2$ represents the case of vinyl polymerization terminated by radical recombination. The polydispersity decreases with increasing values of z.

It should, however, be strongly emphasized that a limited number of parameters cannot define uniquely a molecular weight distribution. This point is illustrated in Fig. I.3, where two systems containing two sharp

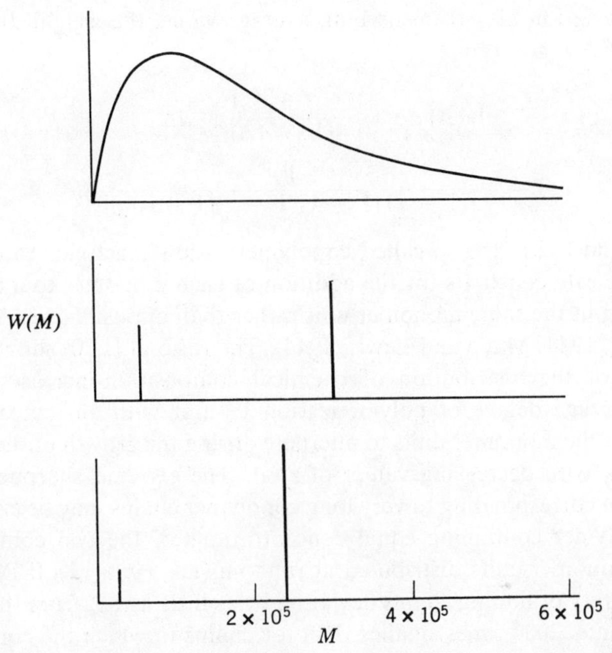

Fig. I.3. Three molecular weight distributions with $\bar{M}_n = 10^5$ and $\bar{M}_w = 2 \times 10^5$.

fractions are shown, which have the same number-average and weight-average molecular weights as the normal distribution shown on top. Thus, a ratio of $\bar{M}_w/\bar{M}_n = 2$ does not, by itself, guarantee that we are dealing with a normal distribution. On the other hand, if the conditions under which a polymer sample was prepared are such that a distribution function with two maxima ("bimodal distribution") seems unlikely, the generalized distribution function (I.28) is likely to be a close approximation to the real distribution of chain lengths.

6. The Distribution of Copolymer Composition

When we deal with copolymers, we must consider the heterogeneity of the sample with respect to chemical composition as well as the polydispersity of molecular weights. This problem may be divided into two parts, considering first the distribution in the composition of chains produced at any given time during a copolymerization and treating subsequently the drift of copolymer composition with polymerization time.

The first problem was solved by Stockmayer (1945), who considered the case of a vinyl copolymerization leading to a normal distribution of polymer chain lengths. If the two comonomers have the same molecular weight, x is the mole fraction of one of the monomers in a given polymer chain, and Δ is the deviation of x from its bulk average value, the weight distribution function of Δ is given by

$$W(\Delta)\, d\Delta = \frac{3\sqrt{A}}{4(1 + A\Delta)^{5/2}} d\Delta$$

$$A = \frac{\bar{P}_n}{2x(1 - x)\,[1 - 4x(1 - x)(1 - r_1 r_2)]^{1/2}} \qquad \text{(I.30)}$$

where r_1 and r_2 are the so-called copolymerization reactivity ratios giving the relative rate constants for the addition of each monomer to a chain end terminating in the same monomer unit rather than opposite one (Alfrey and Goldfinger, 1944; Mayo and Lewis, 1944). The form of (I.30) shows that the sharpness of the distribution of chemical composition increases with the number-average degree of polymerization $\bar{P}_n$ and with an increase in the tendency of the monomer units to alternate during the growth of the polymer chains (i.e., with decreasing values of $r_1 r_2$). The extreme sharpness of the distribution corresponding to very long copolymer chains may be exemplified by a copolymer containing equal concentrations of the two comonomers, with the monomer units distributed at random (i.e., $r_1 r_2 = 1$). If $\bar{P}_n = 1000$, the probability of finding chains deviating by as little as 1% from the average composition is 2000 times smaller than for chains in which the composition corresponds to the bulk average. However, the ratio of these probabilities drops to 17 if the chains contain, on the average, only 100 monomer units.

In practice, copolymer samples will have much broader distributions, since the comonomers usually differ in their reactivity, so that the more reactive one is depleted more rapidly in the course of the polymerization. The drift in the composition of the monomer mixture will then lead to a corresponding drift in the composition of the copolymer. This effect has been analyzed by Skeist (1946) and by Meyer and Lowry (1956). We shall not concern ourselves here with the details of this development, but we should note that in some cases the distribution function of chemical compositions obtained when copolymerization is carried to high conversions may have two maxima.

C. RELATION OF THE STUDY OF MACROMOLECULES IN SOLUTION AND IN BULK

At the outset of a discussion of the properties of macromolecules in solution, it is helpful to gain perspective by considering the relation of studies of solution properties and studies of the properties of macromolecules in bulk.

There is, first of all, the obvious fact that the possibilities of studying a polymer which cannot be brought into solution are very severely limited. All fractionation procedures designed to separate materials of varying molecular weight or varying composition and all chromatographic analyses of molecular weight distributions require prior dissolution of the sample. Also, dissolution of the polymer is required for the preparation of single crystals of polymeric substances and for the preparation of specimens in which the individual macromolecules are deposited, separated from each other, on some suitable surface for study by electron microscopy. It is no exaggeration to say that the existence of substances of very high molecular weight could not have been demonstrated conclusively if these substances could not have been dissolved.

1. Molecular Weight and Polydispersity

One of the important aims in the study of polymer solutions was the development of a number of techniques to determine the mean molecular weight of the sample. It is also possible to characterize the distribution of molecular weights by measurements carried out on a solution of a polydisperse sample without resort to a tedious and time-consuming fractionation procedure. Similar analyses can be carried out to characterize either samples with a continuous distribution of chemical composition (such as may occur in synthetic copolymers) or samples containing several well-defined macromolecular species (characteristic of specimens of biological origin).

We alluded in Section I.A1 to the possibility of determining molecular weights of macromolecules from information obtained by x-ray crystallography for the very restricted class of high molecular weight substances represented by the crystalline globular proteins. Even in this case, however, this alternative to molecular weight determination is of only academic interest, since the crystallographic studies require a tremendous effort as compared with that needed for a conventional molecular weight determination of a dissolved polymer. Of greater practical interest is the possibility of deriving molecular weights from measurements on the images obtained with the electron microscope (Hall, 1960). However, in using this method one has to exercise special care in the preparation of specimens to avoid artifacts due particularly to molecular aggregation (Hayes et al., 1959). The limits of resolution obtainable with the electron microscope seem to restrict this method, for the present, to the molecules of globular proteins and to helically coiled molecular particles, such as those of nucleic acid. Hall and Doty (1958) have shown that electron micrographs may be evaluated in terms of the distribution of molecular weights. However secure we may feel in the use of the theoretical interpretation of data obtained on polymer solutions, the old adage "seeing is believing" is still applicable to some extent, and it is comforting, therefore, that measurements on images representing the individual molecules are in agreement with results obtained in a seemingly less direct manner.

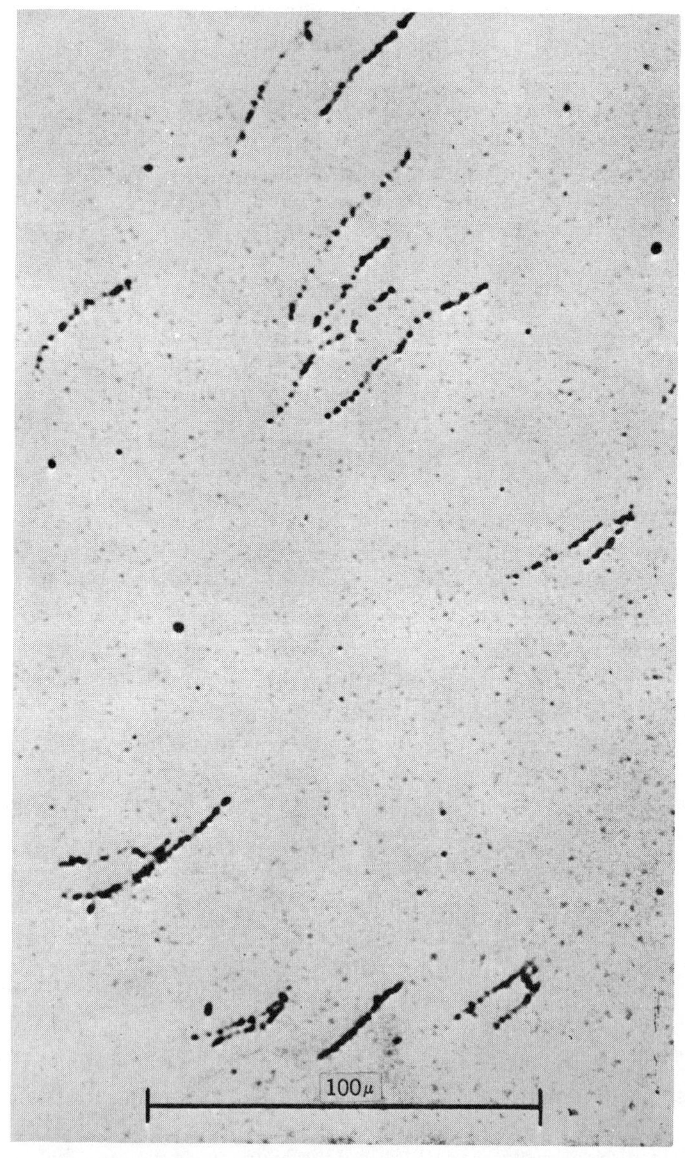

Fig. I.4. Autoradiograph of DNA from T-2 bacteriophages labeled with tritiated thymine (11.2 Ci. / mM). The autoradiographic exposure was 63 days. Courtesy of Dr. J. Cairns.

The electron microscopic method has also the important advantage that it allows us to study chains which are so long that hydrodynamic forces lead easily to breaks in the molecular backbone. This may become a particularly serious problem with deoxyribose nucleic acid (DNA) (Hershey and Burgi, 1960; Levinthal and Davison, 1961), and it is certain that the longest DNA chains occurring in living organisms could not be studied by the solution methods without extensive degradation. Kleinschmidt et al. (1962) developed techniques for the electron microscopic investigation of the DNA spilling from a particle of the T-2 bacteriophage, and they showed that the DNA in this virus exists as a single double-stranded chain with a length estimated at $46 \pm 4\ \mu$. A corresponding study of two bacteria (Kleinschmidt et al., 1961) seemed to indicate that even in these organisms the DNA exists as a single chain.

These extremely long chains may also be studied by autoradiography. In this method the organism is grown on a material containing a radioactive isotope so that the DNA is labeled. A very dilute solution of the DNA is dried on a support which is brought into intimate contact with a photographic emulsion. If sufficient time is allowed so that a large number of radioactive atoms in every DNA molecule of the sample have decomposed, if the molecule is stretched out, and if its dimensions are large compared to the resolving power of the emulsion, the images of the distintegrating atoms will outline the shape of the molecule. The beautiful pictures obtained by Cairns (1961) with the use of this method on DNA preparations from the T-2 bacteriophage are shown in Fig. I.4. The images correspond to a molecular length of 52 μ, in good agreement with the electron microscopy results cited above. When the DNA from the bacterium *Escherichia coli* was investigated in this manner, its length was determined as 400 μ (Cairns, 1962).

2. The Shape of Macromolecules

All of our knowledge concerning the shapes of macromolecules was originally derived from a study of these substances in dilute solution. The methods for obtaining such information constitute one of the main subjects of this book. Nevertheless, knowledge obtained in studies of polymers in bulk has been helpful in a variety of ways and has originated several concepts necessary in the interpretation of solution properties. When we consider in Chapter III the behavior of polymer chains which may be represented by flexible coils, we shall see that the treatment of the resistance to expansion of the coil by osmotic forces is exactly analogous to the treatment of the resistance of a rubber to elongation by an applied mechanical stress. Here, again, observations of measurable retractive forces on rubber samples which we can see give us added confidence in reasoning about the behavior of chain molecules, which we can deduce only indirectly. Moreover, the detailed consideration of the conformations of linear chain polymers which exist in solution as "random coils" owes much to the crystallographic determination of the conformations assumed by these or similar chains when they are packed in a crystal lattice.

A much more important contribution to our understanding of the behavior of dissolved macromolecules has been made by crystallography in connection with the phenomenon of "helix-coil transitions" of synthetic polypeptides and polynucleotides, which we shall discuss in Section III. C. These transitions introduce an entirely new concept of "one-dimensional crystallization" into physical chemistry, and it is doubtful whether a detailed understanding of them could have been obtained without information based on x-ray diffraction analysis of these ordered structures in the crystalline state.

The spectacular progress achieved in recent years in applying the technique of x-ray crystallography to single crystals composed of the large molecules of globular proteins has made it possible to derive the detailed shape of these molecules. Recent reviews (North and Phillips, 1969; Dickerson and Geis, 1969; Blow and Steitz, 1970) list about 20 proteins whose structures are known to varying degrees of precision, and the list may be confidently expected to grow fairly fast. The astonishing amount of information obtained from high-resolution maps of electron density distributions is illustrated in Fig. I.5, which shows the shape of the chain backbone in the enzyme carboxypeptidase A (Lipscomb et al., 1969). Compared with this detailed structure, any results which may be hoped for from the analysis of the properties of protein solutions must appear extremely crude, since they can yield only a small number of parameters characterizing the shape of the dissolved particle. Nevertheless, the following arguments lead us to the conclusion that the success of the crystallographic method has not rendered the study of protein solutions obsolete:

(a) Solution of the structure of a protein crystal from x-ray diffraction data represents a formidable undertaking (requiring typically the efforts of large research teams over several years), and there is as yet no assurance that the task can be accomplished in every case. By comparison, the task involved in the measurement of solution properties and an analysis of the data is relatively minor and well worth while for whatever information it may yield.

(b) In spite of the evidence (to be discussed in Section III.C.6) that the molecular shapes of proteins in the crystalline state are often fair representations of the shapes of these molecules in solution, we cannot be sure that this conclusion has general validity. In fact, we know that some proteins in solution may undergo conformational transitions which may be of crucial importance to their biological function. In any case, if the molecular shape is different in solution and in the crystalline state, we must regard the property

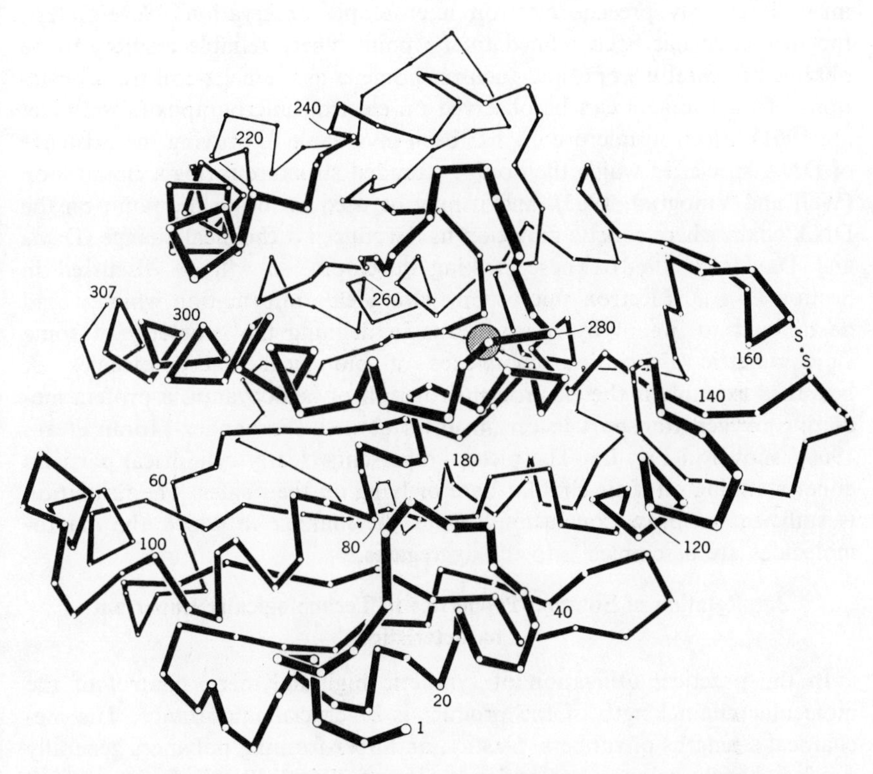

Fig. I.5. Folding of the polypeptide backbone of carboxypeptidase A. The circle near the center represents a zinc atom coordinated to three of the amino acid residues.

of the molecule in solution as more significant in terms of its function in a living organism. The crystallographic and solution methods, therefore, complement one another, with one giving us incomparably more detailed information, while the other is more closely related to the system which we ultimately wish to understand.

(c) There are a number of problems (apart from the analysis of complex protein mixtures) which can be clarified only by solution studies. These include, for instance, association equilibria of proteins with each other or with small molecules and ions, and most studies of rate processes which are excluded by the very long times required for data collection in the crystallographic method.

Electron microscopy, can also contribute to our knowledge of molecular shapes. However, it is important to realize that surface tension may result in serious distortions of molecular shapes during the drying of a solution, which must necessarily precede electron microscopic observation. Nevertheless, the technique has been refined to the point where reliable results can be obtained in careful work, and such phenomena as the helix-coil transformations of nucleic acid can be observed on electron micrographs (Kisselev et al., 1961). Electron microscopy has been invaluable in proving the existence of DNA species in which the double-stranded structure forms a closed loop (Weil and Vinograd, 1963), and it may be used to locate the point on the DNA chain where genetic mutation has produced a chemical change (Davis and Davidson, 1968). These striking developments will be discussed in Section III.C.3. Electron microscopy also yields information which would be difficult to obtain by other means in defining the geometry of some stoichiometric association complexes of biological macromolecules. A beautiful example is the electron micrograph of hemocyanin, a protein mediating oxygen transport in certain invertebrates (Fernandez-Moran et al., 1966), shown in Fig. I.6. The picture represents clearly cylindrical particles either standing on their circular base or lying on their sides. The resolution is sufficient to draw conclusions about the manner in which the macromolecules are assembled into the aggregates.

3. Relation of Solution Properties to Technologically Important Characteristics

In the practical utilization of synthetic high polymers, control of the molecular chain length of the product is of crucial importance. The mechanical strengths of rubbers, plastics, and fiber-forming polymers generally decline sharply below a molecular weight of 20,000–30,000. For very high molecular weights, the mechanical properties tend to approach an asymptotic limit and become independent of a further chain length extension, but such high molecular weight materials are extremely viscous at the elevated temperatures at which polymers are being processed into useful shapes. The industrial polymer chemist must, therefore, control the molecular weight of his material, with a view both to its properties under the conditions of its technological utilization and to its processability characteristics. He will aim at as close a reproducibility of polymer chain length as he can obtain. With a sample polydisperse with respect to molecular weight, useful properties depend often, to a first approximation, on the weight-average molecular weight. Nevertheless, other molecular weight averages are useful indices for the purposes of quality control, since for samples prepared by any given procedure the shape of the molecular weight distribution function is fixed, so that various averages are related by a constant ratio.

The details of the molecular weight distribution function are of less importance to the mechanical characteristics of a polymer in bulk, but they should affect them to some extent (Bueche, 1962b; Ferry, 1970b). For any

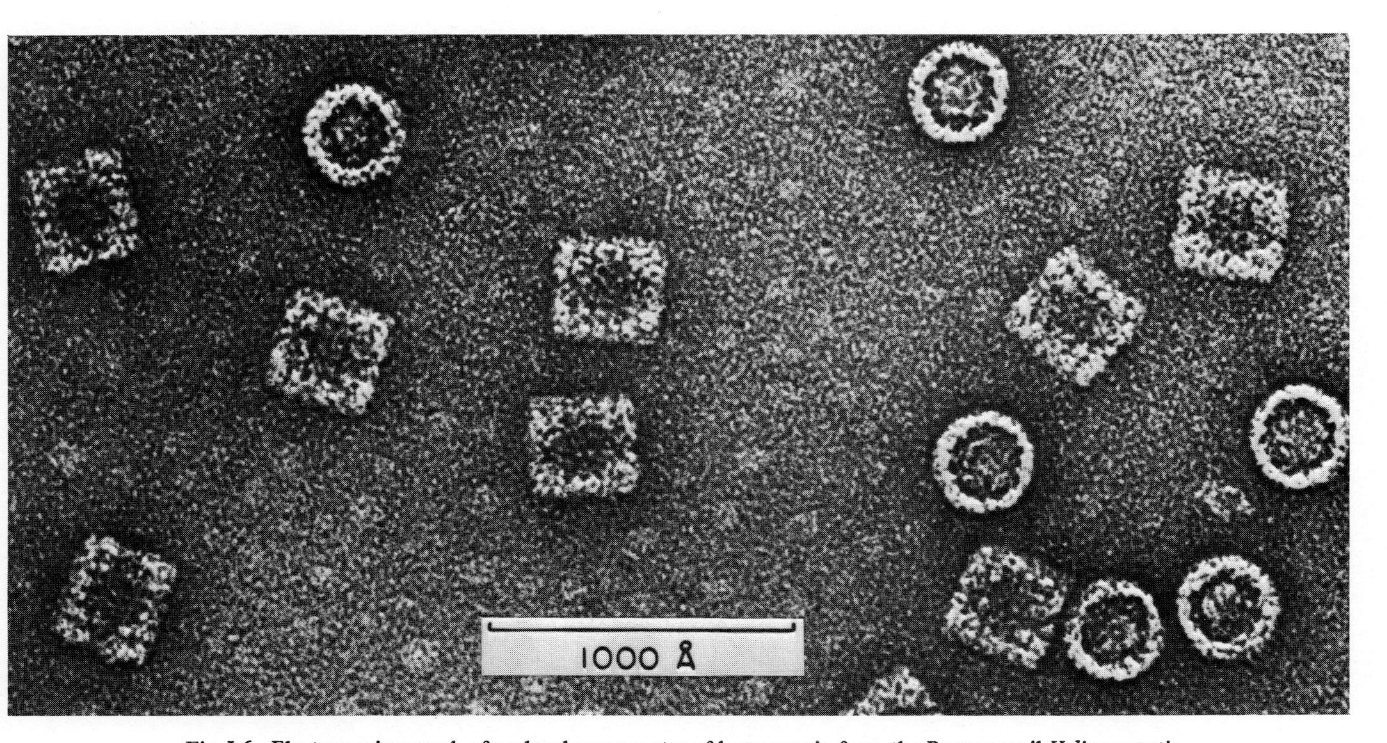

Fig. I.6. Electron micrograph of molecular aggregates of hemocyanin from the Roman snail *Helix pomatia*.

given weight-average molecular weight, theory leads to the prediction that the elastic component of the viscoelastic response will be accentuated by an increasing polydispersity of the sample. Careful experimental studies of the effect of molecular weight distribution on polymer bulk properties are surprisingly scarce. The significance of such data is greatly enhanced if relatively monodisperse samples (such as those prepared by the techniques described in Section I.B.3) are available as standards of comparison. Outstanding studies of this type have been reported by McCormick et al. (1959) and by Rudd (1960). As for copolymers, it is generally desirable that they be as homogeneous with respect to chemical composition as possible, since variations in chemical composition may lead to a tendency toward phase separation, resulting in mechanical weakness.

We should realize, however, that some properties of crucial importance to the behavior of polymers in bulk have little or no relation to their behavior in solution. For instance, the useful properties of a rubber depend rather critically on a relatively low potential energy barrier which has to be overcome when the molecule changes its shape—yet it is not easy to assess the height of this barrier from the properties of the rubber in solution. Other technologically important polymer properties which cannot be predicted from the behavior of polymer solutions include the temperature of the glass transition, the crystallizability, and the kinetics of the crystallization process under some given set of conditions. As an example, let us consider poly(hexamethylene adipamide) (commercial nylon 66) and a polyamide resulting from a condensation of adipic acid with a mixture of pentamethylene diamine and heptamethylene diamine. The properties of the dissolved polyamides will probably be indistinguishable, but the behavior of the two materials in bulk will be strikingly different, since the regular spacing of the amide linkages in nylon 66 leads to a highly crystalline product, while the random distribution of these linkages in the copolymer renders crystallization impossible.

In the case of vinyl polymers, a high degree of stereoregularity is a precondition to crystallizability. Most solution properties are rather insensitive to differences in "tacticity." Even though in later discussions nuclear magnetic resonance spectroscopy of polymer solutions will be shown to be, in some cases, the most powerful method currently available for the quantitative evaluation of tacticity, a stereoregular structure does not in itself ensure that a given polymer will crystallize.

These examples are offered merely as typical of the limits of the information which may be obtained when macromolecules are studied in solution—and usually in highly dilute solution. Such studies aim mainly at the definition of the properties of the isolated macromolecule and cannot be expected to predict phenomena dominated by the enormous viscosity of high polymers in bulk, which frequently prevents the system from attaining a state of thermodynamic equilibrium. They can also not be expected to reveal the geometric possibilities of packing chain molecules into a crystal lattice, which may result in large differences between the bulk properties of crystalline and of amorphous polymers.

Chapter II

THE SOLUBILITY OF MACROMOLECULES

Most high polymers occurring in nature or synthesized in the laboratory are soluble under some conditions. Insoluble polymers are crosslinked structures (e.g., lignin, vulcanized rubber, "cured" phenolic resins) and a few highly crystalline materials such as polytetrafluoroethylene (Teflon) or poly (*p*-phenylene). The insolubility of these substances certainly sets severe limits to their physicochemical study.

Before discussing the special problems encountered in the solubility of high molecular weight solutes, it will be convenient to consider briefly the conditions governing the formation of a solution in binary systems whose components consist of molecules of comparable size. We shall then discuss solubility in binary systems containing a polymeric species and a low molecular weight solvent and finally describe the solubility properties of polymers in multicomponent systems.

A. FORMATION OF SOLUTIONS FROM SPECIES OF LOW MOLECULAR WEIGHT

1. Thermodynamic Considerations

For a system at constant temperature T and constant pressure P the equilibrium state is represented by the state of minimum free energy. The Gibbs free energy function G is defined by

$$G = H - TS = E + PV - TS \tag{II.1}$$

where H is the enthalpy, E the internal energy, V the volume, and S the entropy of the system. If we consider a system of several components in two phases, the condition of equilibrium requires that the free energy remain unchanged if an infinitesimal number of moles δn_i of component i is transferred from phase I to phase II:

$$\left[-\left(\frac{\partial G}{\partial n_i}\right)^{\mathrm{I}}_{T,P,n_j \neq n_i} + \left(\frac{\partial G}{\partial n_i}\right)^{\mathrm{II}}_{T,P,n_j \neq n_i}\right] \delta n_i = 0 \tag{II.2}$$

where the superscripts I and II refer to quantities evaluated in phases I and II, respectively. Using $\bar{X}_i \equiv (\partial X/\partial n_i)_{T,P,n_j\neq n_i}$ for partial molar quantities we may state that equilibrium between two phases requires

$$\bar{G}_i^{\mathrm{I}} = \bar{G}_i^{\mathrm{II}} \tag{II.3}$$

The partial molar free energies $\bar{G}_i$ are customarily referred to as the "chemical potentials." The activity of species i, denoted by a_i, is defined by

$$\overline{\Delta G_i} \equiv \bar{G}_i - \bar{G}_i^0 = \boldsymbol{R}T \ln a_i \tag{II.4}$$

$$\overline{\Delta G_i} = \overline{\Delta H_i} - T\overline{\Delta S_i}$$

$$\overline{\Delta H_i} \equiv \bar{H}_i - \bar{H}_i^0$$

$$\overline{\Delta S_i} \equiv \bar{S}_i - \bar{S}_i^0 \tag{II.5}$$

where $\boldsymbol{R}$ is the gas constant and the superscript ^{0} refers to an arbitrarily chosen standard state. When dealing with nonelectrolytes, we shall invariably use the pure component as its standard state. The activity a_i is related to the "escaping tendency" of component i, as measured by its fugacity f_i, according to

$$a_i = \frac{f_i}{f_i^0} \tag{II.6}$$

The fugacity is equal to the partial vapor pressure P_i in the limit of low pressures, but for finite pressures f_i/P_i deviates from unity in a manner depending on the deviation of the vapor from ideal gas behavior. Specifically, if we represent the equation of state of the vapor by $PV = RT + A_{2g}/V + A_{3g}/V^2 + \cdots$, where A_{2g}, A_{3g}, etc., are the second, third, and higher virial coefficients, we obtain from

$$\boldsymbol{R}T \ln\left(\frac{f_2}{f_1}\right) = \int_{P_1}^{P_2} V\,dP$$

the relation

$$\frac{f}{P} = \exp\left[\frac{A_{2g}P}{(\boldsymbol{R}T)^2}\right] + \cdots \tag{II.7}$$

For the vapors of liquids below their normal boiling points the deviation of f from P is generally quite small. As an example, we may consider carbon tetrachloride, for which $A_{2g} = -171^2$/atm mole^2 at 25°C, where the saturated vapor pressure is 0.15 atm. Under these conditions, $f/P = 0.996$.

It is convenient to interpret the thermodynamic behavior of nonelectrolyte solutions by comparing them with *ideal solutions* in which the activities of all components a_i are equal to their mole fractions x_i. The a_i values found in real solutions are then expressed by

$$a_i = \gamma_i x_i \tag{II.8}$$

where γ_i is referred to as the activity coefficient. We may then rewrite (II.4) as

$$\overline{\Delta G_i} = \boldsymbol{R}T \ln x_i + \boldsymbol{R}T \ln \gamma_i \tag{II.9}$$

where the second term on the right represents the contribution to the chemical potential due to the nonideality of the solution. If an "excess function" is defined as $X^E \equiv X - X_{\text{ideal}}$, the excess chemical potential $\overline{\Delta G_i^E}$ is given by

$$\overline{\Delta G_i^E} = \boldsymbol{R}T \ln \gamma_i \tag{II.10}$$

and the excess free energy, enthalpy, and entropy of mixing in a binary system are related to the partial molar excess functions by

$$\begin{aligned} \Delta G_M^E &= n_1 \overline{\Delta G_1^E} + n_2 \overline{\Delta G_2^E} \\ \Delta H_M^E &= n_1 \overline{\Delta H_1^E} + n_2 \overline{\Delta H_2^E} \\ \Delta S_M^E &= n_1 \overline{\Delta S_1^E} + n_2 \overline{\Delta S_2^E} \end{aligned} \tag{II.11}$$

Since the ideal solution is, by definition, formed from its components without any heat effect, $\Delta H_M^E = \Delta H_M$ and $\overline{\Delta H_i^E} = \overline{\Delta H_i}$. As for the excess entropy of mixing, its meaning will be clarified later when we consider the theory of ideal solutions in more detail.

Utilizing with (II.9) the Gibbs-Helmholtz relation

$$\left[\frac{\partial}{\partial T}\left(\frac{\Delta G}{T}\right)\right]_P = -\frac{\Delta H}{T^2} \tag{II.12}$$

we obtain for the temperature dependence of the activity coefficient

$$\left(\frac{\partial \ln \gamma_i}{\partial T}\right)_{P, n_j \neq n_i} = -\frac{\overline{\Delta H_i}}{\boldsymbol{R}T^2} \tag{II.13}$$

In the representation of thermodynamic quantities of binary systems it is convenient to use plots referring to a total of 1 mole. Figure II.1 shows two typical plots of the free energy of mixing ΔG_M, which may be easily shown to be related to the chemical potentials of the components by

$$\Delta G_M = n_1 \overline{\Delta G_1} + n_2 \overline{\Delta G_2} \tag{II.14}$$

In Fig. II.1*a* the ΔG_M curve has no inflection points, and any system represented by two coexisting phases corresponds to a higher free energy than the single phase formed when the two phases merge. This diagram is then typical of binary systems which form solutions in all proportions. On the other hand, Fig. II.1*b* shows a ΔG_M curve, two points of which have a common tangent. The two phases characterized by x_1' and x_1'' have identical values for $\overline{\Delta G_1}$ and $\overline{\Delta G_2}$,so that they are in equilibrium with each other. It is clear that points representing any one phase system intermediate

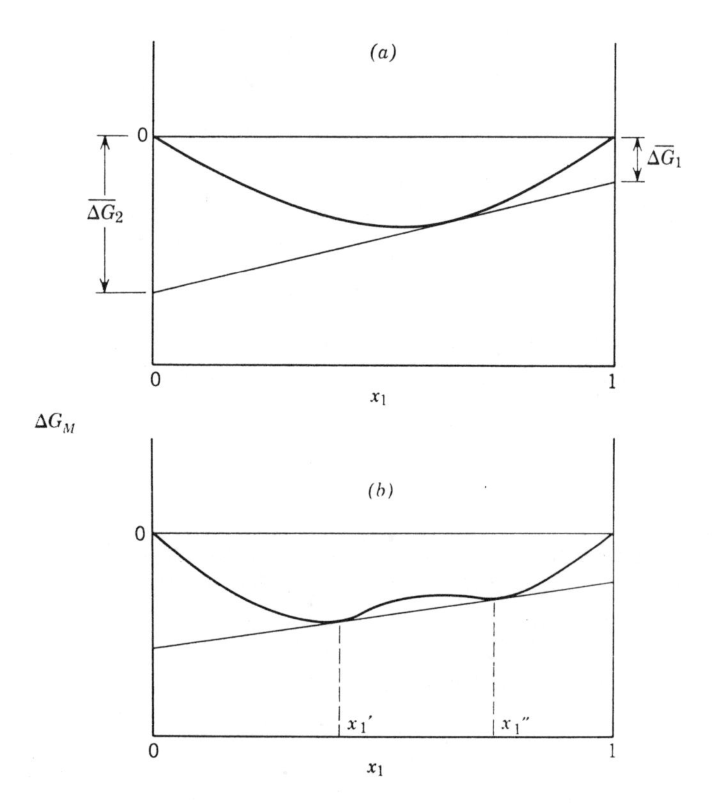

Fig. II.1. Schematic representation of the free energy of mixing. (*a*) System of two components miscible in all proportions. (*b*) System of two components with limited miscibility.

between x_1' and x_2'' correspond to a state of higher free energy than the corresponding point representing a two-phase system with the same overall composition, and any such single phase is, therefore, thermodynamically unstable.

The temperature dependence of the composition of two coexisting phases of a binary system may be obtained from (II.4) and (II.12) as

$$\left(\frac{\partial \ln a_1}{\partial x_1}\right)'_{T,P} \frac{dx_1'}{dT} - \left(\frac{\partial \ln a_1}{\partial x_1}\right)''_{T,P} \frac{dx_1''}{dT} = \frac{-(\overline{\Delta H_1})' + (\overline{\Delta H_1})''}{\boldsymbol{R}T^2} \qquad \text{(II.15)}$$

where ( )′ and ( )″ symbolize quantities evaluated at the compositions x_1' and x_1'', respectively. For the special case of a binary system in which one phase is the pure component 2 at all temperatures (e.g., a crystalline solid which does not form solid solutions with the solvent), (II.15) reduces to

$$\frac{dx_2}{dT} = \frac{(\overline{\Delta H_2})'}{\boldsymbol{R}T^2(\partial \ln a_2/\partial x_2)'_{T,P}} \tag{II.16a}$$

or

$$\frac{d \ln x_2}{dT} = \frac{(\overline{\Delta H_2})'}{\mathbf{R}T^2 [1 + (\partial \ln \gamma_2/\partial \ln x_2)'_{T,P}]} \tag{II.16b}$$

2. The Ideal Solution

Ideal solutions are defined as solutions in which the activity is equal to the mole fraction:

$$a_i = x_i \tag{II.17}$$

for all components and all compositions of the system (Raoult's law). It was once believed that any multicomponent system in which forces between like and unlike molecules are equal (and for which, therefore, the solution process is athermal) will lead to an ideal solution, but we shall see later that the mixing of molecules of different sizes may lead to deviations from ideal solution behavior even if no heat effect accompanies dissolution. We shall begin, therefore, with a consideration of binary systems in which both the molecular size and the forces exerted by the molecules are the same for both components.

The entropy of the pure components is the sum of contributions due to molecular translation, rotation, and vibration. When the components are mixed to form an ideal solution, these entropy contributions may be considered to remain unchanged since molecular interactions have a relatively small effect on translational, vibrational, and rotational motions of the individual molecules. However, a configurational term S_{conf} will now make a contribution to the entropy of the systems, since the solution may contain a large number of distinguishable arrangements of the molecules of the two components. We have, therefore, in good approximation

$$\Delta S_M = S_{\text{conf}} \tag{II.18}$$

The evaluation of S_{conf} is most easily carried out by placing the molecules onto a lattice. If we have N_1 molecules of type 1 and N_2 of type 2, the number of distinguishable arrangements on $N_1 + N_2$ lattice points is

$$\mathscr{W} = \frac{(N_1 + N_2)!}{N_1! N_2!} \tag{II.19}$$

and since all of these configurations represent states of equal energy, the configurational entropy is

$$S_{\text{conf}} = \boldsymbol{k} \ln \mathscr{W} \tag{II.20}$$

where $\boldsymbol{k}$ is the Boltzmann constant and $\mathscr{W}$ is the multiplicity of statistical states. Approximating the factorials in (II.19) by Stirling's approximation formula

$$y! = \left(\frac{y}{e}\right)^y \tag{II.21}$$

we obtain from (II.18), (II.19), and (II.20) for the ideal entropy of mixing

$$\Delta S_M^i = \boldsymbol{k}\left[N_1 \ln\left(\frac{N_1 + N_2}{N_1}\right) + N_2 \ln\left(\frac{N_1 + N_2}{N_2}\right)\right] \tag{II.22a}$$

or

$$\Delta S_M^i = -\boldsymbol{R}\,(n_1 \ln x_1 + n_2 \ln x_2) \tag{II.22b}$$

The ideal partial molar entropy is then obtained by differentiation with respect to n_1 or n_2 [with $x_1 = n_1/(n_1 + n_2)$ and $x_2 = n_2/\,(n_1 + n_2)$] as

$$\overline{\Delta S_1^i} = -\boldsymbol{R} \ln x_1; \qquad \overline{\Delta S_2^i} = -\boldsymbol{R} \ln x_2 \tag{II.23}$$

and since the solution is athermal, so that $\overline{\Delta H_1} = \overline{\Delta H_2} = 0$, we have for the ideal free energy of mixing

$$\Delta G_M^i = -T\,\Delta S_M^i = \boldsymbol{R}T\,(n_1 \ln x_1 + n_2 \ln x_2) \tag{2.24}$$

yielding for the chemical potentials of the two components in an ideal solution

$$\overline{\Delta G_1^i} = \boldsymbol{R}T \ln x_1; \qquad \overline{\Delta G_2^i} = \boldsymbol{R}T \ln x_2 \tag{II.25}$$

Comparison with (II.4) shows that the activities are equal to the mole fractions. This proves the validity of the ideal solution law under the special conditions which we have considered.

The *excess* entropy of mixing is given, after generalizing (II.22b) to a multicomponent system, by

$$\Delta S_M^E = \Delta S_M + \boldsymbol{R} \sum n_i \ln x_i \tag{II.26}$$

and the excess partial molar entropies are

$$\overline{\Delta S_i^E} = \overline{\Delta S_i} + \boldsymbol{R} \ln x_i \tag{II.27}$$

3. Regular Solutions

The concept of a "regular solution" was introduced by Hildebrand (1929) and defined as a solution in which the partial molar entropies of the components are those to be expected from the ideal solution law. From this definition it follows that any deviation from ideal solution behavior in a regular solution is entirely accounted for by the heat of mixing. However, it is obvious that any difference in the forces between similar and dissimilar molecules in a solution (such as will cause the mixing process to be accompanied by the evolution or absorption of heat) should lead also to a deviation from a random distribution of the interacting molecules, that is, to ΔS_M values lower than those predicted for an ideal solution. The concept of a regular solution will, therefore, be a valid approximation only as long as the ordering forces are insufficient to reduce significantly the disorder due to the thermal motion of the molecules. In practice this condition is reasonably well satisfied in many cases. For example, in a solution containing 0.5 mole of benzene and 0.5 mole of cyclohexane at 25 °C the ΔH_M value of 182 cal is small compared to $\boldsymbol{R}T$, so that the error introduced by neglecting differences of the energy corresponding to different configurations in the computation of the configurational entropy of the solution would be expected to be minor.

When Hildebrand first formulated the concept of regular solutions, he assumed that athermal solutions would necessarily follow the ideal solution law. Much later, when the physical chemistry of solutions of high molecular weight substances was subjected to detailed investigation, it became obvious that differences in the molecular size of solute and solvent may lead to a very large deviation from solution ideality even if no heat effect accompanies the formation of the solution. Therefore, the proposal was made (Hildebrand, 1953) to define a regular solution as "one in which thermal agitation is sufficient to give practically complete randomness." This means that the entropy of mixing of a regular solution, defined in this manner, need not be ideal, and we shall see later that it may deviate very far from ideality in systems containing polymeric solutes. The definition has the advantage that a theoretical interpretation of the thermodynamic behavior of a solution may be treated by considering *separately* the entropy of mixing, based on a count of distinguishable configurations, and the heat of solution, based on the assumption of random mixing.

4. Estimation of the Heat of Mixing from Cohesive Energy Densities

When a solution is formed from two pure liquids, similar molecules are separated from each other and contact points are created between dissimilar molecules. It is then qualitatively obvious that the direction of heat flow in the mixing of two components at constant temperature will depend on the relative magnitude of forces operating between like and unlike molecules. A quantitative theory of the energy of mixing is derived most simply for two substances consisting of molecules of equal size which are placed into the cells of a three-dimensional lattice. Let each molecule have z nearest neighbors, and let the energetic interactions, restricted to nearest neighbors, be characterized by an energy w_{ij} required for the separation of a mole of pairs of molecules of species i and j. If we then transfer 1 mole of a species A into a large volume of a species B (so that all molecules of A are completely surrounded by molecules of B), the required energy will be

$$(\Delta E_M)_A^\infty = z\left(\frac{w_{AA}}{2} + \frac{w_{BB}}{2} - w_{AB}\right) \equiv z\,\Delta w \tag{II.28}$$

We see then that such transfers would be athermal if w_{AB} were the arithmetic mean of w_{AA} and w_{BB}. It was pointed out by van Laar and Lorenz (1925), however, that forces between two dissimilar particles will, in general, be more closely approximated by the *geometric mean* of the forces which each of the particles would exert on a second particle of its own kind, so that we should expect for the energetic interaction of dissimilar molecules

$$w_{AB} = \sqrt{w_{AA}w_{BB}} \leqslant \frac{w_{AA} + w_{BB}}{2} \tag{II.29}$$

Hildebrand and Scott (1950) analyzed in detail the physical basis for the assumption expressed in (II.29) and pointed out that the "postulate of the geometric mean" accurately expressed interactions of permanent dipoles and is a close approximation to energetic interactions due to London dispersion forces. It is definitely unsatisfactory for the description of forces between permanent and induced dipoles, but this type of interaction usually makes only a small contribution to the total of cohesive forces between molecules. They concluded, therefore, that the van Laar postulate is justified when considering the mixing of relatively nonpolar molecules. Combining (II.28) and (II.29), it is then predicted that such mixing will be generally endothermic. Since the molar energy of vaporization of species i is given by $\Delta E_i^v = zw_{ii}/2$, we obtain, by combining (II.28) and (II.29),

$$(\Delta E_M)_A^\infty = \Delta E_A^v + \Delta E_B^v - 2\sqrt{\Delta E_A^v\,\Delta E_B^v} = [(\Delta E_A^v)^{1/2} - (\Delta E_B^v)^{1/2}]^2 \tag{II.30}$$

In the general case, where mixing leads to solutions that are not highly dilute, we have to estimate the relative number of A···A, B···B, and A···B nearest-neighbor pairs. This problem may be treated approximately as a "quasi-chemical equilibrium" (Guggenheim, 1935; Fowler and Guggenheim, 1939), so that

$$A\cdots A + B\cdots B \rightleftharpoons 2A\cdots B$$

$$\frac{(A\cdots B)^2}{(A\cdots A)(B\cdots B)} = \tfrac{1}{4}\exp\left(\frac{-\Delta w}{\boldsymbol{R}T}\right) \tag{II.31}$$

In many cases $\Delta w \ll \boldsymbol{R}T$, and we may obtain a reasonable approximation with the assumption that the occupancy of a given lattice cell is independent of the nature of the species occupying neighboring cells. This leads, for the mixing of n_A moles of species A with n_B moles of species B, to the energy of mixing

$$\Delta E_M = z\left[\frac{n_A w_{AA}}{2} + \frac{n_B w_{BB}}{2} - \frac{n_A^2 w_{AA} + n_B^2 w_{BB} + 2n_A n_B w_{AB}}{2(n_A + n_B)}\right] \quad (II.32)$$

or, using the identity of the right-hand side of (II.28) and (II.30),

$$\Delta E_M = \left(\frac{n_A + n_B}{n_A n_B}\right)[(\Delta E_A^v)^{1/2} - (\Delta E_B^v)^{1/2}]^2 \quad (II.33)$$

We may extend the lattice model to a mixture of molecules of different volumes, where each molecule may occupy one or several adjoining lattice cells. If it is assumed that the number of neighboring lattice cells with which a given molecule engages in energetic interactions is proportional to the molecular volume (i.e., that the "molecular surface-to-volume ratio" is the same for all molecular species), a treatment analogous to that outlined above leads to

$$\Delta E_M = V\phi_A\phi_B\left[\left(\frac{\Delta E_A^v}{V_A}\right)^{1/2} - \left(\frac{\Delta E_B^v}{V_B}\right)^{1/2}\right]^2 = V\phi_A\phi_B(\delta_A - \delta_B)^2 \quad (II.34)$$

where ϕ_A, ϕ_B are the volume fractions of the two components. The energy of vaporization per unit volume of a liquid, $\Delta E_i^v/V_i$, is often referred to as the "cohesive energy density" and its square root δ_i as the "Hildebrand solubility parameter."

Relation (II.34) may also be obtained by applying to the liquids the van der Waals equation for nonideal gases or by integrating the intermolecular potentials between pairs throughout the liquid by use of a continuous distribution function (Hildebrand and Scott, 1950, 1962). We see then that within the limitations of this treatment the heat of mixing is predictable from the values of a single parameter assigned to each constituent of the solution. Typical values of δ at 25 °C (in $cal^{1/2}/cm^{3/2}$) are 7.3 for hexane, 7.4 for diethyl ether, 7.5 for octane, 8.2 for cyclohexane, 8.6 for carbon tetrachloride, 9.2 for benzene and chloroform, 9.8 for methylene chloride, and 10.0 for dioxane, carbon disulfide, and nitrobenzene. As the temperature is raised, the increasing intermolecular distance will lead to a significant decrease of the cohesive energy density and the δ parameter.

Hildebrand and Scott have discussed in detail the various factors that limit the scope of the van Laar treatment. We may first note that this treatment applies to constant-volume processes (i.e., the volume of the solution is the sum of the volumes of its components before mixing), whereas in practice the solution process is carried out at constant pressure and leads generally to changes in volume. Hildebrand and Scott suggest that this be taken into account by the relation

$$(\Delta H_M)_P = (\Delta E_M)_V + T\left(\frac{\partial P}{\partial T}\right)_V (\Delta V_M)_P + \cdots \quad (II.35)$$

implying that the estimate of the energy of mixing from the solubility parameters would be exact for an imaginary constant-volume process. Although ΔV_M is in most cases less than 1% of the volume of the system, the large values of $T(\partial P/\partial T)_V$ (typically of the order of about 3000 atm) frequently produce a significant difference between ΔH_M and ΔE_M. For the system benzene-cyclohexane with 0.5 mole of each component, $\Delta V_M = 0.65$ cc, $\Delta H_M = 182$ cal, and $\Delta E_M = 131$ cal (Goates et al., 1959). A much more serious limitation arises because of the different nature of forces between molecules. Whenever relatively polar substances are involved in the mixing process, an athermal solution is no longer assured if the two components have equal cohesive energy densities, but the dipole-dipole forces and the dispersion forces of the components must each separately be equal. If this factor is neglected, the heats of mixing calculated from latent heats of vaporization will, in general, be too small (Hildebrand and Scott, 1950).

5. Deviation from Ideality in Real Solutions of Small Molecules

The simple procedure for estimating the heat of mixing from cohesive energy densities, as described in the preceding section, has been found to have serious limitations. Discrepancies between measured and predicted values have been listed for a considerable number of systems by Brown et al. (1955) and by Mathieson and Thynne (1956), who found ΔE_M to be generally higher than predicted by the δ values of Hildebrand and Scott (1950). These discrepancies were found also in cases where dipole-dipole interaction could not have contributed appreciably to the heat of mixing. Moreover, Brown et al. showed that it is impossible to assign δ parameters to the liquids studied in such a way as to lead to correct predictions for ΔE_M of all liquid pairs.

In addition, difficulties were encountered in explaining the observed excess entropies of mixing. Any finite heat of mixing (positive or negative) would be expected to decrease the randomness of the spatial distribution of the molecular components of the system and lead to a negative ΔS^E. However, in the mixing of nonpolar liquids ΔS^E is typically found to be positive. Such an effect could result when one of the components of the mixture consists of highly symmetrical molecules, so that the pure liquid might be characterized by a partially ordered structure which is destroyed in the mixing process. This interpretation was advanced by Scatchard et al. (1939) for the positive ΔS^E in the benzene-cyclohexane system and by van der Waals and Hermans (1950), who found that ΔS^E was about twice as large when *n*-hexadecane was mixed with the highly branched 2,2,4-trimethylpentane than for the *n*-hexadacane–*n*-heptane system. However, such an interpretation cannot be invoked to explain the positive ΔS^E obtained in the mixing of, for example, carbon tetrachloride with cyclohexane (Goates et al., 1959), illustrated in

Fig. II.2. An increase in the volume of the system accompanying the mixing process will make a positive contribution to the entropy, but this can account only in part for the observed values.

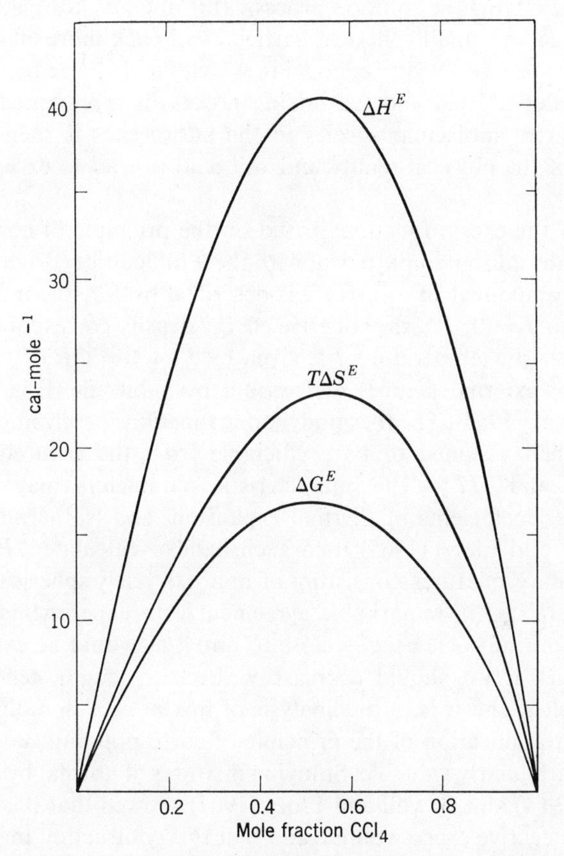

Fig. II.2. Excess thermodynamic functions for the carbon tetrachloride-cyclohexane system at 25°C.

On closer reflection, it becomes clear that a single parameter cannot characterize the interaction of a given molecule with other molecular species, since both the intermolecular forces and the molecular shapes must affect the nature of the dissolution process. This is reflected, for instance, in the lack of a general correlation between heat and volume effects. Although ΔH and ΔV^E have frequently the same sign, since favorable energetic interactions should lead to a contraction of volume, important exceptions have been observed, for example, in the carbon tetrachloride-neopentane system (Mathot and Desmyter, 1953) and with mixtures of aliphatic hydrocarbons

(Desmyter and van der Waals, 1958), where $\Delta E_M > 0$ and $\Delta V^E < 0$. These observations suggest that the efficiency with which dissimilar molecules will pack in a mixture will have a significant effect on the thermodynamic parameters characterizing the solution process. It is obvious, for instance, that a mixture of large and small spherical particles will pack more efficiently than spheres of a single size. A lattice model in which the distance between neighboring molecules is fixed and the mixing process is represented merely by assigning different molecular species to the lattice sites is then a poor representation of the physical reality and will tend to lead to excessively high ΔE_M values.

A theory of the excess functions based on the principle of corresponding states is an ambitious attempt to deal with these difficulties (Prigogine, 1957). Here the ith component of a mixture is described by V_i^*, the molar volume extrapolated to $T = 0$; P_i^*, the cohesive energy density corresponding to V_i^*, and a characteristic temperature T_i^*, given by $P_i^* V_i^* = CRT_i^*$, where $3C$ is the number of external degrees of freedom per molecule (Prigogine et al., 1957; Flory et al., 1964). The thermodynamic functions describing the mixing process are then assumed to be predictable from the reduced quantities V_i/V_i^*, P/P_i^*, and T/T_i^*. The characteristic parameters may be derived from densities, coefficients of thermal expansion, and isothermal compressibilities. Abe and Flory (1965) used such data to calculate ΔH, ΔV^E and ΔS^E for 23 binary mixtures consisting of approximately spherical molecules and obtained results in remarkable agreement with experiments. For these systems, the constant C is usually close to unity, as would be expected. For chain molecules C/V_i^* should decrease with chain length, tending to an asymptotic value, and it is in the analysis of mixtures of homologous chain molecules that application of the principle of corresponding states has been found to be particularly valuable. Studying mixtures of normal hydrocarbons, Flory et al. (1964) and Orwoll and Flory (1967) showed that theory predicts the observed negative excess volume and that this contraction makes a negative contribution to the enthalpy of mixing, which becomes relatively more important with rising temperature. This agrees with experimental evidence that $\partial \Delta H_M/\partial T < 0$ and that mixing of paraffin hydrocarbons actually may become exothermic at high temperatures (Friend et al., 1963). In fact, Orwoll and Flory were able to match closely experimental data obtained under widely varying conditions, using the intrinsic properties of the components of mixtures and a single adjustable parameter describing the difference in the interaction energies between the internal methylene and the terminal methyl residues.

When the energetics of the mixing process are determined by such highly specific interactions as hydrogen bonding and the association of electron donor and acceptor molecules, the dependence of the thermodynamic functions on the composition of the system may become very complex. Two situations may be distinguished. In the first case, one component of the mixture consists of associating molecules, while the other component is nonpolar. This case may be exemplified by the carbon tetrachloride-ethanol system (Barker et al., 1953), Its behavior is shown in Fig. II.3 Here the

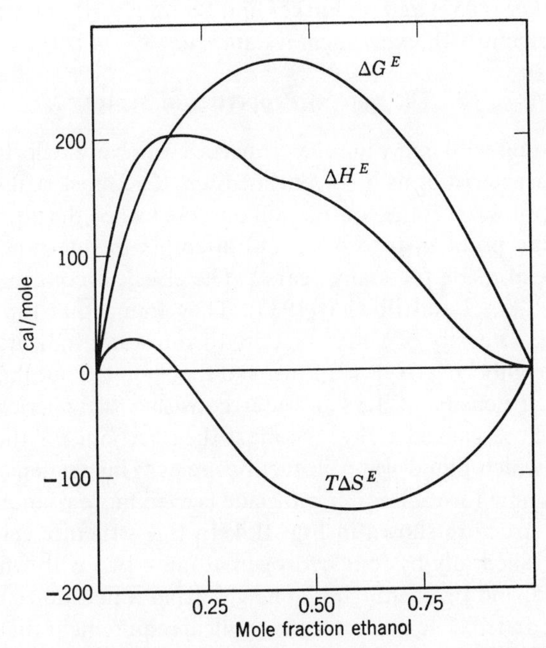

Fig. II.3. Excess thermodynamic functions for the carbon tetrachloride-ethanol system at 45°C.

symmetrical curve representing the excess free energy is quite deceptive in hiding a very complex behavior of both ΔH_M^E and $T\Delta S_M^E$. This complexity is hardly surprising, however, since the alcohol molecules may interact with each other to form a series of open-chain or cyclic association complexes, each species behaving differently on dilution. Also, the strength of the hydrogen bond is not the same in all these species; for instance, the cyclic tetramers are believed to be characterized by stronger hydrogen bonds than the open-chain polymers (Kuhn and Bowman, 1961). The second case involves mixtures in which the forces between the dissimilar molecules, far from being approximated by the rule of the geometric mean as given in (II.29), are actually much stronger than the forces operating between similar molecules of either component of the system. Such specific interaction of the components of a mixture will, of course, lead to exothermic mixing and can be described, in extreme cases, by the formation of a stoichiometric complex. A classical example is the chloroform-acetone system (Staveley et al., 1955), in which neither component can associate with molecules of its own kind but a hydrogen bond forms on mixing. This system is characterized, as would be expected, by very large negative ΔH_M and ΔS_M^E values. A statistical-mechanical theory of the excess functions in such strongly interacting systems advanced by Barker (1952) and by Barker and Smith (1954) appears to provide excellent agreement with experimental data.

6. The Solvent Properties of Water

Water is a liquid with many unique properties which are reflected in various ways in its characteristics as a solvent medium. The most striking anomaly in the behavior of water is the contraction observed when the liquid is warmed from its freezing point of 0 to 4°C, and attempts to interpret this phenomenon have been made for many years.‡ The classical contribution in this field is due to Bernal and Fowler (1933). They found that the angular dependence of the intensity of x-rays scattered from water indicates a distance of 1.38 Å for neighboring oxygen atoms, and they pointed out that this would correspond to a density of 1.84 if water consisted of spherical molecules arranged in a close-packed array. We must then account for the very much larger volume which liquid water in fact occupies. This tendency to take up a very large volume for each water molecule is even more pronounced in ice, which has the structure shown in Fig. II.4. In this structure each oxygen is surrounded tetrahedrally by four hydrogen atoms—two with which it forms a covalent bond, and two belonging to neighboring water molecules to which it is hydrogen bonded. The quantum-mechanical requirement that a hydrogen bond can form between two neighboring water molecules only if a hydrogen lies close to the line connecting their oxygen atoms is reponsible for preventing the molecules from packing as efficiently as they would if their interactions were due solely to dispersion forces and dipole-dipole interactions. According to Bernal and Fowler, liquid water retains to some extent molecular aggregates with the open icelike structure, thus accounting for its low density. This representation is also in good quantitative agreement with the angular distribution of the intensity of scattered x-rays. As the temperature of the liquid is raised, the tendency for icelike aggregates to "melt" contributes a volume contraction which is superimposed on the expansion due to thermal agitation. This concept was considered in greater detail by Frank and Wen (1957), who pointed out that the partial charge transfer

$$\begin{array}{cc} \text{H} & \text{H} \\ {}^{-\delta}| & {}^{+\delta}| \\ \text{—O—H}\ldots & \text{O—H} \end{array}$$

‡As early as 1892, W. K. Röntgen proposed a model to explain various anomalies of water (*Ann. Phys.*, **45**, 91).

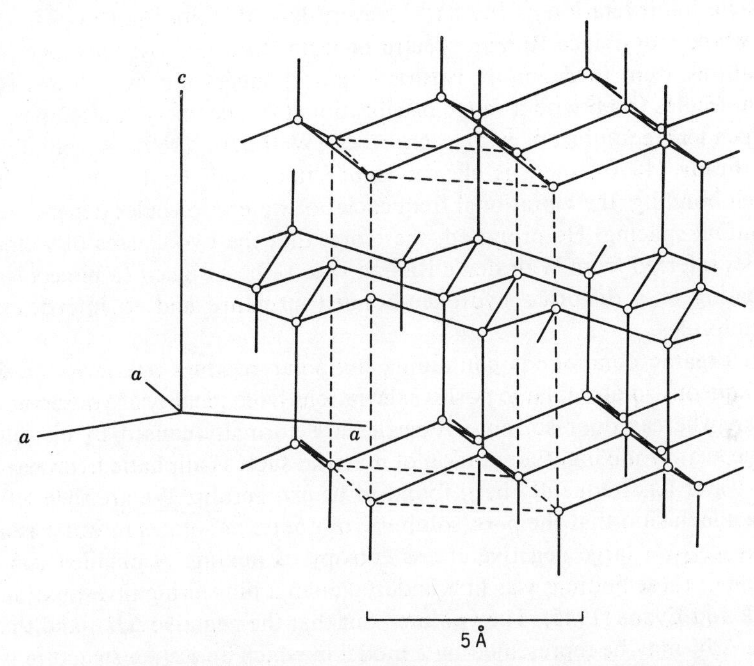

Fig. II.4. Arrangement of Oxygen Atoms in Ice (Owston, 1958).

accompanying the formation of a hydrogen bond:
makes the acceptor molecule a stronger hydrogen bond donor, while the donor molecule becomes a stronger hydrogen bond acceptor. As a result, there will be a tendency for hydrogen bonds to reinforce one another, favoring the formation of relatively large "flickering clusters" which form and dissolve with local energy fluctuations.

Experimental evidence bearing on the structure of water has been summarized in a monograph by Eisenberg and Kauzmann (1969), and a more recent one edited by Franks (1972). The results obtained by vibrational spectroscopy are of crucial importance in this context, since the periods of vibration are short compared to the times required for diffusional motions of the molecules, so that they reflect their "instantaneous" states. It is important, therefore, that Raman or infrared spectra (Wall and Hornig, 1956; Falk and Ford, 1966) contain broad unstructured bands which cannot be accounted for by a *small* number of distinct species, but indicate instead a continuous range of states for the water molecules. An increase in the temperature shifts the maxima of these bands to higher frequencies, indicating a decreasing effectiveness of the average hydrogen bonding. Eisenberg and Kauzmann suggest that the model proposed by Pople (1951), in which the hydrogen-bonded structure of water is subject to variable distortions, offers a plausible interpretation of the data. Neverthless, this does not appear to be the whole story since Raman spectra of water recorded over a range of temperatures seem to define an isosbestic point, suggesting two classes of water molecules (each with a range of vibrational frequencies) in a temperature-dependent equilibrium with each other (Walrafen, 1968), As pointed out by Frank (1970), there is clear evidence that, even in the absence of hydrogen bonding, the vibrational frequencies of water molecules depend on their mutual spacing. He proposed, therefore, that the two classes of water molecules inferred from Walrafen's Raman spectra be assigned to molecules participating in a distorted hydrogen-bonded structure and to interstitial water molecules.

When organic compounds containing nonpolar residues are introduced into an aqueous solution, large positive deviations from ideality are observed. However, whereas poor solvent properties are normally caused by an unfavorable heat of mixing, the solution of material such as aliphatic hydrocarbons in water has frequently been found to be exothermic. We are then left with the conclusion that the poor solubility of nonpolar solutes in water is a consequence of a large negative excess entropy of mixing. A detailed consideration of these findings was first undertaken in a pioneering investigation of Frank and Evans (1945). They pointed out that the negative ΔH_M and the negative ΔS_M^E may be represented by a model in which an icelike structure is stabilized in the neighborhood of nonpolar solutes; thus the heat evolved may be thought of as due to the latent heat of freezing of "icebergs" which represent regions of crystalline order and whose formation, therefore, leads to a loss of entropy. The tendency of nonpolar solutes to aggregate in aqueous media reduces the number of water molecules in their immediate vicinity and leads to "iceberg melting." This provides the driving force toward such aggregation and has been referred to as the "hydrophobic bond" (Kauzmann, 1959). Frank and Evans also noted that the solution of nonpolar hydrocarbons in water frequently leads to a pronounced contraction of volume. Typical thermodynamic data for the solution of liquid hydrocarbons in water (Nemethy and Scheraga, 1962) are listed in Table II.1. The very large negative values of ΔV appear to be inconsistent with the stabilization of an expanded structure similar to that of the usual low-pressure modification of ice. We may circumvent this difficulty by postulating either that a more compact water structure is stabilized in the vicinity of hydrocarbon molecules, or that the solute molecules are incorporated as interstitials in the structured water regions. It is best, however, to remember that the "icebergs" represent only a model for the unusual heat and entropy effects and that there is no reason why they should exhibit properties of known crystalline forms (Holtzer and Emerson, 1969).

TABLE II.1.
Thermodynamic Parameters for the Solution of Liquid Hydrocarbons in Water at 25°C

Substance	ΔG° (kcal/mole)	ΔH° (kcal/mole)	ΔS° (cal/deg. mole)	$\overline{\Delta C}_p$ (cal/deg. mole)	ΔV° (cm³/mole)
Propane	+4.9	−1.8	−22	90	−22.7
Butane	+5.9	−0.8	−22	90	
Benzene	+4.6	+0.6	−13	73	−6.5
Toluene	+5.3	+0.6	−16	85	

A very striking feature characterizing the transfer of hydrocarbon molecules to an aqueous environment is the large value of the partial excess heat capacity $\overline{\Delta C}_p$. For instance, $\overline{\Delta C}_p = 73$ cal / deg mole for benzene is more than twice the molar heat capacity of liquid benzene. These $\overline{\Delta C}_p$ values may be thought of as due to the additional heat requirement for the melting of "icebergs" in the neighborhood of the nonpolar solutes. Since $\overline{\Delta C}_p = \partial \overline{\Delta H}_2/\partial T$, the enthalpy of solution will typically change from a negative value at low temperature to a positive value at elevated temperatures. It follows from (II.15) that the solubility of these hydrocarbons will pass through a minimum at the temperature at which $\overline{\Delta H}_2 = 0$. Such behavior of the solubility-temperature relationship was first observed for aqueous solutions of benzene and other aromatic hydrocarbons (Bohon and Claussen, 1951).

When a substance dissolved in the aqueous medium contains not only

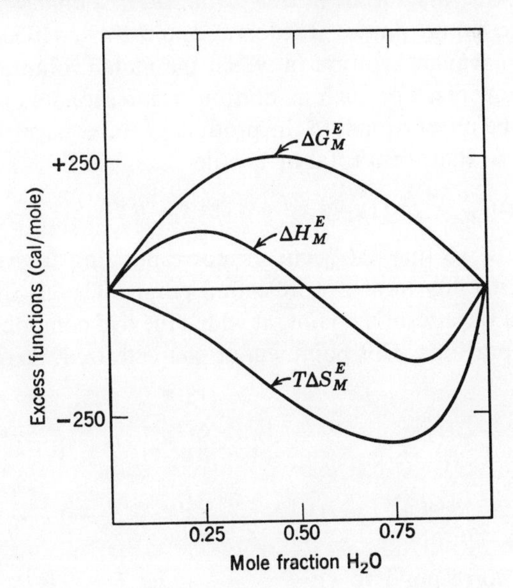

Fig. II.5. Excess thermodynamic functions for the dioxane-water system.

nonpolar residues but also functional groups which may participate in hydrogen bonding, the situation becomes much more complex. We may consider as typical examples the systems water-ethanol (Mitchell and Wynne-Jones, 1953) and water-dioxane (Malcolm and Rowlinson, 1957), for which the excess thermodynamic functions are plotted in Fig. II.5. For both systems there is a positive deviation from ideal solution behavior and a negative ΔS_M^E over the entire concentration range. However, the heat of mixing for the dioxane-water system shows an interesting inversion of sign, which apparently signifies that exothermic "iceberg formation" is the dominant effect only in water-rich media, whereas the endothermic mixing for systems of high dioxane concentration follows the normal pattern for the mixing of two liquids of different cohesive energy densities. Walrafen (1966) has shown that it is possible to distinguish by Raman spectroscopy between solutes which reinforce the hydrogen-bonded structure of water (e.g., sucrose) and solutes which favor the disintegration of this structure (e.g., urea).

7. Phase Equilibria

Limited solubility of liquids in liquids may be the result of an unfavorable heat of mixing or of a large negative value of the excess entropy characterizing the solution process. The first is by far the more frequent case. We may then inquire what magnitude of ΔH_M will lead in a binary system to the coexistence of two liquid phases. Hildebrand and Scott (1950) considered this problem for a regular solution in which the molar volumes are similar for the two components. For such a solution it is reasonable to expect the heat of mixing to be proportional to the product of the concentrations of the two components, so that for a total of 1 mole

$$\Delta G_M = \boldsymbol{R}T(x_1 \ln x_1 + x_2 \ln x_2) + Bx_1x_2 \tag{II.36}$$

We saw in Fig. II.1b that ΔG_M curves corresponding to systems which exhibit limited miscibility have two inflection points. These will coalesce in the limiting case of the consolute point, at which the two components become miscible in all proportions. This point will be characterized, therefore, by the conditions:

$$\left(\frac{\partial^2 \Delta G_M}{\partial x_1^2}\right)_{T,P} = 0, \qquad \left(\frac{\partial^2 \Delta G_M}{\partial x_2^2}\right)_{T,P} = 0$$
$$\left(\frac{\partial^3 \Delta G_M}{\partial x_1^3}\right)_{T,P} = 0, \qquad \left(\frac{\partial^3 \Delta G_M}{\partial x_2^3}\right)_{T,P} = 0 \tag{II.37}$$

These conditions correspond to $x_1 = x_2 = {}^1/_2$ and $B = 2\boldsymbol{R}T_c$, where T_c is the consolute temperature, so that at the consulate point the mixing is endothermic to the extent of $\boldsymbol{R}T_c/2$ per mole. If we use the van Laar theory for estimating the heat of mixing from (II.34), then, for liquids with a molar volume of 100 ml, a critical miscibility temperature T_c of 300° K requires that the solubility parameters δ of the two components differ from each other by 3.5 units. Such large differences in the δ parameters are rare. Nonpolar systems of low molecular weight organic liquids, therefore, generally do not exhibit phase separation, unless one of the components is a fluorocarbon (with very low δ values of 5.6–6.1) or the strongly hydrogen-bonded methanol. Other systems with incomplete miscibility at room temperature are formed by aniline, nitrobenzene, or phenol with normal aliphatic hydrocarbons.

If the mixing process is endothermic, solubility will increase with rising temperature and the critical point will lie at an "upper consolute temperature" above which the components of the system are miscible in all proportions. An example of such a system (Hildebrand, 1953) is given in Fig. II.6*a*.

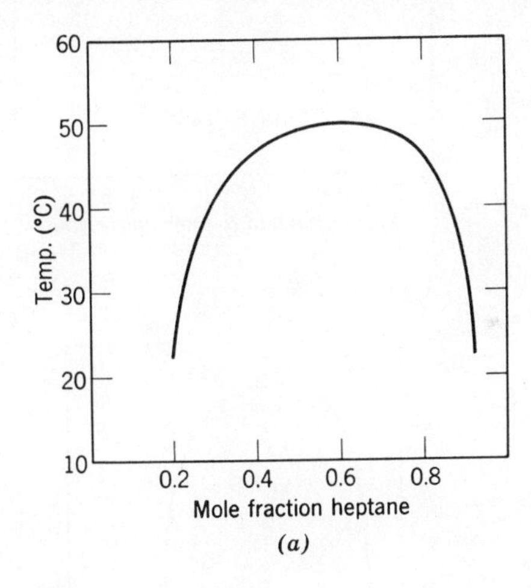

Fig. II.6(*a*) Phase diagram of the heptane-perfluoroheptane system.

It is much more unusual to have a binary system with complete miscibility below a "lower consolute temperature" and separation into two phases above it. The conditions leading to phase diagrams of this type have been discussed by Copp and Everett (1953). A lower consolute temperature can occur only if mixing is exothermic and the excess entropy of mixing has a large negative value, so that ΔG_M^E attains the value required for phase separation. These conditions are satisfied in some aqueous solutions due to the "iceberg formation" discussed in the preceding section, and Fig. II.6*b*, representing the phase diagram of the *N*-methylpiperidine-water system (Flaschner, 1908), is a typical example of this case. Finally, it is possible for a binary system to

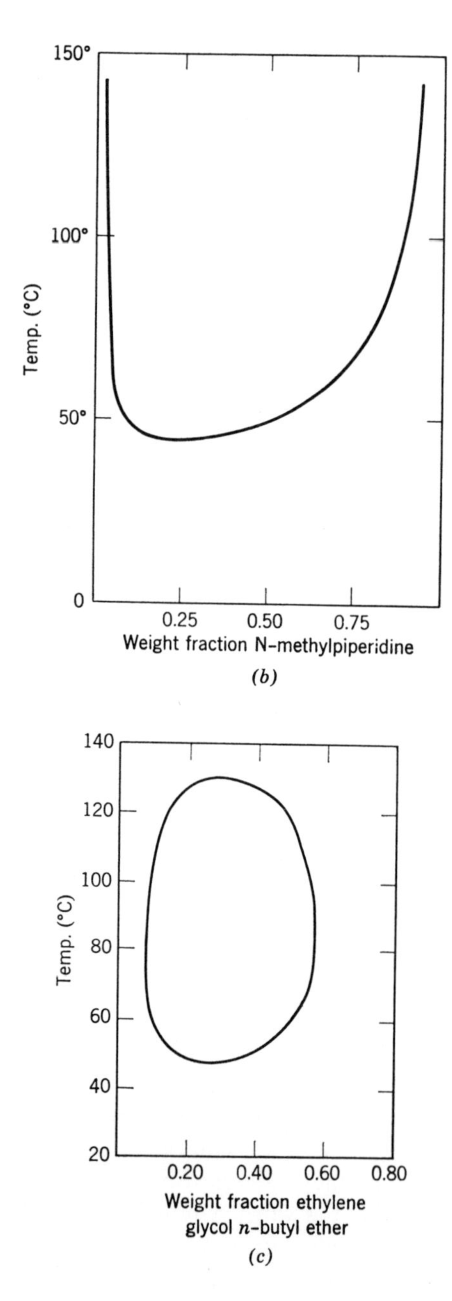

Fig. II.6 (continued) (*b*) Phase diagram of the water-*N*-methylpiperidine system. (*c*) Phase diagram of the water-ethylene glycol mono-*n*-butyl ether system.

have both an upper and a lower consolute temperature, so that coexistence of two liquid phases is observed only in a limited temperature range and the phase diagram is characterized by a closed loop. The thermodynamic requirement for this phenomenon is a change of sign of ΔH_M from a negative value at the lower critical solution temperature to a positive value at the upper critical solution temperature. This requires that $d\,\Delta H_M/dT = \Delta C_p$ have a large positive value, a condition which is fulfilled, as we saw in Table II.1, for solutions of "iceberg-forming" solutes in water. Barker and Fock (1953) have pointed out that phase diagrams with a closed loop might be expected, in general, for systems in which interactions between unlike molecules are repulsive for most relative orientations but attractive for a few. A typical phase diagram of this type, for the system ethylene glycol monobutyl ether-water shown in Fig. II.6*c*, was studied by Cox and Cretcher (1926), who listed a number of other systems with similar behavior.

8. Solubility of Crystalline Solids

The general relation for the solubility of a crystalline solute was given in (II.16b). Here the partial molar heat of solution $(\overline{\Delta H_2})'$ may be considered as the sum of the latent heat of melting and the ΔH corresponding to the mixing of 1 mole of the molten solute with a large volume of an almost saturated solution. If the molten solute forms an ideal solution with the solvent, this heat of mixing vanishes, as does also the term $(\partial \ln \gamma_2/\partial x_2)_{T,P}$. We may then integrate (II.16b) from any temperature T to the melting temperature T_m, and, disregarding the variation of the latent heat of melting ΔH_m with T, we obtain an estimate of the "ideal solubility":

$$\ln x_2 = -\left(\frac{\Delta H_m}{R}\right)\frac{T_m - T}{T_m T} \tag{II.38}$$

We see then that the ideal solubility of a crystal is related to its melting point and its latent heat of melting. But these two quantities are not independent of each other, since T_m is the ratio of the heat to the entropy of melting, so that T_m usually increases with an increase in ΔH_m. For similar compounds, ΔH_m depends on the efficiency of packing in the crystal lattice and tends to rise with an increasing symmetry of the molecule. This is the cause, for instance, of the general rule that *para*-disubstituted benzenes have higher melting points and lower solubilities than their *ortho* and *meta* isomers. Data typifying this effect are listed in Table II.2.

An instructive illustration of the principles outlined above is provided by *p*-phenylene oligomers and their derivatives. Whereas octa (*p*-phenylene) melts above 500 °C and is virtually insoluble at room temperature, a derivative carrying a methyl group on each phenyl residue melts at 254 °C and has a

TABLE II.2.
Thermodynamic Data for the Melting of the Isomeric Dinitrobenzenes and Their Ideal Solubilities

	ortho	*meta*	*para*
ΔH_m (cal/mole)	5430	4150	6720
ΔS_m (cal/deg-mole)	13.9	11.4	15.0
T_m (°K)	390	363	447
Ideal solubility (x_2) at 303°K	0.136	0.319	0.027

solubility in toluene at 20 °C of 13 g/l (Kern et al., 1960). This may be ascribed to both less efficient packing in the crystal (reducing ΔH_m) and a higher value of ΔS_m.

B. BINARY SYSTEMS CONTAINING MACROMOLECULES

1. Deviation of Polymer Solutions from Ideal Solution Behavior

In classical treatments of the thermodynamic behavior of solutions it was generally assumed—implicitly or explicitly—that deviations from ideal solution behavior must be associated with a finite heat of solution. However, when physical chemists began to investigate the properties of systems containing high molecular weight components, such systems showed extremely large deviations from the behavior to be expected of ideal solutions, even in cases where the heat of mixing of a polymer with a low molecular weight solvent was negligible. We may use as an extreme example a vulcanized automobile tire swollen with 1 gram of benzene. By the classical definition, the tire is a single molecule, since any two atoms in it are connected by a series of covalent bonds. We have then a system containing 1 rubber molecule and 8×10^{21} benzene molecules. Since the heat of mixing of rubber and benzene is extremely small, the system might have been expected to behave ideally, so that the vapor pressure of the benzene swelling the tire should have been indistinguishable from that of pure benzene. We know that this is far from being the case and that the vapor pressure of benzene under those conditions will be very small indeed.

In studying the thermodynamic behavior of a binary system containing a polymeric and a low molecular weight component, it is most convenient to focus attention on the thermodynamic activity of the small molecules. The large negative deviations from ideal solution behavior imply a large negative value of $\overline{\Delta G_1^E}$, and since this is found even when the enthalpy of dilution is zero or positive, we must conclude that polymer solutions are characterized by large positive excess entropies of dilution. A theory of this effect may be developed by the methods of statistical mechanics, but before discussing the results obtained in this manner we shall try to obtain a qualitative feeling for the origins of the effect.

Most polymer molecules may be represented as flexible chains. If such chains are sufficiently long, the shape of their backbones may be likened to the random flight path of a particle undergoing Brownian motion and is then commonly referred to as a "random coil." At extreme dilutions, each one of these chains can assume a large number of shapes (conformations), and the probability that any one chain exists at a given time in a given shape will be independent of the shapes assumed by all the other chains. In the pure amorphous polymer the chain molecules are just as flexible as in solution and may be assumed to be able to exist in a similar number of conformations. However, it is obvious that now these molecular shapes are not independent of each other; the shape of each molecular chain must be correlated with the shape assumed by its neighbors so as to fill the available space. It is this restraint which is eliminated when a molecular chain is transferred from the pure polymer phase to a dilute solution, and this accounts for the characteristic positive ΔS_M^E values of solutions of chain molecules. In considering the problem more closely, we shall be led to distinguish two ranges of concentration in systems containing chain molecules. In relatively dilute solution, the loose molecular coils will only occasionally interpenetrate, and the frequency of such interference will depend on the size of the individual coil. In this range we may, therefore, suspect that ΔS_M^E is sensitive to the polymer chain length. At higher concentrations the total available volume is much less than the sum of the volumes enclosed by the twisting chain molecules, and therefore, the molecular chains must necessarily become heavily intertwined. In this range we might suspect that the interference with the shape of a given chain, due to the presence of other chain molecules, will depend on the fraction of the volume occupied by these chains. It should be rather independent of the occasional occurrence of chain ends so that it should not depend significantly on the molecular weight of the polymer.

A quantitative theory of the change in conformational entropy produced by the mixing of flexible chain polymers with a solvent of low molecular weight was formulated by Flory (1942) and Huggins (1942a,b,c,), who evaluated the number of distinguishable ways in which N_1 solvent molecules with a molar volume V_1 and N_2 polymer chains with a molar volume V_2 can be placed on a lattice so that each lattice site is occupied by either a solvent molecule or one of the V_2/V_1 segments of a polymer chain. The crucial point in the calculation is the assumption that, in placing a given chain segment on the lattice, which already contains previously placed chains, the probability of occupancy of a lattice site may be approximated by the overall fraction of occupied sites. This approximation is clearly untenable in very dilute solutions, where molecular coils, representing high local concentrations of chain segments, are separated by regions of pure solvent. However, the assumption of the Flory-Huggins theory is reasonable in the concentration range in which the chains interpenetrate each other, so that the density of chain segments is uniform, on the molecular scale, throughout the system, and it is in this range that the theory has been eminently successful. Moreover the result of the Flory-Huggins treatment was later found to be most useful in the theoretical interpretation of the extension of individual chain molecules.

Using a more concise procedure than that employed originally by Flory and by Huggins, we may obtain the partial molar excess entropy of the solvent in a solution containing flexible chain molecules in the following manner. We start with the chains at infinite dilution, so that each segment may occupy z lattice sites adjoining the preceding chain segment. If the ith polymer chain is now transferred to a solution in which other chains occupy a volume fraction $\phi_{i'}$ the number of positions for the chain segment is reduced, on the average, to $z(1 - \phi_i)$ if the probability of finding a chain segment of another polymer is considered to be uniform over the entire system. The excess entropy resulting from this restraint is then $\boldsymbol{R} \ln [z(1 - \phi_i)/z] = R \ln (1 - \phi_i)$ per mole of chain segments. (Note that this is independent of z, characterizing the flexibility of the chain molecules.) Averaging this quantity for a process in which ϕ_i increases from 0 to ϕ_2, the volume fraction of polymer in the solution, we have for the transfer of n_2 moles of polymer chains (i.e., n_2V_2/V_1 moles of polymer chains segments)

$$\Delta S^E = -\left(\frac{\boldsymbol{R}n_2V_2}{V_1}\right)\left(\frac{\phi_1 \ln \phi_1}{\phi_2} + 1\right)$$

$$= -\boldsymbol{R}\left(n_1 \ln \phi_1 + \frac{n_2V_2}{V_1}\right) \tag{II.39}$$

Substituting $\phi_1 = n_1V_1/(n_1V_1 + n_2V_2)$ and differentiating with respect to $n_{1'}$ we obtain for the partial molar excess entropy of the solvent

$$\overline{\Delta S_1}^E = -\boldsymbol{R}(\ln \phi_1 + \phi_2) \tag{II.40}$$

This result is independent of the length of the chains, as a necessary consequence of the model employed, that is, the assumed uniform distribution of the chain segments over the volume of the system.

In relatively dilute systems the mole fraction x_2 of solute molecules may be approximated by ϕ_2V_1/V_2, so that the ideal partial molar entropy of the solvent is

$$\overline{\Delta S_1}^i = -\boldsymbol{R} \ln\left(1 - \frac{\phi_2V_1}{V_2}\right)$$

$$= R\left[\frac{\phi_2 V_1}{V_2} + 1/2\left(\frac{\phi_2 V_1}{V_2}\right)^2 + \cdots\right] \tag{II.41}$$

Since $\overline{\Delta S_1} = \overline{\Delta S_1}^i + \overline{\Delta S_1}^E$, we obtain for athermal solutions in which $ln\ a_1 = -\overline{\Delta S_1}/\boldsymbol{R}$ and in which higher powers of V_1/V_2 are negligible

$$\ln a_1 = \ln \phi_1 + \left(1 - \frac{V_1}{V_2}\right)\phi_2$$

$$= -\phi_2\left(\frac{V_1}{V_2}\right) - \frac{\phi_2^2}{2} - \frac{\phi_2^3}{3} - \cdots \tag{II.42}$$

This may be compared to the value for dilute ideal solutions:

$$\ln a_1^i = -\phi_2\left(\frac{V_1}{V_2}\right) - 1/2\left(\frac{\phi_2 V_1}{V_2}\right)^2 - 1/3\left(\frac{\phi_2 V_1}{V_2}\right)^3 - \cdots \tag{II.43}$$

In classical physical chemistry, thermodynamic measurements on solutions usually emphasized the determination of the molecular weight from colligative properties of dilute solutions, where the effect observed is proportional to $\ln a_1$. A comparison of (II.42) and (II.43) shows that, at a typical solute concentration of $\phi_2 = 0.01$, $\ln a_1$ differs from its ideal value by factors of only 1.01 and 1.10 for $V_2/V_1 = 2$ and 20, respectively, so that it is not surprising that effects due to an excess configurational entropy of dilution should have been missed until solutions of very long chain molecules became the subjects of investigation. On the other hand, deviations from solution ideality, observable as reduced vapor pressures of the solvent, become very obvious in the range of composition where most of the volume is occupied by the larger molecules even if V_2/V_1 has relatively low values. For instance, at $\phi_2 = 0.99$ the Flory-Huggins theory predicts a decrease of a_1 below the ideal solution value by factors of 1.2, 2.2, and 6.4 for $V_2/V_1 = 2$, 5, and 20, respectively.

In addition to the deviation from ideality caused by the chainlike nature of the solute molecules, we have to consider contributions to ΔH_M and ΔS_M^E due to solute-solvent interactions. These effects are short range (except for the long-range Coulombic interactions in solutions of polyelectrolytes) and should, therefore, be proportional to the "number of contact points" between solute and solvent. The concentration dependence of this somewhat nebulous quantity should not be altered appreciably when chain segments are joined together to a macromolecule, and we may assume that the contribution to ΔG_M^E due to nearest neighbor contacts is approximately proportional to $V\phi_1\phi_2$. We have seen from (II.34) that this is the form predicted by the van Laar theory for ΔE_M, and the dependence of the heat of mixing on the composition of the system should have a similar form also in cases where this theory does not apply (e.g., in systems in which mixing is exothermic.) In binary systems of small molecules, in which molecular interactions are governed largely by dispersion forces, an analogous concentration dependence holds also for ΔS_M^E (see the carbon tetrachloride-cyclohexane system, Fig. II.2). However, inspection of Fig. II.3 and especially Fig. II.5 shows that ΔG_M^E approaches the predicted parabolic behavior even in systems with very complex interactions which lead to striking deviations from this concentration dependence for ΔH_M^E and ΔS_M^E.

Using the concept of a regular solution, we may treat the free energy of mixing as being made up additively from contributions due to configurational probability and a free energy arising from nearest-neighbor interactions. The latter are characterized by the "Flory-Huggins interaction parameter" χ, which specifies, in units of $\boldsymbol{RT}$, the excess free energy for the transfer of a mole of solvent molecules from the pure solvent to the pure polymer phase. With the initial state involving a solvent and a disordered polymer phase, the Flory-Huggins treatment leads to

$$\Delta G_M = \boldsymbol{RT}(n_1 \ln \phi_1 + n_2 \ln \phi_2 + n_1 \chi \phi_2) \tag{II.44}$$

We obtain then for the activity of the solvent

$$\ln a_1 = \frac{\overline{\Delta G_1}}{\boldsymbol{RT}} = \ln \phi_1 + \left(1 - \frac{V_1}{V_2}\right)\phi_2 + \chi\phi_2^2 \tag{II.45}$$

which yields, on expansion of the $\ln \phi_1 = \ln(1 - \phi_2)$ term,

$$\ln a_1 = \frac{-\phi_2 V_1}{V_2} - (\tfrac{1}{2} - \chi)\phi_2^2 - \tfrac{1}{3}\phi_2^3 \cdots \tag{II.46}$$

It is instructive to compare this result with one which would be obtained if the entropy of dilution had the ideal value given in (II.23). The solvent activity would then be given by

$$\ln a_1 = -\frac{\phi_2 V_1}{V_2} - \frac{1}{2}\left(\frac{\phi_2 V_1}{V_2}\right)^2 - \frac{1}{3}\left(\frac{\phi_2 V_1}{V_2}\right)^3 - \cdots + \chi\phi_2^2 \tag{II.47}$$

In this case a large value of V_2/V_1 would make mixing impossible if χ had an appreciable positive value. Thus, endothermic mixing of high molecular weight polymers with solvents is possible only because of the conformational entropy gained by flexible chain molecules in the process of dilution.

It should be emphasized that a large ratio in the molar volumes of solute and solvent does not necessarily lead to a pronounced deviation from solution ideality in athermal solutions. Thus, calculations of the excluded volume effect in solutions of large molecules behaving as rigid spheres lead to a solvent activity given, for athermal solutions, by (Huggins, 1948)

$$\ln a_1 = -\frac{\phi_2 V_1}{V_2} - 4\left(\frac{V_1}{V_2}\right)\phi_2^2 - 10\left(\frac{V_1}{V_2}\right)\phi_2^3 - \cdots \tag{II.48}$$

so that the relative magnitudes of the leading term and the terms depending on the higher powers of solute concentration remain unchanged as the molecular volume of the solute is increased.

The validity of the Flory-Huggins theory for solutions of flexible chain molecules may be tested by plotting $\ln (a_1/\phi_1) - [1 - (V_1/V_2)]\phi_2$ against ϕ_2^2. According to (II.45), such a plot should be linear and have a slope χ. In Fig. II.7 experimental data obtained for four typical systems are plotted in this

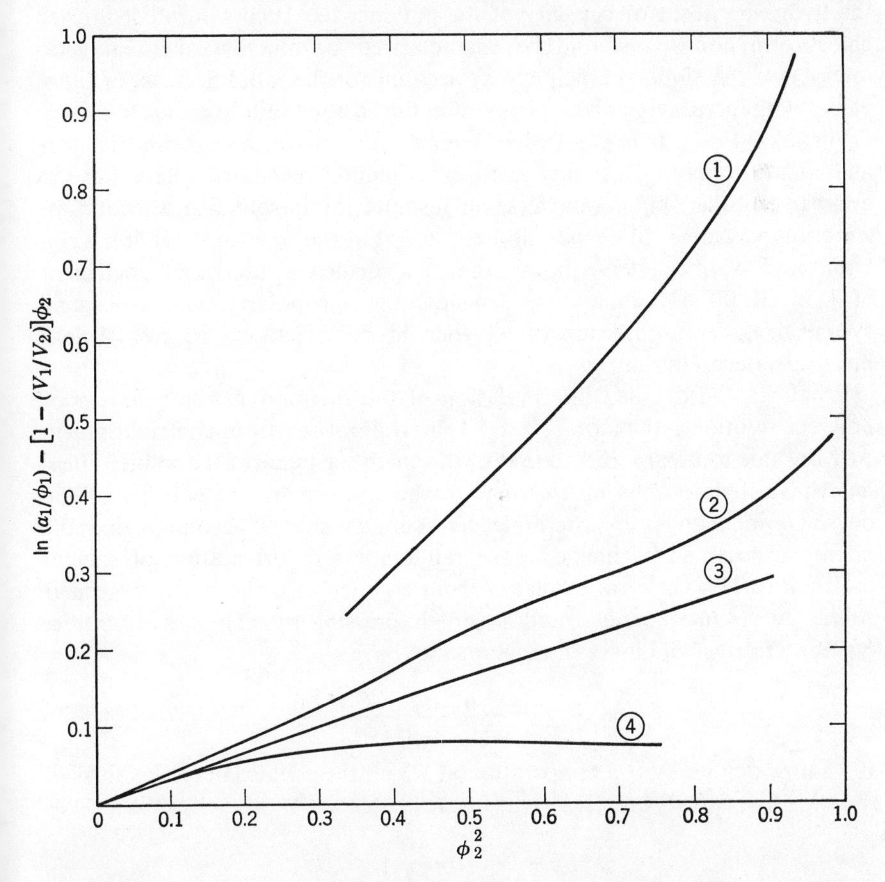

Fig. II.7. Comparison of Flory-Huggins theory with experimental data. (1) Polyisobutene-benzene. (2) Hevea rubber-benzene. (3) Polybutadiene-benzene. (4) Polystyrene-chloroform.

manner. For Hevea rubber in benzene (Gee and Treloar, 1942) and polybutadiene in benzene (Jessup, 1958), two systems in which mixing is nearly athermal, the predicted linearity of the plot is in fairly good agreement with experiment. For polyisobutene in benzene, which is a poor solvent for this polymer (Jessup, 1958), the plot has an upward curvature, a feature which seems to be typical of very poor solvents in general (Booth et al., 1957). On the other hand, the plot for the polystyrene-chloroform system (Bawn and Wajid, 1956) is characterized by a strong downward curvature. This is a system in which one would expect exothermic mixing as a result of hydrogen bonding of the acidic hydrogen of chloroform with the aromatic nuclei of polystyrene (Creswell and Allred, 1962) in systems rich in polystyrene. Once the hydrogen acceptor capacity of the polymer has been saturated, further chloroform addition should be characterized by much weaker energetic interactions. A similar principle may account for the rapid decrease of the χ values with increasing polymer content in the acetone-cellulose nitrate system (Takenaka, 1957). It is possible, of course, to improve the agreement of an analytical expression for ln a_1 with experimental results by increasing the number of adjustable parameters, for instance, by introducing a second interaction parameter to be used as a coefficient of the ϕ_2^3 term in (II.46). Krigbaum and Geymer (1959) showed that this coefficient, unlike the coefficient of $\frac{1}{3}$ in (II.46), is temperature dependent in the polystyrene-cyclohexane system. It is, doubtful, however, whether this coefficient can be given a clear physical interpretation.

Whatever the detailed interpretation of the thermodynamic behavior of polymer solutions, the term $(\frac{1}{2} - \chi)\,\phi_2^2$ in (II.46) arises from contributions to $\overline{\Delta G_1^E}/RT$ due to binary interactions of the chain segments of the solute. These may have their origins in a change in the conformational entropy of the polymer, in changes in intermolecular contact energy accompanying the mixing process, in a change in the randomness of orientation of solvent molecules when they are displaced from contact with the macromolecular solute, in volume changes, and so on. Expressing by asterisks quantities resulting from such binary interactions, we have

$$\tfrac{1}{2} - \chi = -\frac{(\overline{\Delta H_1^*} - T\,\overline{\Delta S_1^*})}{RT\phi_2^2} \tag{II.49}$$

If we now denote by Θ a temperature at which the coefficient of ϕ_2^2 vanishes, then $\overline{\Delta H_1^*} = \Theta\,\overline{\Delta S_1^*}$, and (II.49) may be rewritten as

$$\tfrac{1}{2} - \chi = \Psi\left(1 - \frac{\Theta}{T}\right)$$

$$\Psi = \frac{\overline{\Delta S_1^*}}{R\phi_2^2} \tag{II.50}$$

The use of the two parameters Θ and Ψ, as suggested by Flory (1953), has largely supplanted the interaction parameter χ, which describes the behavior of a given polymer-solvent system at a single temperature. We shall see in Chapters III and IV that solutions in Θ-solvents, exhibiting pseudoideality at low concentrations of the solute, play an important role in the interpretation of the behavior of flexible chain molecules. It should be understood, however, that $\overline{\Delta H_1^*}/\phi_2^2$ and $\Delta S_1^*/\phi_2^2$ can be assumed to be temperature independent only within a relatively narrow temperature range, so that the linear dependence of χ on $1/T$ is subject to the same limitation. We shall return to this point when discussing polymer-solvent phase equilibria in Section II.B.3.

2. Phase Equilibirium in Systems of a Polymer and Poor Solvents

The general theory of phase equilibria in systems containing long chain molecules and low molecular weight solvents is rather complex, and it is questionable whether a theoretical model lending itself to mathematical analysis can take account of all the factors involved. It is much simpler to consider the limiting case of a two-component system in which the mutual solubility of the components is very low, so that each phase contains one component in highly dilute solution, while the other component is virtually in its standard state. The condition for equilibrium is then

$$\bar{G}_i^0 = \bar{G}_i^0 + \boldsymbol{R}T \ln x_i + \bar{G}_i^E$$

$$x_i = \exp\left[\frac{-\bar{G}_i^E}{\boldsymbol{R}T}\right] \quad \text{(II.51)}$$

The $\bar{G}_1^E$ characterizing solvent molecules in very slightly swollen polymer will be due to nearest-neighbor interaction effects between solvent molecules and polymer segments. The chain length of the polymer will affect it only insofar as the medium will have different properties in the immediate vicinity of chain ends. For high polymers the concentration of chain ends will be very low, and $\bar{G}_1^E$ will approach the limiting value characteristic of solvent molecules in polymer of infinite molecular weight. We conclude that the equilibrium swelling of polymers with small solvent molecules will not depend appreciably on the chain length of the polymer.

The situation is quite different with $\bar{G}_2^E$, characterizing the chain molecules in dilute solution. This quantity contains contributions from both nearest-neighbor interactions of solvent molecules with polymer segments and the increase of conformational entropy accompanying the separation of flexible chains from each other. Both contributions should be proportional to chain length, so that the solubility of polymers in poor solvents should fall off with their molecular weight as indicated by

$$x_2 = \exp(-\mathrm{A}M_2) \quad \text{(II.52)}$$

where A is a characteristic constant.†

†An unexplained exception to the expected decrease of polymer solubility as the chain length is extended has been reported for the amylose-water system. The solubility passes, in this case, through a pronounced minimum (Kerr, 1949; Husemann et al., 1958), which has been located at degrees of polymerization of 75–80 (Pfannemüller et al., 1971).

Results based on the Flory-Huggins theory are in agreement with these qualitative considerations. Setting $\ln a_1 = 0$ and $\phi_2 \to 1$ in (II.45), we obtain for the volume fraction of solvent contained in the swollen polymer in equilibrium with a poor swelling agent

$$\phi_1 = \exp\left(-1 - \chi + \frac{V_1}{V_2}\right) \tag{II.53}$$

Similarly, from the activity of the polymeric species:

$$\ln a_2 = \frac{1}{RT}\left(\frac{\partial G_M}{\partial n_2}\right)_{n_1} = \ln \phi_2 + \left(1 - \frac{V_2}{V_1}\right)\phi_1 + \frac{V_2}{V_1}\chi\phi_1^2 \tag{II.54}$$

we obtain, by setting $\ln a_2 = 0$, $\phi_1 \to 1$, for the volume fraction of polymer in a solution phase which is in equilibrium with very slightly swollen polymer

$$\phi_2 = \exp\left[-\left(\frac{V_2}{V_1}\right)\left(\chi + \frac{V_1}{V_2} - 1\right)\right] \tag{II.55}$$

This expression has the form postulated in (II.52).

3. The Consolute Point

In our discussion of the mutual solubility of liquids in Section II. A.7 we saw that the consolute point (at which the difference between two coexisting phases vanishes) requires $(\partial^2 \Delta G_M/\partial x_1^2)_{T,P} = 0$ and $(\partial^3 \Delta G_M/\partial x_1^3)_{T,P} = 0$. We may write these conditions in the form $\partial \ln a_1/\partial\phi_2 = 0$, $\partial^2 \ln a_1/\partial\phi_2^2 = 0$ and use, for $\ln a_1$, the expression given in (II.45). This leads to the prediction that the critical point is characterized (Flory, 1942) by

$$(\phi_2)_{\text{crit}} = \frac{1}{1 + \sqrt{V_2/V_1}} \tag{II.56}$$

$$\chi_{\text{crit}} = 1/2 + \sqrt{V_1/V_2} + V_1/2V_2 \tag{II.57}$$

This result depends on the applicability of the Flory-Huggins expression for the free energy of mixing at the relatively high dilution corresponding to the consolute point of high polymer solutions. We shall see in Chapter IV that a different approach becomes necessary when dealing with systems in which the polymer coils occupy discrete regions of space. However, it seems reasonable to assume that at the critical point the chains will interpenetrate sufficiently to justify the assumption of the uniform concentration of chain segments, on which the Flory-Huggins theory is based. On the other hand, it can be shown that results (II.56) and (II.57) are quite sensitive to the dependence of $\ln a_1$ on ϕ_2^3. In the Flory-Huggins theory this dependence is constant for all solvent media [see (II.46)], but Krigbaum and Geymer (1959) have shown that the coefficient of ϕ_2^3 tends to vanish at the consolute point and this should introduce an error in the theory of the critical point as outlined above. This applies particularly to the value of $(\phi_2)_{crit}$, which is found experimentally to be appreciably higher than predicted (Shultz and Flory, 1952).

Relations (II.56) and (II.57) have two features which should be noted. First, the critical polymer concentration, for high molecular weight solutes, occurs at $\phi_2 \approx \sqrt{V_1/V_2}$, that is, the consolute point will occur in a very dilute system if the polymer chains are very long. This means that a slight variation in the quality of the solvent may result in a shift from complete miscibility to a very low equilibrium solubility of the polymer. Also, since even the more concentrated of the two phases in equilibrium close to the consolute point may be highly dilute, it may separate as a viscous liquid, rather than as a gel. This phenomenon, termed "coacervation" by Bungenberg de Jong and Kruyt (1930), occurs usually as a result of attractive interactions of two polymers, but it has also been observed in systems containing a single polymeric species (Turska and Łaczkowski, 1957). Second, the critical value of the interaction parameter χ approaches a limiting value of $\frac{1}{2}$ as V_2/V_1 becomes very large. This result is an obvious consequence of (II.46). As the chain length of the solute approaches infinity, the first term on the right of this expression vanishes and the transfer of polymer molecules into the pure solvent requires $\lim c_2 \to 0\ (d \ln a_1/d\phi_2) < 0$. We have then, in the limit of extremely long polymer chains, a sharp transition from miscibility in all proportions, when $\chi < \frac{1}{2}$, to complete insolubility in media characterized by $\chi > \frac{1}{2}$. According to (II.50), the limiting condition $\chi = \frac{1}{2}$ corresponds to $T = \Theta$. This condition may be attained either by a variation in the temperature for solvents of constant composition or by a variation in the composition of a mixed solvent medium at constant temperature.

Combination of (II.50) and (II.57) leads to

$$\frac{1}{T_c} = \frac{1}{\Theta}\left[1 + \frac{1}{\Psi}\sqrt{V_1/V_2} + V_1/2V_2)\right] \tag{II.58}$$

Shultz and Flory (1952, 1953) found plots of $1/T_c$ against $\sqrt{V_1/V_2} + V_1/2V_2$ to be indeed linear, as predicted, defining the two characteristic parameters Θ and Ψ. Their values for these parameters for a number of systems are listed in Table II.3. For two of the systems, poly(acrylic acid) in dioxane and polymethacrylonitrile in butanone, the value of Ψ is negative. This makes the polymer dilution an exothermic process, and the solubility must then decrease with rising temperature. In these cases T_c *decreases* with increasing chain length of the polymer and Θ represents, therefore, the *highest* temperature at which very high molecular weight polymer and solvent are miscible in all proportions.

The preceding discussion is based on the implicit assumption that the entropy of mixing and the heat of mixing are temperature independent.

TABLE II.3.
Thermodynamic Data for Polymer-Solvent Systems from the Dependence of the Consolute Temperature on the Polymer Chain Length

Polymer	Solvent	Θ (°K)	Ψ
Polystyrene	Octadecanol	474	1.30
Polystyrene	Cyclohexanol	358.4	1.51
Polystyrene	Cyclohexane	307.2	1.056
Polystyrene	Ethylcyclohexane	343.2	0.875
Polyethylene	Nitrobenzene	503	1.09
Polyisobutene	Diisobutyl ketone	331.1	0.65
Poly (methyl methacrylate)	Heptanone-4	305	0.61
Polydimethylsiloxane	Phenetole	358	0.69
Polydimethylsiloxane	Butanone	298.2	0.43
Cellulose tricaprylate	Dimethylformamide	413	0.39
Cellulose tricaprylate	3-Phenylpropanol-1	323	0.21
Poly (acrylic acid)	Dioxane	302.2	−0.31
Polymethacrylonitrile	Butanone	279	−0.63

Although this may be a reasonable approximation when dealing with data obtained within relatively narrow temperature limits, such a constancy of ΔS_M and ΔH_M cannot be assumed to hold over a wide temperature range. In particular, when ΔH_M changes its sign, we may obtain two different critical solution temperatures T_c for polymers of a given molecular weight

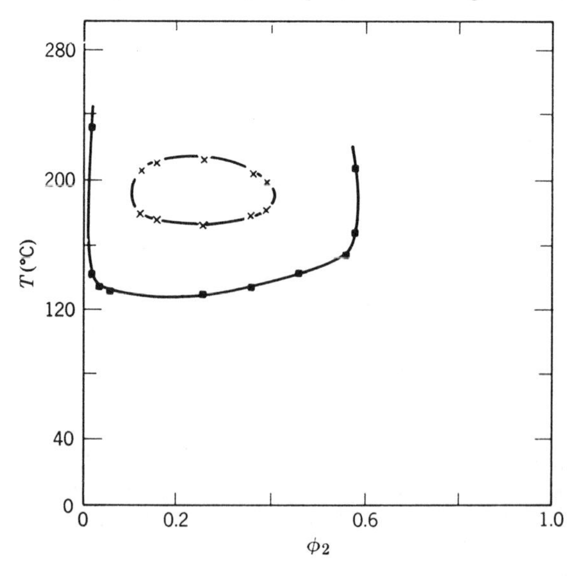

Fig. II.8. Phase diagram for the poly(ethylene oxide)-water system. (■) $\bar{M} = 5000$; (×) $\bar{M} = 3000$.

or, extrapolating T_c to an infinite molecular weight, two different Θ-temperatures. The appearance of the phase diagram will depend on the sign of $d\,\Delta H_M/dT$. If $d\Delta H_M/dT > 0$. we may observe phase separation within a limited temperature interval, while the polymer and solvent are miscible in all proportions below and above this temperature range. This behavior, analogous to that of the system depicted in Fig. II.6*c*, was observed by Malcolm and Rowlinson (1957) with the poly (ethylene oxide)-water system (Fig. II.8) Such phase diagrams are not unexpected for aqueous solutions in which ordering of the water molecules in the vicinity of organic solutes makes a negative contribution to ΔH_M and a large positive contribution to $d\,\Delta H_M/dT = \Delta C_p$ (see Table II.1), so that the solution process, exothermic at low temperatures, may become endothermic as the temperature is raised.

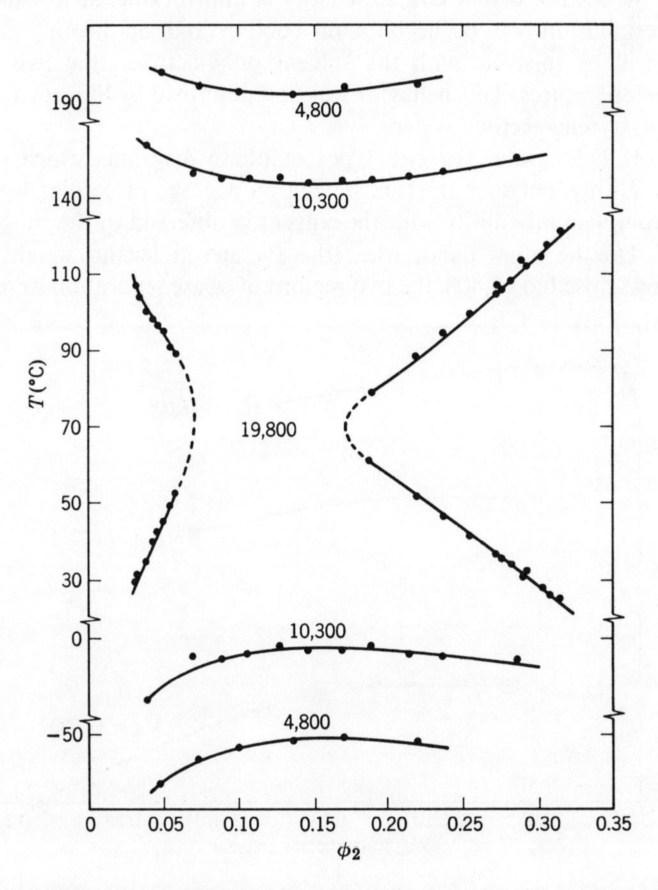

Fig. II.9. Phase diagram for the polystyrene-acetone system. The numbers indicate the molecular weights of the fractions.

However, it is more common to have $d\,\Delta H_M/dT < 0$. We have already alluded to this pattern of behavior in discussing mixtures of paraffin hydrocarbons in Section II. A.5. The mixing of long-chain molecules with substances of low molecular weight leads generally to improved packing, which makes a negative contribution to ΔH_M; this contribution tends to increase as the temperature is being rasied. The driving force toward polymer dissolution therefore acquires, eventually, a negative temperature coefficient and a point may be reached at which phase separation occurs. Patterson (1967) has reviewed systems in which polymers have been found to precipitate from solution on heating, and even if we are cautious in accepting the view (Patterson, 1969) that this must happen for all polymer solutions, there can be no doubt that this behavior is quite widespread. If the polymer is dissolved in a medium in which the solute-solvent contact energy is unfavorable, at low temperatures, precipitation will occur both on cooling and on heating and the polymer will be miscible with the solvent only between the two critical solution temperatures. This behavior was first described by Siow et al. (1972) for the polystyrene-acetone system.

Figure. II.9 illustrates the two types of phase diagrams which may be obtained. With a polymer fraction having an average molecular weight of 10, 300, complete miscibility with the solvent is observed in the range –2 to +144 °C. On the other hand, when the average molecular weight of the polymer was raised to 19,800, the two regions of phase separation were found

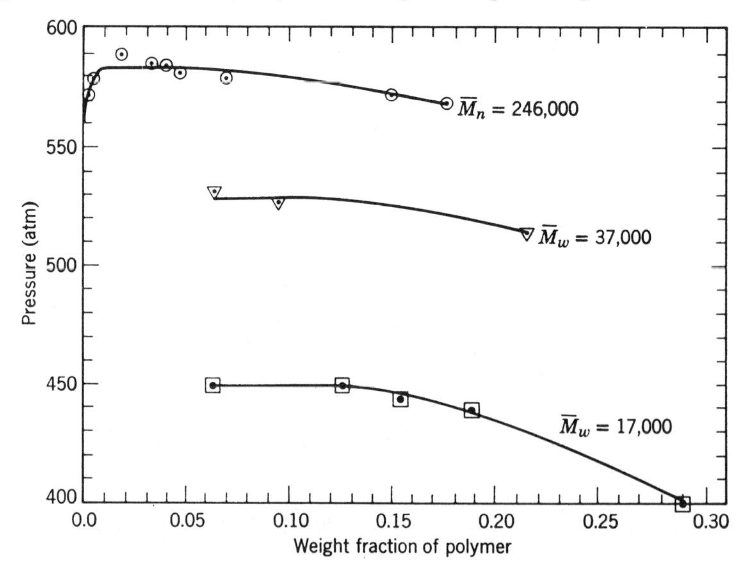

Fig. II.10. Pressure-composition phase diagram for the polyethylene-*n*-propane system at 110°C.

to overlap so that even at 70 °C, the temperature at which polymer-solvent interation was optimized, systems with a polymer volume fraction from $\phi_2 = 0.07$ to $\phi_2 = 0.17$ were characterized by the coexistence of a polymer-rich and a polymer-poor phase. Other systems with a similar behavior have been described by Saeki et al. (1973).

Gases above the critical temperature may also provide a solvent medium for polymeric solutes. In this case the cohesive energy density is very sensitive to pressure, and it is then of interest to determine the pressure-composition phase diagram at a given temperature. Figure. II.10 gives plots of the data obtained by Ehrlich and Kurpen (1963) for the polyethylene-propane system using three fractions of the polymer. Such phase diagrams define a "critical solution pressure" (which is a function of temperature), above which the gas and the polymer are miscible in all proportions.

4. Anisotropic Solutions of Macromolecules

If a solution contains solute molecules, which may be represented by stiff, elongated rods, significant deviations from a random orientation of the rodlike particles will persist even at high dilution. In that case the solute particle will have a very large excluded volume, and yet the entropy gained in transferring it from an ordered to a disordered phase will be relatively small. Under these circumstances the transfer of a solute molecule from a dilute to a more concentrated phase may result in a net increase of entropy, so that the system will separate into two phases even if mixing is athermal (Onsager, 1949; Flory, 1956). The more concentrated phase should contain partially oriented solute molecules and may be expected to exhibit optical anisotropy. Such a phenomenon was, in fact, observed by Bernal and Fankuchen (1941) and by Oster (1950) with solutions of tobacco mosaic virus, where the solute is a cylindrical particle of 150 Å diameter and 3000 Å length. The appearance of the anisotropic phase occurs in this case at a solution concentration of about 2%, and if the virus particles vary in length, the longer ones will be accumulated in the concentrated phase.

The interpretation of the data on tobacco mosaic virus solutions is undoubtedly complicated by the fact that the rods bear ionic charges, and the same difficulty arises in interpreting the formation of an anisotropic phase in solutions of ribonucleic acid (Spencer et al., 1962). It is of special interest, therefore, that synthetic un-ionized polymers in organic solvent media exhibit similar phenomena. Such observations were first made on optically active synthetic polypeptides (Robinson, 1956; Robinson et al., 1958) in solvent media which, as we shall see (Section III. C.1), stabilize a helical conformation of the polypeptide chain backbone so that the polymer behaves like a rather rigid rodlike particle. The anisotropic liquid phase is first formed at a critical concentration, which decreases to a limiting value as the chain length of the polymer is extended. When the solution concentration is raised above the critical point, the volume fraction of the anisotropic phase is gradually increased, while the compositions of the two phases remain invariant, until all of the system becomes anisotropic. The striking appearance of the anisotropic phase under a polarizing microscope is shown in Fig. II.11. The pattern of equidistant dark bands is interpreted by a structure in which rigid polypeptide helices are arranged at right angles to an axis so that each rodlike polypeptide is tilted by a characteristic small angle with respect to the preceding one (rotation occurs always in the same sense because of the asymmetry of the polypeptide helix). Such a spatial arrangement is characteristic of cholesteric liquid crystals. The spacing of the microscopically observed bands is then interpreted as representing half of the pitch of the periodic array of the macromolecules. This pitch depends on the nature of the solvent medium, and in any given solvent it is inversely proportional to the square of the concentration of the anisotropic phase.

Although the appearance of an anisotropic phase in solutions containing polypeptide chains in the helical conformation is in qualitative accord with the predictions of Flory's theory for solutions containing rodlike solute particles, the phase equilibria reported by Robinson show some important quantitative differences from the expected behavior. In particular, the concentrations of the isotropic and anisotropic phases are surprisingly similar to one another, and the limiting concentration of the isotropic phase does not decrease to very low values for chains of high molecular weight, as would be expected. A careful study of poly (γ-benzyl-L-glutamate) in dimethylformamide (a solvent favoring the helical conformation of the polypeptide) has also shown that the solution remains isotropic even in poorly solvating media, in which theory would predict separation of the anisotropic phase (Goebel and Miller, 1970). It is possible that these discrepancies between theory and experimental behavior result from the imperfect rigidity of the polypeptide helices, but no theory exists which would allow us to estimate how the behavior of a solution of rodlike particles would be modified if these particles acquired some flexibility.

The formation of anisotropic liquid phases has also been observed in solutions of block copolymers. For instance, solutions of block copolymers of styrene with ethylene oxide (Sadron, 1963) or butadiene (Vanzo, 1966) exhibit iridescence characteristic of liquid crystals. At high solution concentrations, the structures of such systems may be studied by x-ray diffraction; the results of such studies are discussed in section V.D.

A most spectacular alternative technique for demonstrating such structures has been reported by Gallot and Sadron (1971). They used methyl methacrylate as solvent to form anisotropic solutions of block copolymers of butadiene or isoprene and polymerized the methyl methacrylate to transform

Fig. II.11. Appearance of poly (γ-L-benzyl glutamate) solution in methylene chloride under crossed nicols. Solution concentration, 12 g/100 ml; molecular weight of polymer, 70,000. The spacing of the dark lines in the anisotropic phase is 14μ. Courtesy of Dr. C. Robinson.

the solution into a rigid glass. Sections of such specimens were stained with osmium tetroxide, which binds to the areas rich in double bonds so that they can be visualized by electron microscopy. The electron micrographs revealed patterns of equidistant bands or of dark circles in hexagonal array, corresponding to a lamellar structure of the solution or a structure in which the polydiene blocks are concentrated in hexagonally packed cylinders.

5. Some Typical Examples of Polymer-Solvent Interaction

The discovery that heat may be evolved in the solution of nonpolar polymers with nonpolar solvents has illustrated rather dramatically the limitations of the simple theory of van Laar and Lorenz (1925) on which we based the discussion in Section II.A.4. Nevertheless, estimation of the heat of mixing from cohesive energy densities is a useful approximation for nonpolar compounds at ordinary temperatures. However, if (II.34) is to be used for a system in which one component is polymeric, some method must be found to estimate δ for the polymer, since its energy of vaporization is obviously not measurable.

A variety of methods have been suggested toward that objective. It can be assumed, for instance, that the swelling of lightly crosslinked rubber (Gee, 1942; Scott and Magat, 1949) or polystyrene (Boyer and Spencer, 1948) with a series of solvents involves only nonspecific interactions, so that ΔH for the process is either zero or positive. In that case the maximum swelling would correspond to athermal mixing and to equal values of δ for the two components of the system. This approach works well enough with rubber, but in the case of polystyrene a number of anomalies have been observed. The same assumption that an athermal solvent is the best solvent is the basis for the method of Alfrey et al. (1950), in which the maximum on a plot of intrinsic viscosities (see Chapter. VI) as a function of the cohesive energy density of the solvent is assumed to indicate that solvent and polymer have identical CED values. Walker (1952) estimated the δ value of polyacrylonitrile from the values of a group of its low molecular weight analogs, and Small (1953) assumed that $\delta V = \sum \delta_i V_i$, where δ_i and V_i are contributions to the solubility parameters and the molar volumes characteristic of various groups comprising the molecule. A number of investigators have shown that the cohesive energy ΔE^v can be represented as the sum of group contributions, and the values proposed by them are listed in Table II.4. Using the heat of mixing as a rough measure of ΔG_M^E, we obtain from (II.34), (II.44), and (II.57) at the critical point for polymers of very long chain length

$$V_1(\delta_2 - \delta_1)^2 = \frac{RT_c}{2} \tag{II.59}$$

This relation indicates that, for a typical solvent with a molar volume

$V_1 = 100ml$, polymer-solvent systems will be miscible in all proportions around room temperature if the difference in the δ parameters is less than about 1.7. We see that the large gain in conformational entropy, characterizing the dilution of flexible chains, permits the mixing to be endothermic to a considerable extent before the onset of phase separation. We should also note that the maximum value of $\delta_2 - \delta_1$ which will allow complete miscibility is inversely proportional to the square root of the molar volume of the solvent.

We have referred (p. 44) to the application of the theory of corresponding states to the heat of mixing of nonpolar chain molecules with their low molecular weight analogs (Prigogine et al., 1957; Flory et al., 1964). This theory predicts a contraction on mixing, and this contraction is expected to make a negative contribution to ΔH_M. In practice, measurement of the heat of solution of high polymers is quite difficult since the dissolution is a slow process requiring continuous stirring. As a result, corrections for heat leakage from the calorimeter to the surroundings and for the energy dissipated in stirring the solution assume particular importance, if reliable values for the small heats of mixing are to be obtained. Frequently a problem also arises in regard to the proper interpretation of the experimental data. Many polymers exist at ordinary temperatures in the glassy state. In this state the volume of the polymer is larger than it would be at thermodynamic equilibrium, and expansion of the polymer beyond the equilibrium state corresponds to an enthalpy higher than the equilibrium value. The measured ΔH_M will, therefore, be less positive (or more negative) than for the mixing of an equilibrium polymer sample with a given solvent, and this should be taken into account in comparing experimental results with theoretical predictions. This complication occurs in the interpretation of the heat of solution of polystyrene (Hellfritz, 1951; Schulz et al., 1955), but it may be circumvented by measuring heats of dilution of polystyrene solutions, where the initial and final systems are in their equilibrium states. Tompa (1952) found that the dilutions of polystyrene solutions in both benzene and toluene are exothermic. With polyisobutene, which is at ordinary temperatures well above its glass transition point (−70°C), a significant deviation of the polymer from its equilibrium state is unlikely, and the heats of solution may be directly compared with theoretical predictions. Delmas et al. (1962a, b) measured ΔH_M for polyisobutene solutions in a homologous series of paraffins and found, as predicted by theory, that it contained a negative term inversely proportional to the square of the chain length of the solvent.

When we consider polymers carrying highly polar, strongly interacting groups, the problem of interpreting solubility behavior becomes much more complex. Schuerch (1952) studied the solubility properties of lignin and pointed out that solvents characterized by the same value of the Hildebrand

TABLE II.4.
Group Contributions to the Cohesive Energy[a]

Group	ΔE^v (cal/mole) Dunkel (1928)	Bunn (1955)	Hayes (1961)	Holftyzer and van Krevelen (1970)
$-\overset{\mid}{\underset{\mid}{C}}-$	(−1,750)	(−1,500)	—	−1,600
$-\overset{\mid}{C}H-$	−380	(−440)	—	100
$-CH_2-$	990	680	990	1,000
$-CH_3$	1,780	1,700	—	2,300
$-CH(CH_3)-$	(1,400)	1,360	1,700	(2,400)
$-C(CH_3)_2-$	1,810	1,900	2,840	(3,000)
$-\overset{\mid}{C}{=}CH-$	(610)	(700)	—	400
$-CH{=}CH-$	1,980	1,700	1,790	2,000
$-C(CH_3){=}CH-$	2,390	2,400	2,740	(2,700)
Phenyl	7,380	5,400	—	6,800
p-phenylene	—	3,900	5,700	—
$-O-$	1,630	1,000	1,630	1,500
$-OH$	7,250	5,800	—	—
$-\overset{O}{\overset{\|}{C}}-$	4,270	2,660	—	—
$-\overset{O}{\overset{\|}{C}}-O-$	(3,820)	2,900	3,380	3,200
$-\overset{O}{\overset{\|}{C}}-OH$	8,970	5,600	—	—
$-O-\overset{O}{\overset{\|}{C}}-O-$	—	—	—	—
$-\overset{O}{\overset{\|}{C}}-\overset{H}{\overset{\mid}{N}}-$	16,200	8,500	10,680	14,500
$-O-\overset{O}{\overset{\|}{C}}-\overset{H}{\overset{\mid}{N}}-$	—	8,740	6,280	—
$-S-$	—	2,200	—	—
$-CN$	—	—	(6,200)	6,800
$-Cl$	3,400	2,800	—	3,100

[a] The values are calculated from experimental data at 298°K, except for those of Bunn which are based on heats of vaporization at the normal boiling point. This procedure is, in principle, undesirable since ΔE^v is temperature-dependent. Values in parentheses are not given in the literature cited but are derived from primary data.

solubility parameter δ may differ widely in their solvent power. He correlated such differences with the ability of the solvent to accept a hydrogen bond. The question then arises how such hydrogen-bonding capabilities should be characterized in a quantitative manner. Badger and Bauer (1937) proposed that the O – H stretching frequency should be a linear decreasing function of the strength of the hydrogen bond, and Schuerch found that he could make a reliable prediction of lignin solubility on the basis of the δ parameter of the solvent combined with its ability to shift the O – D stretching frequency of CH_3OD (Gordy, 1941). A careful study of excess functions in mixtures of poly (ethylene oxide) or poly (propylene oxide) with chloroform (Kershaw and Malcolm, 1968; Malcolm et al., 1969) provides data for systems in which polymer-solvent interaction is dominated by hydrogen bonding. These systems are characterized by a very large evolution of heat during polymer dissolution, which in the case of the PEO-$CHCL_3$ system attains a maximum value of 12 cal/cm^3. The statistical-mechanical theory of Barker (1952), which has been found to describe well systems of strongly interacting small molecules, seems to be equally applicable to these polymer solutions.

An extensive study on cellulose nitrate by Crowley et al. (1966) showed that even a combination of the Hildebrand δ and the hydrogen-bonding parameter is insufficient for a prediction whether or not this polymer will be soluble in a given solvent. However, such predictions could be made reliably if the dipole moment of the solvent was added as a third parameter. A large dipole moment appears also to be important for solvents of polyacrylonitrile (Phibbs, 1955). A group of "active" solvents, which are frequently effective in dissolving such strongly interacting polymers, include dimethylformamide, *m*-cresol, dichloroacetic acid, dimethylsulfoxide, cyclic tetramethylene sulfone, and ethylene carbonate. All of these have very high dipole moments. Phibbs points out that succinic anhydride, with a dipole moment of 4.2 D, is a solvent for polyacrylonitrile, whereas the closely related acetic anhydride, with a dipole moment of 2.5 D, is a nonsolvent. Probably the other cyclic molecules among the active solvents, such as tetramethylene sulfone, ethylene carbonate, and γ-butyrolactone, also have dipole moments much higher than those of their linear analogs. Hexamethylphosphoramide is another powerful solvent which dissolves polyacrylonitrile (Dickey et al., 1949) and other solvent-resistant polymers such as poly (vinylidene chloride) (Coover and Dickey, 1956) and poly [3,3-bis (chloromethyl)-oxetane] (Campbell, 1958).

Exothermic solution of polar polymers accompanied by a negative excess entropy leads in some cases to polymer precipitation when the system is heated to moderate temperatures. Two systems of this type, poly (acrylic acid)-dioxane and polymethacrylonitrile-butanone, are listed in Table II.3, and cellulose nitrate solutions in alkyl acetates have been found to behave in a similar fashion (Doolittle, 1946). The nature of the interactions leading to exothermic mixing is sometimes quite obscure. For instance, it used to be believed that the evolution of heat characterizing the dissolution of poly-(vinyl chloride) in solvents such as tetrahydrofuran involved hydrogen bonding of the tertiary hydrogens in the polymer to the ether oxygen. However, it was found that model compounds with primary, secondary, and tertiary chlorines have a similar behavior and that an evolution of heat is observed even on mixing tetrahydrofuran with carbon tetrachloride, so that this interpretation has had to be abandoned (Pouchlý and Biroš, 1969). We should also note that the enthalpy of dilution may change sign as the composition of the system is altered. Such behavior was reported by Moore and Shuttleworth (1963) for the cellulose acetate-acetone system, which was found to be characterized by an endothermic dilution when the polymer content was low, while the dilution of concentrated solutions led to a large evolution of heat. This behavior may be related to the tendency of cellulose acetate to form a molecular complex with acetone, which has been observed as a distinct crystalline phase (Katz and Weidinger, 1932).

The interaction of copolymers with solvents depends on the distribution of the monomer residues in the macromolecular chains. If monomer residues A and B are distributed at random and if the heat of solution is relatively small, a simple argument (Stockmayer et al., 1955) leads to a Flory-Huggins interaction parameter $\chi = f_A \chi_A + f_B \chi_B - f_A f_B \chi_{AB}$, where f_A, f_B are the volume fractions of the two monomer residues in the copolymer, χ_A and χ_B are the interaction parameters for the homopolymers, and χ_{AB} characterizes the mixing of the two monomer residues in the copolymer in bulk. Usually the mixing of the monomer residues will be energetically unfavorable, so that $\chi_{AB} > 0$ and the reduction of contact between these residues will increase the driving force toward solution. For instance, the Θ-temperatures of polystyrene and of poly (methyl methacrylate) in cyclohexanol are very similar (81.8 and 71.4 °C, respectively), but for styrene-methyl methacrylate random copolymers the Θ-temperature may be as much as 20 °C lower than for either homopolymer (Kotaka et al., 1969). If two monomer residues are very different in polarity, a binary copolymer may actually be soluble in a medium which would be a strong precipitant for both homopolymers. Chuang and Morawetz (1973) have observed such behavior with an acrylamide copolymer containing 58 mole % styrene, which was found to be soluble in methanol.

The solubility behavior of block copolymers has been studied by Kotaka et al. (1971), using the styrene-methyl methacrylate system. They characterized it by the following generalizations: (a) copolymers containing one block of each kind (S-M) are soluble in media which dissolve either homopolymer, although the medium would be a precipitant for the other component; (b) copolymers of the M-S-M type are usually soluble in poly-

(methyl methacrylate) solvents, but not in polystyrene solvents, which precipitate poly(methyl methacrylate). We shall see later that the remarkable ability of triblock polymers to remain in solution, although the medium cannot solvate the central block, is most important for the study of hydrophobic polypeptides in an aqueous environment. Finally, a graft copolymer will dissolve in solvents for the grafted side chains even if the medium is a strong precipitant for the chain backbone (Gallot et al., 1962). We may visualize such systems as containing compact, poorly solvated particles which are prevented from aggregating by a "halo" of the solvated side chains.

Water is usually the solvent of choice for polymers with a high concentration of hydrogen-bonding groups such as poly (vinyl alcohol), polyacrylamide, or polysaccharides. With these materials the search for an alternative solvent medium is frequently quite difficult. Poly(vinyl alcohol) has been reported to be soluble in solvents such as piperazine, formamide, and diethylenetriamine; polyacrylamide dissolves in morpholine, and amylose in ethylenediamine and dimethylsulfoxide (Meyerson, 1966).

The special properties of water as a solvent medium are due in part to the effect of the solute on the structure of the solvent. We discussed in section I.A.6 the tendency toward "iceberg formation" when nonpolar solutes are introduced into water, resulting in heat evolution but a large negative excess entropy. These effects manifest themselves frequently in a reduction of the solvent power of water with rising temperature, such as was observed, for example, for poly(methacrylic acid) (Silberberg et al., 1957). Sometimes the aqueous polymer solution will precipitate on heating. This has been observed with vinyl alcohol-acetate copolymers (Nord et al., 1951), methylcellulose (Heymann, 1935; Uda and Meyerhoff, 1961), and other water-soluble polymers. However, the solvent power of water does not always decrease with rising temperature. In the case of polyacrylamide, for instance, Silberberg et al. (1957) found that water became a thermodynamically better solvent as the temperature was raised. In any case, the interaction of water molecules with macromolecular solutes will be strongly localized at specific sites, so that the solvent-solute interaction parameter would be expected to vary widely with the composition of the system. To cite an extreme case, nylon has a very high affinity for water as long as the water concentration is low (Starkweather, 1959), yet the solubility of water in the polymer is quite limited and the solubility of nylon in water is too low to be detected. Obviously, in a case such as this we have to consider separately the interaction of the water with the highly polar amide groups and with the nonpolar portions of the polymer solute.

Another property of water which is of crucial importance to its characteristics as a solvent is its ability to provide an ionizing medium of high dielectric constant. This accounts for its striking solvent power for polymeric electrolytes. For instance, poly(sodium acrylate) or poly(vinylpyridinium chloride) have a high affinity for water and are miscible with it in all proportions, but are highly insoluble (and little swollen) even in methanol. We must assume that the lowering of the dielectric constant results in almost complete association of the fixed charges of the polymer with the counterions and that the mutual interactions of the ion pairs are too strong to be disrupted by the forces of solvation. The same point may be made with respect to the mutual interactions of the dipoles in polymeric ampholytes, particularly proteins (Cohn and Edsall, 1943), although hydrazine, ethylenediamine, and anhydrous hydrogen fluoride, among others, are also protein solvents (for a review see Singer, 1962). With synthetic amphoteric polymers, the dipolar ion structure will make a large contribution only if the uncharged base has a higher hydrogen-ion affinity than the anionic base. For instance, in a copolymer of vinylpyridine and methacrylic acid:

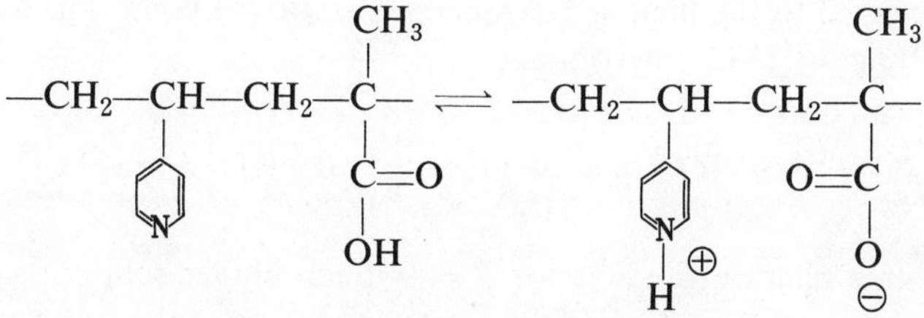

$$-CH_2-\underset{\underset{\text{(pyridyl, N)}}{|}}{CH}-CH_2-\overset{CH_3}{\underset{\underset{OH}{|}}{\underset{C=O}{|}}}{C}- \rightleftharpoons -CH_2-\underset{\underset{\text{(pyridyl, N}^{\oplus}\text{–H)}}{|}}{CH}-CH_2-\overset{CH_3}{\underset{\underset{O^{\ominus}}{|}}{\underset{O=C}{|}}}{C}-$$

the equilibrium favors the uncharged species, while in a methacrylic acid copolymer with dimethylaminoethyl methacrylate:

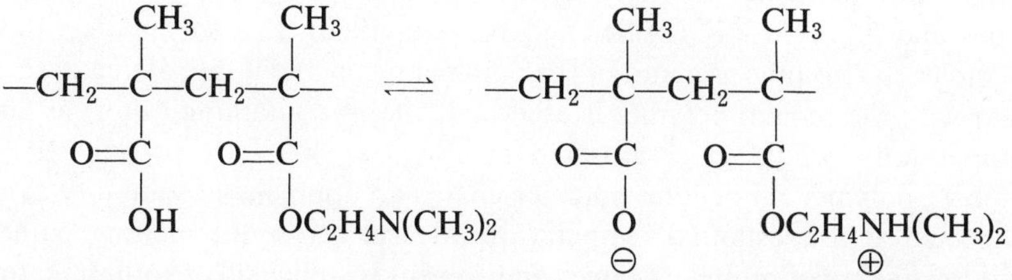

$$-CH_2-\overset{CH_3}{\underset{\underset{OH}{|}}{\underset{O=C}{|}}}{C}-CH_2-\overset{CH_3}{\underset{\underset{OC_2H_4N(CH_3)_2}{|}}{\underset{O=C}{|}}}{C}- \rightleftharpoons -CH_2-\overset{CH_3}{\underset{\underset{O^{\ominus}}{|}}{\underset{O=C}{|}}}{C}-CH_2-\overset{CH_3}{\underset{\underset{OC_2H_4NH(CH_3)_2^{\oplus}}{|}}{\underset{O=C}{|}}}{C}-$$

the dipolar ion form predominates. Alfrey and Morawetz (1952) found that the first polymer was insoluble in water, whereas the second dissolved easily. This difference is due to the different extent of ionization of the two substances in their isoelectric forms.

6. The Solubility of Crystalline Chain Polymers

We saw in Section II. A.8 that the solubility of a crystalline solute is intimately related to its melting point. For crystalline chain molecules, the melting point $T_m = \Delta H_m/\Delta S_m$ approaches a limiting value for samples of very high molecular weight (Flory and Vrij, 1963). This value increases with the magnitude of the energetic interactions between the chains (which increases ΔH_m) and with increasing chain stiffness (which reduces ΔS_m). However, ΔH_m is also quite sensitive to the efficiency with which the chains are packed in the crystal lattice. Thus isotactic polypropylene has ΔH_m = 2.37 kcal/mole of monomer units, appreciably higher than the value of 1.84 kcal/mole for polyethylene (Miller and Nielsen, 1961).

In considering the phase equilibrium between a crystalline polymer and its saturated solution, we may assume in first approximation that the effect of the swelling of the polymer phase may be neglected. Spontaneous transfer of the chain molecules from the crystallite to the solution requires then that the sum of the free energies, corresponding to melting and the dilution of the melt by the solvent, be negative. If we neglect the temperature dependence of the latent heat of fusion in applying the Gibbs-Helmholtz relation to the free energy of fusion and if we use (II.54) for the activity of the polymeric species, referred to the melt as the standard state, the condition which must be met (Huggins, 1942c) becomes

$$\frac{\Delta H_m(T_m - T)}{T_m} + RT\left[\ln \phi_2 + \left(1 - \frac{V_2}{V_1}\right)\phi_1 + \left(\frac{V_2}{V_1}\right)\chi\phi_1^2\right] < 0 \qquad \text{(II.60)}$$

Although this relation predicts for $\chi = 0$ much higher solubilities than the "ideal solubilities" given by (II.38) (due to the excess configurational entropy of dilution), these solubilities still become extremely low for very long chains. It is instructive to illustrate the consequences of (II.60) on a typical example. For isotactic polystyrene, $\Delta H_m = 2.15$ kcal/mole of monomer residues and $T_m = 513\,°\text{K}$ (Miller and Nielsen, 1960). If a solvent has the same molecular volume as a unit of the polymer chain and if the mixing of the solvent with the molten polymer is athermal, chains containing as few as 20 monomer units would give saturated solutions at 300 °K with only 10^{-6} volume % polymer.† For long polymer chains an appreciable solubility can be expected at a substantial temperature interval below the melting point only if the mixing of molten polymer and solvent is sufficiently exothermic to compensate for the heat absorbed in the melting process. With a material such as polyethylene, the specific solute-solvent interactions required for strongly exothermic mixing cannot be achieved, and it is not surprising that this crystalline polymer is insoluble at low temperatures in all solvent media. On the other hand, it is reasonable that crystalline polymers in which hydrogen bonding contributes a large fraction of the energy of the crystal lattice (e.g., polyamides and polypeptides) should be soluble in media known to be

†In this estimate we have neglected the difference between T_m for a low polymer and the limiting T_m value for very high polymers.

strong hydrogen bond acceptors (e.g., dimethylformamide) or hydrogen bond donors (e.g., cresol). A compilation of ΔH values for hydrogen bonds formed by phenol (Joesten and Drago, 1962) gives -3.2 kcal/mole for ethyl acetate, acetone, or acetonitrile as the acceptors, -5.0 kcal/mole for diethyl ether, -6.1 kcal/mole for dimethylformamide, and -9.2 kcal/mole for triethylamine. The heat of melting for nylon 66 is 11 kcal/mole of the repeat unit (Miller and Nielsen, 1960); it may be concluded that this polymer will be soluble only if the repeat unit can participate in at least two more hydrogen bonds in solution than in the crystal lattice. This condition is apparently satisfied when nylon 66 is dissolved in solvents such as *m*-cresol, formic acid, or fluorinated alcohols (Middleton and Lindsey, 1964). It is less easy to see how the mixing of poly(vinylidene chloride) with solvents such as hexamethylphosphoramide and tetramethylene sulfoxide can be sufficiently exothermic to render this highly crystalline and high-melting polymer soluble below 30 °C (Wessling, 1970).

In practice, several factors may increase the solubility of crystalline polymers above the values predicted by the considerations discussed above, which apply to relatively large, perfectly formed crystals. The solubility of all crystalline substances increases with their degree of subdivision, and with high polymers crystalllized in bulk the crystallites are usually very small. Moreover, a variety of "molecular imperfections" will tend to reduce the crystal stability. We may cite two examples of such an effect: (a) isotactic vinyl polymers probably never attain perfect stereoregularity, and a relatively small content of chain segments with the "wrong" steric configuration will weaken considerably the stability of the crystal lattice; (b) branched polyethylene as produced in the oxygen-catalyzed high-pressure process, precipitates from solution at a substantially lower temperature (Myers, 1954) than the linear polymer (Griffith and Rånby, 1959). Finally, we should realize that many crystalline polymers are very easily supercooled from the melt to an amorphous product, which may be quite easily dissolved to yield solutions that are metastable with respect to the crystalline polymer. The polymer will eventually precipitate from such systems, but since nucleation of a polymer crystal is a slow process, the metastable solutions may persist over extended periods of time. Isotactic polystyrene and poly(vinylidene chloride) are typical examples of crystalline polymers which behave in this manner. The stability of the metastable solutions is strongly dependent on the nature of the solvent medium. For instance, isotactic polystyrene crystallizes readily from trimethylbenzene and more slowly from xylene, while solutions in toluene are stable for very long periods (Blais and Manley, 1966). When the systems are seeded with polymer crystals, the crystal growth rate may be increased by several orders of magnitude by using a polymer solution in a thermodynamically poor solvent (Keith et al., 1970).

7. Thermally Reversible Gelation

One of the most striking phenomena observed with macromolecular solutions is the thermally reversible gelation of gelatin. Dilute aqueous solutions of this protein, which have a viscosity differing little from that of pure water, are transformed within a narrow temperature range into an elastic substance whose mechanical properties suggest a three-dimensional crosslinked network structure. Since this transformation is rapidly reversible, it clearly does not involve the formation or breaking of covalent bonds, and the quasi-crosslinkages in the gel network must be due to secondary valence forces.

Detailed interpretation of the phenomenon emerged only slowly over several decades of study. Trunkel (1910) and Smith (1919) found that gelatin solutions undergo a pronounced increase in optical activity in the temperature range in which the setting or melting of gels is observed. Since this change in optical activity is observed also in solutions which are too dilute to form a gel network (Harrington and von Hippel, 1961a, b), it must reflect a change in the molecular properties of the protein that is independent of the gelation process. More concentrated gelatin gels have been found to exhibit sharp x-ray diffraction patterns, so that the system must contain extended regions with crystalline order (Herrmann et al., 1930; Katz et al. 1931; Gerngross et al., 1932). The dependence of the "melting point" T_m of the gel on the molecular weight M and the concentration c of the gelatin was studied by Eldridge and Ferry (1954). Assuming that the formation of a quasi-crosslinkage from two interacting regions along the polymer chains may be represented formally as an association equilibrium of the type $2A \rightleftharpoons A_2$, and that T_m corresponds to the temperature at which a sufficient number of quasi-crosslinkages have formed to result in a network structure extending through the entire volume of the system, they obtained the relation

$$-\left(\frac{d \ln c}{dT_m}\right)_M = \frac{\Delta H_{cl}}{RT_m^2} \tag{II.61}$$

where ΔH_{cl} is the enthalpy characterizing the formation of an average crosslinkage. The values obtained for ΔH_{cl} were extremely large (up to -220 kcal/mole), indicating that a single crosslinkage must involve a fairly large number of amino acid residues in the protein chain. It was also found that the T_m values and the ΔH_{cl} derived from them depend on the temperature at which the gelatin solution was allowed to set to a gel.

The elucidation of the structure of collagen, the protein from which gelatin is derived, has provided a clearer understanding of the nature of this solution-gel transformation. The collagen structure will be discussed at greater length in the next chapter; here we should note only that three polypeptide chains are twisted around each other to form a rodlike molecular particle. At temperatures above T_m the structure disintegrates to randomly coiled isolated chains. If the temperature is now reduced below T_m at appreciable protein concentrations, reconstitution of the original three-stranded particles becomes a very unlikely process, since different regions of any one chain will find different partners to associate with. Figure II.12 gives a schematic representation of the collagen → gelatin solution → gelatin gel

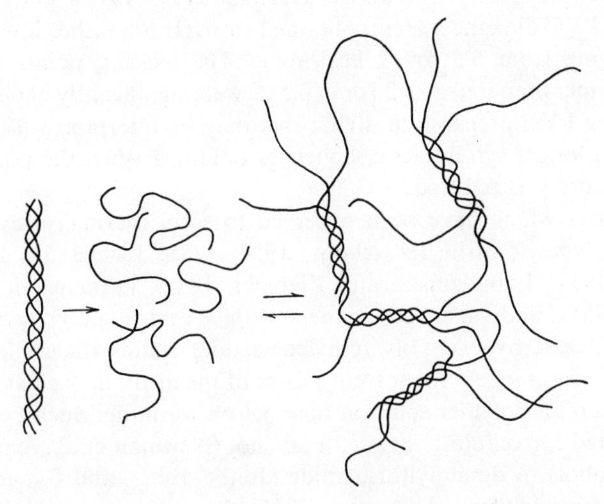

Fig. II.12. Schematic representation of the collagen → gelatin solution → gelatin gel transformation.

transformations. Once a continuous network is formed, the mobility of the chains is greatly restricted and an extension of the associated regions (or the formation of new quasi-crosslinkages) becomes a very slow process. It is probable that all gelatin gels should be regarded as nonequilibrium systems, owing their metastability to the difficulty of reconstituting the collagen particles.

A number of acidic polysaccharides occurring in living organisms also exhibit solution-gel transformations under various conditions. Investigations of fibers drawn from two of these materials, carrageenan (Anderson et al., 1969) and hyaluronic acid (Dea et al., 1973), have revealed a high degree of crystalline order, and the crystallographic evidence shows that two of the polysaccharide chains are wound into a double-helical structure. Hence the gel formation seems to involve a process quite analogous to that encountered with gelatin.

Solutions of some synthetic polymers have also been observed to gel, although the phenomenon usually requires higher concentrations than in the case of gelatin and the melting of the gels occurs over a much wider temperature range. Takahashi et al. (1972) found that poly (vinyl chloride) (PVC) gels exhibit two x-ray diffraction rings which are weak but sharp, suggesting that the quasi-crosslinkages are formed by microcrystallites probably containing syndiotactic sequences of the polymer chains. They reported that $1/T_m$ is linear in the logarithm of the product of the concentration and the molecular weight of the polymer, as predicted by the analysis of Eldridge and Ferry referred to above. Harrison et al. (1972), studying the gelation of the PVC-dioxane system, obtained from (II.61) rather low values for ΔH_{cl}, ranging from 6.5 to 11 kcal/mole. The melting points of gels containing polymer prepared at -20 or at 0.5 °C were significantly higher than those containing PVC prepared at 40 °C; this may be interpreted as a consequence of the longer syndiotactic sequences obtained when the polymerization temperature was reduced.

Other polymers which have been observed to form thermally reversible gels include polyacrylonitrile (Bisschops, 1954, 1955; Jost, 1958; Labudzińska et al., 1967; Labudzińska and Ziabicki, 1970), gutta percha (Van Amerongen, 1951), and block copolymers with a central poly (propylene oxide) block flanked by two poly (ethylene oxide) chains (Schmolka and Bacon, 1967). In cases where the solvent power of the medium decreases with rising temperature, a polymer solution may gel on warming. Such behavior has been reported for cellulose nitrate in ethanol (Newman et al., 1965), for poly (vinyl alcohol) in dimethylformamide (Jones, 1962), and for aqueous solutions of methylcellulose (Heymann, 1935). In some cases, gels exhibit syneresis, that is, they tend to contract with separation of pure solvent on prolonged standing.

Many of the systems discussed above contain either crystallizable polymers or polymers with regions which may be incorporated into crystallites, and a number of polymers other than gelatin and poly (vinyl chloride) exhibit a sharp x-ray diffraction pattern in the gel state (Katz and Derksen, 1932; Derksen and Katz, 1932; Jost, 1958; Labudzińska et al., 1967). Yet there can be no doubt that crystallizability is not a necessary condition for the formation of quasi-crosslinkages leading to thermally reversible gel networks. This has been amply demonstrated in studies on aqueous solutions of homopolymers and copolymers of *N*-acrylylglycinamide (Haas et al., 1967, Haas et al., 1970a, b) and *N*-methacrylylglycinamide (Haas et al., 1971). However, the heat of crosslink formation as estimated from relation (II.61) or from differential thermal analysis is quite low and suggests that a single hydrogen bond is involved. The very high ΔH_{cl} values characterizing gelatin cannot be duplicated in a synthetic polymer unless crosslink formation results from a cooperative interaction between a large number of residues in the interacting chains. Haas and MacDonald (1970) demonstrated that ΔH_{cl} values even higher than those characterizing gelatin can be achieved with poly (γ-benzyl-

L-glutamate) in xylene solution. As we shall see in the next chapter, this polypeptide assumes in poor solvent media a helical conformation in which the chain behaves as a very slightly flexible rod, and particles of this kind can apparently interact with one another very efficiently to form the gel network.

8. Application of the Concept of "Solution" to Extremely Large Molecules

The question may arise whether there is an upper limit to the size of a particle that may be considered to be in "true solution." A generation ago a rather arbitrary dividing line was drawn between solutions in the classical sense and "colloid solutions" comprising systems in which the disperse phase had characteristic dimensions between about 0.01 and 1μ. Such systems could be of various types, comprising solutions of macromolecules, solutions in which a number of the solute molecules associated into micellar aggregates, and finally "sols" of metals, sulfur, and so forth in which the size of the particles held in suspension bore no relation to the size of their constituent molecules. In recent years the emphasis has been increasingly on the essential unity of physical chemistry and on the interpretation of "colloid phenomena" in terms of general physicochemical principles. From that point of view we may define a solution as a system of several components forming a single phase. A phase is defined, in turn, as a physically and chemically distinct, mechanically separable portion of the system. However, here we are facing an obvious ambiguity in trying to decide what constitutes "mechanical separability." If passage of a liquid through a filter paper results in the separation of suspended particles, we have no doubt in considering them a separate phase. Yet, membranes of various pore dimensions are known, making it possible to separate a liquid first from "dissolved" macromolecules and eventually even from solute molecules as small as sucrose. In considering a series of filtrations through increasingly tight filters, it will be impossible to define, between suspensions and solutions, a demarcation line which will not be arbitrary.

A possible pragmatic criterion could be based on the extent to which the components of a system will be separated from each other at equilibrium by the gravitational field of the earth. If a particle of mass m_2 and density ρ_2 is suspended in a liquid of density ρ_1, the gravitational pull on it will be, after allowing for the buoyancy effect, $m_2\boldsymbol{g}(\rho_2 - \rho_1)/\rho_2$, where $\boldsymbol{g}$ is the acceleration due to the gravitational field. The ratio of the concentrations of the particles at two heights, h_1 and h_2, will then be given by the Boltzmann distribution law as

$$\frac{c_2}{c_1} = \exp\left[\frac{-m_2\boldsymbol{g}(\rho_2 - \rho_1)(h_2 - h_1)}{\rho_2 \boldsymbol{k}T}\right] \tag{II.62}$$

If we then allow some limiting value for c_2/c_1 as being compatible with the concept of a "solution," the weight of the particle considered to be dissolved will depend critically on an arbitrary choice of the dimensions of the system and will be highly sensitive to the relative densities of solute and solvent, so that a definition of "solutions" based on such a criterion cannot be considered to have any fundamental significance. With $h_2 - h_1 = 10$ cm, $(\rho_2 - \rho_1)/\rho_2 = 0.2$, and $T = 300\,°K$, a particle weight corresponding to a molecular weight of 10^6 would be sufficient to produce a 10% concentration difference, while a solute with a molecular weight of 30×10^6 would be, at equilibrium, 10 times as concentrated at the bottom as at the top of the system. Such a situation hardly seems to be compatible with the classical concept of a "phase." However, solutions such as those discussed above appear much more homogeneous in practice, since approach to an equilibrium distribution of the solute in the gravitational field of the earth is extremely slow.

Whatever the nomenclature which we may want to adopt, it is significant that the same experimental methods used for the study of solutions of macromolecules may be adopted, for instance, for aqueous systems containing virus particles (Cummings and Kozloff, 1960; Davison and Freifelder, 1967), which we might hesitate to describe as "dissolved." In the field of synthetic polymers, spherical particles with dimensions in the range of 0.1–1 μ may be prepared by emulsion polymerization in the presence of a crosslinking agent, and sols of these "microgels" may also be studied by techniques similar to those employed in the investigation of linear polymers in solution (Shashoua and Beaman, 1958).

C. THREE-COMPONENT AND MULTICOMPONENT SYSTEMS CONTAINING MACROMOLECULES

1. The Effect of Polydispersity on Polymer Solubility

The phase rule requires that in a two-component system with two condensed phases the composition of the phases should be uniquely determined at a given temperature and pressure. This means, in particular, that the phase composition should be unaffected by any variation in the relative volume of the coexisting phases. When we deal with a polymer which is a single molecular species, this principle is valid and the concentration of a solution in equilibrium with the polymer phase does not increase as more undissolved polymer is added. This behavior is encountered in practice only with certain biological macromolecules. Typical data are those of Butler (1940), showing that, after purification of chymotrypsinogen, crystals of the protein are in equilibrium with solutions of a unique concentration. However, we should be careful in using a well-defined solubility as an ultimate proof that a sample consists of identical molecules. Genetic mutation in living organisms leads to a variability in the composition of a given kind of protein which frequently involves the replacement of a single amino acid residue in the macromolecule. Molecules of such variants can often replace one another in the crystal lattice, and if such replacement may occur at random, then the phase equilibrium of a system containing a mixture of protein variants will be indistinguishable from that of a system containing a single molecular species of the protein.

The situation will be quite different in a two-phase system containing macromolecules of varying chain length, such as are obtained in the preparation of synthetic polymers. We may then apply expressions (II.54) for the activity of the polymeric species to the distribution of chains of any given length between the polymer gel phase and the solution phase. Denoting by ϕ_i and ϕ_i' the concentrations of species i in solution and in the gel, respectively, and setting $\ln a_i = \ln a_i'$, we obtain

$$\ln\left(\frac{\phi_i}{\phi_i'}\right) = \left(1 - \frac{V_i}{V_1}\right)(\phi_1' - \phi_1) + \frac{V_i}{V_1}\chi[(\phi_1')^2 - \phi_1^2]$$

$$\approx \frac{V_i}{V_1}\{(\phi_1 - \phi_1') - \chi[\phi_1^2 - (\phi_1')^2]\} \qquad \text{(II.63)}$$

so that the distribution coefficient is an exponentially decreasing function of the chain length, as would be expected from the general considerations discussed in Section II.B2.

The molecular heterogeneity of the solute affects also the theory of the critical miscibility point. The use of relations (II.37) for the location of the critical point from analytical expressions describing the free energy of mixing of binary systems is not justified for systems containing polydisperse polymers, since the molecular weight distribution of the polymeric component will, be different in the two coexisting phases. The problem has been analyzed in detail by Stockmayer (1949), who concluded that the location of the critical point should depend on an average lying in general between the weight-average and the z-average molecular weight of the solute. For most molecular weight distributions encountered in practice, Stockmayer expects $\bar{M}_w$ to be the controlling factor. This conclusion is consistent with the experimental data of Shultz and Flory (1952).

2. Solubility of Polymers in Mixed Organic Solvents

Frequently a polymer is more soluble in a mixture of two liquids than in either component of the mixture. A systematic study of this phenomenon was first undertaken by Gee (1944), who pointed out that it is possible to match the cohesive energy density (or the Hildebrand δ parameter) of a polymer by mixing in suitable proportions solvents characterized by high and low δ values. For instance, a styrene-butadiene copolymer with $\delta = 8.1$

is insoluble both in pentane ($\delta = 7.1$) and in ethyl acetate ($\delta = 9.1$) but will dissolve in a 1–1 mixture of the two solvents. However, the behavior of mixed solvents frequently cannot be accounted for in this manner. Often, small concentrations of a liquid, which is decidedly a nonsolvent for a polymer, will solubilize it in a second nonsolvent because the cosolvent can satisfy hydrogen-bonding requirements of functional groups carried by the polymer. We have observed, for example, that styrene copolymers with carboxylic acid comonomers may be insoluble in typical polystyrene solvents unless a small amount of an alcohol or an amine is added to the solvent medium. Another principle appears to be involved in the long-known phenomenon that cellulose nitrate is soluble in ethanol-ether mixtures and the recent report that addition of alcohols as cosolvents renders poly (methyl methacrylate) soluble in carbon tetrachloride (Deb and Palit, 1973). In these two cases, hydrogen bonding of the alcohol to the ester groups serves, apparently, to reduce dipole-dipole interactions opposing polymer dissolution. Finally, we may cite the case of a 2-vinylpyridine-methacrylic acid copolymer, which was insoluble in methanol but dissolved in methanol mixtures containing either benzene or water (Alfrey and Morawetz, 1952). Since benzene is less polar and water more polar than methanol, cosolvent action must be of a highly specific type in this case.

The relative importance of cohesive energy density and specific solvation is illustrated in an interesting manner in the studies of cellulose nitrate solubility carried out by Doolittle (1944, 1946). He dissolved polymer samples in a homologous series of solvents and titrated with a nonsolvent, noting the solvent concentration at the precipitation point corresponding to an arbitrarily chosen polymer concentration. Some typical results are reproduced in Fig. II.13; they show that the concentration of the solvating groups required to hold the cellulose nitrate in solution approaches a limiting value for the higher members of the homologous series of solvents. For low molecular

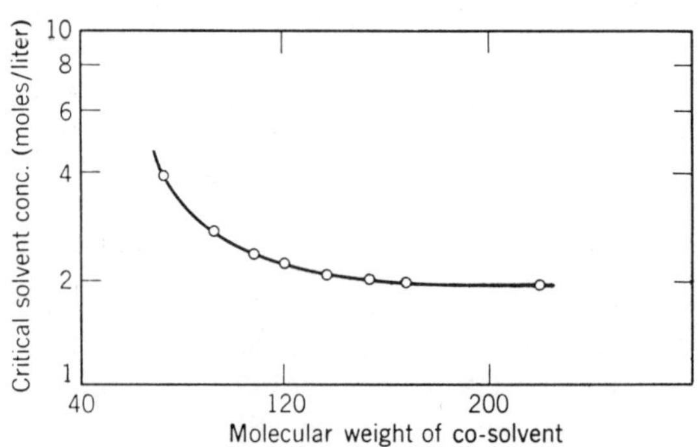

Fig. II.13. Concentration of *n*-alkyl acetate cosolvents required to dissolve cellulose nitrate in toluene. Polymer concentration, 0.5 g/100 ml.

weight solvents the required concentration of solvating groups increases. This effect may be understood by considering separately the paraffinic "tail" and the polar "head" of the solvent molecule, which may be thought of as more or less firmly attached to the macromolecular solute. As the paraffinic portion is shortened. the cohesive energy density of the polymer-cosolvent complex will increase and a more polar medium will be required for dissolution of this complex.

Gee (1944) cites an interesting case where a polymer will precipitate when two of its solutions in different solvent media are mixed. Cellulose acetate will dissolve both in aniline and in glacial acetic acid but not in a mixture of these two solvents, since the mutual interaction of the acidic and the basic solvent is much stronger than the interaction of either of them with the polymeric species.

3. Solubility in Aqueous Electrolyte Solutions

The solubility of organic nonelectrolytes in water is usually reduced by the addition of neutral salts. This "salting-out" was once thought to result from the formation of the hydration shells of the ions, leading to a reduction in the number of water molecules available for the solvation of nonelectrolyte solutes. However, the phenomenon is more complex, since the relative "salting-out efficiency" of different salts varies widely with the nonelectrolyte solute (Long and McDevit, 1952). Polymeric nonelectrolytes behave in a manner similar to that of low molecular weight solutes, and their water solubility is frequently reduced sharply by salt addition. Occasionally, systems containing polymeric nonelectrolytes, water, and salt separate into two solution phases, one containing most of the polymer and the other most of the salt (Albertsson, 1958).

There exist, however, some important cases where water-insoluble polymers dissolve in aqueous solutions of electrolytes. Cellulose is soluble in solutions of Cu(II)-ammonia or Cu(II)-amine salts (Stamm, 1952), ferric ammonium tartrate (Jayme and Lang, 1955), and salts containing the cadmium-ethylenediamine complex (Jayme and Neuschäffer, 1957). In the case of the cupric ammonia solutions, the solvation of the cellulose has been accounted for by the formation of cupric complexes with the glucose residues (Reeves, 1949), and an analogous mechanism is undoubtedly responsible for the solvent properties of the ferric and cadmium salts. Similar complexation appears to be involved with concentrated solutions of strong acids, which dissolve cellulose (Stamm, 1952) and polyacrylonitrile (Binder, 1960). The existence of the cellulose-nitric acid and the cellulose-perchloric acid complexes has been proved crystallographically (Andress, 1928; Andress and Reinhardt, 1930). A similar phenomenon has been observed with poly(ethylene oxide), which is insoluble in methanol but is solubilized in the presence of potassium iodide (Lundberg et al., 1966). The nuclear magnetic resonance spectrum of such solutions indicates that it is the potassium ion which is bound to the polymer (Liu, 1969), and the coordination of the cation to the ether oxygens can be assumed to be analogous to that in poly (ethylene oxide)-$HgCl_2$ complexes, which have been characterized crystallographically (Iwamoto et al., 1968; Yokoyama et al., 1969).

The characteristic ability of strong (> 55 wt. %) aqueous solutions of certain neutral salts of metals such as lithium and calcium (which do not have a particularly strong tendency to form complex ions) to dissolve such water-insoluble polymers as cellulose (Stamm, 1952), silk fibroin (von Weimern, 1926; Ambrose et al., 1951), poly (L-serine) (Bohak and Katchalski, 1963), or polyacrylonitrile (Rein, 1938; Stanton et al., 1954) seems to fall into a different category. The solvent power of the salts depends strongly on both the cation and the anion, increasing with the polarizability of the anion. Stanton et al. give the data reproduced in Table II.5, where the cations increase and the anions decrease in solvent power from the top to the

TABLE II.5.
Aqueous Salt Solutions Dissolving Polyacrylonitrile
(All Cations Associated with Anions at the Same or Higher Level)

Cations	Anions
K^+, Fe^{3+}, Pb^{2+}	
NH_4^+, Cd^{2+}, Al^{3+}	SCN^-
Na^+, Ba^{2+}	I^-
Ca^{2+}, Li^+, Mg^{2+}	Br^-
Ga^{3+}, Sb^{3+}, In^{3+}, Tl^{3+}, Sn^{4+}	Cl^-
Zn^{2+}, Ag^+, Ni^{2+}, Co^{2+}, Mn^{2+}	NO_3^-
	SO_4^{2-}

bottom of the table. A salt is a solvent for polyacrylonitrile if the anion is placed at least as high as the cation; thus all salts containing K^+, Fc^{3+}, Pb^{2+}, or SO_4^{2-} are nonsolvents. Fairly frequently, simple organic compounds also form crystalline adducts with some of these "solvent salts," and a cellulose-lithium thiocyanate complex has been characterized by x-ray diffraction (Katz and Derksen, 1931).

With ionizable polymers, water solubility generally increases with the net charge density of the polymeric species. For instance, poly (2-vinylpyridine) or isotactic poly (acrylic acid) are water insoluble in the un-ionized state, but dissolve on partial neutralization. With amphoteric polyelectrolytes, particularly proteins, the solubility is generally at a minimum when the polymer is isoelectric and increases on addition of acid or base. Figure II.14 gives a schematic representation of the various equilibria in a system which contains

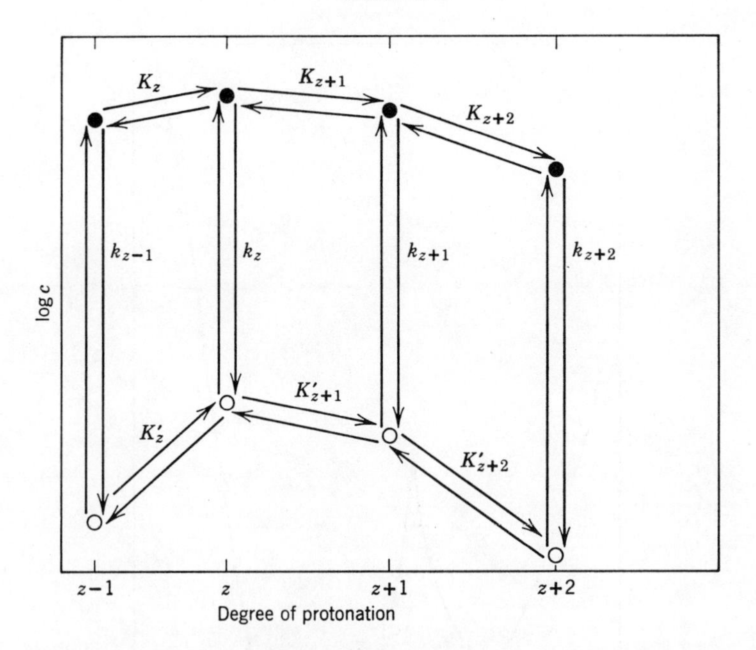

Fig. II.14. Distribution of species with different degrees of protonation for ionizable macromolecules in two coexisting phases.

an ionizable macromolecule in two coexisting phases, with K_z and K'_z denoting ionization constants and $k_z = c_z/c'_z$ distribution coefficients of the various species between the two phases. Since $\log(c_{z-1}/c_z) = \text{pH} - \text{p}K_z$ and $\log(c'_{z-1}/c'_z) = \text{pH} - \text{p}K'_z$, we have $\text{p}K'_z - \text{p}K_z = \log(k_{z-1}/k_z)$. Thus, any variation in the distribution coefficients with the state of ionization is a consequence of differences in the ionization equilibria in the two coexisting phases. Rupley (1968) has pointed out that the crystal structure of globular proteins is sufficiently loose to permit a random distribution of species with different states of ionization in the crystal lattice. Under these conditions the above analysis is applicable, and the pH dependence of protein solubility may be interpreted in terms of differences in the average degree of protonation of the protein molecules in solution and in the crystals. A particularly interesting situation is observed with horse carbonylhemoglobin, which can exist in two crystalline modifications characterized by different ionization equilibria. As a result, the solubility-pH relation exhibits two minima and a sharp cusp at the pH of phase transition (Fig. II.15).

The effect of electrolyte addition on the solubility of proteins has been considered in detail in the monograph by Cohn and Edsall (1943) and may be understood by an extension of the theory formulated by Kirkwood (1934)

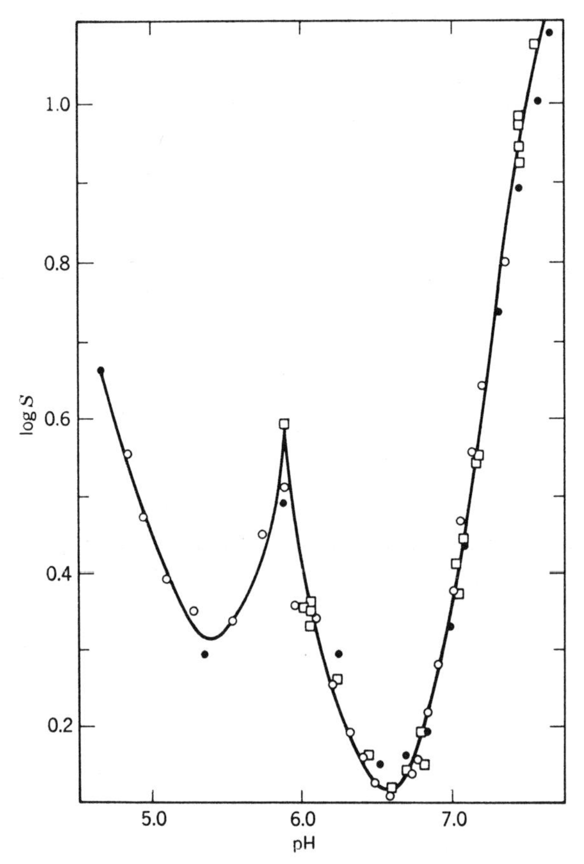

Fig. II.15. The pH dependence of the solubility of horse carbonylhemoglobin (Rupley, 1968).

for dipolar ions. Even though these species carry no net charge, they attract each other because of dipole-dipole interactions. A second reason for the mutual attraction of amphoteric species in isoelectric solution is a fluctuation in the state of ionization of the individual molecules. These fluctuations are not independent of one another but tend to place, at any given time, net charges of opposite sign on neighboring ampholytes (Kirkwood and Shumaker, 1952). Both the dipole (or multipole) interaction and the charge fluctuation theory predict the logarithm of the solubility to increase linearly with the square root of the ionic strength. The magnitude of the effect would be expected to be quite sensitive to the distribution of the ionizable groups on the surface of the protein molecule. Some proteins, designated as globulins, have extremely low solubilities in water in the absence of simple electrolytes.

Protein solubilities in water are also increased by addition of dipolar ions, which may raise substantially the dielectric constant of the medium and in this way suppress electrostatic interactions between solute molecules. The magnitude of this effect varies within rather wide limits. For instance, the addition of glycine increases the solubility of lactoglobulin by a much larger factor than that of hemoglobin (Cohn and Ferry, 1943).

At high concentration, the effect of added salt on protein solubility becomes highly dependent on the nature of the electrolyte, ranging from strongly precipitating salts, such as ammonium sulfate, to strongly solvating salts, exemplified by lithium bromide and calcium thiocyanate. The protein solubility S at a salt concentration c_s may be related to its solubility S_o in pure water by

$$\log\left(\frac{S}{S_o}\right) = K_s c_s \tag{II.64}$$

where K_s, positive or negative, follows a trend similar to that indicated in Table II.5. Experiments with model compounds of low molecular weight, such as diketopiperazine (Meyer and Klemm, 1940), acetyltetraglycyl ethyl ester (Robinson and Jencks, 1965), or adenine (Robinson and Grant, 1966) revealed salting-in and salting-out behavior quite similar to that observed with proteins. Some of the K_s values determined for acetyltetraglycine ethyl ester are listed in Table II.6. It is important to note, however, that the K_s

TABLE II.6.
Effect of Salts on the Solubility of Acetylglycine Ethyl Ester

Salt	K_s	Salt	K_s
$NaClO_4$	–0.33	KBr	–0.023
LiI	–0.28	NaBr	0
NaSCN	–0.25	LiCl	+0.021
NaI	–0.23	NH_4Cl	+0.035
KI	–0.21	KCl, NaCl	+0.046
LiBr	–0.17	$(NH_4)_2SO_4$	+0.45
NH_4Br	–0.11	Na_2SO_4	+0.48

values for a given salt vary from one protein to another, so that salting-out may be employed for efficient protein separation procedures.

Addition of a salt to an aqueous polyelectrolyte solution may also lead to the formation of two liquid phases, one of which contains most of the polyelectrolyte. This phenomenon was studied in great detail by Bungenberg de Jong (1937), who described the separation of a viscous, protein-rich liquid phase when salt was added to certain aqueous protein solutions. Systems containing alkali poly (vinylsulfonates) and alkali halides have been studied by Eisenberg and Mohan (1959). If the system contains two different cations, one will tend to be enriched in the polyelectrolyte solution, so that the system may be considered a "liquid ion exchanger," with the polymer phase concentrating potassium over sodium or ammonium ions.

4. Incompatibility of Polymer Solutions

More than half a century ago, the strange observation was reported that dilute aqueous solutions of gelatin and agar would not mix but would form two coexisting phases (Beijerinck, 1910). Each of these phases was found to contain almost exclusively one of the polymeric solutes. A similar phenomenon was later demonstrated on a large number of ternary systems (Dobry and Boyer-Kawenoki, 1947; Kern and Slocombe, 1955; Kern, 1956; Albertsson, 1958). Mutually compatible polymers are relatively rare (Bohn, 1966), and cellulose nitrate is unusual in being compatible with a variety of other polymers (Petersen et al., 1969). Krause (1972) has reviewed data on ternary systems containing two polymers which do or do not lead to phase separation and has compared critically the experimental evidence with theoretical expectation. A typical phase diagram of two incompatible polymers and a common solvent is that determined by Kern for the polystyrene-poly (*p*-chlorostyrene)-benzene system shown in Fig. II.16. It may be seen that

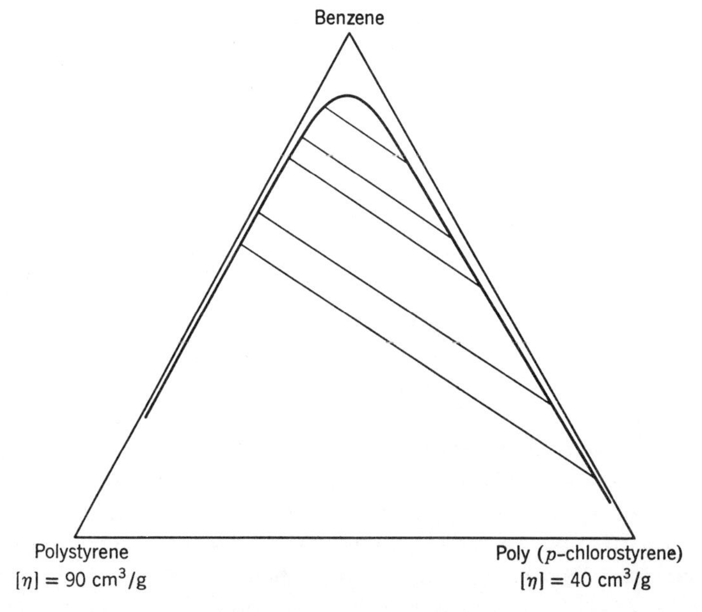

Fig. II.16. Ternary phase diagram of benzene-polystyrene-poly(*p*-chlorostyrene) at 25° C. The intrinsic viscosities were measured in benzene solution. (The use of intrinsic viscosity as a measure of the length of flexible chain molecules will be discussed in Section C2 of Chapter VI.)

benzene, which is miscible in all proportions with either polymer alone, can dissolve only slightly more than 2% of a mixture of equal weights of the two polymers without phase separation. Poly(*p*-chlorostyrene) has a negligible solubility in a phase containing 4% of polystyrene. Kern used a polystyrene sample with a substantially longer chain length than that of the poly(*p*-chlorostyrene), and this accounts for the asymmetry of the phase diagram. It is generally found that the mutual incompatibility of polymers increases rapidly with an increase in their molecular weight.

The thermodynamic causes of this phase separation phenomenon are not difficult to visualize. If two dilute solutions of low molecular weight solutes A and B in the same solvent medium are mixed, the system forms a single phase, since the gain in entropy will outweight even an unfavorable energy of mixing. However, if the solutions contain initially long-chain molecules composed of a large number of A or B units, respectively, the entropy of mixing per unit weight becomes negligible, while the energy of mixing per unit weight, depending on the number of contact points between dissimilar chain segments, remains nearly the same as for the low molecular weight analog. The solutions of the chain molecules will, therefore, resist mixing if contacts between dissimilar chain segments involve an expenditure of energy. Kern (1958) has shown how sensitive the phase separation phenomenon is to small changes in the structure of the polymers: poly (vinyl chloride) solutions, which do not mix with solutions of poly (methyl methacrylate), were found to be miscible with solutions of polymers of ethyl, propyl, butyl, and isobutyl methacrylate.

A very interesting application for the coexistence of two dilute polymer solution phases containing the same solvent medium was demonstrated by Albertsson (1970, 1971). Although the two solutions are very similar, the transfer of a third kind of macromolecule from one phase to the other may be characterized by a significant change of energy. These macromolecules may, therefore, accumulate in one of the two phases, and this phenomenon can be utilized for fractionation procedures. Sasakawa and Walter (1972) studied distribution coefficients for proteins between aqueous solutions of dextran and poly (ethylene oxide) (PEO) and found them to be sensitive to molecular size, pH, and added salts. When deoxyribose nucleic acid (DNA) is dissolved in aqueous PEO, no second liquid phase is formed, but the exclusion of DNA from the PEO domains induces it to form aggregates which sediment rapidly in the ultracentrifuge and are spectroscopically distinct from normal dilute DNA (Lehrman, 1971; Jordan et al., 1972).

5. Emulsions of Systems Containing Two Polymeric Solutes

In speaking of emulsions, the traditional colloid chemist refers to systems which contain two phases whose main components are water and an organic solvent ("oil"), respectively. We distinguish between "oil-in-water" and "water-in-oil" emulsions, depending on the phase which is continuous. The system is generally metastable, with the driving force toward the reduction of the interface diminished by an "emulsifying agent" consisting of molecules which have highly polar and nonpolar portions.

The mutual incompatibility of dilute solutions of two different polymers in organic solvent media makes it possible to obtain a different type, that is, an "oil-in-oil" emulsion. The emulsifying agents in this case are block or graft copolymers containing subchains similar to the polymers in the two solution phases. The same principle which leads to the segregation of the homopolymers in these two phases then forces the copolymer molecules to occupy the interface so that the subchains can extend into regions where they encounter only similar polymeric species.

Emulsions of this type were first observed by Merrett (1954), who polymerized methyl methacrylate in the presence of Hevea rubber to obtain graft copolymers. He fractionated his product by the addition of methanol to a benzene solution and encountered considerable difficulty with the formation of stable emulsions consisting presumably of benzene-rich and benzene-poor solutions, each containing one of the two homopolymers. Similar problems have been recorded by other workers in the purification of block or graft copolymers.

A specific study of the use of graft copolymers as emulsifying agents for solutions of two homopolymers in a single solvent was carried out by Molau (1965), and the subject has been extensively reviewed in a monograph on block copolymers (Molau, 1970). Molau found that the addition of an increasing amount of the graft copolymer reduces the size of the droplets in the discontinuous phase and that the emulsifying agent must have a molecular weight in excess of 10^5. He believes that the emulsions are metastable, but they could well represent systems of true thermodynamic equilibrium.

An interesting approach to the study of systems of this type has been reported by Molau and Wittbrodt (1968), who stabilized solutions of polystyrene and polybutadiene in monomeric styrene by a styrene-butadiene block copolymer. The styrene was subsequently polymerized, and the product was examined by electron microscopy after selective staining of the polybutadiene regions. Although the structure of the system would tend to change during the polymerization, this difficulty may possibly be overcome if the polymerization is sufficiently rapid.

Although "water-in-water" emulsions have not been described, there is no reason why they would not form in an analogous fashion in systems containing water, water-soluble homopolymers of monomers A and B, and an A_n-B_m block copolymer.

6. Ternary Systems with Strongly Interacting Polymers

We saw in the preceding section that even polymers of very similar structure may be highly incompatible, whenever their mixing is endothermic. The corollary of this situation is the very high affinity of polymers for one another, if their mixing is accompanied by an evolution of heat. In such cases the mixing of two polymer solutions in the same solvent medium will result typically in the separation of a concentrated phase containing both polymers in a ratio such as to maximize their energetic interaction. This behavior is typical, in particular, of systems containing both positively and negatively charged polyions. Fuoss and Sadek (1949) showed that the formation of the association complex between polycations and polyanions may be observed trubidometrically in solutions as dilute as $10^{-6}N$. Maximum turbidity corresponds to equivalence between the cationic and anionic charges, so that turbidometric titrations provide an unusually sensitive tool for the analytical determination of polyions. Terayama (1952) carried out similar titrations, using as the end-point indicator various dyes, such as toluidine blue, which show a striking color change in solutions containing an excess of polyanions over polycations.

When two proteins are contained in a solution at a pH lying between their isoelectric points, one species will bear a positive and the other a negative charge, so that they will tend to precipitate one another. This has long been recognized as an inherent limitation in any protein fractionation, and procedures have been developed to take advantage of such interactions for the

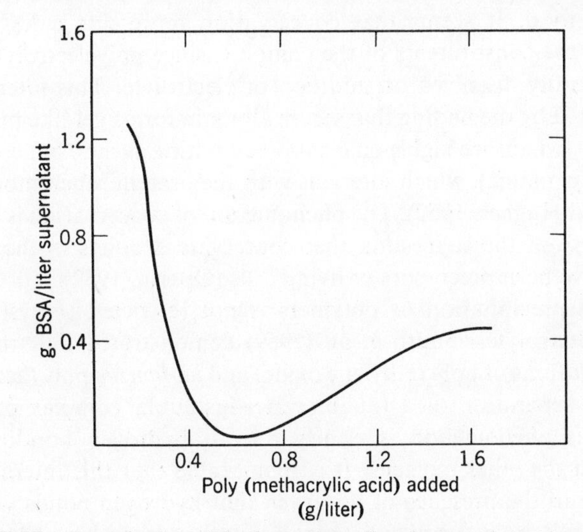

Fig. II.17. Precipitation of bovine serum albumin by poly(methacrylic acid).

separation of groups of proteins from complex mixtures. These groups may then be subdivided by refractionation at higher ionic strength or in the presence of dipolar ions, that is, under conditions which minimize electrostatic interactions between the macromolecular solutes (Cohn et al., 1950). The solubility relations in a system containing a polymeric acid and a protein below its isoelectric point are illustrated in Fig. II.17 on the poly (methacrylic acid)—bovine serum albumin system. It may be noted that an excess of the polymeric acid tends to redissolve the precipitate, implying the formation of negatively charged complexes. The solubility curves are independent of the chain length of the polymeric acid, suggesting that the phase equilibrium depends merely on the ratio of the positive and negative charges carried by the two macromolecules. The composition of the precipitate depends on the relative amounts of the cationic and anionic species in the supernatant, as was shown both in protein precipitation with synthetic polyacids (Morawetz and Hughes, 1952) and in a detailed study of the precipitation of ovalbumin by nucleic acid (Björnesjö and Teorell, 1945). As would be expected, the solubility of the complex is at a minimum when it is isoelectric (Bungenberg de Jong, 1949).

Sometimes the concentrated phase, forming in systems containing cationic and anionic polymers, separates, not in the form of a flocculent precipitate, but rather as a highly viscous fluid. This phenomenon, termed "complex coacervation," was subjected to extensive studies (Bungenberg de Jong and Kruyt, 1930; Bungenberg de Jong, 1937). Even so, the conditions required for coacervation, rather than the formation of gel-like precipitates, are not clearly understood. It seems that coacervation represents a less intimate interaction of the constitutents of the complex, since polyelectrolyte coacervates are generally dissolved on addition of electrolyte. This interpretation is also supported by the finding that serum albumin forms gel-like precipitates with polyacids (which are highly efficient precipitating agents) but coacervates with poly (vinyl amine), which interacts with the protein much more weakly (Morawetz and Hughes, 1952). The phenomenon of coacervation is of special interest because of the suggestion that coacervate droplets in the primeval ocean may have been precursors of living cells (Oparin, 1957).

The mutual precipitation of polymers is not restricted to systems containing polyelectrolytes. Smith et al. (1959) demonstrated that the mixing of aqueous solutions of poly (ethylene oxide) and *un-ionized* poly (acrylic acid) results in the separation of a tough, water-insoluble complex of the two polymers. This phenomenon is clearly due to hydrogen bonding of the carboxyls with the ether oxygens. It is remarkable that this interaction can be so effective in the presence of water, so that hydrogen bonds with water have to be broken to form hydrogen bonds between the polymers. The observation can be explained only in terms of a cooperative phenomenon, that is, it is the multiplicity of interacting groups arranged at corresponding spacings in the two interacting molecular chains which accounts for the stability of the complex.

Two very striking observations have been reported in cases where stereoisomeric forms of the same polymer form association complexes resulting in a sharp reduction of solubility. Watanabe et al. (1961) found that the mixing of dilute solutions of isotactic and syndiotactic poly (methyl methacrylate) in good solvent media resulted in instantaneous gelation. The gel gave a characteristic sharp x-ray diffraction pattern and had a well-defined melting point, which was quite insensitive to the ratio in which the two polymeric components were mixed. The data suggested that a stoichiometric complex is formed from the two stereoisomeric polymers. Observations bearing on the nature of this complex will be presented in Chapter VIII. A similar phenomenon was described by Yoshida et al. (1962), who found that the mixing of dimethylformamide solutions of poly (γ-methyl-L-glutamate) and poly-(γ-methyl-D-glutamate) results in the precipitation of an optically inactive material, whatever the ratio in which the optically active polymers are mixed. We shall see in the next chapter that poly (γ-methyl-L-glutamate) chains exist in dimethylformamide in the form of right-handed helices, and left-handed helices would similarly characterize the chains obtained from the enantiomorphic monomer. A mixture of equal numbers of right- and left-handed helices packs more efficiently in the crystal lattice than do helices of one sense only (Mitsui et al., 1967), and this may well account for the small solubility of the racemic polymer mixture. These observations suggest that effects due to stereochemical complementarity, so characteristic of specific interactions of proteins (e.g., the formation of enzyme—substrate and antigen—antibody complexes), are not necessarily restricted to macromolecules produced in living organisms.

Chapter III

CONFIGURATION AND CONFORMATION OF CHAIN MOLECULES

A large proportion of the problems arising in the study of macromolecules involves considerations of molecular geometry. This subject may be logically subdivided into two parts. First, we have to concern ourselves with the manner in which the atoms constituting the macromolecule are joined, that is, with the possibility of chemical isomerism. Second, once a chain molecule has been defined in chemical terms, we have to consider its flexibility, which will allow it, in general, to assume a wide variety of shapes. As we shall see later, these two properties may be interdependent.

Several types of isomerism may be distinguished in chain molecules involving, for instance, the distinction between linear and branched molecular chains, copolymers with different sequential arrangements of the monomer units, and macromolecular stereoisomerism. The subject of stereoisomerism of molecular chains (the "chain configuration") has become particularly important since the discovery of catalytic systems by which this variable may be controlled. (For a review of early work in this area see Gaylord and Mark, 1959.) We shall discuss in this chapter the definition and representation of macromolecular stereoisomerism and reserve for later chapters the experimental techniques by which it may be studied in polymer solutions.

The high flexibility of long-chain molecules is one of their most characteristic properties. It is reflected in the macroscopic phenomenon of rubber-like elasticity exhibited by polymers above their glass transition temperature, provided that the slippage of the chain molecules past one another is prevented by chain entanglements or by chemical crosslinking. Calorimetric experiments have shown that adiabatic extension of typical rubbers is attended by a temperature rise, and thermodynamic analysis of this observation leads to the conclusion that the retractive force of the deformed rubber must be due to a tendency of the chain molecules to return to a state of higher entropy (i.e., an extension of higher probability). The theoretical and experimental work in this field has been summarized in an excellent monograph by Treloar (1958).

The principles governing the equilibrium extension of a flexible chain molecule in dilute solution are analogous to those determining its elastic behavior in bulk. The role of a mechanical stress, which brings about an extension of a bulk specimen, is played by osmotic forces, which expand the dissolved chain molecule. In both cases, the forces are counteracted by the reluctance of the molecular chain to pass from a more probable to a less probable extension. An attempt to construct a theory that would interpret and predict the probability distribution of molecular shapes in a system containing dissolved chain molecules may be subdivided into three parts. First, the concept of the flexibility of the polymer chain should be defined in terms of its molecular structure. We may then explore the statistical problem of the distribution of shapes in model chains characterized by a given contour length and a flexibility parameter. At this stage the chains are treated as mathematical lines of zero volume. Finally, we have to account for the perturbations in the shapes of the molecules which are brought about by their spatial requirements and by energetic interactions dependent on the solvent medium, which may favor or hinder contacts of chain segments with each other.

The polymer chains to which the above remarks apply may assume a wide variety of "randomly coiled" shapes, none of which is of any special importance. However, a restricted class of linear chain molecules is able, in solution, to assume precisely defined conformations corresponding to helically wound, rodlike structures. This behavior is typical of some proteins, the nucleic acids, and their synthetic analogs. The transition from the random coil form of a chain to the helical conformation may be considered a one-dimensional analog of crystallization, and the principles underlying such phenomena transcend, therefore, the specialized interests of the polymer chemist. It appears, moreover, that only large molecules with precisely defined spatial relationships, such as result from the ordered conformations of proteins and nucleic acids, can exhibit the high degree of specificity of molecular interactions which is indispensable for the existence of the life process. This consideration has undoubtedly been responsible for the massive effort expended to achieve a detailed understanding of conditions conducive to the stabilization of ordered conformations in solutions of polypeptides and polynucleotides. The additional problem of defining the forces responsible for the folding of polynucleotide and polypeptide chains consisting of ordered sections and sections with no structural regularity into the unique "tertiary" structure of some nucleic acids (Section III. C. 3) and of globular proteins (Section III C. 5) is only very imperfectly understood at the present time.

A. THE STEREOISOMERISM OF CHAIN MOLECULES

During the growth of a typical vinyl polymer, an asymmetric center is created every time a monomer molecule adds to the propagating chain end.

This introduces the possibility of stereoisomerism, since the new asymmetric center may have the same configuration as the asymmetric center that is its immediate predecessor or the opposite configuration. For instance, in a growing polystyrene chain, the addition of a monomer unit may take the two alternative routes represented below:

$$\cdots CH_2{-}\overset{\phi}{\underset{H}{C}}{-}CH_2{-}\dot{C}\begin{matrix}\diagup H\\ \diagdown \phi\end{matrix} + CH_2{=}CH\phi \begin{cases}\rightarrow \cdots CH_2{-}\overset{\phi}{\underset{H}{C}}{-}CH_2{-}\overset{\phi}{\underset{H}{C}}{-}CH_2\dot{C}\begin{matrix}\diagup H\\ \diagdown \phi\end{matrix}\\[2ex] \rightarrow \cdots CH_2{-}\overset{\phi}{\underset{H}{C}}{-}CH_2{-}\overset{H}{\underset{\phi}{C}}{-}CH_2{-}\dot{C}\begin{matrix}\diagup H\\ \diagdown \phi\end{matrix}\end{cases}$$

Since these two alternative modes of monomer addition are available for each monomer unit, a polymer prepared under conditions which do not discriminate between them is composed of a very large number of stereoisomeric forms. This is the nature of the product obtained, for instance, when styrene is polymerized in the presence of free-radical initiators, and the multiplicity of stereoisomeric forms in such polymers is the cause of their inability to crystallize (Flory, 1953a).

In 1955, Natta and his collaborators made the unexpected discovery that it is possible to polymerize such typical vinyl derivatives as propylene, 1-butene, and styrene with special catalyst systems to obtain highly crystalline polymeric products. An analysis of the x-ray diffraction pattern from such polymer samples suggested that the chain backbone must contain long sequences of monomer units in which the carbons carrying the chain substituents have identical steric configurations, and such polymer chains were named "isotactic" (Natta et al., 1955). It is most suggestive to represent the steric configuration of a polymer chain by a convention popularized by Natta. In this representation the atoms comprising the chain backbone lie in a plane and form the zig-zag corresponding to maximum chain extension compatible with the bond angles. Such a representation of a section of an isotactic polystyrene chain is shown in Fig. III.1*a*. We should note the following properties of such a chain:

(a) On translation of a monomer unit by a distance equal to its length in the direction of the chain axis, it will be exactly superimposed onto the next monomer unit.

(b) The methylene groups lie in a plane of symmetry, so that the steric relation of two adjoining monomer units is that characteristic of a *meso* structure.

(c) A backbone carbon carrying the chain substituents is not truly asymmetric, since two of its valences are bound to sections of the polymer which may differ in length but are identical in the neighborhood of the carbon atom under consideration. Such carbons are called "pseudoasymmetric." Rotation around the axis represented by the dotted line changes a chain in which the substituents lie above the plane containing the chain backbone to one in which the substituents lie below that plane. Therefore, we may *not* distinguish between *d* and *l* isotactic sequences.

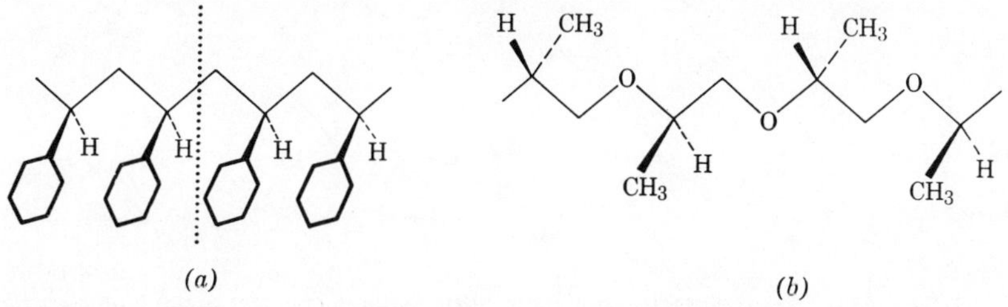

(*a*) (*b*)

Fig. III.1. Typical isotactic polymers. (*a*) Isotactic polystyrene. (*b*) Isotactic poly-(propylene oxide).

When the monomer unit in the chain backbone contains an odd number of atoms, the geometrical relationship in an isotactic sequence is distinctly different. A typical case of this kind is isotactic poly (propylene oxide) (Price and Osgan, 1956), represented schematically in Fig. III.1*b*. In this case superposition of a monomer unit onto the next one along the polymer chain requires translation by a distance equal to the length of the unit and rotation by 180° around the chain axis. As a result, the chain substituents come to lie alternately above and below the plane containing atoms of the extended chain backbone. In poly (propylene oxide) the carbons carrying the chain substituents are truly asymmetric, since the chain is joined to them on one side through an oxygen atom and on the other side through a methylene group. We may, for instance, define the configuration of the asymmetric carbons by specifying that, when viewing the fully extended chain from a direction such that an oxygen precedes the asymmetric center, the direction of the bond to the next methylene carbon has to be rotated clockwise (or counterclockwise) by 120° to be superimposed on the position previously occupied by the methyl group.

Later studies resulted in the discovery of polymerization conditions leading to a regular alternation of the configuration of the pseudoasymmetric carbon atoms. This type of stereoregulation was first described by Natta and Corradini (1956) for 1,2-polybutadiene and poly (vinyl chloride), as depicted in Figs. III.2*a* and III.2*b*, and the term "syndiotactic" was suggested to

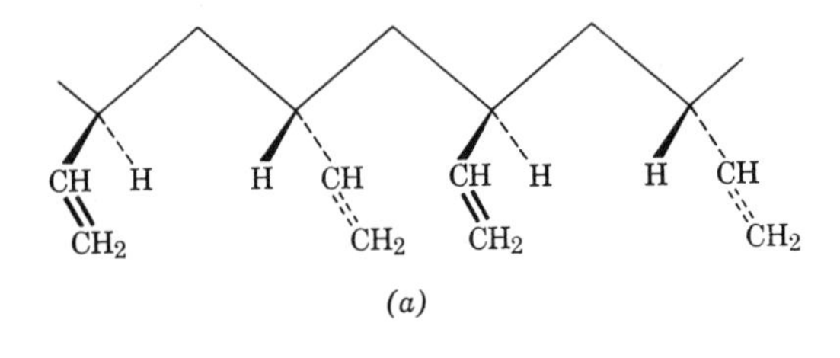

(*a*)

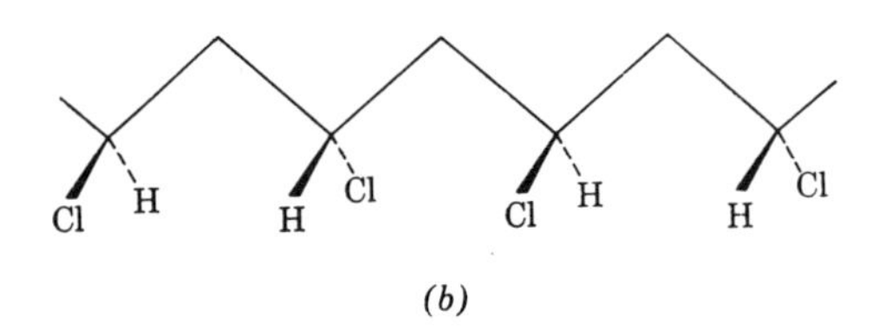

(*b*)

Fig. III.2. Typical syndiotactic polymers. (*a*) Syndiotactic 1,2-polybutadiene. (*b*) Syndiotactic poly(vinyl chloride).

describe it. In a syndiotactic chain it is impossible to obtain superposition of two nearest-neighbor monomer units by combinations of translations and rotations; such superposition is obtained only onto next-to-nearest neighbors by a translation of the monomer unit. Specific catalysts leading to syndiotactic polymers seem to be less common than those by which isotactic chains are produced, but they have been reported, for example, for poly (methyl methacrylate) (Fox et al., 1958) and for polypropylene (Zambelli et al., 1967b). In addition it is frequently possible to approach the syndiotactic structure by carrying out free-radical polymerizations at low temperatures, since the activation energy involved in chain growth seems to be generally significantly lower if monomers add to the chain so as to produce pseudoasymmetric centers with alternating configurations. This effect has been described for poly(vinyl chloride) (Fordham et al., 1959) and for poly(isopropyl acrylate) (Garrett et al., 1959). In the free-radical polymerization of methacrylic acid, both the temperature and the solvent affect the stereoregularity of the product. A highly syndiotactic polymer may be obtained at $-80\,^{\circ}\mathrm{C}$ in the proper solvent medium (Lando et al., 1970).

The phenomenon of stereoregulation might not have been discovered were it not for the striking ability of many stereoregular polymers to crystallize. Nevertheless, some polymers known to be highly stereoregular cannot be made to crystallize. It was found, for instance, that poly (*p*-iodostyrene), derived from isotactic polystyrene (Braun, 1959), and poly(vinyl acetate), derived from crystalline poly(vinyl formate) (Fujii et al., 1962), are not crystallizable. In many other cases crystallization of highly stereoregular polymers is a very slow process and has been achieved only after a long empirical search for suitable experimental conditions. Conversely, in the case of poly(vinyl alcohol) the small size of the hydroxyl substituent enables the polymer to crystallize even if it is not stereoregular. It is then obvious that criteria other than crystallizability are needed to study the phenomenon of stereoregulation.

There are still other reasons why such additional techniques for the characterization of polymeric stereoisomerism are required. On one hand, the concept of an isotactic and a syndiotactic chain represents the ideal of perfect stereoregularity, which, in practice, can never be quite attained, even under the most favorable conditions. On the other hand, it has been proposed that we describe as "atactic" all polymer samples which "do not possess steric order of the units or possess tacticities of such a complex character that it cannot be understood by a simple intellect "(Natta and Danusso, 1959). Obviously, this definition reflects the lack of interest of the crystallographer in all chains in which the regularity of the distribution of configurations is too low to allow crystallization. This distribution may, in fact, deviate to different extents from perfect randomness, and we should aim to describe it so that the nature of the distribution of pseudoasymmetric centers will be characterized quantitatively for all cases, ranging continuously from perfect order to complete randomness. This problem was discussed first by Coleman (1958) and later by Bovey and Tiers (1960), Miller and Nielsen (1960), and Miller (1962). In the simplest case, the probability σ^* that a newly formed pseudoasymmetric center in the growing chain has the same configuration as the preceding center is independent of the configuration of the rest of the chain. We may then use the single parameter σ^* to predict the probability of any sequence of steric configurations. For instance, if we consider a sequence of three pseudoasymmetric centers and denote arbitrarily the configuration of the central one by $+$, the probability of an isotactic $(+++)$, syndiotactic $(-+-)$, and heterotactic $(++-$ or $-++)$ triad will be $(\sigma^*)^2$, $(1 - \sigma^*)^2$, and $2\sigma^*(1 - \sigma^*)$, respectively. If we have a spectroscopic method for distinguishing such triads (p. 232–235), we can decide whether their relative frequency is compatible with a chain growth mechanism in which the steric configuration is governed by a single parameter. For example, if we find that dyads of pseudoasymmetric centers are with equal probability isotactic and syndiotactic, $\sigma^* = 0.5$ and if isotactic, syndiotactic, and heterotactic triads constitute 25%, 25% and 50%, respectively, of the total, the single parameter σ^* controls the stereosequence distribution. On the other hand, if the polymer is found to be diisotactic (Huggins et al. 1962) with a sequence: $++--++--++--++--$ then *all* the triads are heterotactic, although isotactic and syndiotactic dyads still occur with equal frequency. In this case, the chain growth mechanism is obviously influenced by the configuration of the penultimate unit, and two probabilities have to be specified: $\sigma_i^* = 0$ (a + unit is never added to a ++ sequence) and $\sigma_s^* = 1$ (a + unit is always added to a −+ sequence). There are also indications that the nature of the propagating chain end may change during the synthesis of a polymer (Coleman and Fox, 1963). This can lead, in an extreme case, to the formation of a stereoblock polymer:

$$++++++++-+-+-+-+-$$

which will be characterized by a high content of isotactic and syndiotactic and a low content of heterotactic triads.

A different type of stereoisomerism could arise, in principle, in the case of polymer chains containing residues of monomers such as maleic anhydride, maleimide, acenaphthylene, or vinylene carbonate:†

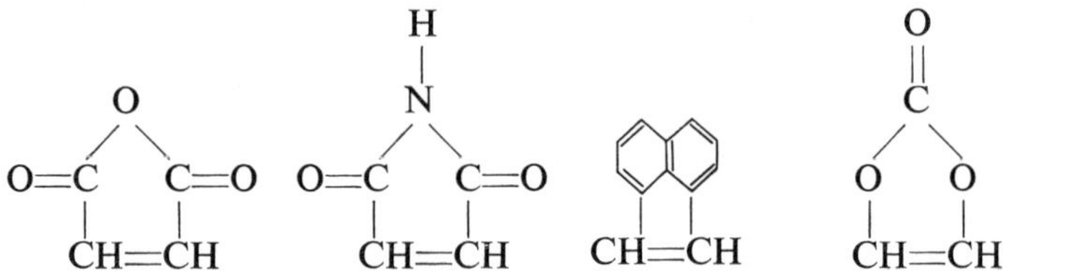

Maleic anhydride	Maleimide	Acenaphthylene	Vinylene carbonate
O=C(–O–)C=O, CH=CH	O=C(–NH–)C=O, CH=CH	CH=CH	O=C(–O–)(–O–), CH=CH

One could envisage the bonds linking the residue into the chain backbone as being either on the same side or on opposite sides of the five-membered ring, but molecular models indicate that only the trans linkage is possible.

When macromolecules are produced in the metabolic processes of living organisms, the chemical stereospecificity of enzymatic catalysis leads typically to perfectly stereoregulated chain molecules. Apart from proteins and nucleic acids, to be discussed more fully in section III.C, this category includes a variety of polysaccharides; the structures of some of them are shown in Fig. III.3. Synthetic macromolecules of a uniquely defined configuration may be obtained from optically active monomers if no bond attached to the asymmetric center is involved in the polymerization reaction and no new asymmetric centers are formed in the polymerization. This is the case in the preparation of high molecular weight polypeptides from *N*-carboxy-α-amino acid anhydrides (Katchalski, 1951; Katchalski et al., 1964), of polyesters from optically active acids (Kleine and Kleine, 1959), of polyamides from optically active lactams (Overberger and Jabloner, 1963), and so forth. Polymers with atropisomeric groups in the side chains or in the backbone of

†The polymerization of maleic anhydride is difficult, and this monomer is usually encountered only in copolymers; the other three cyclic monomers form homopolymers quite easily.

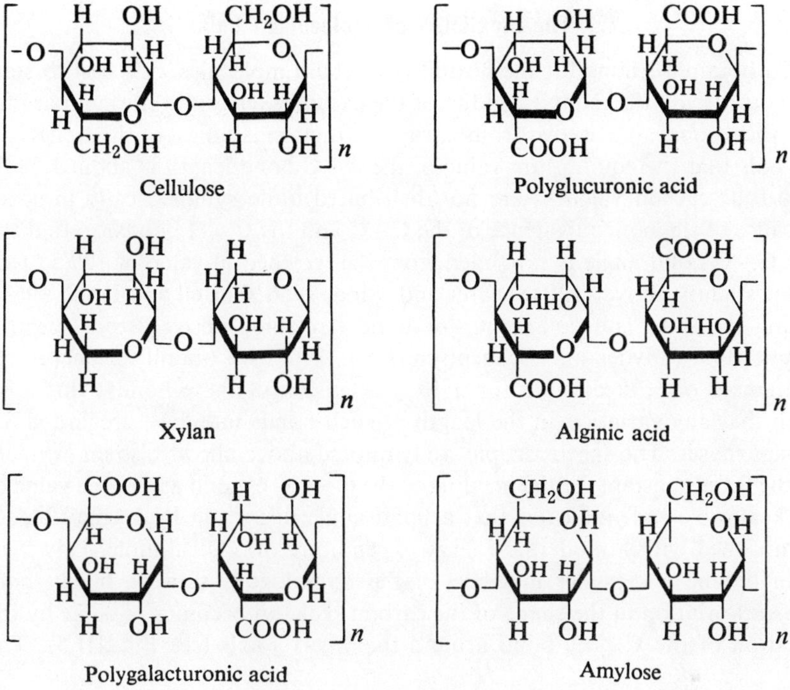

Fig.III.3. Structures of some typical polysaccharides.

the macromolecular chain (i.e., groups whose asymmetry depends on highly hindered rotation around a single bond between two aromatic rings) have been reported by Schulz and Jung (1966, 1968) and are represented in Fig. III.4. For a general discussion of the preparation of optically active polymers, the reader is referred to reviews by Schulz and Kaiser (1965), Pino (1965), and Goodman et al. (1967).

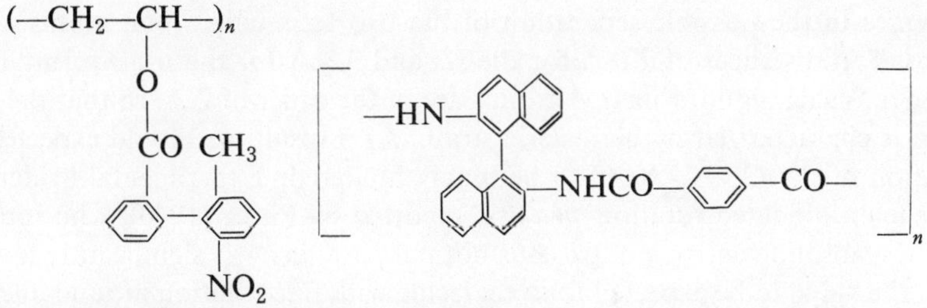

Fig. III.4. Polymers with atropisomeric groups.

B. GEOMETRY OF FLEXIBLE CHAINS

1. The Flexibility of Molecular Chains

To obtain a feeling for the flexibility of chain molecules, it is best to start by considering the relative rigidity of the carbon skeleton in normal paraffin hydrocarbons. We know from x-ray diffraction studies of hydrocarbon crystals that the equilibrium value of the C—C bond length is about 1.53 Å. The four carbon valences are not distributed quite symmetrically in space because of the nonequivalence of the C—C and the C—H bonds, so that the C—C—C bond angle is expanded from the tetrahedral value of 109.5° to a value slightly above 112° (Shearer and Vand, 1956; Bartell and Kohl, 1963). Using the force constant for C—C bond stretching derived from spectroscopic data (Snyder and Schachtschneider, 1965) we obtain an energy requirement of 60 kcal/mole for a 10% extension of these bonds; this is so high that any variation in the length of such bonds may be disregarded for our purposes. The spectroscopic study quoted above allows also an estimate of the force constant for the bending of the C—C—C bond angle; the value of 65 kcal/mole rad^2 indicates that a bond angle distortion by 5° may be accomplished with a moderate energy expenditure of 500 cal/mole. However, with butane and higher members of the homologous series a much more drastic variation in the shape of the carbon skeleton becomes possible by the rotation of the C_3-C_4 bond around the C_2-C_3 axis (see Fig. III.5). The

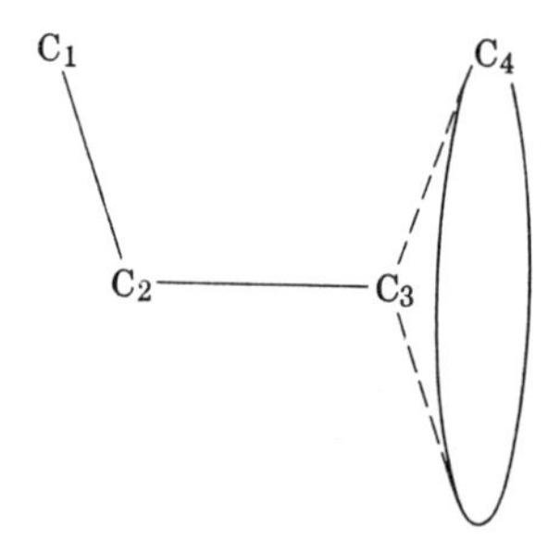

Fig. III.5. Rotation around the C_2-C_3 bond in *n*-butane.

extremes in the possible separation of the two terminal carbon atoms correspond to distances of 2.6 Å for the *cis* and 3.8 Å for the *trans* form. The van der Waals radii for methyl groups are of the order of 2 Å, so that the *cis* form is characterized by high steric strain. As a result we should expect the rotation of the C_3-C_4 bond to be highly hindered. Experimental evidence for such a hindered rotation was first reported by Pitzer (1940), who found that the absolute entropy of gaseous normal paraffins was significantly lower than the value to be expected for a molecule with free rotation around all the valence bonds. He showed that the discrepancy can be accounted for if the potential energy barriers hindering the rotation are considered. The variation in the shape of the molecule because of rotation around a covalent bond ("molecular conformation") is visualized most easily by the schematic method introduced by Newman (1956). In this representation the bond between the second and third carbons of *n*-butane lies in the line of sight. The conformation of the carbon skeleton is then characterized by an "internal angle of rotation," ϕ, by which the bond is rotated from the *trans* conformation. Figure III.6 shows the three "eclipsed" forms corresponding to ϕ values of $+60°$, $-60°$, and $180°$ in which steric strain is maximized by close approach of the terminal carbons to each other or to a hydrogen atom attached to the next-to-nearest carbon atom. Also shown are the three staggered conformations, the *gauche* (g^+ and g^-) forms, $\phi = 120°$ and $\phi = -120°$, and the *trans* (t) form, $\phi = 0°$. The *trans* conformation, corresponding to a maximum separation of the bulky methyl groups, would be expected to be the lowest energy form.

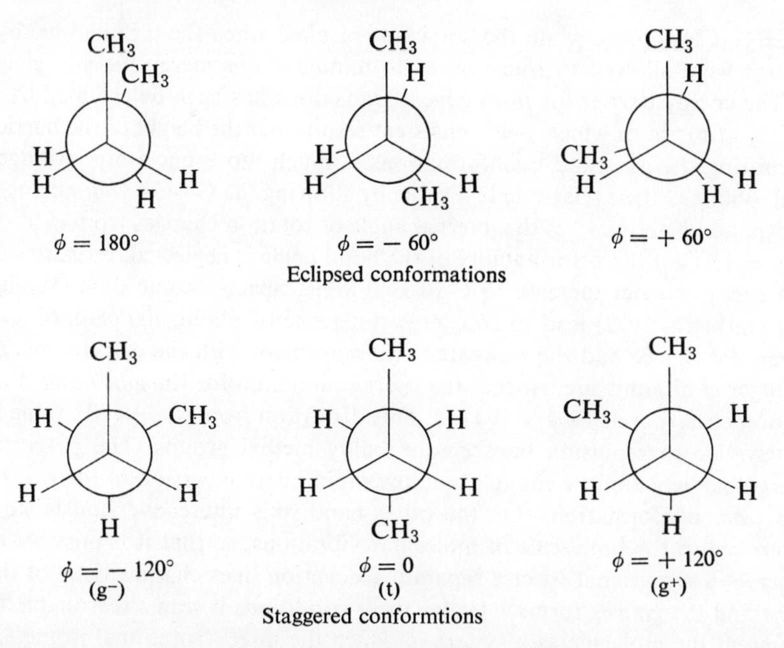

Fig. III.6. Newman projections of *n*-butane conformations.

Theoretical treatment of the energy of *n*-butane as a function of ϕ must take account not only of the van der Waals interactions of nonbonded atoms, but also of interactions between the bonding electron orbitals, which resist any departure from the staggered conformation. In addition the deformability of bond angles may be important in relieving steric strain. Figure III.7 represents the results of the calculations of Allinger et al. (1967). The dotted line was obtained assuming a staggered conformation for the

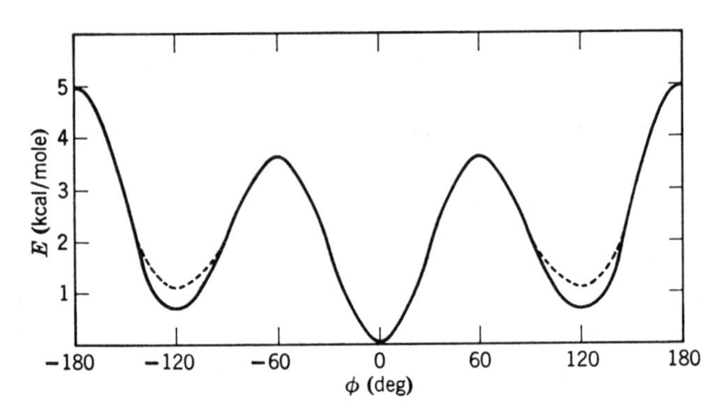

Fig. III.7. The relative energy of *n*-butane as a function of the angle of rotation around the 2,3-bond.

$-CH_2-CH_3$ group, while the solid line resulted when the terminal methyl groups were allowed to rotate so as to minimize the energy for any given ϕ. The energy barrier for *trans-gauche* transitions has been determined by a variety of methods which yield consistent results, but the height of the barrier separating the g^+ and g^- conformations is much more uncertain. Allinger et al. obtained their relatively low value by allowing the C—C—C bond angle to expand by about 4° as the internal angle of rotation changes from $\phi = 0°$ to $\phi = 180°$; if the deformability of the bond angle is neglected, estimates of this energy barrier increase to 6–10 kcal/mole. Spectroscopic data (Woller and Garbisch, 1972) lead to an energy difference of about 700 cal/mole between the *gauche* and the *trans* states (in agreement with the calculations of Allinger et al.) and suggest that the energy minimum for the *gauche* conformation corresponds to $\phi = \pm 114°$. This distortion from the $\pm 120°$ value is clearly due to repulsion between the bulky methyl groups. The potential energy barriers are low enough to allow rapid interconversion of the *gauche* and *trans* conformations. On the other hand, this interconversion is slow compared to the time scale of molecular vibrations, so that it is possible to observe in vibrational spectra separate absorption lines characteristic of the *trans* and the *gauche* forms.† Under these conditions it seems reasonable to represent the molecule as a system in which the three "rotational isomers," with ϕ values centered at 0° and about $\pm 120°$, are treated as separate chemical species in equilibrium with each other, while the eclipsed conformations are transition states in the conversion of one rotational isomer into

†Kohlrausch (1932) compared the Raman spectra of ethyl halides with those of *n*-propyl halides and oberved the doubling of certain absorption lines in the propane derivative, which he interpreted correctly as due to rotational isomerism. His proposal was incorrect in detail, since he thought that the eclipsed *cis* form was one of the conformations contributing to the spectrum.

another. This point of view, in which hindered rotation is taken into account in terms of discrete rotational isomers, was introduced into the study of chain molecules with most fruitful results by the Leningrad school of investigators, whose work has been summarized by Volkenstein (1963). It is instructive (Volkenstein, 1963a) to compare rotational isomerism in saturated paraffins with the *cis-trans* isomerism of 1,2-disubstituted ethylenes. Rotation around the double bond is much more difficult (with an energy barrier of the order of 40,000 cal/mole), and interconversion of the isomers is, therefore, sufficiently slow to allow their isolation as pure chemical entities. But the difference between rotation around single and around double bonds is only quantitative and appears more important merely because of our common experimental time scale. There is no reason why short-lived rotational isomers of molecules containing single bonds only should not be treated in the same manner as equilibrium mixtures of the *cis—trans* isomers of 1,2-disubstituted ethylene derivatives.

Yet another principle emerges in the consideration of the possible conformations of the next member of the homologous series, *n*-pentane. A description of the shape of the carbon skeleton now involves the specification of two internal angles of rotation, so that assignment of the discrete values of 0°, +120°, and −120° to these angles leads to the four distinct molecular shapes tt, tg^- (equivalent to tg^+, g^-t, and g^+t), g^-g^- (or g^+g^+), and g^+g^- (or g^-g^+). These arrangements are represented in Fig. III.8 in a manner such that the central three carbon atoms lie in the plane of the paper, while lines with variable thickness and dashed lines represent bonds pointing toward or away from the observer. The conformation g^+g^- may be seen to result in a very small distance between the terminal carbon atoms (about 2.5 Å), and Taylor (1948) has pointed out that the potential energy of this form must be extremely high. Thus, although conformations g^+ and g^- are equivalent in butane, conformations g^-g^- and g^+g^- in pentane represent very different potential

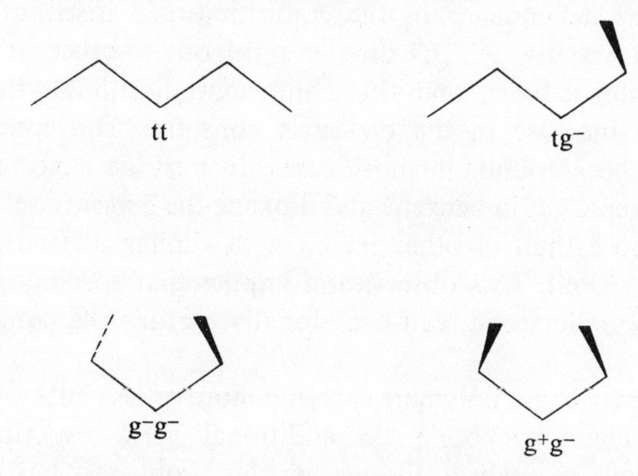

Fig. III.8. Conformations of the carbon skeleton in *n*-pentane.

energies, and we may conclude, in general, that the potential energy of a given conformation of a chain molecule may not be evaluated by summing the contributions characteristic of the individual angles of internal rotation. In particular, with a long polyethylene chain (see Lifson, 1959) one should, in principle, take account of correlations of internal angles of rotation beyond the nearest-neighbor pairs. However, such a model would lead to prohibitively difficult calculations, and it is reasonable to accept an approximate model in which (a) g^+g^- and g^-g^+ sequences are excluded; (b) for all other combinations the difference in the potential energy of g and t bonds is assumed to be independent of the conformation of the neighboring bonds. The stiffness of the chain will increase with an increase of the difference in the energy characterizing *gauche* and *trans* bonds, and it will decrease with an increase in temperature, which will tend to equalize the occupancy of all accessible chain conformations.

Before discussing more complicated chain molecules, we should point out that considerations of steric interference are occasionally insufficient to account for the relative stability of the various rotational isomers of small molecules. Volkenstein (1963b) cites spectroscopic data showing that the *gauche* conformation is the low-energy form of *n*-propyl chloride and *n*-propyl bromide, and similar conclusions were drawn from electron diffraction studies of gaseous *n*-butyl chloride (Ukaji and Bonham, 1962). The nature of the forces responsible for the reversal of the usual order of stability in rotational isomers has not been fully clarified. In cases where the rotational isomers differ appreciably in their dipole moments, the more polar form should be stabilized in the liquid state as compared to the gaseous state because of dipole-dipole interactions of the closely spaced molecules. Thus, Mizushima et al. (1949) estimated from spectroscopic data that gaseous 1,2-dichloroethane at 25°C contains 25% of the *gauche* isomer, while in the liquid state 57% of the molecules are in the *gauche* conformation. Similar reasoning may be applied to the dependence of the conformational distribution on the solvent medium. The two C—Cl dipoles repel one another in the *gauche* form of 1,2-dichloroethane, and this conformation should, therefore, be stabilized by an increase in the dielectric constant. The conformational distribution has been found, in most cases, to vary as expected with the nature of the solvent, but in benzene and dioxane the *gauche* conformation is favored much more than in other media with similar dielectric constants (Oi and Coetzee, 1969). This observation implies that specific interactions, which are poorly understood, can occasionally perturb the conformational equilibrium.

When we consider chain polymers carrying more or less bulky substituents attached to the chain backbone, the additional steric restictions become much more difficult to analyze. In tackling this problem it has been an inestimable advantage that the geometry of the molecular chain may be obtained by x-ray diffraction analysis of crystalline polymers. Bunn and Holmes (1958) have pointed out that the conformation of chain molecules in the crystalline state should, in general, be close to the form corresponding to the lowest energy of the isolated chain, being apparently affected to only a minor extent by intermolecular forces. This principle has been greatly strengthened by the work of Natta et al. (1962) and De Santis et al. (1963), who calculated the energies of various chain molecules as a function of the internal angles of rotation on the basis of estimated van der Waals repulsion energies of chain substituents. They found striking agreement in a number of cases between the predicted lowest-energy form and the conformation assumed by the chain molecule in the crystalline state. Nevertheless, some exceptions exist to this broad generalization. For instance, Natta et al. (1959) found quite different conformations for crystalline isotactic polystyrene and for isotactic poly (*p*-fluorostyrene), and this discrepancy must clearly be due to effects produced by a changing chain conformation on the efficiency with which the chains can be packed in the crystal lattice. However, such anomalies would be expected only in cases where two chain conformations have very similar energies, so that a more favorable crystal lattice energy would give a decisive advantage to a form which would be slightly less stable in the isolated chain. This situation is probably rare, and molecular chain conformations in crystalline polymers may be used as points of departure in speculations about the preferred shape of these molecules in solution.

Before presenting the crystallographic evidence, let us consider briefly the extent of steric interference to be expected in a chain segment of two monomer units with all possible combinations of *trans* and *gauche* bonds. The various ways in which two monomer units may be arranged in isotactic

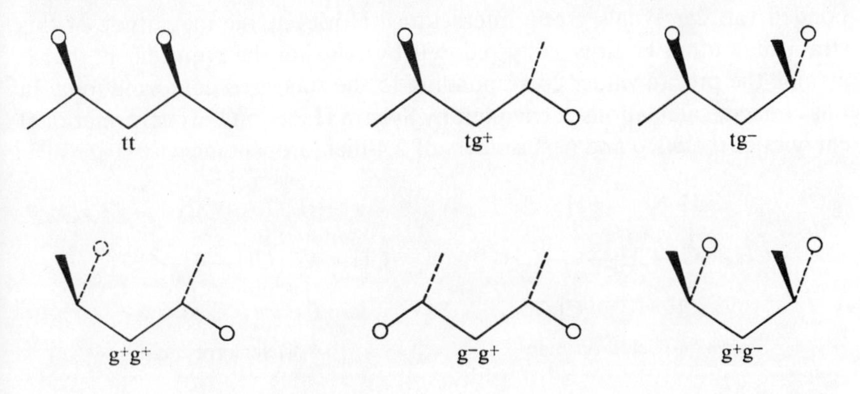

Fig. III.9. Staggered conformations in The Chain backbone in a sequence of two monomer residues of an isotactic vinyl polymer. The circles symbolize chain substituents.

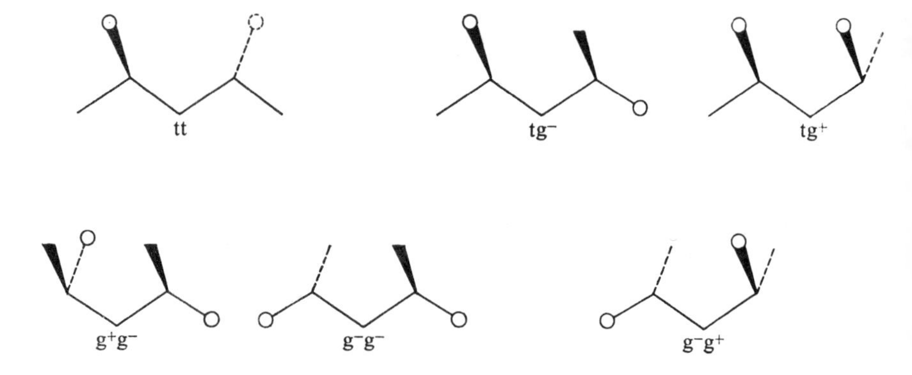

Fig. III.10. Staggered conformations in The chain backbone in a sequence of two monomer residues of a symdiotactic vinyl polymer.

and syndiotactic chains are represented schematically in Figs. III.9 and III. 10. In defining the sign of g we use the following convention: (a) the tt rotational isomer is represented with the first chain substituent on the left in front of the plane defined by the chain backbone; (b) in viewing a bond in the chain backbone from the left, a clockwise rotation of the backbone carbon lying further from the observer is taken as positive. The figures show that, with the chain substituents, the conformations tg^+ and tg^- or g^-g^+ and g^+g^- are no longer equivalent. For the isotactic sequence, only tg^+, shown in Fig. III.9 (as well as its mirror image g^-t), is free from steric strain. For the syndiotactic sequence, tt and g^-g^- (or tt and g^+g^+ in the mirror image of the chain represented in Fig. III.10) are unstrained; all other rotational isomers would be expected to have high potential energies because of non-bonded van der Waals group interactions. However, the magnitude of this strain may often be drastically reduced by relaxing the requirement that ϕ assume the precise values corresponding to the staggered conformations. In this context, calculations carried out by Sykora (1968) on the conformational energies of the *meso* and ($\pm$) isomers of 2,4-dichloropentane:

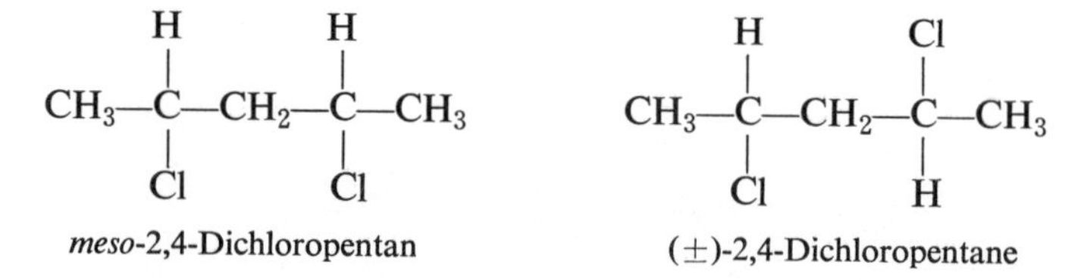

are highly instructive. For instance, in the *meso* compound the $tg^+ \rightarrow tt$ transition is estimated to require 18.6 kcal/mole if the *trans* form is restricted to $\phi = 0$, but only 2.9 kcal/mole if the internal angles of rotation of the tt form can adjust themselves to the most favorable values, that is, $\phi_1 = +28°$, $\phi_2 = -12°$. Sykora did not consider the possibility of bond angle distortions, and these should lead to further reduction in the energy of conformational transitions.

In crystalline polyethylene the polymer chains assume the planar zig-zag form of the all-*trans* conformation (Bunn, 1939), which would be expected to correspond to the potential energy minimum. The crystal structures of a large number of isotactic vinyl polymers have been determined by Natta and

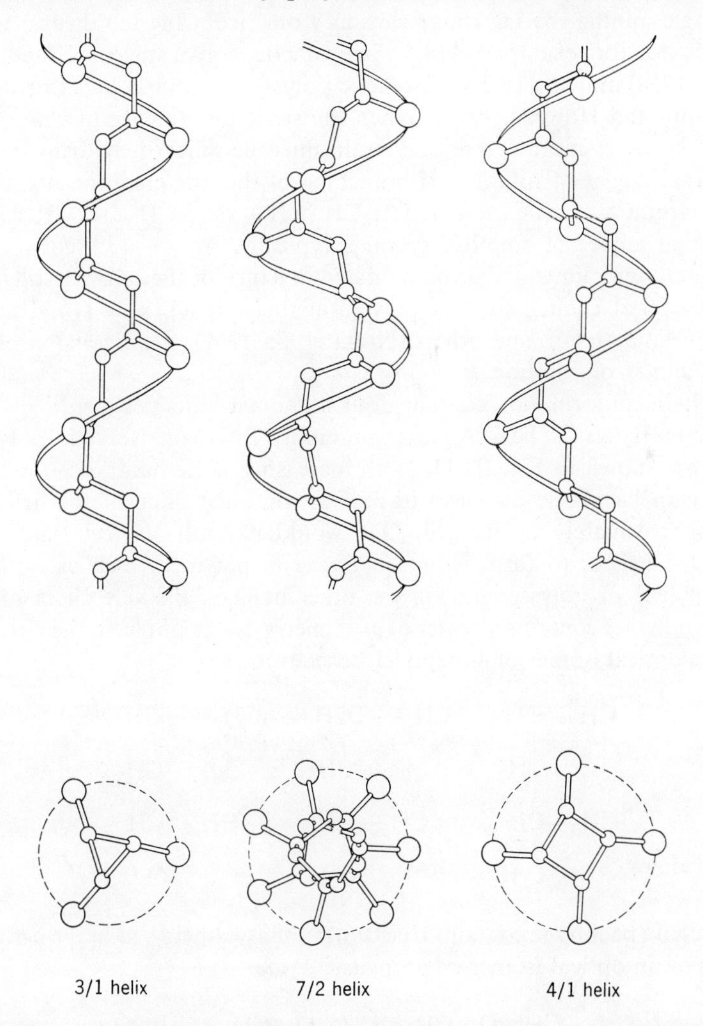

Fig. III.11. Helical conformations of isotactic vinyl polymers in the crystalline state.

his collaborators (Natta, 1957; Natta et al., 1959). In a number of chains in which the substituents have moderate spatial requirements [such as isotactic $(-CH_2—CHR—)_n$, where R = $—CH_3$, $—C_2H_5$, $—C_3H_7$, $—C_6H_5$, $—OCH_3$,] the stable conformation of the chain backbone has a regular alternation of *trans* and *gauche* bonds ($\phi_1 = 0°$, $\phi_2 = 120°$), generating a helix in which three monomer units are contained in one turn. This conformational sequence is the one predicted as corresponding to the least steric strain on the basis of the crude representations of Fig. III.9. Such steric interference of the chain substituents as is present in these polymers is relieved by expansion of the bond angle joining the backbone carbon atoms from the tetrahedral angle of 109.5° to, for example, 114.5° in isotactic polypropylene (Bunn and Holmes, 1958) and to 116.5° in isotactic polystyrene (Natta and Corradini, 1955; Bunn and Howells, 1955). When the steric interference of chain substituents becomes more severe, the strain must be relieved by distortion of the internal angles of rotation. If branching of the side chain occurs at the second carbon atom [i.e., R = $—CH_2CH(CH_3)_2$ or $—CH_2CH(CH_3)C_2H_5$] the internal angles of rotation become, typically, $\phi_1 = -13°$, $\phi_2 = 110°$, and $3\frac{1}{2}$ monomer units are accommodated in a turn of the helix; in still more highly hindered chains, such as poly (*o*-methylstyrene), poly (vinyl cyclohexane) (Natta, 1960), and others (Natta et al., 1959), the distortion of the internal angles of rotation is doubled to $\phi_1 = -26°$, $\phi_2 = 100°$, and the helical chain conformation contains four monomer units per turn.

The three types of helices most commonly found in isotactic polymer crystals are shown in Fig. III.11. With increasing steric hindrance the helix of the chain backbone increases in radius, while the pitch of the helix remains approximately unchanged. One would naturally expect right- and left-handed helices to form with equal ease in polymers such as isotactic polypropylene or polystyrene. On the other hand, if the side chain of the isotactic polymer contains a center of asymmetry, for example, in the polymer of a pure optical isomer of 4-methyl-1-hexene:

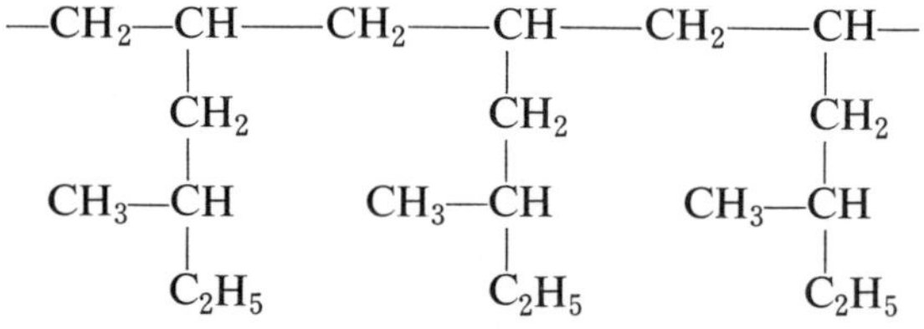

```
—CH2—CH——CH2——CH——CH2——CH—
      |             |             |
     CH2           CH2           CH2
      |             |             |
 CH3—CH        CH3—CH        CH3—CH
      |             |             |
     C2H5          C2H5          C2H5
```

or if the chain backbone contains true centers of asymmetry, as in the isotactic polymer of an optical isomer of propylene oxide:

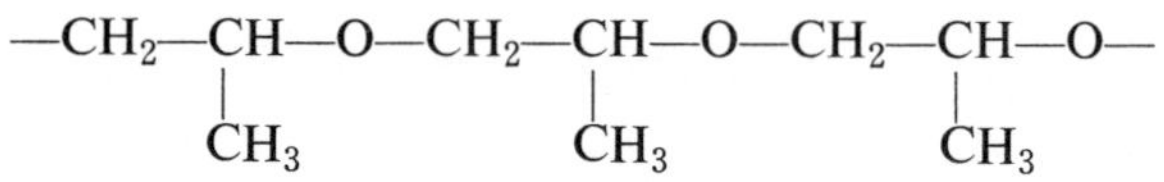

```
—CH2—CH—O—CH2—CH—O—CH2—CH—O—
      |            |            |
     CH3          CH3          CH3
```

then a right- or left-handed helix will, in general, be energetically preferred (De Santis et al., 1963). In the case of syndiotactic polymers the qualitative estimates based on the steric relations represented in Fig. III.10 are again found to predict correctly that conformations close to the all-*trans* planar structure will generally be favored (Natta and Corradini, 1956). However, we have seen that the $g^{\pm}g^{\pm}$ sequence is also strain-free in a syndiotactic pair. Longer sequences of *gauche* bonds are sterically forbidden, but the $(ttg^{\pm}g^{\pm})_n$ chain conformation in syndiotactic polypropylene has an energy very similar to that of the all-*trans* chain (Natta et al., 1962). In fact, syndiotactic polypropylene exists in different crystalline modifications characterized by one or the other chain conformation (Natta, 1960; Natta et al., 1964; Corradini et al., 1967).

If the chain backbone contains also atoms other than carbon, dipole-dipole interactions may make significant contribution to the conformational energy. This factor (Uchida et al., 1956) stabilizes in polyoxymethylene, $(—CH_2O—)_n$, a distorted all-*gauche* conformation with $\phi = 102.5$ (Tadokoro et al., 1960) (in contrast to the all-*trans* polyethylene chain), but the spatial requirements of the ether oxygen and the methylene group are sufficiently different to favor the helical polyoxymethylene chain even if the dipole interactions are neglected (De Santis et al., 1963). With polycycloxabutane, $(—CH_2CH_2CH_2O—)_n$, three crystalline forms with chain conformations $(tt)_n$, $(t_3g^+t_3g^-)_n$, and $(t_2g_2^{\pm})_n$ are known (Tadokoro et al., 1967); here again the conformational energies must be sufficiently similar so that small differences in crystal lattice energies determine the conformation assumed by the chain molecule in the crystalline state.

For chains with two bulky substituents of the type $(—CH_2—CR_1R_2—)_n$ all conformations represent appreciable strain, so much, in fact, that such molecules would probably be considered impossible from inspection of conventional molecular models. Yet many such polymers do exist and are quite stable since the steric strain may be relieved by bond angle distortions. This possibility introduces a serious problem into the interpretation of x-ray diffraction data from crystalline polymers. The fiber diagrams from synthetic polymers have a much lower information content than x-ray diffraction data from single crystals, such as are formed by compounds of low molecular weight, and the interpretation of these fiber diagrams in terms of a chain conformation requires a priori assumptions of bond lengths and bond angles. Corradini et al. (1963) introduced an ingenious method to overcome this difficulty. Dicarboxylic acids form hydrogen-bonded "pseudochains," yet yield single crystals suitable for x-ray analysis. Thus, if the chemical structure of such a dicarboxylic acid has features analogous to those of some given polymer chain, its x-ray analysis will yield precise data on bond lengths, bond angles, and internal angles of rotation. In a typical application of this principle, Benedetti et al. (1970) used 2,2,4,4-tetramethyladipic acid:

```
    CH3        OH···O   CH3      CH3        OH···O   CH3      CH3       OH···
    |         /      \\  |        |         /      \\  |        |        /
···-C-CH2-C          C-C-CH2-C-CH2-C          C-C-CH2-C-CH2C
    |         \\      /  |        |         \\      /  |        |        \\
    CH3        O···HO   CH3      CH3        O···HO   CH3      CH3       O···
```

"Pseudochain" of 2,2,4,4-tetramethyladipic acid

to study the geometry of the —$C(CH_3)_2$—CH_2—$C(CH_3)_2$—CH_2— grouping in polyisobutene. They found that the repulsion between the four methyl groups expanded the bond angle between the methylene and the two quaternary carbons to the unusual value of 122.6°. Allegra et al. (1970) have shown that the polyisobutene conformation predicted on the basis of this unusual expansion of the bond angle is in good agreement with crystallographic data on this polymer.

In poly (methyl methacrylate), where the asymmetric centers carry two different bulky substituents, the reported crystal structures suggest that the isotactic chain forms a helix with five monomer residues per turn. This is generated, according to Tadokoro et al. (1970), by the conformational sequence $\phi_1 = 0$, $\phi_2 = 72$. However, their analysis of the x-ray data assumed for the C—CH_2—C bond angle a value of 114°, which seems much too low in view of the results cited above for methylene joined to two quaternary carbon atoms.

When a chain molecule is transferred from a perfect crystal to a dilute solution, restraints imposed on its shape by factors influencing the efficiency of packing in the crystal lattice are eliminated. This will make it possible for the internal angles of rotation to change to values leading to an irrational number of monomer units in a turn of the helix. There is also no longer a need for the bond angle θ or the internal angles of rotation ϕ to have well-defined values; they may vary within a range, leading to a flexibility of the macromolecular conformation. Finally, the tendency to increase the entropy of the system will make a number of the bonds in the chain backbone assume conformations of higher energy, leading to "kinks" in the regular arrangement characteristic of the macromolecule in the crystalline state. Figure III.12 shows how a single kink produced in this manner in an all-*trans* poly-

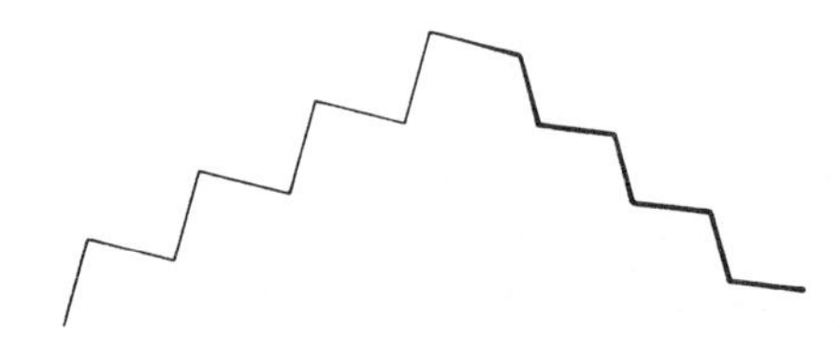

Fig. III.12. Conformation of polyethylene chain backbone with one *ganche* bond.

ethylene chain leads to a drastic change in the end-to-end distance of the molecule. Since the ratio of the probability of two conformations is given by $\exp(-\Delta E_{\mathrm{conf}}/\boldsymbol{R}T)$, where ΔE_{conf} is the energy of the conformational transition, chain flexibility resulting from a multiplicity of conformational states requires the existence of ΔE_{conf} with values which are not large compared to $\boldsymbol{R}T$. We should emphasize that a chain with high steric strain may still be highly flexible if several strained conformations are characterized by comparable energies. This point may be illustrated by analogy to a pair of low molecular weight compounds. Although the *trans-gauche* transition in *n*-butane requires an energy of about 700 cal/mole (Fig. III.7), the corresponding *anti* and *gauche* forms of the highly hindered 2,3-dimethylbutane (Fig. III.13) have, according to experimental data (Szasz and Sheppard, 1949) and theoretical calculations (Allinger et al., 1968), no significant difference in energy. Finally, chains in which no conformational transitions of the backbone are possible will still retain some flexibility because of the low energy requirements for oscillations of the internal angles of rotation around their most favored values and the deformability of the bond angles.

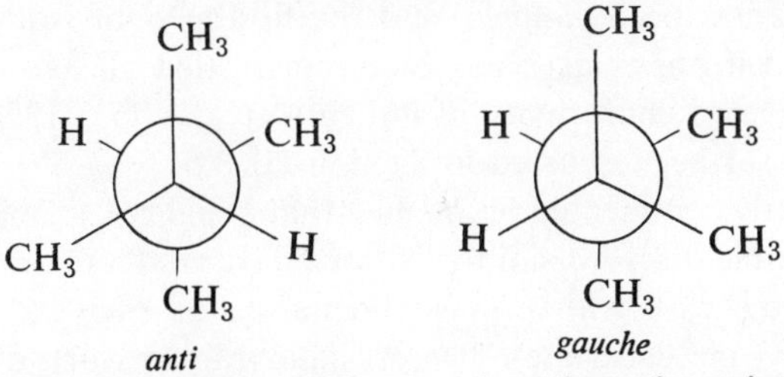

Fig. III.13 Newman projection of the *anti* and *gauche* conformations of 2,3-dimethylbutane.

Very extensive efforts to gain a theoretical understanding of the flexibility of chain molecules have been reviewed by Volkenstein (1963a), by Birshtein and Ptitsyn (1966), and by Flory (1969). In these investigations the energy associated with a given value of the internal angle of rotation ϕ and of pairs of consecutive values ϕ_i, ϕ_{i+1} were estimated by summing contributions from nonbonded van der Waals interactions, electrostatic interactions of bonding electrons (which, as we have noted, resist any departure from the staggered conformation), and dipole-dipole Coulombic interactions. Here we shall comment on only a few cases.

With syndiotactic vinyl polymers, spatial interference of the chain substituents favors sequences tt and $g^{\pm}g^{\pm}$. Since longer sequences of *gauche* bonds are sterically forbidden, the conformation of the dissolved chain should be of the type

$$\cdots(\mathrm{tt})_p(\mathrm{gg})(\mathrm{tt})_q(\mathrm{gg})(\mathrm{tt})_r\cdots$$

where g^+g^+ and g^-g^- are equally probable. The flexibility of the chain will depend on the energy requirement of a tt → gg transition, which may depend in turn on the nature of the chain substituent.

The question of the flexibility of isotactic vinyl polymers is much more controversial. The sequences $(tg^+)_n$ and $(g^-t)_n$, as defined in Fig. III.9, generate helices of opposite sense; and if all other conformational sequences were rigorously excluded, the dissolved chains could deviate from a rigid helical structure only by assuming a conformational sequence of the type

$$\cdots(g^-t)_p(tg^+)_q(g^-t)_r\cdots$$

However, such a sequence contains g^+g^- junctions, which correspond to a very high potential energy, so that one would be led to expect few breaks in the helical structure. Experimental data suggest a much higher flexibility of isotactic polymer chains than is predicted from this model. This has led Flory et al. (1966) and Flory (1970) to assert that the experimentally determined flexibility must be the result of imperfect stereoregularity of the "isotactic" samples employed. However, this view is by no means universally accepted, and it has been claimed that the flexibility of isotactic chains is greatly increased if one relaxes the requirement that all the ϕ angles correspond to staggered conformations (Allegra et al. 1963, 1973). This controversy will be further discussed in Section III.B.5.

We have already commented on the fact that a highly "crowded" molecule may be very flexible if several conformations have comparable energies. This is, apparently, the case with polyvinylidene chains such as polyisobutene and poly (methyl methacrylate). The crystallographic work of Benedetti et al. (1970), which demonstrated the pronounced expansion of the C—C—C bond angle linking methylene to two quaternary carbons, has placed our understanding of such chains on a more reliable foundation, and Allegra et al. (1970) found for polyisobutene four combinations of ϕ_i, ϕ_{i+1} with comparable energies.

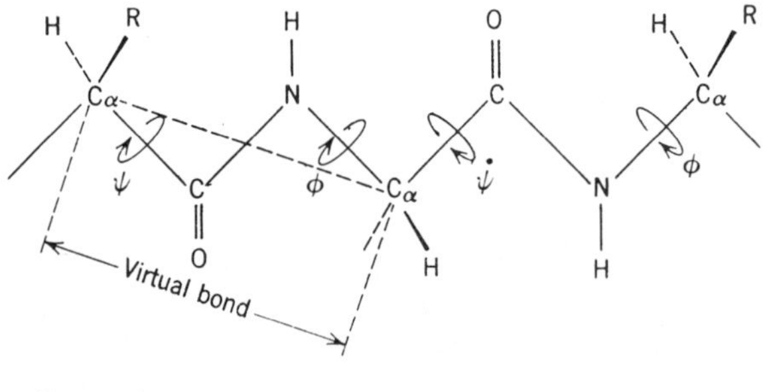

Fig. III.14. Convention used to specify the conformation of polypeptide chains. A "virtual bond" indicates the distance between neighboring α-carbons, which is not subject to change.

The flexibility of polypeptide chains—either synthetic, containing identical amino acid residues, or derived from a protein by disrupting its specific structure has been the subject of intensive study (Ramachandran and Sasisekharan, 1968). A segment of the chain with two amino acid residues is represented in Fig. III.14. Since the amide C—N bond has partial double-bond character, the C_α—C′—N—C_α sequence is coplanar. The amide group also exists invariably in the *trans* form, so that the distance between two neighboring C_α atoms is invariant. The conformation of the chain may then be described by assigning the values of two angles of internal rotation for every amino acid residue. The convention by which these angles are specified is shown in Fig. III.14, with ϕ and ψ representing rotations around the C_α-N and the C_α-C′ bonds, respectively. Figure III.15, due to Nemethy et al. (1966), represents the combinations of ϕ and ψ which are sterically allowed for glycyl peptides, and it shows that the restrictions become more and more

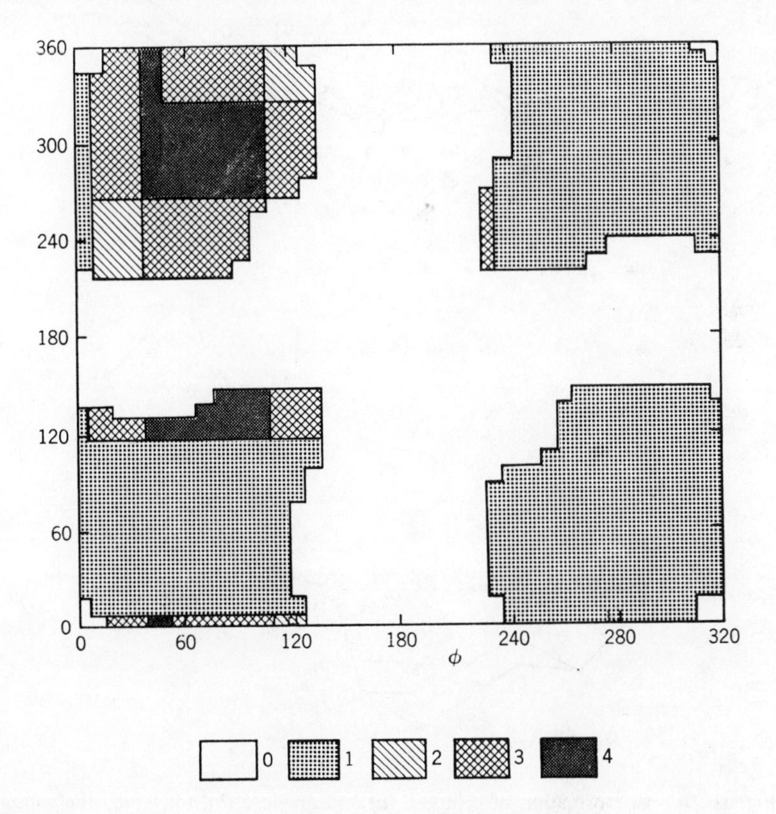

Fig. III.15. Sterically allowed conformations of glycyl peptides: O, forbidden for all peptides; 1–4, allowed for glycyl glycine; 2–4, allowed for glycyl-L-alanine; 3–4, allowed for homologs of glycyl-L-alanine with unbranched side chain; 4, glycyl-L-valine.

severe as the R group changes from R = H to the bulkier R = CH_3, R = CH_2CH_2X, and R = CH_2—$CH(CH_3)_2$.

The flexibility of polyglucoside chains is highly dependent on the manner in which the anhydroglucose residues are joined to one another. This is apparent from Fig. III.16, which represents two consecutive monomer residues in the cellulose and amylose chains (Rao et al., 1969; Yathindra and Rao, 1970). In these polymers the anhydroglucose ring may be regarded as rigid, so that chain flexibility depends on rotation around the two bonds joining the glucosidic oxygen to two of these rings. This rotation is severely restricted in both chains (Rao et al., 1969; Brant and Dimpfl, 1970; Yathindra and Rao, 1970), but the nature of this restriction corresponds to a much greater expansion of the cellulose chain.

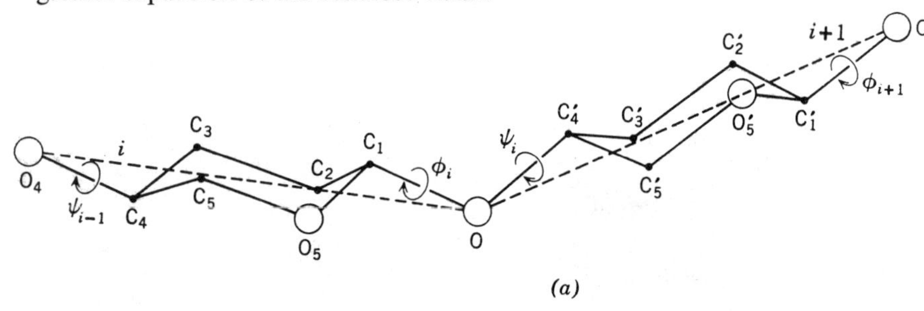

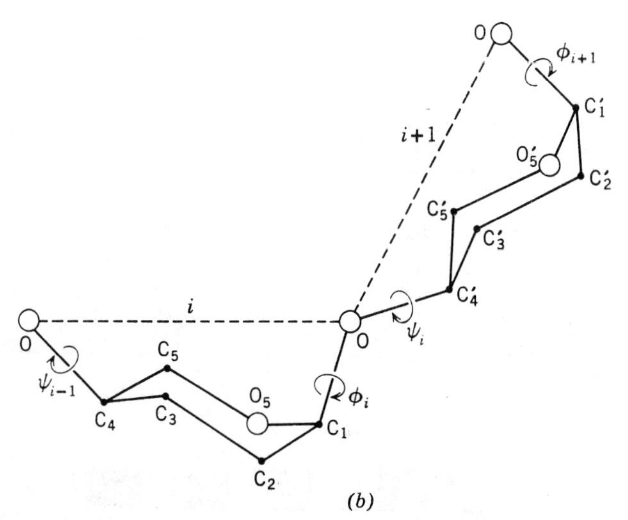

Fig. III.16. Conformation of cellulose (*a*) and amylose (*b*), for which the rotational angles are defined as $\phi = 0$, $\psi = 0$. The convention specifying the direction in which ϕ, ψ are rotated is also shown. The dashed lines represent "virtual bonds" connecting monomer residues.

In concluding this section it should be stressed that we have explored the question of the flexibility of polymer chains only from the point of view of accessibility of different shapes, without considering the kinetic problem, that is, how fast these changes of shape may come about. The rate of such transitions will, of course, be governed by the height of the energy barriers separating the various accessible conformations. This subject will be discussed in Section IX.A.1. We shall touch on it also in Chapter V when considering nuclear magnetic resonance spectroscopy of polymer solutions and in Chapter VI when we introduce the concept of the "internal viscosity" of chain molecules.

2. Statistics of Random Flight Chains

Having considered the factors which determine the flexibility of molecular chains, we now have to explore the shape which a chain molecule of a given length and flexibility will assume in solution. For a long flexible chain the number of distinguishable shapes will be very large, and it is clearly impossible to describe such chain molecules in terms of the probability distribution of the individual conformations in which the position of each atom is specified. Even if such a result could be obtained theoretically, it could not be checked against experimental data which yield one parameter (or at most two), characterizing the average shape of the dissolved macromolecules. A frequently employed parameter is the distance h between the ends of the molecular chains, which is described by a probability distribution function $\mathscr{W}(h)$, defined so that

$$\int_0^\infty \mathscr{W}(h)\, dh = 1$$

For many purposes, the significant quantity is the mean square end-to-end displacement $\langle h^2 \rangle$, given by

$$\langle h^2 \rangle = \int_0^\infty h^2 \mathscr{W}(h)\, dh \qquad \text{(III.1)}$$

Although $\langle h^2 \rangle$ is easily visualized, no rigorous method exists for measuring this quanity. The parameter characterizing the expansion of macromolecules, which may be derived unambiguously from experimental data, is the mean square radius of gyration $\langle s^2 \rangle$, defined by

$$\langle s^2 \rangle = \frac{1}{Z} \sum_1^Z \langle r_i^2 \rangle \qquad \text{(III.2)}$$

where Z is the number of chain elements and $\langle r_i^2 \rangle$ is the mean square distance of the ith element from the center of gravity of the chain. Values of $\langle s^2 \rangle$ are obtained from light scattering measurements (Chapter V); they are frequently converted into $\langle h^2 \rangle$ by assuming that the two quantities are related by the same factor as in the "random flight" model of the molecular chains. The random flight chain, usually taken as the point of departure in theoretical treatments of the shape of flexible chain molecules, is an idealized model in which links represented by mathematical lines of zero volume are "freely jointed" so that all angles in space between successive links are equally probably (Kuhn, 1934). This implies, of course, that changes in the shape of the chain are not accompanied by any changes in energy. Let us consider first the one-dimensional case, where Z links of length b may be added with equal probability in the positive and negative directions along the x axis of the coordinate system. If Z_+ is the number placed in the positive direction, the probability distribution of Z_+ will be

$$\mathscr{W}(Z_+, Z) = \frac{(\frac{1}{2})^Z Z!}{Z_+!(Z - Z_+)!}$$

Starting at the origin, the coordinate of the chain end will be $x = b(2Z_+ - Z)$ and the probability of x will be given by

$$\mathscr{W}(x) = \frac{(\frac{1}{2})^Z Z!}{(Z/2 + x/2b)!(Z/2 - x/2b)!} \tag{III.3}$$

For large values of Z, the probability of finding Z_+/Z very different from $\frac{1}{2}$ will become minute, so that we need to consider $\mathscr{W}(x)$ only in the range where x is very small compared to Zb, the length of the fully extended chain. We may use the Stirling approximation to the factorials, expand the result in powers of x/Zb, and retain only the first term of the expansion to obtain the continuous probability distribution function

$$\mathscr{W}(x) = \frac{\exp(-x^2/2Zb^2)}{b\sqrt{2\pi Z}} \tag{III.4}$$

If we consider the three-dimensional case of a random flight chain, the reasoning outlined above will be applicable to the probability distribution of the x coordinate of the chain end. The average component of the length of the chain links in the x direction is $b_x = b/\sqrt{3}$, so that the corresponding component of the end-to-end displacement has a probability distribution function

$$\mathscr{W}(h_x)\, dh_x = \frac{\exp(-3h_x^2/2Zb^2)}{b\sqrt{2\pi Z/3}}\, dh_x \tag{III.5}$$

Since $h^2 = h_x^2 + h_y^2 + h_z^2$, the probability of finding the chain end in a given location of the coordinate system will then be

$$\mathscr{W}(h_x, h_y, h_z)\, dh_x\, dh_y\, dh_z = \frac{\exp(-3h^2/2Zb^2)}{(b\sqrt{2\pi Z/3})^3}\, dh_x\, dh_y\, dh_z \tag{III.6}$$

Usually we shall be interested only in the value of the end-to-end displacement h rather than the precise location of the chain ends. If one chain end is at the origin of our coordinate system and the distance to the other end lies between h and $h + dh$, this end will lie within a spherical shell of area $4\pi h^2$ and thickness dh. The probability of such a chain will then be

$$\mathscr{W}(h)\,dh = \frac{\exp(-3h^2/2Zb^2)}{(b\sqrt{2\pi Z/3})^3}\,4\pi h^2\,dh \tag{III.7}$$

The most probable value of h, corresponding to $d \ln \mathscr{W}(h)/dh = 0$, will be denoted by h^*, and its value is given for the distribution function (III.7) by

$$(h^*)^2 = \frac{2Zb^2}{3} \tag{III.8}$$

while the mean square chain end displacement obtained by combining (III.1) and (III.7) becomes†

$$\langle h^2 \rangle = Zb^2 \tag{III.9}$$

In Fig. III.17 the distribution function (III.7) is represented by a plot of $\langle h^2 \rangle^{1/2}\,\mathscr{W}(h)$ against $h/\langle h^2 \rangle^{1/2}$. We may note that, even for a relatively short chain of 100 segments, h^* is only 8% of the full contour length of the chain.

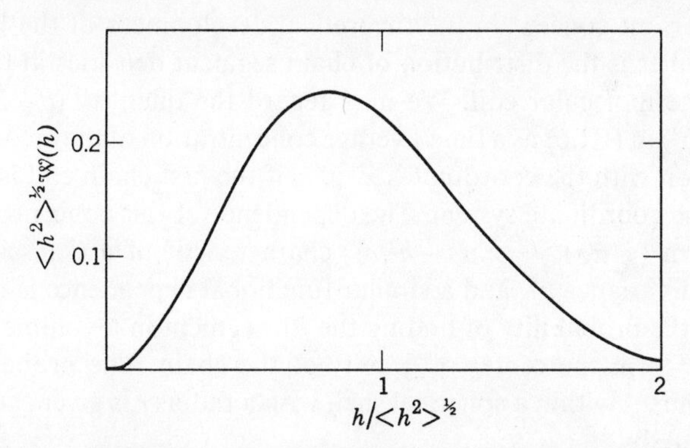

Fig. III.17. Probability distribution of chain end displacements in a random flight chain.

The probabilities of extensions much larger than h^* fall off rapidly to extremely small values. Thus, for instance, $\mathscr{W}(3h^*)/\mathscr{W}(h^*) \approx 10^{-3}$ and $\mathscr{W}(5h^*)/\mathscr{W}(h^*) \approx 10^{-9}$. We see then that for long chains the probabilities

†The relation given in (III.9) may also be obtained in the following simple manner. The vector **h** characterizing the chain end displacement is the sum of the vectors $\mathbf{b}_i$ of the chain elements. Thus $\langle h^2 \rangle = \langle \Sigma \mathbf{b}_i \cdot \Sigma \mathbf{b}_i \rangle = Zb^2 + \langle \Sigma \mathbf{b}_i \mathbf{b}_j \rangle_{j \neq i}$. For a freely jointed chain any value of $\mathbf{b}_i \mathbf{b}_j$ is equally likely to be positive and negative, so that $\langle \Sigma \mathbf{b}_i \mathbf{b}_j \rangle_{i \neq j}$ must vanish.

become negligible long before the chain end separation approaches the length of the fully extended chain.

The arguments used to obtain the distribution function of the separation of the chain ends may be used with equal justification for the distance between any given pair of chain elements. Thus the displacement between the ith and the jth link will be given, provided that $|j - i|$ is sufficiently large, by

$$\langle h_{ij}^2 \rangle = b^2 |j - i| \tag{III.10}$$

Such a distribution of chain elements has been shown by Debye (1946) to lead to a mean square radius of gyration

$$\langle s^2 \rangle = \frac{Zb^2}{6} \tag{III.11}$$

The approximations used in the above derivations restrict the validity of the distribution functions (III.6) and (III.7) to chain end separations much smaller than Zb, the length of the fully extended chain. This is apparent from the form of these functions, which give a finite probability for all values of h up to infinity, while values exceeding Zb can have no physical significance. However, for chains consisting of a large number of links, $\mathscr{W}(h)$ will drop to very low values long before h approaches Zb, so that this deficiency of the distribution function given by (III.7) is without practical importance.

A very important concept in the theoretical development of the behavior of chain molecules is the distribution of chain segment densities in the space occupied by the molecular coil. We may regard the quantity $(b\sqrt{2\pi Z/3})^{-3}$ $\exp(-3h^2/2Zb^2)$ in (III.6) as a time-average concentration of the second chain end at a location with the coordinates x, y, z, if the first chain end is fixed at the origin of the coordinate system. The dependence of this concentration on h is of the form $(\sqrt{\pi\sigma})^{-3/2}\exp(-h^2/\sigma^2)$ characteristic of a Gaussian error function with a variance σ^2, and a similar functional dependence is obtained if we consider the probability of finding the jth segment in a volume element at a distance r from the center of gravity of the chain. The probability of finding this segment within a spherical shell with a radius r is given, according to Isihara (1950), by

$$\mathscr{W}(j,r)\,dr = (\pi\sigma_j)^{-3/2}\exp\left(\frac{-r^2}{\sigma_j^2}\right)4\pi r^2\,dr$$

$$\sigma_j^2 = \frac{2\langle h^2 \rangle}{9}\frac{j^3 + (Z-j)^3}{Z^3} \tag{III.12a}$$

The total number of chain segments to be found within the spherical shell is then obtained by intergrating $\mathscr{W}(j, r)$ over all j values;

$$\mathscr{W}(r)\,dr = \int_{j=0}^{j=Z} \mathscr{W}(j, r)\,dj\,dr \tag{III.12b}$$

This integration was carried out by Debye and Bueche (1952), and their results are compared in Fig. III.18 with the $\mathscr{W}(r)$ corresponding to a Gaussian distribution of chain segment desnites. The discrepancy is most pronounced at large distances from the center of the coil, where the chain segment density decays much less sharply than would be the case in a Gaussian distribution.†

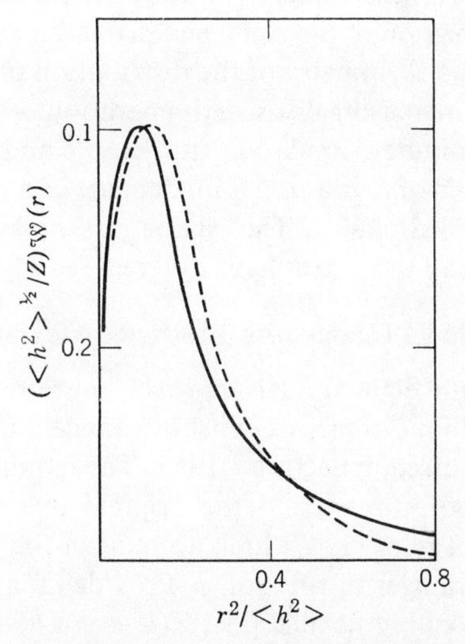

Fig. III.18. Segment distribution in a random flight chain. The dashed line corresponds to a Gaussian distribution of segment densities: the solid line represents the distribution calculated by Debye and Bueche (1952) without consideration of the excluded volume effect.

In spite of this difference, mathematical convenience has led many theoreticians to assume a Gaussian distribution of chain segment densities when analyzing the behavior of dissolved chain molecules.

Although the chain segment distribution is spherically symmetrical around the center of gravity of the chain when averaged over a period of time, no such symmetry would be expected for the instantaneous segment distribution. According to Kurata et al., (1960) the chain molecules occupy an ellipsoidal volume, with the dimension larger in the direction of the chain end displacement than in a direction perpendicular to it. The deviation from spherical symmetry is characterized most conveniently by comparing the mean square radius of gyration around axes running through the center of gravity of the coil parallel to the chain end displacement or at right angles to that direction. These quantities are then given, for moderate deviations of the chain end displacement from its rms value, by

†We shall see later (pp. 137) that the spatial interference of chain segments leads to further distortions of the Gaussian distribution.

$$\langle s_{\parallel}^2 \rangle = \frac{Zb^2}{36}\left(1 + \frac{3h^2}{Zb^2}\right)$$

$$\langle s_{\perp}^2 \rangle = \frac{Zb^2}{36} \qquad \text{(III.13)}$$

Even with h^2 at its average value, $h^2 = Zb^2$, the extension of the chain is twice as long in the direction of the chain end displacement as in the direction at right angles to it. The asymmetry of the distribution of the chain segments increases sharply with increasing chain and separation.

More recently, a computer simulation study (Šolc and Stockmayer, 1971), showed that the ellipsoid of revolution is inadequate as a model for the overall shape of randomly coiled chains. This shape ressembles more closely to a cake of soap, in which all three axes have different lengths.

3. Model Chains with Restricted Flexibility

To make the random flight model approach more realistically to conditions existing in a chain molecule, we must introduce the various restrictions to chain flexibility discussed in Section III.B.1. The requirement of a constant bond angle θ was incorporated into the original theory by Kuhn (1934), while Oka (1942) and Taylor (1947) took account of the unequal probability of the various internal angles of rotation ϕ. Provided that the chain is sufficiently long and is reasonably flexible [i.e., $Z(1 + \cos\theta) >> 1$; $\langle\cos\phi\rangle$ not too close to unity] and that the probability distribution of internal angles of rotation remains unchanged when the sign of ϕ is reversed,

$$\langle h^2 \rangle = Zb^2 \frac{1 - \cos\theta}{1 + \cos\theta} \cdot \frac{1 + \langle\cos\phi\rangle}{1 - \langle\cos\phi\rangle} \qquad \text{(III.14)}$$

For a tetrahedral bond angle $\cos\theta = -\frac{1}{3}$, so that the bond angle restriction doubles $\langle h^2 \rangle$. A chain with tetrahedral bond angles, a fraction f_t of *trans* bonds, and equally probable g^+ and g^- has then, according to (III.14),

$$\langle h^2 \rangle = \frac{2Zb^2(1 + 3f_t)}{3(1 - f_t)} \qquad \text{(III.15)}$$

Both (III.14 and III.15) are based on the assumption that internal angles of rotation are independent of each other. The discussion in Section III.B.1 made it clear, however, that the ϕ values in successive bonds of real chains are governed by complex correlations which result from the finite size of the atoms and the forces acting between them in any given short segment of the molecular chain. We may generate model chains composed of mathematical lines in which such additional restraints are taken into account, and mathematical techniques for calculating the dependence of the end-to-end displacement as a function of the length of the chain have been developed (Flory,

1969). No matter what the nature of the restraints due to bond angles, internal angles of rotation, and their correlations, $\langle h^2 \rangle$ will still be proportional to Z for sufficiently long chains with the "characteristic ratio"

$$C_\infty = \lim_{Z \to \infty} \left(\frac{\langle h^2 \rangle}{Zb^2} \right) \tag{III.16}$$

giving a measure of the chain flexibility. Here we define b and Z as the length and number of bonds in chains in which rotation can take place around each bond of the chain backbone, while chains containing rigid elements in their backbones are described in terms of "virtual bonds." This is the case, for example, for polypeptides, in which b is the distance between two consecutive α-cabons (Fig. III.14) or in polyglucosides, where it refers to the distance between two consecutive glucosidic oxygens (Fig. III.16). The parameter C_∞ is related to a concept advanced by Kuhn (1934) in his classical paper. He suggested that the real chain with fixed bond angles and various restrictions to internal rotation around the bonds of the chain backbone be substituted for by an "equivalent chain" consisting of a smaller number Z_s of "statistical chain elements" of length b_s which are freely jointed to each other. The real and equivalent chains have the same contour length and chain end separation, that is, $Zb = Z_s b_s$, $\langle h^2 \rangle = Z_s b_s^2$, and this is achieved by setting $b_s = C_\infty b$, $Z_s = Z/C_\infty$.

Occasionally it is necessary to deal with chains so stiff that the contour length is no longer very large compared to the length of Kuhn's statistical chain element. In such cases the equivalent random chain with its long rigid links and abrupt kinks leads to faulty conclusions, and it may be preferable to use in its place the model of a "wormlike chain," as proposed by Kratky and Porod (1949). In their model chain, adjoining links subtend an angle deviating only slightly from 180° and there are no barriers to internal rotation. Thus the direction of successive links has a slowly decreasing correlation with the direction of the first link of the chain. Kratky and Porod showed that the mean cosine of the angle subtended by the directions of the first and the last link of the chain decays as exp $(-L/a)$, where L is the contour lenth and a is a "persistence length" characterizing the rigidity of the chain. The mean square chain end separation (Kratky and Porod, 1949) and the mean square radius of gyration (Benoit and Doty, 1953) are given by

$$\langle h^2 \rangle = 2a\left\{L - a\left[1 - \exp\left(\frac{-L}{a}\right)\right]\right\} \tag{III.17}$$

$$\langle s^2 \rangle = a\left(\frac{L}{3} - a\left\{1 - \frac{a}{2L} + \frac{2a^2}{L^2}\left[1 - \exp\left(\frac{-L}{a}\right)\right]\right\}\right) \tag{III.18}$$

which reduces for very long chains ($L >> a$) to $\langle h^2 \rangle = Z_s b_s^2 = b_s L$ and

$\langle s^2 \rangle = \frac{1}{6} b_s L$ if the persistence length is set at half the length of the statistical chain element.

A very interesting alternative development of the concept of a wormlike chain has been proposed by S. E. Bresler and Ya. I. Frenkel (as quoted by Landau and Lifshitz, 1958). The energy expended in the bending of a rod with cylindrical symmetry and a length L may be represented by

$$E = \frac{B}{2} \int_{l=0}^{l=L} \rho^{-2}(l)\, dl \tag{III.19}$$

where B is a parameter characterizing the resistance of the rod to bending, and $\rho(l)$ is the radius of curvature at a distance l from the end of the rod. When such a rod is suspended in a fluid, it will be distorted by Brownian motion so that the mean square end-to-end distance will be given by (III.17) with

$$a = \frac{B}{kT} \tag{III.20}$$

The concept of wormlike chains is then obviously applicable to cases where the chain length is not very large compared to the persistence length. Such situations arise with chain molecules in which no conformational transitions are possible in the chain backbone, so that the chain flexibility is due merely to small variations of the internal angles of rotation around the energetically preferred values and to distortion of bond angles. This applies particularly to cellulose derivatives and to polymer chains which exist in solution in helical conformations, such as polypeptides, deoxyribose nucleic acid (DNA), and polyisocyanates. Such structures commonly have persistence lengths of the order of 500 Å, and it is instructive to interpret this value in terms of (III.19) and (III.20). For a cylindrical rod with a radius r and a modulus of elasticity Y, $B = \pi r^4 Y/4$. Thus a rod with a radius of 2.5 Å and the observed persistence length would have a modulus of 10^{11} dynes/cm^2, an order of magnitude below the value characterizing steel.

4. Random Flight Models of Branched Chain Molecules

The statistical treatment of the extension of branched chain molecules has been presented in some detail by Zimm and Stockmayer (1949). It is instructive to consider first their result for a "star-shaped" molecule which has f^* branches radiating from one point. Macromolecules of this type have been prepared by Schaefgen and Flory (1948) by adding monomer units to a f^*-functional initiator species. Zimm and Stockmayer derived for the mean square radius of gyration of such structures consisting of freely jointed segments

$$\langle s^2 \rangle = Zb^2 \sum_{i=1}^{i=f^*} \left(\frac{Z_i^2}{2Z^2} - \frac{Z_i^3}{3Z^3} \right) \tag{III.21}$$

where Z_i is the number of links in the ith chain. For the special case where all the branches are of equal length $Z_b b$, this becomes

$$\langle s^2 \rangle = Z_b b^2 (\tfrac{1}{2} - \tfrac{1}{3} f^*) \tag{III.22}$$

Thus, for a constant length of the branches and increasing functionality of the branch point, $\langle s^2 \rangle$ will approach a limiting value. This is physically understandable since the location of the branch point will approach, with increasing functionality, more and more closely to the center of gravity of the structure. If we consider two of the branches as constituting the "chain backbone," we find that a transformation of this linear structure into a star-shaped one with equal length of branches may increase $\langle s^2 \rangle$ by a factor of as much as $\frac{3}{2}$ as f^* becomes large. On the other hand, we may characterize the effect of branching on the radius of gyration by comparing two structures with the same total number of links. If $\langle s^2 \rangle$ refers to the branched chain and $\langle s_u^2 \rangle$ to the unbranched reference chain, then $\langle s^2 \rangle$ must necessarily be smaller than $\langle s_u^2 \rangle$. For a star-shaped molecule with branches of equal lengths $\langle s^2 \rangle / \langle s_u^2 \rangle = (3/f^*) - [2/(f^*)]^2$. We may note that this ratio will increase toward unity as the relative lengths of the branches become more unequal.

Treatment of the general case of structures with n^* branchpoints with a functionality f^* is much more complex, and any calculation must be based on a model which defines the distribution of branch points and branch lengths. Zimm and Stockmayer used a random distribution of branch points, such as would be obtained in a polycondensation reaction in a mixture of bifunctional and f^*-functional reagents. Defining the parameter $g_{f*}^*(n^*)$ by

$$g_{f*}^*(n^*) = \frac{\langle s^2 \rangle}{\langle s_u^2 \rangle} \tag{III.23}$$

they obtained a result which is approximated for $n^* > 5$ by

$$g_3^*(n^*) \approx \frac{3}{2}\left(\frac{\pi}{n^*}\right)^{1/2} - \frac{5}{2n^*}$$

$$g_4^*(n^*) \approx \frac{1}{2}\left(\frac{3\pi}{n^*}\right)^{1/2} - \frac{2}{3n^*} \tag{III.24}$$

It should be stressed that the model on which this result is based is not applicable to graft polymers. Any application of statistical analyses of the shapes of branched macromolecules to specific cases must be based on a detailed knowledge of the factors which determine the kinetics of the chain branching process.

5. The "Unperturbed Chain Dimensions"

In the statistical treatments discussed so far, the model chains were represented by mathematical lines of zero volume. The spatial requirements of the chain backbone and the chain substituents were taken into account only insofar as they determine chain flexibility, but no restriction was placed on the return of the chain to a point arbitrarily close to one occupied by a previous chain segment. When we take the spatial interference of chain segments into account, the probability distribution of chain end displacements (or of radii of gyration), arrived at previously, will have to be modified. In addition, we need to consider the forces operating between the various components of the system. In a good solvent, in which nearest-neighbor polymer-solvent contacts are thermodynamically favored over polymer-polymer contacts, the polymer chain segments will effectively repel one another, while an effective attraction of the chain segments will characterize poor solvent media. This may be taken formally into account by a suitable adjustment of the "excluded volume" of a chain segment, which will increase for repulsive and decrease for attractive segment interactions. A positive value of the excluded volume will lead to an expansion of the molecular coil, since the fraction of conformations which has to be excluded will decrease with an increasing separation of the chain ends. The theory of this "excluded volume effect" will be the subject of the next section of this chapter.

We may, however, envisage a thermodynamically poor solvent in which chain segments attract one another just enough to compensate for the effect of the physically occupied volume of the chain. In that case the excluded volume effect will vanish, and the chain should behave, as pointed out by Flory (1949, 1953b), according to predictions based on mathematical model chains of zero volume, provided that the short-range restraints to chain flexibility are taken properly into account. The elimination of binary interaction effects between segments of a single chain implies inevitably that binary interactions between solute molecules make no contribution to deviation from solution ideality, so that $\frac{1}{2} - \chi$ in (II.46) must be zero. Solvent media satisfying this condition are called, following Flory, Θ-solvents, and chain dimensions in such media are referred to as "unperturbed." The unperturbed mean square chain end displacement and mean square radius of gyration are designated by $\langle h_0^2 \rangle$ and $\langle s_0^2 \rangle$, respectively. They may be found by extrapolation of osmotic or light scattering data (see Chapter. IV and V) to a solvent composition or temperature at which the second virial coefficient disappears. We may then determine the characteristic ratio from $\langle h_0^2 \rangle / Zb^2$ in Θ-solvents using light scattering or viscosimetric data (as explained in Chapters V and VI) and compare the result to C_∞ values estimated on the basis of short-range interactions to be expected in chains of a given kind.

The mathematical techniques used to predict values of the characteristic ratio C_∞ have been described in detail by Flory (1969). We may distinguish two categories of chain molecules. In the first one, there is an uninterrupted interdependence between the internal angles of rotation around consecutive bonds of the chain backbone. Polyethylene, poly(ethylene oxide), and polymers of vinyl derivatives are characteristic examples of such chains.† In the second category, pairs of internal angles of rotation are strongly interdependent, but the values assumed by one pair of these angles are quite independent of those in neighboring pairs. Characteristic examples of such chains are polypeptides (Fig. III.14) and polyglucosides (Fig. III.16).

In the case of polyethylene the chain flexibility is determined by two factors; that is, the energy requirements for t → g and $g^{\pm}g^{\pm} \rightarrow g^{\pm}g^{\mp}$ transitions, respectively. Two experimental values have to be matched: the characteristic ratio C_∞ at a given temperature, and the relative change of $\langle h_o^2 \rangle$ with changing temperature, $d \ln \langle h_o^2 \rangle / dT$. According to Abe et al. (1966), a satisfactory agreement with experimental data is obtained by assigning an energy requirement of 400 ± 100 cal/mole for the *trans-gauche* transition and by assuming that the energy of the chain is raised by an additional 2000 ± 250 cal/mole for every $g^{\pm}g^{\mp}$ sequence. It is interesting that the difficulty with which $g^{\pm}g^{\mp}$ sequences are obtained leads to a very pronounced chain expansion, with C_∞ increasing at 140 °C from 3.4 to 6.8 (Flory, 1971). Since t → g transitions are endothermic and lead to a chain contraction, $d \ln \langle h_o^2 \rangle / dT$ would be expected to be negative; experimental data yield values of $-1.1 \pm 0.1 \times 10^{-3}$ (°C)$^{-1}$.

Poly(ethylene oxide), $(\text{—}CH_2CH_2O\text{—})_n$, provides another instructive case. Here the repeat unit consists of three atoms in the chain backbone, and we must differentiate between rotations around —O—C— and —C—C— bonds. Mark and Flory (1965) have proposed the following model to account for $C_\infty = 4.0$ (at 40 °C) and $d \ln \langle h_o^2 \rangle / dT = +0.23 \times 10^{-3}$ (°C)$^{-1}$.

(a) for rotation around the —O—C— bond, the t → g transition is even more endothermic than in polyethylene, requiring an estimated 900 cal/mole.

(b) rotation around the —C—C— bond leads to a repulsion of the two C—O dipoles in the *gauche* state, which should be, therefore, energetically unfavorable. In spite of this, the fitting of experimentally observed C_∞ and the positive temperature coefficient of chain expansion require the *gauche* state to be favored by an estimated 400 cal/mole.

† In the case of polymers with quaternary carbons in the chain backbone (such as polyisobutene), correlations between internal angles of rotation around up to four consecutive bonds must be considered (De Bolt and Flory, 1975). For a treatment of poly(methyl methacrylate), see P. R. Sundararajan and P. J. Flory, *J. Am. Chem. Soc.*, **96**, 5025 (1974).

(c) Conformations of the type $g^{\pm}g^{\mp}$ are highly unfavorable in the $-CH_2CH_2OCH_2CH_2-$ sequence, but require no additional energy in the sequence $-OCH_2CH_2OCH_2-$.

The situation with vinyl polymers is much more complex because of the effect of stereoisomerism. We commented in Section III.B.1 on the nature of the "breaks" in a regular conformational sequence which will determine the flexibility of syndiotactic and isotactic chains. In syndiotactic chains, C_∞ will depend on the probability q_g that a tt unit is followed by a monomer residue with the $g^{\pm}g^{\pm}$ conformation. According to Allegra et al. (1963), $C_\infty = 3/q_g\,(2 - q_g)$. They obtained $q_g = (\sqrt{5} - 1)/2$ for the special case where the tttt $\rightarrow$ tt$g^{\pm}g^{\pm}$ transition is thermoneutral so that $C_\infty \approx 3.5$. This should be a good estimate for syndiotactic polypropylene, since crystalline forms with an all-*trans* and a $(ttg^{\pm}g^{\pm})_n$ conformation are both known, so that they must correspond to a similar potential energy. If the side chain is more extended than $-CH_3$, the side-chain conformation has to be correlated with that of the chain backbone if steric interference is to be avoided (Flory et al., 1966), and this factor would be expected to modify the value of q_g. For pure isotactic polymers, the theory of chain extension is much more controversial, since Flory et al. (1966) have given an estimate of the energy associated with junctions of right- and left-handed helical chain segments which differs widely from that of Allegra et al. (1963, 1973) and Luisi (1972).

Unfortunately, the extension of stereoregular polymers has often been studied on samples whose stereoregularity has not been adequately characterized. Inagaki et al. (1966) found $C_\infty = 4.7$ for predominantly syndiotactic polypropylene at 135°, as against $C_\infty = 5.0$ for the atactic species. As for isotactic polymers, only Heatley et al. (1969) studied the extension of a polymer after a careful spectroscopic determination of its stereoregularity. They obtained $C_\infty = 4.73$ for a polypropylene containing only 2% of randomly distributed racemic dyads. This value is even lower than $C_\infty = 7.48$ (Kinsinger and Hughes, 1963) and $C_\infty = 6.78$ (Nakajima and Saijyo, 1968), reported for samples of more uncertain stereoregularity. According to Flory et al. (1966), such a high flexibility would require 7–10% of racemic dyads, but Abe (1970) and Luisi (1972) find it consistent with the high degree of stereoregularity indicated by the spectroscopic data. According to Allegra et al. (1973), potential energy calculations for pure isotactic polypropylene lead to chain extensions in close agreement with experimental data if a continuous distribution of the internal angles of rotation is used in place of the rotational isomeric state approach. Luisi found that the chain extension should increase slightly for isotactic polymers of the higher linear olefins, while a dramatic increase in chain stiffness should result when the side chain is branched close to the chain backbone. For polymers with asymmetric centers in their side chains, one sense of the helix will be heavily favored, even if the energetic difference between the two helical forms is quite small, if junctions of right- and left-handed helices have a large energy requirement.

The characteristic ratio C_∞ for atactic vinyl polymers seems to be generally in the range of 5–10. Wherever atactic polymers were compared with polymers considered isotactic by the usual criteria of solubility and crystallinity, the isotactic chains were found to be somewhat more extended. However, the difference in the unperturbed dimensions was generally surprisingly small, leading, for instance, to an increase in C_∞ by about 15–20% for polystyrene (Krigbaum et al., 1958). In the case of poly(pentene-1), C_∞ was 14% larger for the isotactic species at 32.5 °C, but was identical for the isotactic and the atactic polymer at 150 °C (Moraglio and Gianotti, 1969).

Unperturbed chain dimensions of vinylidene polymers are in the same range as those of polymers of vinyl derivatives. For polyisobutene at 24° C, $C_\infty = 6.6$ (Fox and Flory, 1951) and values in the range 6.9 ± 0.5 were reported for atactic poly(methyl methacrylate) (Schulz and Kirste, 1961; Fox, 1962). The high degree of "crowding" of these chains does not lead to increased chain expansion, since the similarity of the energy corresponding to several conformations leads to high chain flexibility (Allegra et al., 1970). An extreme case of this kind is polyacenaphthylene:

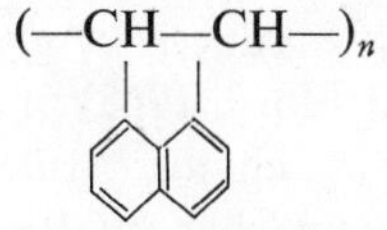

(—CH—CH—)$_n$

Polyacenaphthylene

This polymer has remarkably small unperturbed dimensions, with $C_\infty = 8.6$, actually slightly smaller than $C_\infty = 10.2$ in the much less hindered polystyrene. Barrales-Rienda and Pepper (1967) argued that this could be accounted for by assuming that it is equally probable for the monomer residue to be incorporated into the chain by linkages *cis* and *trans* to the five-membered ring. However, molecular models indicate that a *cis* linkage is most unlikely, and the behavior of the chain appears then to be governed by the relative probability of isotactic and syndiotactic dyads.

"Ladder polymers," in which two chains are linked to each other by closely spaced crosslinkages, would be expected to form highly rigid structures. Methods for the preparation of such materials have been reviewed by Overberger and Moore (1970). A polysiloxane of this type with a structure represented as follows:

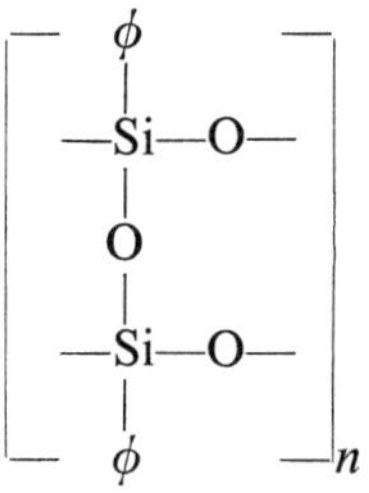

```
 ┌   φ    ┐
 │   |    │
 │ —Si—O— │
 │   |    │
 │   O    │
 │   |    │
 │ —Si—O— │
 │   |    │
 └   φ    ┘n
```

has been claimed by Brown (1963). Frye and Klosowski (1971) have disputed Brown's conclusion that this polyphenylsiloxane has a ladder structure, but extensive studies of this material by Tsvetkov (1972) indicate a rigidity which would be expected of a ladder polymer. His data correspond to C_∞ values in the range of 50–120, marking a dramatic increase over C_∞ values of 5.7–7.6, variously estimated for single chain polysiloxanes (Flory, 1969).

A few words are in order concerning unperturbed dimensions of natural polymers and their derivatives. In media in which the "native" structure of proteins (Section III. C.5) is destroyed, so that the molecular chain behaves as a random coil, $C_\infty = 3.6 \pm 0.7$ (Lapanje and Tanford, 1967), where the "virtual bond" (see Fig. III.14) has a length of 3.8 Å. Theoretical calculations for polypeptides composed of L-amino acid residues with unbranched side chains led Brant et al. (1967) to an estimate of $C_\infty \approx 9$. The chain extension is expected to be sharply reduced by glycine or proline residues (Miller et al., 1967; Schimmel and Flory, 1968) and slightly increased by amino acid residues with branched side chains (Miller and Goebel, 1968), but the experimental C_∞ values are somewhat smaller than those predicted on the basis of the protein compositions. With polyglucosides one observes an enormous difference in chain extension, depending on the manner in which the anhydroglucose residues are joined to one another. Goebel and Brant (1970) determined C_∞ values of 5.3 and 6.4 for two amylose derivatives where the length of the virtual bond (see Fig. III.16) is 4.25 Å. A theoretical analysis of the extension of this chain shows that C_∞ is highly sensitive to the bond angle at the glucosidic oxygen (Brant and Dimpfl, 1970). The cellulose chain (with a virtual bond length of 5.45 Å) is much more highly extended. Theoretical estimates of C_∞ lead to values in the range 40–80, depending on the bond angle at the glucosidic oxygen. This result seems to be in agreement with experimental indications (Yathindra and Rao, 1970). In both the amylose and the cellulose chains C_∞ has an unusually high temperature dependence.

The question now arises whether the flexibility of a given polymer chain (as characterized by C_∞) is a function of temperature only, being independent of the nature of the solvent medium. In principle, solvation can perturb the relative energy of the various conformations accessible to a segment of a chain molecule, and the theory of such modifications of chain flexibility has been discussed by Lifson and Oppenheim (1960). The most straightforward characterization of the role of the solvent in determining chain flexibility depends on a comparison of unperturbed dimensions in different Θ-solvents. One would suspect that C_∞ would be particularly susceptible to variation if the chain backbone contains polar bonds. Hexene-1-polysulfone is a material of this type with Θ-solvents either more or less polar than the polymer, and the mean square chain dimensions differ by about 15% in the two types of Θ-media (Bates and Ivin, 1967). Similar but smaller variations were found for polydimethylsiloxane, which is less expanded in Θ-media more polar than the polymer (Crescenzi and Flory, 1964). A different problem arises when we aim at a characterization of C_∞ in a good solvent medium, in which the excluded volume effect leads to an expansion of the chain over its unperturbed dimensions. Differences in chain flexibility may be demonstrated by spectroscopy sensitive to chain conformation, and this method has shown that poly(ethylene oxide) has very different conformations in water and in nonpolar solvents (Liu and Parsons, 1969; Assarson et al., 1969). Alternatively, extrapolation procedures to be described in Chapter VI may be used to estimate the expansion of the chain over its unperturbed dimensions. (This approach is necessary in cases where the polymer would either crystallize or assume a helical conformation in poor solvent media, so that the randomly coiled chain can be studied only in good solvents.) Flory et al. (1958) studied solutions of cellulose nitrate in this manner and concluded that the flexibility of the molecular chains is strongly affected by the nature of the solvent. Dondos and Benoit (1971) have summarized other data on the variation of unperturbed dimensions of polymers in different solvent media. Such variations are particularly large [e.g., leading to 40% variations of C_∞ for poly(2-vinylpyridine)] when a Θ-medium is obtained by mixing a good with a poor solvent, and in that case C_∞ tends to increase if $\Delta G^E > 0$ for the mixing of the two cosolvents. Variations of C_∞ are smaller with polymers dissolved in a single solvent. The effect is easily interpretable when the polymer may associate by hydrogen bonding with a solvent molecule. In that case, the bulk of the side chain is effectively increased, and this would be expected to lead to a modification of the conformational distribution.

6. Effect of Long-Range Interactions on the Shape of Flexible Chains

It was pointed out in the preceding section that restrictions to chain flexibility, arising from bond angle requirements and short-range interactions between the atoms of a molecular chain, determine the "unperturbed dimensions" observed in Θ-solvents. In better solvent media, chain segments will be characterized by a finite "excluded volume" (a concept to be examined more closely in Chapter IV), and the elimination of all conformations in which the excluded volumes of any two segments overlap will lead to an expansion above the unperturbed dimensions.

A very revealing procedure for the study of this effect involves the generation of chains on some suitable lattice, rejecting all those in which a lattice point is occupied more than once. The probability distribution of the end-to-end distances is then analyzed and compared with the distribution which would have been obtained if the self-intersecting chains had not been eliminated. The method is illustrated in Fig. III.19 on chains of six segments

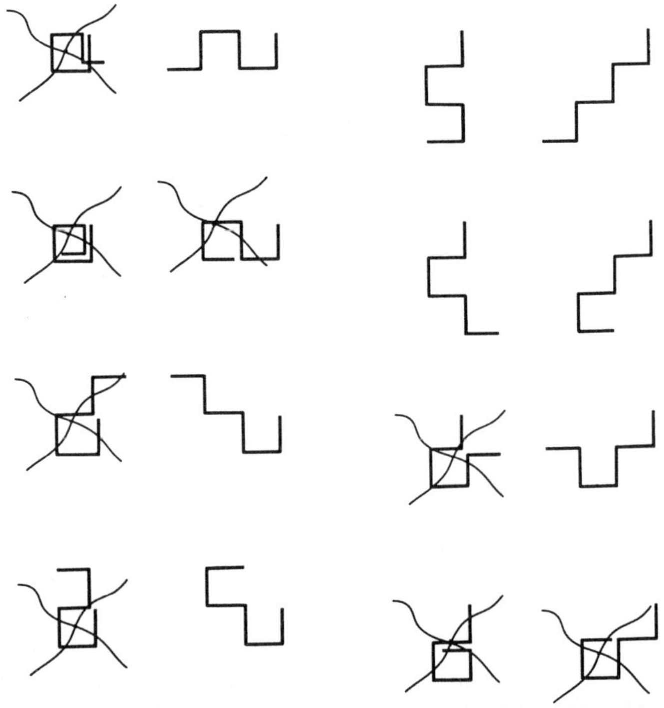

Fig. III.19. Two-dimensional analogy to the excluded volume effect with chains on a square lattice.

placed on a plane square lattice with the positions of the first two segments fixed and the requirement of a right angle between two successive segments. We see that of the 16 possible chains only 8 are not self-intersecting—if one more segment were added, the fraction of nonintersecting chains would decrease to $^{13}/_{32}$. If the length of one step is taken as unity, the average end-to-end distance of all the six-segment chains is 2.05; this increases to 3.1 after elimination of the intersecting chains. The main problem encountered in a study of this type is the rapid decrease in the fraction of nonintersecting chains with increasing Z, the number of chain segments, so that it is difficult to retain a reasonable statistical sample by the time the model chains have grown to a size analogous to the length of the molecular chains of high polymers. Although the attrition of chains is somewhat less severe in three dimensions than on a plane lattice, Wall et al. (1954) found that the longest nonintersecting chain generated on a cubic lattice in 140,000 trials had only 121 segments. On a tetrahedral lattice (which is most pertinent to the problem of a polymer chain with a hydrocarbon backbone) the fraction of nonintersecting chains decays by about 4% for each additional segment added, so that only 1 chain out of 10^7 attains a length of 400 segments. Although the problem of generating the chains and analyzing the size of those without intersection points can be programmed for an electronic computer, the number of chains which must be started to retain a sufficient sample of the nonintersecting variety is so large that the speed of even the fastest computers is insufficient. The problem may be largely overcome, however, by approximate methods in which "successful" chains are added to each other; Wall and Erpenbeck (1959) were able in this manner to generate statistical samples of chains up to $Z = 800$. They found that the mean square end-to-end distance of the nonintersecting chains is proportional to $Z^{1.18}$, as compared to unperturbed random walk chains, for which $\langle h^2 \rangle$ is proportional to Z. It is also possible to design, as an added refinement, a computer program for chains placed on a tetrahedral diamond lattice with a bias discriminating between segments added in the *trans* and in the *gauche* conformations (Wall et al., 1962).

In a thermodynamic analysis of the excluded volume effect, we start with the assumption that the chain flexibility is independent of the solvent medium. We have seen that this approximation may not be strictly correct, but it is amply justified since the increase in chain dimensions of a polymer coil with increasing solvent power of the medium is usually much larger than the variation of unperturbed dimensions. If the effective excluded volume of these chain segments is positive, the chain will tend to expand so as to reduce the free energy of mixing G_M of the chain with solvent molecules. On the other hand, the number of conformations consistent with a given chain end displacement decreases as h is expanded beyond h^* and the chain therefore resists such expansion. An analogous entropic retractive force is encountered in the elasticity of a rubber network and is customarily expressed in terms of an "elastic free energy" G_{elastic}. The equilibrium expansion of a chain with interacting segments will then correspond to that value of h at which the osmotic swelling force is balanced by the retractive force of the elastic chain. Using as the characteristic parameter the factor $\alpha_e = h^*/h_0^*$, by which the most probable chain end displacement is expanded over the value in the unperturbed random flight chain, the equilibrium extension must sotisfy the condition

$$\frac{\partial G_M}{\partial \alpha_e} + \frac{\partial G_{\text{elastic}}}{\partial \alpha_e} = 0 \qquad \text{(III.25)}$$

In formulating a theory which would predict the dependence of α_e on chain length, Flory (1949, 1953b) used a model in which the continuous unperturbed chain was substituted by a cloud of unconnected segments whose concentration was spherically symmetrical and was a Gaussian function of the distance from the center of gravity. It was assumed, moreover, that the excluded volume effect would expand all chain dimensions by the same factor, so that the Gaussian character of the segment distribution would be maintained. On this basis the elastic retractive force becomes

$$\frac{\partial G_{\text{elastic}}}{\partial \alpha_e} = 3\boldsymbol{k}T(\alpha_e - \alpha_e^{-1}) \qquad \text{(III.26)}$$

To simplify the derivation, we shall substitute for the Gaussian chain segment distribution a model in which the chain segments are uniformly distributed within an "equivalent sphere" with a molar volume $V_e = V_{e0}\alpha_e^3$. Within this sphere, the chain segments occupy a volume fraction $\phi_{2l} = V_2/V_{e0}\alpha_e^3$ and the number of solvent molecules is $N_1 = V_{e0}\alpha_e^3/V_1(1 - \phi_{2l}) \approx V_{e0}\alpha_e^3/V_1$, since ϕ_{2l} is rather low within the region of the swollen molecular coil. We have then

$$\frac{\partial G_M}{\partial \alpha_e} = \left(\frac{\partial G_M}{\partial N_1}\right)\left(\frac{\partial N_1}{\partial \alpha_e}\right) = \frac{(3V_{e0}\alpha_e^2/V_1)\overline{\Delta G_1}}{\boldsymbol{N}} \qquad \text{(III.27)}$$

The value of $\overline{\Delta G_1}$ is estimated from the Flory-Huggins theory [(II.45) and (II.46)], assuming that only contributions of binary segment interactions to the excess chemical potential of the solvent need be considered. Thus, $\overline{\Delta G_1}/\boldsymbol{N} = -\boldsymbol{k}T(\frac{1}{2} - \chi)\,\phi_{2l}^2 = -\boldsymbol{k}T(\frac{1}{2} - \chi)V_2^2/V_{e0}^2\alpha_e^6$, so that

$$\frac{\partial G_M}{\partial \alpha_e} = \frac{-3\boldsymbol{k}T\,(\frac{1}{2} - \chi)\,V_2^2}{V_1 V_{e0}\alpha_e^4} \qquad \text{(III.28)}$$

Substituting (III.26) and (III.28) into (III.25), we obtain

$$3\boldsymbol{k}T\,(\alpha_e - \alpha_e^{-1}) - \frac{3\boldsymbol{k}T\,(\frac{1}{2} - \chi)\,V_2^2}{V_1 V_{e0}\alpha_e^4} = 0$$

$$\alpha_e^5 - \alpha_e^3 = (\tfrac{1}{2} - \chi)\left(\frac{V_2^2}{V_1 V_{e0}}\right) \qquad \text{(III.29)}$$

We shall now substitute $V_2 = M_2/\rho_2$, where ρ_2 is the density of the polymer, and choose V_{e0} so that the equivalent sphere has the same radius of gyration as the real chain in its unperturbed state, $V_{e0} = N(4\pi/3)(5/18)^{3/2}\langle h_0^2\rangle^{3/2}$. Relation (III.29) may then be written as

$$\alpha_e^5 - \alpha_e^3 = \frac{1.63}{V_1 N \rho_2^2}\left(\frac{M_2}{\langle h_0^2\rangle}\right)^{3/2}(\tfrac{1}{2} - \chi)\, M_2{}^{1/2} \qquad \text{(III.30)}$$

where $M_2/\langle h_0^2\rangle$ may be described alternatively by $M_0/C_\infty b^2$. Here M_0 is the molecular weight per bond (or virtual bond) of the chain backbone, and b is the length of this bond.

The result is not very sensitive to the details of the distribution of the segment densities as long as they remain proportional to $1/\alpha_e^3$. With Flory's assumption of a Gaussian distribution, only the numerical coefficient of (III.30) is slightly altered to 1.72. The most important feature of the result is the prediction that the expansion factor α_e increases indefinitely with increasing chain length. For very long chains, where $\alpha_e^5 \gg \alpha_e^3$, we should then expect α_e to be proportional to $M_2^{0.1}$ so that h would increase as the 0.6 power of the chain length. Note also that the right side of (III.30) has V_1 in the denominator; thus the excluded volume effect should decrease in importance with an increase in the volume of the solvent molecules. In particular, when a polymer chain is embedded in a medium of other very large molecules, its expansion should correspond to the unperturbed dimensions at high dilution in a Θ-solvent. This interesting conclusion has been experimentally substantiated (Kirste et al., 1972; Benoit et al., 1973; Ballard et al., 1973).

The assumption that the swelling of the polymer chain beyond its unperturbed extension can be represented by an increase of all distances by the same factor α_e is equivalent to replacement of the original random flight chain with Z_{s0} elements of length b_{s0} by a new random flight chain whose elements are expanded to $b_{s0}\alpha_e^2$ and reduced in number to Z_{s0}/α_e^2. This representation has been subjected to a critical analysis by Krigbaum (1955) and by Ptitsyn (1959). They point out that the excluded volume effect must lead to the most pronounced changes in the central region of the molecular coil, where the segment density is highest. As a result, the distribution function of chain end displacements tends to become much sharper than in the unperturbed random flight chain. One consequence of this sharpening is a decrease in the ratio $\langle h^2\rangle/\langle s^2\rangle$ below the value of 6 [(III.9) and (III.11)] predicted for random flight chains. The modified treatment led Krigbaum (1955) to conclude that the ratio $(\alpha_e^5 - \alpha_e^3)/M_2{}^{1/2}$ should not be molecular weight independent, as implied by (III.30), but should pass through a shallow maximum as the chain length is increased.

Another criticism advanced against Flory's theory of the expansion of molecular coils concerns the assumption of spherical symmetry of chain segment distribution. We saw from (III.13) that the coil is, in fact, rather elongated and that increasing separation of the chain ends increases the coil dimensions in the direction parallel to the chain end displacement, but not the dimensions in a direction perpendicular to it. Kurata et al. (1960) suggested, therefore, a model in which the chain segments are distributed uniformly within an equivalent ellipsoid of revolution, chosen so as to give the same values for the principal radii of gyration, according to (III.13), as correspond to a Gaussian chain. The volume of such an ellipsoid is proportional to $(1 + 3\alpha_e^2)^{1/2}$, and a treatment analogous to that outlined above for the equivalent sphere model leads to a chain expansion due to the excluded volume effect given, in terms of our notation, by

$$\alpha_e^3 - \alpha_e = \frac{5.4\,(M_2/N\langle h_0^2\rangle)^{3/2}\,(\frac{1}{2} - \chi)\,M_2^{\ 1/2}}{V_1\rho_2^2\,[1 + (1/3\alpha_e^2)]^{3/2}} \qquad \text{(III.31)}$$

In comparing (III.31) with (III.30), the most important difference concerns the asymptotic behavior of α_e. Whereas the equivalent sphere model predicts that α_e will become proportional to the 0.1 power of the chain length for very long chains, the ellipsoidal model leads asymptotically to α_e proportional to $M_2^{1/6}$.

The above theories all assume that it is legitimate to treat the excluded volume with the use of models in which the chain character of the macromolecule is taken into account only in the estimate of the elastic retractive force, while the excess free energy of mixing is approximated by a value to be expected for a cloud of disconnected segments. To judge the validity of this assumption is difficult, and it is important, therefore, that Zimm et al. (1953) and Fixman (1955) were able to treat the excluded volume problem with the use of a "pearl necklace" model in which the continuity of the molecular chain is taken into account. If β is the mutually excluded volume between two statistical chain elements, then βZ_s^2 is twice the volume excluded by all binary segment interactions. The extension of the molecular coil may be characterized from (III.7) by

$$\lim_{h\to 0} \frac{\mathscr{W}(h)\,dh}{4\pi h^2\,dh} = \left(\frac{3}{2\pi Z_s b_s^2}\right)^{3/2} \equiv \frac{1}{V_c} \qquad \text{(III.32)}$$

where V_c is a volume in which two independent chain segments would have the same probability of colliding with one another as the two terminal segments of the molecular chain. It is then generally agreed that the expansion coefficient α_e is a function of the dimensionless ratio $\mathbf{z} = \beta Z_s^2/V_c$, so that with the substitutions $6\,\langle s_0^2\rangle = Z_s b_s^2$ and $Z_s = Z/C_\infty$ we obtain

$$\alpha_e^2 = 1 + a_1\mathbf{z} - a_2\mathbf{z}^2 + \cdots$$
$$\mathbf{z} = (4\pi\,\langle s_0^2\rangle)^{-3/2}\,\beta\left(\frac{Z}{C_\infty}\right)^2 \qquad \text{(III.33)}$$

The dependence of the excluded volume parameter β on the properties of the solvent may be expressed as $\beta = (2V_s^2/NV_1)(\frac{1}{2} - \chi)$, where $V_s = V_2/Z_s$ is the molar volume of the statistical chain element. At temperatures not too far removed from Θ, we may substitute $(\frac{1}{2} - \chi) = \Psi(1 - \Theta/T)$ [cf. (II.50)] and obtain $\beta = \beta_0 (1 - \Theta/T)$. The ratio $\langle h^2 \rangle / \langle s^2 \rangle$ increases slowly with $\mathbf{z}$ (Yamakawa, 1971); if we define $\alpha_e^2 = \langle s^2 \rangle / \langle s_0^2 \rangle$, then $a_1 = 134/105$ (Zimm et al. 1953), $a_2 = 2.082$ (Yamakawa et al., 1966). Unfortunately, the slow convergence of the power series in (III.32) restricts a precise theoretical evaluation of α_e^2 to $\mathbf{z} \ll 1$, that is, the immediate vicinity of the Θ-point. The experimental evaluation of $\mathbf{z}$ utilizes thermodynamic measurements which yield, as we shall see (p. 192), the value of β_0. Berry (1966) obtained by this method $\alpha_e^2 - 1 = f(\mathbf{z})$ for polystyrene in two solvent media over a range of temperatures, and his data are well fitted (Fig. III.20) by the relation of Flory and Fisk (1966):

$$\alpha_e^2 - 1 = 0.648\bar{\mathbf{z}}\,[1 + 0.97(1 + 10\bar{\mathbf{z}})^{-2/3}]$$
$$\bar{\mathbf{z}} = \frac{\alpha_e^3}{\mathbf{z}} \tag{III.34}$$

In all of the preceding discussion, the expansion of the molecular coils was treated for an isolated chain molecule without taking account of the consequences of polymer-polymer interactions. The result of theories based on such models should, strictly speaking, be valid only in the limit of infinite dilution of the chain molecules. At finite concentration, the equilibrium coil

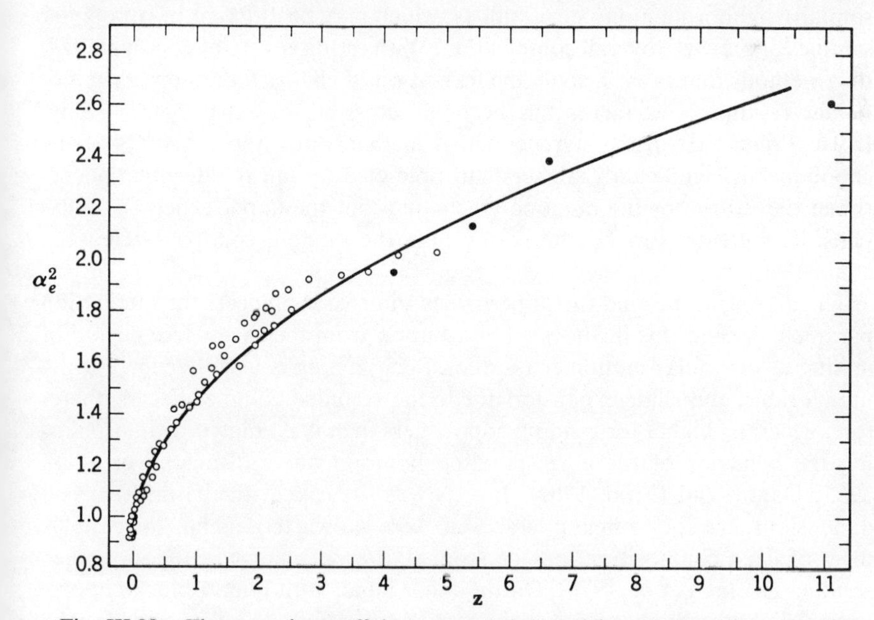

Fig. III.20. The expansion coefficient α_e as a function of the excluded volume parameter $\mathbf{z}$. Experimental points by Berry (1966); the solid line represents eg. (III.34).

expansion corresponds to the minimum of free energy, which depends on the number of contacts between polymer segments no matter whether these segments belong to the same or to different chains. The expansion of the molecular coil, which reduces the intramolecular interaction of chain segments, increases the probability of interaction of different chain molecules with one another. We should expect, therefore, the driving force toward chain expansion to decrease with an increasing concentration of the chain molecules, so that the molecular coils would tend to shrink as increasing numbers of them are being placed in a given volume. This effect was first described by Weissberg et al. (1951), and statistical-mechanical theories of it have been formulated by Fixman (1960) and by Grimley (1961). It could be studied, in principle, by examining neutron scattering in solutions containing a low concentration of a deuterated polymer and increasing concentrations of the same polymer containing the light hydrogen isotope. This technique has been described by Kirste et al. (1972), Benoit et al. (1973), and Ballard et al. (1973).

7. Conformation of Copolymers

The unperturbed dimensions of copolymers are frequently assumed to be characterized by a C_∞ value which is linear in the mole fraction of the components. There is no reason, however, why such a simple relation should hold. Moreover, the unperturbed dimensions should reflect the frequency of dissimilar neighboring monomer residues, which may be different in copolymer samples of the same overall composition. With the discovery of copolymerization methods that favor heavily the formation of chains with alternating monomer residues, this factor has become accessible to experimental study. It was found that in the styrene-methyl methacrylate system the C_∞ of the copolymer is significantly higher than predicted by linear interpolation between the values for the homopolymers and that the unperturbed chain expansion is greater for the alternating than the random copolymer (Kotaka et al., 1970).

The expansion beyond the unperturbed dimensions reflects the interaction parameter χ, and this includes a contribution from the excess free energy of mixing of dissimilar monomer residues. Such mixing is usually energetically unfavorable, and chain expansion due to the excluded volume effect is, therefore, generally higher for random copolymers than is estimated by interpolating the behavior of the corresponding homopolymers (Stockmayer et al. 1955; Danon and Derai, 1969). Just as was the case with the unperturbed dimensions, the long-range interactions were shown to depend, for copolymers of fixed composition, on the sequential arrangement of the monomer residues (Kotaka et al. 1970). On the other hand, with binary block copolymers $\langle h^2 \rangle$ of the chain is close to the sum of $\langle h^2 \rangle$ values to be expected for the constituent blocks, suggesting that interactions between dissimilar monomer residues do not perturb appreciably the overall chain expansion (Dondos et al., 1967).

With solutions of graft copolymers, it is possible to add nonsolvents which will collapse either the side chains or the chain backbone. Such systems were first studied using Hevea rubber grafted with poly (methyl methacrylate), and it was found that films cast from solutions in which the chain backbone was extended were rubbery, whereas media in which only the side chains were solvated yielded glassy films like poly(methyl methacrylate) (Merrett, 1954, 1957). Similar effects are observed with styrene-butadiene-styrene block copolymers. In this case, films cast from the copolymer solution may be stained with osmium tetroxide (which deposits on the butadiene-rich regions), and the dependence of the film structure on the solvation of one or the other block in the casting solution may be observed by electron microscopy (Miyamoto et al., 1970).

C. SPECIFIC CONFORMATIONS

1. Ordered Conformations of Polypeptide Chains

In our discussion of the flexibility of dissolved chain molecules (Section B.III.1) we started with the assumption that the extended or helical conformation assumed by a polymer in the crystalline state represents the lowest energy form of the isolated macromolecule. If rotational isomerism of the chain backbone involves relatively small changes of energy, the entropy gain associated with an increasing number of accessible conformations will lead to the occupancy of a very large number of conformations, and the dissolved macromolecules will then be characterized by a high degree of flexibility. On the other hand, as the energy requirement for rotational isomerism of the chain is increased, the molecule will approximate more and more closely to the conformation of lowest energy. Eventually, when the energy needed to produce a kink in the conformation characteristic of the crystalline polymer is very high compared to kT, the probability of finding such a kink in a chain of a given length may become negligible, so that the chain will have in solution a conformation very similar to that existing in the crystalline state.

This type of behavior was first observed with some synthetic polypeptides. The experimental evidence is of many different kinds, including the estimation of molecular dimensions from light scattering data, frictional properties of the dissolved macromolecules, characteristic spectral shifts, changes in optical activity, and changes in reactivity. Discussion of the applicability of these techniques to the study of the transition from a random coil to an ordered conformation will be left for later chapters. Here we shall concern ourselves only with the factors which determine the nature of these conformations and their stability in solution.

The structures encountered in crystalline polypeptides were clarified by Pauling and his collaborators. Their proposals were supported by extensive data on the crystal structures of amino acids and peptides which served to establish reliable values for bond lengths, bond angles, and the geometry of the peptide group. The atoms of this group:

```
        O
        ‖
       C       C
      /  \    /
     C     N
           |
           H
```

were found to be invariably coplanar and to exist in the *trans* form. The hydrogen-bonding potential of the polypeptide chain could then be satisfied either intramolecularly, stabilizing a helical conformation, or intermolecularly between extended chains, to form a "pleated sheet" structure.

In considering the various possible helical forms, Pauling et al. (1951a) were guided by the assumption that the structure should represent a potential energy minimum for the isolated chain. This required the hydrogen bonds N—H ··· O to have a length of 2.72 Å, with the hydrogen atom lying close to the connecting line between the nitrogen and oxygen atoms. Such restrictions could be satisfied by only two types of helices, and x-ray data obtained on poly (γ-methyl glutamate), poly(γ-butyl glutamate), and poly(L-alanine) (Bamford et al., 1951; Pauling and Corey, 1951b; Brown and Trotter, 1956) are, in fact, consistent with a conformation for the polypeptide backbone which is very close to the "α-helix" postulated by Pauling and Corey (1951a) on theoretical grounds. This helical structure, represented in Fig. III.21, contains 18 amino acid residues in five turns with a pitch of 5.4 Å per turn. In the extended form of the polypeptide chain, commonly referred to as the β form, the chains may lie parallel or antiparallel to one another, as represented in Fig. III.22. As seen in that figure, the parallel arrangement does not allow the hydrogen bonds to be linear, and this has led Pauling and Corey (1951*c*) to propose a slight deviation from the fully extended form which eliminates most of this difficulty. Their structure leads to repeat distances of 6.5 and 7.0 Å for the parallel and antiparallel pleated sheets, as against 7.23 Å, corresponding to fully extended polypeptide chains. These values may be compared with a repeat of 6.89 $\pm$ 0.02 Å, reported by Brown and Trotter (1956) for the β form of poly (L-alanine).

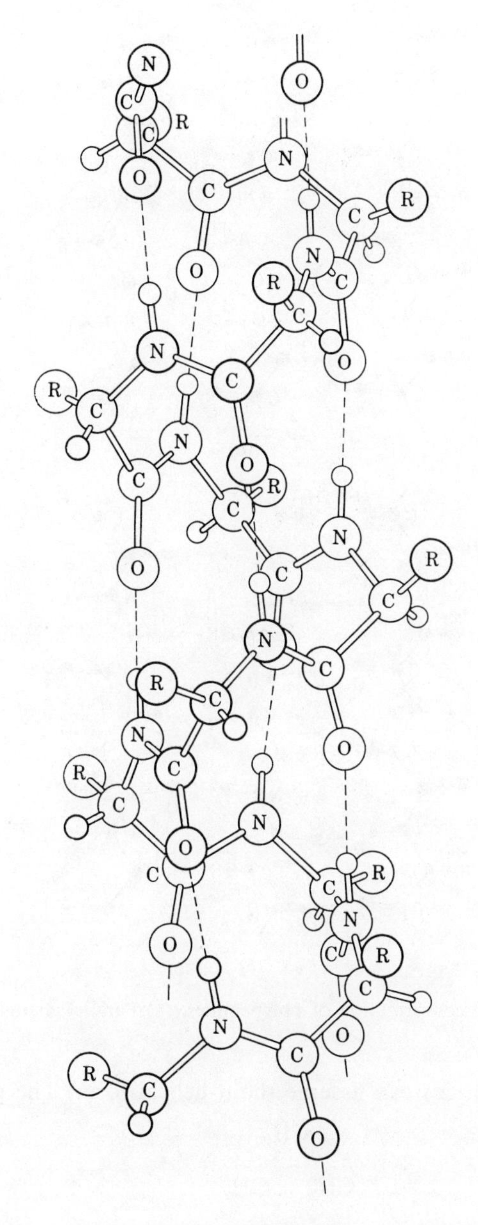

Fig. III.21. The Pauling-Corey α-helix (Corey and Pauling, 1955).

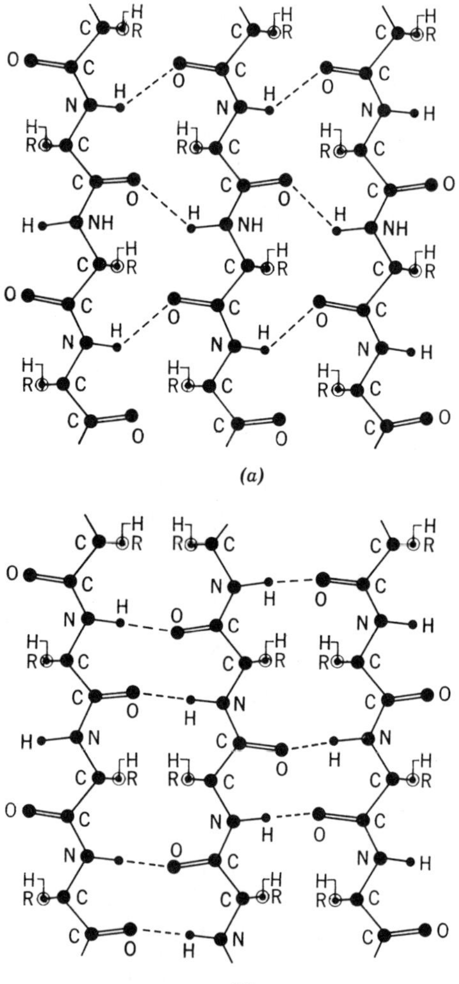

(a)

(b)

Fig. III.22. The β conformation of polypeptides. (*a*) Parallel chains. (*b*) Antiparallel chains.

Not all polypeptides can assume the α-helical form. The prolyl residue:

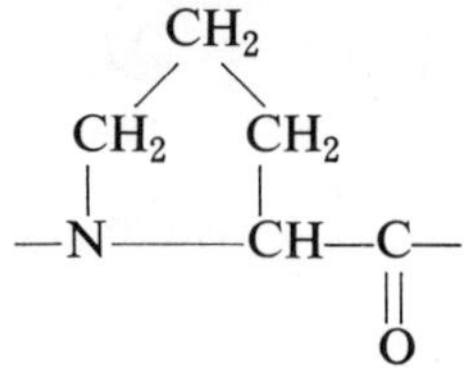

```
            CH2
           /   \
        CH2     CH2
         |       |
       —N————————CH—C—
                    ‖
                    O
```

in which the peptide nitrogen is part of a pyrrolidine ring, is excluded on steric grounds, although, as we shall see, poly(L-proline) can form helices of a different type. Polyglycine has also never been observed to form an α-helix, although no steric restraints can exist in this case. It has been suggested that the thermodynamic instability of the α-helical form of polyglycine is a consequence of the high flexibility of this polypeptide. Thus, a transition from the helix to a random coil has been estimated (Nemethy et al., 1966) to lead to an entropy increase, per amino acid residue, which is 2.5 eu larger than for poly(L-alanine) and 5 eu larger than for polypeptides with branched side chains, such as poly(L-valine). Finally, it was found (Blout et al., 1960) that polypeptides in which the amino acid residues carry electronegative atoms attached to the β-carbon of the side chains (i.e., the oxygen of serine and threonine and the sulfur of cysteine) exist invariably in the β form.

The existence of helical conformations of synthetic polypeptide molecules in dilute solution was first inferred by Doty et al. (1954) on the basis of viscosimetric data. They found that poly(γ-benzyl-L-glutamate) molecules behaved like rodlike particles if dissolved in chloroform saturated with formamide, while results obtained in dichloroacetic acid solution had the usual pattern compatible with a flexible coil model for the polymer chain. Viscometric and light scattering data obtained with poly(γ-benzyl-L-glutamate) in "helicogenic" media were interpreted by Doty et al. (1956) as consistent with particle lengths close to 1.50 Å per amino acid residue, corresponding to the α-helix of Pauling and Corey (1951a). Similar conclusions were arrived at on the basis of x-ray scattering studies (Saludjian and Luzzati, 1966). In spite of these results, it is not certain that the α-helix describes correctly the helical structure observed in dilute solutions. This problem has been studied in great detail by a group of researchers in Strasbourg, using a variety of experimental methods. In a summary of their work, Benoit et al. (1967) show that the interpretation of experimental data is sufficiently ambiguous to make it difficult to assign an exact value to the length of the projection of a monomer residue on the axis of the helix. This parameter may possibly exceed the classical value of 1.50 Å and be as large as 2.1 Å, corresponding to a helix with 10 monomer residues in three turns. In any case, there can be no doubt that the helical structure has some flexibility. It is best described by the wormlike chain model (Section III.B.3), and the persistance length has been estimated as 875 $\pm$ 100 Å (Moha et al., 1964).

The helical conformation of dissolved polypeptide molecules can be stable only if the solvent-solute interactions lie within rather narrow, well-defined limits. The solvent must interact sufficiently strongly with the side chains attached to the polypeptide backbone to prevent aggregation of the macromolecules. Yet the solvation of the polypeptide must not be too strong, or else the intramolecular hydrogen bonds, stabilizing the helical conformation, will be disrupted. If the polypeptide in the solid state assumes the helical conformation and if the helices are disrupted in dilute solution by a given solvent, it would be expected that the helical conformation becomes stable as the activity of the solvent is reduced at higher solution concentrations. Such an effect has indeed been observed with poly(γ-benzyl-L-glutamate) in dichloroacetic acid solution, where the α-helix is formed at concentrations exceeding 20% (Frenkel et al., 1967). This transition, which may be observed by infrared spectroscopy, is accompanied by the appearance of the anisotropic phase characteristic of solutions containing helical polypeptides.

It is also reasonable to assume that the mutual interactions of the side chains attached to the polypeptide backbone will contribute significantly to the stability of the helical form. The absence of such a side chain (i.e., the introduction of a glycine residue into the polypeptide) has the effect of reducing substantially the α-helix stability, as demonstrated in studies of copolypeptides containing the glycyl residues at regular spacings along the polymer chain (Fraser et al., 1967). Variations in helix stability, characteristic of individual polypeptides, may be studied conveniently by following the conformational transitions through a series of solvent media of increasing hydrogen-bonding potential (e.g., $CHCl_3 < CHCl_2COOH < CF_3COOH$). In this way Fasman (1962b) found that poly(L-leucine) formed the most stable helices, which remained intact in dichloroacetic acid solution and could be completely disrupted only in dichloroacetic-trifluoroacetic acid mixtures containing at least 60% TFA. Since our interest in the conformational transitions of synthetic polypeptides is motivated largely by the analogy in the behavior of these chain molecules and that of the polypeptide chains in proteins, the stability of helical conformations in an aqueous medium may be considered most significant. As would be expected, the mutual repulsion of ionized groups attached to a polypeptide will tend to disrupt the helical structures. This effect has been observed with poly(α-L-glutamic acid) (Doty et al., 1957) and with poly(L-lysine) (Applequist and Doty, 1962), which exist as random coils when their ionizable groups are charged, but assume a helical conformation in the pH range where the charge density on the polypeptide chain is low. In general, water-soluble polypeptides will have the stability of their helical conformations stabilized by the introduction of comonomers with strongly hydrophobic side chains, as illustrated by the behavior of copolypeptides of L-glutamic acid with L-leucine (Fasman et al., 1962).

Two approaches have been used to study the effect of an aqueous environment on the stability of the α-helical form of water-insoluble polypeptides. The first method uses block copolymers in which the central hydrophobic block is held in solution by the hydrophilic blocks at the two ends of the chain. Such block copolymers have been prepared to study the conformation of poly (L-alanine) (Gratzer and Doty, 1963; Ingwall et al., 1968), poly-

(L-valine) (Epand and Scheraga, 1968), poly(L-leucine) (Ostroy et al. 1970), and poly(L-phenylalanine) (Auer and Doty, 1966) in aqueous solution. The stability of the helical form of poly(L-alanine) is surprisingly high, considering that the side chains are too short to be in contact with one another. As would be expected, hydrophobic interactions of the side chains of poly-(L-leucine) lead to a further pronounced stabilization of the helical structure in the aqueous medium. With poly(L-phenylalanine), the helical form retains its stability even in media which destroy completely the native structure of globular proteins, for example, 8*M* urea solutions at temperatures up to 95 °C. A second method for studying the helix-forming tendency of various α-amino acids in aqueous solution, when their homopolypeptides are either water insoluble or unable to form helices stable in an aqueous environment, has been described more recently (Platzer et al., 1972; Hughes et al., 1972). In this method a "host polypeptide," which forms a stable α-helix in aqueous solution, is modified by incorporation of "guest" α-amino acid residues in the polypeptide chain. The change in the stability of the helical form is then extrapolated to yield thermodynamic parameters characterizing the helix-coil equilibrium in the limit when all the "host" α-amino acids have been replaced by "guest" α-amino acid residues.

In the original description of the α-helix by Pauling et al. (1951a), the question whether a polypeptide chain composed of L-amino acid residues would form right-handed or left-handed helices was left open. During the intervening years, the study of optical activity and its dependence on the wavelength of the light used was developed into a powerful method for investigating helical conformations. This subject will be discussed at some length in Chapter V; here we should note only that it provides an experimental method by which the question of the sense in which the helix is wound may be resolved. Most of the polypeptides derived from L-amino acids form right-handed α-helices. On the other hand, some exceptions to this generalization provide an interesting insight into the extent to which the mutual interactions of side chains can be the determining factor in the choice between two backbone conformations which differ only slightly in energy. Whereas poly(γ-benzyl-L-glutamate) forms right-handed helices, the sense of the helix is reversed in poly(β-benzyl-L-aspartate), in which the ester group is brought closer to the polymer backbone by the elimination of a methylene residue (Karlson et al., 1960; Bradbury et al., 1961). However, the preference of this polypeptide for the left-handed helix is clearly rather marginal, since a slight perturbation, such as *p*-nitro, *p*-cyano, *p*-chloro, and *p*-methyl substitution of the aromatic residue—a long distance from the polypeptide backbone—suffices to reverse the helical sense (Goodman et al., 1963; Hashimoto and Aritoni, 1966):

```
   C6H5                                                        NO2
    |                        C6H5                               |
   CH2                        |                               C6H4
    |                        CH2                                |
    O                         |                               CH2
    |                         O                                 |
   C=O                        |                                 O
    |                        C=O                                |
   CH2                        |                               C=O
    |                        CH2                                |
   CH2                        |                               CH2
    |                         |                                 |
(—C—CH—N—)n              (—C—CH—N—)n                     (—C—CH——N)n
  ||    |                  ||    |                         ||      |
  O     H                  O     H                         O       H
```

Poly(γ-benzyl-L-glutamate) Poly(β-benzyl-L-aspartate) Poly(β-*p*-nitrobenzyl-L-aspartate)

A similarly delicate balance of factors determining the sense of the helix exists in the poly(β-alkyl- L-aspartates), since the methyl ester is left-handed, whereas the helices of the ethyl, propyl, and isopropyl esters are right-handed (Bradbury et al., 1968a). Moreover, with some polypeptides of this type the helical sense depends on the solvent medium (Toniolo et al., 1968) and the temperature (Bradbury et al., 1968a, b).

The dependence of the stability of the helical conformation on temperature reflects the relative importance of several factors to the heat associated with helix-coil transitions. In the case of poly(β-benzyl-L-aspartate) in *m*-cresol solution, the stability of the helix is reduced by raising the temperature (Bradbury et al., 1961), indicating that solvation of the peptide groups in the randomly coiled polymer is less exothermic than the formation of the intramolecular hydrogen bonds in the helical conformation. A similar labilization of the helix at higher temperatures was observed with poly(α-L-glutamic acid) in a water-dioxane mixture (Doty et al., 1957), but for poly(γ-benzyl-L-glutamate) in ethylene dichloride-dichloroacetic acid the helix stability increased at elevated temperatures (Doty and Yang, 1956). Helix formation is in this case energetically unfavorable, but it is being brought about by an increase in entropy. This must be caused by the release of dichloroacetic acid accompanying the desolvation of the polypeptide. A more complicated pattern of behavior was observed by Fasman et al. (1962) with copolypeptides of glutamic acid and leucine in aqueous solution, where the helix first melts out and then reforms as the temperature is being raised. These phenomena obviously reflect a reversal in the relative magnitudes of heat effects due to hydrogen bonding and to the formation of "hydrophobic bonds" (cf. Sections II.A.6 and II.B.5).

A particularly interesting case is poly(L-proline). This chain may be represented as follows:

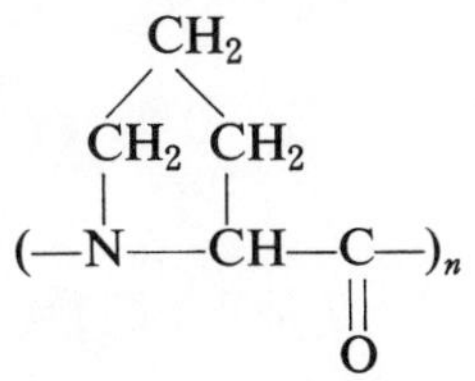

$$
\begin{array}{c}
\quad CH_2 \\
\quad / \quad \backslash \\
\quad CH_2 \quad CH_2 \\
\quad | \qquad\quad | \\
(\text{—N——CH—C—})_n \\
\qquad\qquad \| \\
\qquad\qquad O
\end{array}
$$

so that the polymer backbone lacks the peptide hydrogens and cannot form intramolecular hydrogen bonds. Nevertheless, poly(L-proline) may exist in rigid helical conformations. In fact, two such conformations with helices wound in opposite directions can be stabilized in appropriate solvent media. Downie and Randall (1959) and Katchalski and his collaborators (Steinberg et al., 1960) came to the conclusion that poly(L-proline) I, the form stable in propanol containing 10% acetic acid, contains right-handed helices with the two α-carbons attached to a peptide bond in the *cis* conformation. In poly-(L-proline)II, which is the stable form in most solvent media (water, acetic acid, etc.), the two neighboring α-carbons lie *trans* to one another and the chains are wound in left-handed helices, which are almost 70% longer than those of form I. The geometries of these two helices in solution are analogous to those which have been characterized by x-ray diffraction analysis for the two crystalline forms of poly(L-proline), although the complete identity of conformations assumed in solution and in the crystalline state cannot be taken for granted (Carver and Blout, 1967). The poly(L-proline)II helix has been shown to have 3 amino acid residues per turn with a repeat of 9.36 Å (Cowan and McGavin, 1955). For poly(L-proline)I a helix with 10 residues in three turns and a repeat of 19 Å has been suggested (Traub and Shmueli, 1963).

Figure III.23 shows the arrangement in the two chain conformations. Both the *cis* and the *trans* forms have a carbon atom *cis* to the carbonyl oxygen (in contrast to peptide bonds of the type C_α—CO—NH—C_α, where a *cis-trans* transition replaces the α-carbon with a hydrogen atom), and this explains why *N*-acylproline has no pronounced preference for one of the two isomeric states. The energies of the two helical forms are also very similar (Steinberg et al., 1960), so that small changes in the solvation of the macromolecule produce a complete shift of the equilibrium from one helix to the other. However, a junction of the two helical conformations represents a state of very high energy (estimated at 7 kcal/mole), and the equilibrium state of poly(L-proline) in mixed solvents is characterized, therefore, by a helix-helix transition occurring sharply within a narrow range of the composition of the solvent medium (Ganser et al., 1970). Because of the steric restrictions to rotation around the bonds in the backbone of poly(L-proline), the chain does not exist in the random coil form in nonelectrolyte solutions. However,

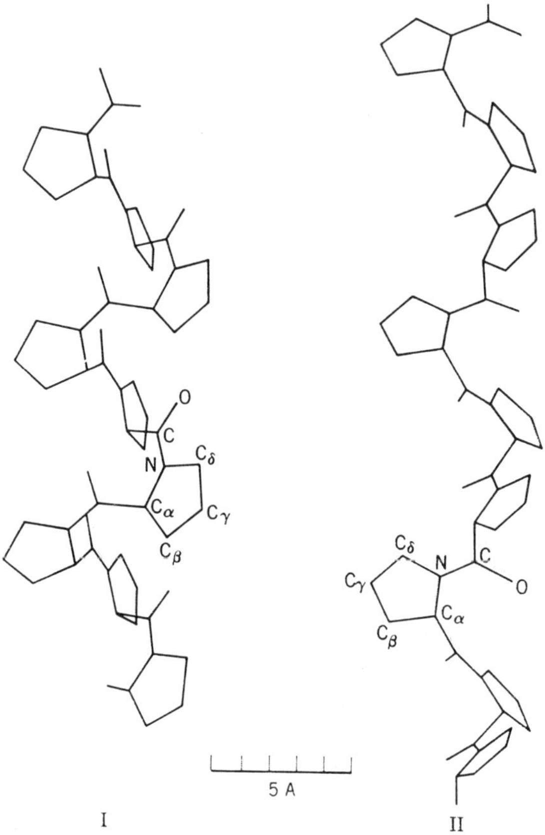

Fig. III.23. Conformations of poly (L-proline) I and poly(L-proline)II (Harrington and von Hippel, 1961b.)

certain "solvent salts," such as lithium bromide, apparently produce a change in the geometry of the chain, so that it become highly flexible (Harrington and Sela, 1958). This effect has been studied in detail by Schleich and von Hippel (1969), who found a relative effectiveness of different salts similar to that cited in Section II.C.3. They proposed that polarizable anions bind to the imide nitrogen and that this leads to an extension of the bond length between the α-carbon and the carbonyl carbon in the poly(L-proline) chain. Schleich and von Hippel argued that this bond extension would increase the allowed range for the internal angle of rotation ψ and thus would greatly increase the flexibility of the helical structure. However, spectroscopic evidence (Torchia and Bovey, 1971; Dorman et al., 1973) favors the original suggestion of

Harrington and Sela, who believed that in solvent salt solutions the poly-(L-proline) chains contain relatively short sections of left- and right-handed helices. This result would be expected of association of the peptide bonds with the ions, which reduces substantially the energy requirement for a junction of the two helical structures.

The case of poly(L-proline) is important in that it proves that stabilization through hydrogen bonding is not a requirement for the existence of stable helical conformations of dissolved chain molecules. In this context, it is interesting that poly(*N*-methyl-L-alanine):

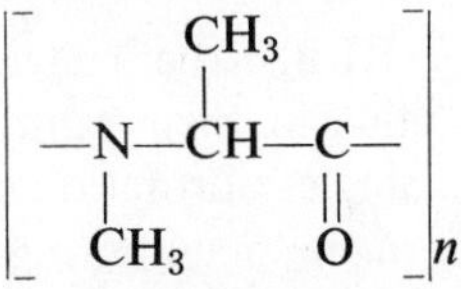

$$\left[-\underset{\substack{| \\ CH_3}}{N}-\overset{\substack{CH_3 \\ |}}{CH}-\underset{\substack{\| \\ O}}{C}- \right]_n$$

an analog of poly(L-proline) in which the pyrrolidine ring is missing, is also capable of maintaining a helical structure in solution (Goodman and Fried, 1967). However, in this case the chain backbone can assume only the *trans* conformation around the peptide C-N bond, and this implies that a replacement of a methyl group by the $—CH(CH_3)—$grouping *cis* to the carbonyl oxygen reduces the steric strain.

We have seen that the β structure of polypeptides, as proposed by Pauling and Corey (1951c), involves intermolecular hydrogen bonding of the extended molecular chains to form an infinite "pleated sheet." This would, of course, imply the insolubility of such a structure. Nevertheless, incontrovertible spectral evidence has been obtained that polypeptides can exist in solution in the β form. For instance, the un-ionized form of poly(L-lysine) exists in the α-helical form at low temperatures, and its solutions exhibit spectral changes characteristic of $\alpha \rightarrow \beta$ transition on heating (Rosenheck and Doty, 1961). The tendency of the polymer to precipitate on standing unless the solution is very dilute is the expected consequence of intermolecular aggregation, but at low solution concentrations the size of the molecular aggregate is apparently limited and the particle can remain dissolved. Sarkar and Doty (1966) claimed that poly(L-lysine) forms the β structure in dilute solution without evidence of molecular aggregation, and they reasoned that the chains must form hairpin loops so that antiparallel extended sections may form intramolecular hydrogen bonds. This conclusion was disputed by Davidson and Fasman (1967), who found that both the stability of the β structure and the rate of its formation are concentration dependent, indicating intermolecular association. Goodman et al., (1971) observed similarly a $\alpha \rightarrow \beta$ transition with an increase of the concentration of poly(L-alanine).

Davidson and Fasman found that poly(L-lysine) exhibits at high pH and low temperature a reversible transition between the α-helix and a random coil; at high temperatures the α-helix is transformed through the random coil intermediate to the antiparallel β form, and this transition is not reversed on cooling. With polypeptides which cannot exist in the helical form, a transition from a random coil to the β form may occur with a change of the solvent medium (Fasman and Blout, 1960). Our ability to characterize the properties of the β structure in solution is particularly important because of the role played by this structure in the conformation of many globular protein molecules.

We discussed in Section III.B.1 the restrictions which determine the flexibility of polypeptide chains in their random coil form. Although the pair of characteristic internal angles of rotation ϕ and ψ is independent of the values assumed by this pair in the neighboring peptide residue, nonbonded interactions between non-nearest-neighbor residues become important if ϕ and ψ are to have a unique pair of values, generating a helical conformation. Conformational analyses dealing with this problem have been carried out using hard sphere models for the atoms of the chain molecule (Ramachandran et al., 1966) or a variety of more sophisticated models in which van der Waals attractions between nonbonded atoms (De Santis et al., 1965), torsional energies, and dipole-dipole interactions are also taken into account (Ooi et al., 1967; Yan et al., 1968, 1970). Such analyses have led to some remarkable successes. Ooi et al., showed that poly(L-valine) should form a stable α-helix, although no α structure had ever been observed in the crystalline state. This conclusion was later confirmed by Epand and Scheraga (1968). The temperature dependence of the sense of α-helices observed by Bradbury et al., (1968a, b) with some poly(L-aspartates) could be accounted for by the reasonable procedure of assigning to the effective radii of the hydrogen atoms values which increase with rising temperature (Lotan et al., 1969). This procedure leads to the prediction that the stability of left-handed, as compared to right-handed helices, should increase as the temperature is raised—a conclusion that agrees with experimental evidence.

Striking success has been reported in accounting for reversals of the helical sense of closely related polypeptides, for example, from the left-handed poly-(β-methyl-L-aspartate) to the right-handed propyl ester (Yan et al., 1968), or from the right-handed poly(β-p-chlorobenzyl-L-aspartate) to the left-handed *o*- and *m*-chlorobenzyl esters (Yan et al., 1970). However, these successes should perhaps be accepted with some caution. Frequently, helix reversal makes contributions of opposite sign to the various components of the change of energy, and these have to be known to a high precision for a reliable prediction whether the net change is exothermic or endothermic. The conformational analysis increases rapidly in complexity with an extension of the length of side chains attached to the polypeptide backbone. It would also be expected that the requirement of strict periodicity would be much less stringent for parts of the side chains situated at some distance from the polypeptide backbone. If this is the case, entropies arising from such "side-chain disorder" should be taken into account in estimating the free-energy difference of the two helical forms. Finally, it seems unlikely that solvation effects are taken adequately into account by considering the dependence of dipole-dipole interactions on the "effective dielectric constant" of the medium. Krimm and Venkatachalam (1971) have shown that the hydrogen bonding of a water molecule to the imide group of poly(L-proline) changes in a striking manner the dependence of the energy on the conformation of the macromolecule and is responsible for the stability of the left-handed poly(L-proline)II helix in aqueous media. Similar effects would be expected to influence the conformations of other polypeptides in water solution.

2. The Coiled-Coil Structures of Collagen and Myosin

The poly(L-proline)II helix, discussed in the preceding section, is of special interest because of its relation to the structure of collagen, the characteristic protein of skin and tendon. The polypeptide chain of this substance contains extended regions with a remarkably periodic sequence of amino acid residues, which may be represented by $(\text{Gly—X—R})_n$, where Gly = glycine, X is frequently proline, and R stands for any of the remaining amino acids. The regular spacing of glycine residues and the high proline (or hydroxyproline) content favor the formation of a complex structure in which three left-handed helices, very similar to those of poly(L-proline)II, are wound around each other to form a right-handed helical aggregate (Ramachandran and Sasisekharan, 1961; Rich and Crick, 1961; Ramachandran, 1967). It is interesting that, in winding the three chains around each other in a sense opposite to that in which the individual helices are wound, nature is using the structural principle which leads to the strongest product in the manufacture of rope. Amino acid sequence studies have shown that the precise periodic placement of glycine residues is observed in long sections of the polypeptide chains of collagen (Bornstein, 1967; Butler, 1970), but that no such regularity exists near the chain ends (Kang and Gross, 1970). There also appears to be a tendency for amino acid residues with ionizable side chains to be segregated in specific regions of the particle. Evidence bearing on the structure of collagen as it exists in solution has been beautifully reviewed by von Hippel (1967) and by Traub and Piez (1971).

Destruction of the ordered structure of the collagen, brought about typically by an increase in the temperature of aqueous solutions, does not necessarily lead to complete separation of the three constituent strands, since interchain covalent bonds exist in most collagen samples. The process is usually not reversible, since conditions favoring the formation of the triple helix from the disordered coils also lead to intramolecular association of sections of folded chains and to more extensive chain aggregation (Harrington and Rao, 1970). The reconstitution of the original triple helix from the denatured state can be accomplished, however, if the temperature is such as to favor rapid equilibration between the helical and the random coil states (Beier and Engel, 1966), if the native collagen is treated with a reagent which leads to the formation of a number of covalent links between the constituent chains (Veis and Drake, 1963), or if such crosslinks exist in the natural material.

The stability of the collagen structure is usually characterized by the "melting temperature" T_m, that is, the temperature at which the helix-coil transition has proceeded to half completion. This T_m value has been found by many workers to increase with the content of proline and hydroxyproline residues (von Hippel, 1967). It had been widely assumed that this effect is a consequence of an increasing rigidity of the chain, so that less entropy would be gained by the destruction of the triple helix. However, calorimetric measurements by Privalov and Tiktopulo (1970) led to the unexpected finding that the entropy of melting generally *increases* with an increasing content of imino acid residues in a collagen sample and that the higher melting point results from an increasing energy requirement for the disruption of the triple helix.

Since it is possible to synthesize polypeptides with a periodic repetition of the sequence of monomer residues, the assumption that the triple helix formation is a consequence of such a periodicity can be tested experimentally. Work in that direction has shown that synthetic polypeptides of the (Gly—Pro—R)$_n$ type have a tendency to associate in solution and exhibit a temperature dependence of their spectral properties which is similar to that characterizing the formation of the collagen triple helix (Millionova et al., 1963; Millionova, 1964; Engel et al., 1966; Oriel and Blout, 1966; Brown et al., 1969). Studies on such model polymers have also shown that the stability of the triple helix is highly sensitive to factors other than the regular spacing of glycine residues and the imino acid content. For instance, a (Gly—Pro—Ala)$_n$ chain forms much more stable collagen-like structures than (Gly—Ala—Pro)$_n$ (Doyle et al., 1971).†

The coiled-coil structure arrived at, apparently, by the evolutionary

†Okuyama *et al.* (1972) have synthesized the oligopeptide (ProProGly)$_{10}$ which forms in solution a triple helix analogous to that of collagen. Single crystals can be grown from this material and these yield on x-ray crystallographic analysis much more detailed and unambigous structural data than can be obtained from collagen fibers. On the other hand, we cannot be certain that the structure of the collagen triple helix is identical with that of its much more regular analog.

process to provide the tensile strength of collagen is also utilized in myosin, one of the characteristic components of muscle tissue. Molecules of myosin from the skeletal muscles of mammals contain two chains with a molecular weight of about 2×10^5. The major portion of the chains has the α-helical conformation, while one end of each chain is folded into a globular form. The two α-helices are wound around one another to form a rodlike structure with a length of 1350 Å (Lowey et al., 1969). Some smaller proteins seem to be rather tightly associated with the long chains in the myosin particle. The extensive work on myosin has been reviewed by Perry (1967). A similar structure is consistent with the solution properties of paramyosin, another muscle protein (Lowey et al., 1963), and crystallographic data indicate that the two α-helices form a loosely coiled superhelix with a pitch of 180 Å (Cohen and Holmes, 1963). Although the two strands of the superhelix are somewhat crosslinked by disulfide bonds, elimination of these crosslinks does not seem to affect the stability of the coiled-coil structure (Olander, 1971).

To provide for the stability of the coiled-coil structure of myosin, evolutionary development has apparently utilized the same principle which we have already encountered in describing the three-stranded helix of collagen, that is, a remarkable periodicity in the spacing of strongly interacting amino acid residues along the constituent polypeptide chains. But whereas this cohesion was provided for collagen by hydrogen bonding of glycine residues, the coiled-coil of myosin is stabilized by hydrophobic interactions of nonpolar side chains appended to every seventh unit of the polypeptides. This was inferred by Cohen and Holmes on the basis of their x-ray data, and their brilliant prediction has been substantiated by an analytical determination of the amino acid sequence (Sodek et al., 1972).

3. Ordered Structures in Solutions of Nucleic Acids and Synthetic Polynucleotides

We have seen that the results of crystallographic studies were of crucial importance in providing the impetus for a search of helical conformations in dissolved polypeptides. In a similar fashion, most of the vast volume of work which has been carried out on nucleic acid solutions was undertaken under the stimulus of the proposal of a crystal structure for deoxyribose nucleic acid (DNA) by Watson and Crick (1953). Research on the nucleic acids has derived much of its glamor from the crucial role played by these substances in the transmission of hereditary characteristics of living organisms. Beautiful accounts of the dramatic discoveries implicating nucleic acids with some of the central problems of biology were given by Crick (1963), Watson (1963), and Wilkins (1963) in their Nobel prize lectures.

A segment of the molecular chain of DNA is represented as follows:

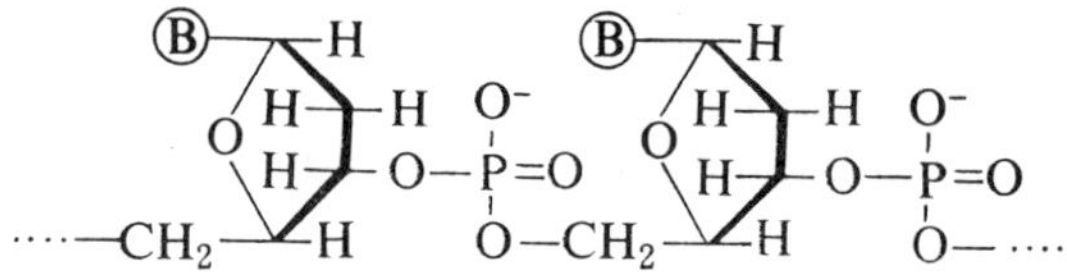

```
      Ⓑ──┐─H              Ⓑ──┐─H
       /H─┤─H  O⁻           /H─┤─H  O⁻
      O   │    |           O   │    |
       \H─┤─O─P=O           \H─┤─O─P=O
·····─CH₂──┘─H  |       ······CH₂──┘─H  |
                O─CH₂                   O─·····
```

The chain molecule is a copolymer of four types of units differing in the residues, denoted by Ⓑ, which are selected from the two purine bases, adenine (A) and guanine (G), and the two pyrimidine bases, thymine (T) and cytosine (C). Although the overall base composition varies within wide limits in DNA samples derived from different sources, the adenine content always equals that of thymine and the guanine content that of cytosine (Chargaff, 1950, 1955). This equivalence was of crucial importance in leading Watson and Crick to their DNA model on the basis of rather limited crystallographic data. They pointed out that the hydrogen bonding of A + T and of G + C leads to structures of almost identical dimensions and that the x-ray diffraction pattern of crystalline DNA may be accounted for if two antiparallel, intertwined chains have base sequences such that adenine in one chain is always juxtaposed to thymine in the other chain and guanine is similarly paired with cytosine. The "spiral staircase" model of the resulting double helix is represented in Fig. III.24. Later extensive crystallographic analyses (Langridge et al., 1960a,b) proved the Watson-Crick model to be correct in its essential features. The geometry of the A + T and G + C pairs emerging from this work is depicted in Fig. III.25. The plane of the base pairs is almost perpendicular to the axis of the helix, which contains 10 base pairs in a repeat distance of 34 Å. The stacking of the conjugated double-bond system of the base residues at the short distance of 3.4 Å from each other allows strong interaction of their π-electrons.

In favorable solvent media, DNA exists to a large extent in the double-helical conformation. The evidence on which this conclusion is based was reviewed by Steiner and Beers (1961) and by Marmur et al. (1963). The weight per unit length of the scattering particle, deduced from low-angle x-ray scattering (Luzzati et al., 1961b), also corresponds to the Watson-Crick model of the double helix. On the other hand, hydrodynamic and light scattering data show that the chain extension does not increase quite in proportion to the molecular weight of the particle. Hearst and Stockmayer (1962) have shown that the hydrodynamic behavior of DNA samples of varying chain lengths is well fitted by the wormlike chain model, so that it appears that all DNA has the double-helical structure, but that this structure is characterized by a slight flexibility. This picture is also in accord with

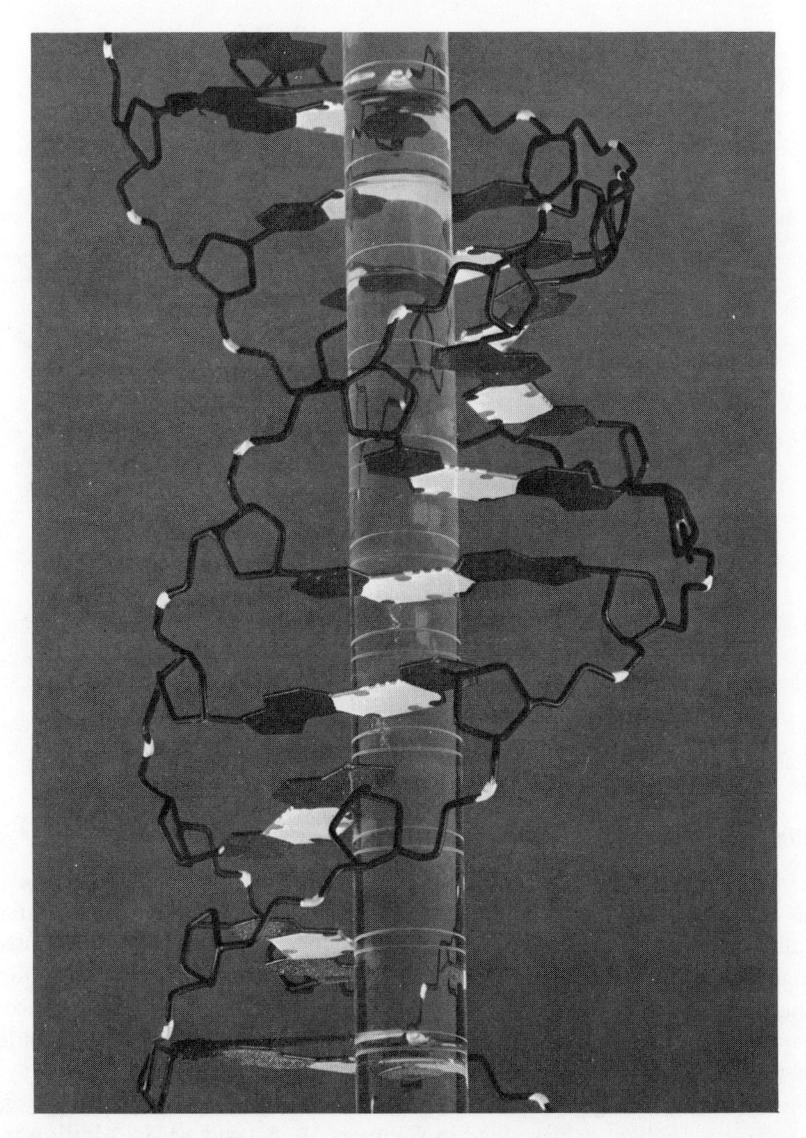

Fig. III.24. Model representation of the DNA double helix Courtesy of Professor N. Davidson.

the appearance of DNA images obtained by electron microscopy. Estimates of the persistence length of the DNA double helix have been subject to considerable variation (Schmid et al., 1971). Hydrodynamic data have been interpreted as indicating a doubling of the persistence length from 330 to

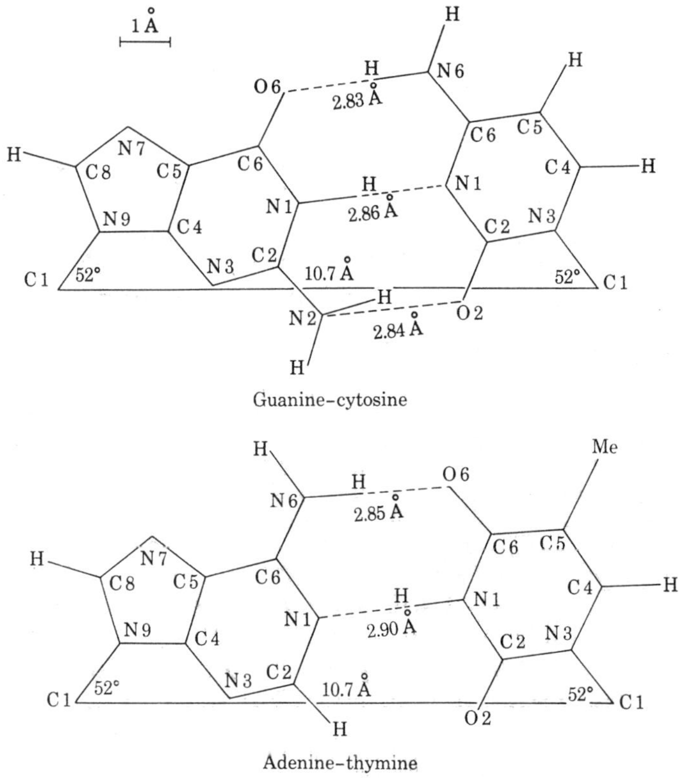

Fig. III.25. Geometry of the hydrogen-bonded guanine-cytosine and adenine-thymine pairs.

660 Å as the NaCl concentration was reduced from 1*M* to 0.005*M* (Hearst et al., 1968). In principle, changes in the excluded volume effect and in short-range interactions could both contribute to a change in the hydrodynamic behavior, but data are not available to assess their relative importance. Transitions in which the double helix is partially destroyed or the strands are completely separated from one another may be followed by spectroscopy, light scattering, changes in frictional properties, or equilibrium centrifugation in a density gradient as described in Chapters IV, V, and VI. In addition to these physicochemical methods, a unique criterion of the integrity of the native helical structure is provided by the biological activity of certain DNA preparations. Avery et al. (1944) discovered that contact of certain bacteria with DNA solutions may lead to transformation of the hereditary characteristics of the microorganisms. This "transforming activity," which is lost by any "denaturation" of the nucleic acid, may be used as a most sensitive tool

for the determination of the fraction of molecules present in intact double helices (Marmur and Lane, 1960; Dove and Davidson, 1962b).

Numerous factors affect the stability of the double helix in solution. Formation of the ordered structure is exothermic, so that the helices tend to "melt" when the temperature of DNA solutions is raised. Among the forces stabilizing the native form, hydrogen bonds and dipole-dipole interactions between the purine and pyrimidine residues stacked up in the double helix (De Voe and Tinoco, 1962) should lead to evolution of heat, whereas hydrophobic bonding would be expected to be endothermic. The role of the hydrogen bonds is reflected in the denaturation of the double helix at low pH, where the purine and pyrimidine bases are protonated (Cox and Peacocke, 1957), while denaturation by organic solvents (Herskovits, 1962) illustrates the importance of hydrophobic bonding in stabilizing the native structure. As would be expected, the high charge density due to the ionized phosphate residues along the DNA chain is a source of instability for the helical conformation. As a result, the addition of moderate concentrations of electrolytes should stabilize the native form of DNA, and this has been found to be the case with such salts as the halides of the alkali and the alkali earth metals (Dove and Davidson, 1962a; Schildkraut and Lifson, 1965). The "melting temperature" T_m appears to be approximately linear in the logarithm of the concentration of alkali metal cations. In a typical case, T_m is raised from 36 to 82°C by raising the concentration of sodium ions from 0.0003 to $0.1N$, and an increasing salt concentration tends also to narrow the temperature range for the helix-coil transition. Some divalent ions, which form specific complexes with the phosphate groups of the DNA backbone (e.g., Mg^{2+}), are particularly effective in stabilizing the helical conformation. The nucleic acid behaves as if it formed a stoichiometric complex with these cations, and the T_m of such complexes is high even at very low ionic strength. Under all conditions the helix-coil transition takes place in a remarkably narrow temperature interval, with 90% of the change occurring typically within less than 10°C.

Samples of DNA derived from different organisms may vary widely in the ratio of A—T and G—C base pairs, with the composition ranging approximately from 25 to 75% G—C. This composition has been found to be strongly reflected in the stability of the helical form, which increases sharply with increasing G—C content. Marmur and Doty (1962) measured T_m values for 40 samples of DNA and found that in solutions containing $0.2N$ Na^+ their data were closely fitted by $T_m = 69.3 + 41x_{GC}$, where x_{GC} is the fraction of G—C base pairs. Whatever may be the physical basis for the higher stability of double helices rich in G—C pairs (Pauling and Corey, 1956; Pullman and Pullman, 1959; De Voe and Tinoco, 1962), the phenomenon makes it possible to study the uneven distribution of nucleotides in a given species of DNA. For instance, electron microscopy of a bacteriophage DNA heated to progressively higher temperatures showed that the denaturation is almost complete in one half of the molecule before the double helix in the other half begins to melt (Inman, 1967).

Although the ease with which DNA solutions may be denatured has been known for a long time, it was demonstrated only much later (Marmur and Lane, 1960; Doty et al., 1960) that the process may be reversed under appropriate conditions. A detailed consideration will show that this is a truly remarkable fact. Typical samples of DNA may consist of chains with some 10,000 monomer units and, if the sequence of the base residues is random, any arbitrary sequence of four or even five residues will be likely to occur many times. We should, therefore, expect a high probability that two chains encountering one another will find short matching sequences in sites other than the unique site that allows pairing of the entire chain molecules. Obviously, this correct matching would have a negligible probability if association of the base pairs were not reversible, so as to allow the "melting out" of complexes which have formed in the wrong locations. It is not surprising, therefore, that "renaturation" of heat-denatured DNA, in which the two macromolecular strands have been fully separated from one another, requires very slow cooling through the temperature region of T_m in which the association process is easily reversible, so that the most stable complex—corresponding to the longest sequence of correctly matched base residues—can be arrived at. It is also essential that the DNA in solution be derived from a single species, or at least from very closely related organisms (Schildkraut et al., 1961, 1962b), so that formation of a hybrid double helix involves only a small number of regions in which the base residues facing one another are mismatched. Davis and Davidson (1968) were able to identify by electron microscopic observation the locus at which the two strands of a hybrid DNA are not complementary to one another.

Geiduschek (1962) showed that renaturation is easily achieved even in systems containing widely different DNA species if the denaturation did not lead to a complete separation of the DNA strands, so that they remain "in register" in a small number of loci. This is ensured most conveniently by producing a few covalent crosslinks between the two molecular chains (Geiduschek, 1961).

A particularly interesting phenomenon is the occurrence of circular DNA. This was first found to be formed from the nucleic acid of certain bacteriophages in which each of the two constituent polynucleotide strands extends some distance beyond one end of the double helix. The two single-stranded ends are complementary to each other and may associate to form a circular structure (McHattie and Thomas, 1964; Hershey and Burgi, 1965). Moreover, an enzyme has been discovered which will catalyze the covalent closure of the two polynucleotide rings (Gellert, 1967). This sequence of events is represented schematically in Figs. III.26*a*, III.26*b*, and III.26*c*. A number of biological sources have now been found which contain circular double-helical DNA. As the nature of the solvent is being varied, the double helix tends to be wound more or less tightly, causing the ring to be twisted into a superhelix as indicated in Fig. III.26*d*, with the number of twists depending on the solvent medium (Vinograd et al., 1968). The ring closure of the DNA tends to stabilize the double-helical structure, and since the intertwined closed rings are topologically prevented from separating in strongly denaturing media, renaturation is quantitative and rapid. Piko et al. (1968) found electron microscopic evidence that DNA can also exist in the form of two interlocked rings (Fig. III.26*e*), and similar observations have since been

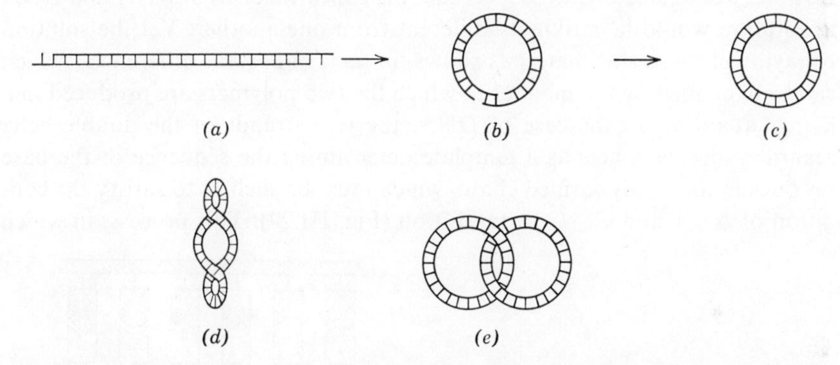

Fig. III.26. Schematic representation of circular DNA. (*a*) DNA double helix with complementary chain ends extending beyond the double-helical structure. (*b*) Formation of circular structure by interaction of complementary chain ends. (*c*) Covalent closure of rings. (*d*) Formation of superhelix. (*e*) Two interlocked circular DNA molecules.

made on DNA isolated from a variety of sources (Wang, 1973). An extensive summary of work on circular DNA has been prepared by Helinski and Clewell (1971).

Ribonucleic acid (RNA) consists of chain molecules with segments of the following type:

```
       (B)—H              (B)—H
      /  H—OH  O⁻        /  H—OH  O⁻
     O   H—O—P=O        O   H—O—P=O
     \           |      \           |
---CH2—H         O—CH2——H           O---
```

where three of the base residues (adenine, guanine, cytosine) are identical with the bases contained in DNA, while the fourth base, uracil (U), is very closely related to the base thymine found in DNA:

```
          O                              O
          ‖                              ‖
          C                              C
        /   \                          /   \
     HN       C—CH3                 HN       CH
      |       ‖                      |       ‖
    O=C       CH                   O=C       CH
        \   /                          \   /
          N                              N
          |                              |
       Thymine                         Uracil
```

Neither the substitution of ribose for deoxyribose nor that of uracil for thymine would lead one to suspect that the conformations of RNA and DNA in solution would be strikingly different from one another. Yet the solution behavior of these two materials shows, in fact, important differences, which can be explained by the manner in which the two polymers are produced in a living organism. In the case of DNA, the two strands of the double helix separate, and each acts as a template determining the sequence of the base residues in the newly formed chain, which must be such as to satisfy the condition of A—T and G—C juxtaposition (Fig. III.27). This process, in which

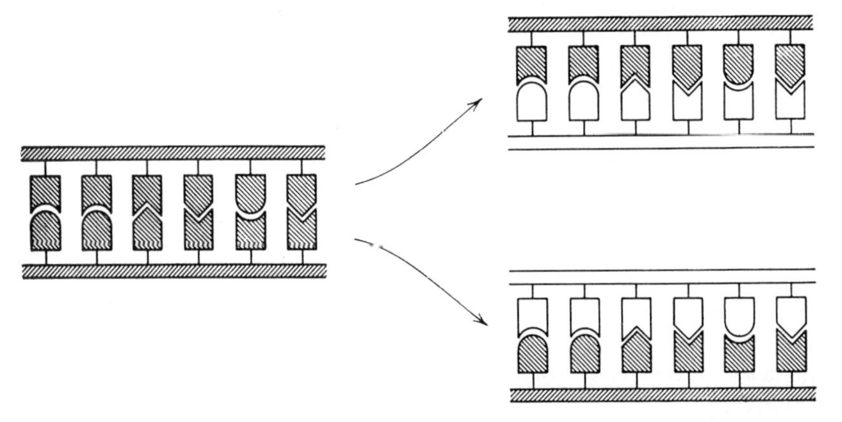

Fig. III.27. Schematic representation of DNA replication. The four kinds of appendages attached to the chain represent the two purine and two pyrimidine bases, with two and two complementary to one another. The shaded chains separate, and each guides the synthesis of a new chain (unshaded) with a complementary sequence of bases.

an equal number of complementary chains are produced with a sequence of monomer units identical with that of the original macromolecule, may be demonstrated in a cell-free system, as shown by the elegant experiments in Kornberg's laboratory (Kornberg, 1959, 1961). The biosynthesis of RNA, however, leads only in the rare case of some viruses to the synthesis of two complementary chains which may associate to a double helix (Gomatos and Tamm, 1963; Tomito and Rich, 1964; Sato et al., 1966). Most samples of RNA have an adenine content differing from that of uracil and a guanine content different from that of cytosine (Magasanik, 1955), so that they cannot be composed of two complementary strands.

Nevertheless, there is considerable evidence for the formation of some ordered conformations along part of the length of the RNA molecular chain (Doty et al., 1959; Steiner and Beers, 1961). The evidence is based on characteristic spectral shifts, changes in optical activity, anomalies in the titration behavior, and changes in reactivity. The picture which emerges suggests a partial formation of double-helical structures and a very gradual transition to the fully disordered form when the temperature is raised, or when other variables are changed in a direction that would be expected to favor denaturation. Since an interaction of two chains may be excluded, the chain must bend back on itself in a hairpin-like manner, so that short complementary sequences can be twisted into double-helical sections with a geometry quite similar to that of DNA. This concept, originally advanced on the basis of physicochemical observations (Fresco et al., 1960), was greatly strengthened when it became possible to isolate, as pure chemical species, so-called transfer ribonucleic acids, which are involved in the addition of a specific amino acid to the growing polypeptide in protein biosynthesis. They have relatively short chains (with about 80 base residues), and complete sequences of nucleotide residues have been determined for a considerable number of these molecules. Although these sequences vary greatly from species to species, all of them, when arranged in the form of a "cloverleaf" (Fig. III.28), allow A—U and C—G residue pairing in corresponding regions of the chain (Zachau, 1969). In view of the large number of molecules for which this generalization holds, we must assume that the "cloverleaf" represents correctly the manner in which the chain is divided into double-helical sections and unpaired loops. X-ray crystallographic analysis at a resolution of 3 angstrom has substantiated this "secondary" structure and has revealed the three-dimensional geometry of one of these molecules. The top and bottom limbs of the "cloverleaf" were found to extend at right angles to each other, while the right and left limbs wind about the hinge so that the molecule appears L-shaped (Kim et al., 1974). Studies on nucleotide sequences in ribosomal RNA reveal also complementary sequences, so that these molecules seem to be characterized by a similar folded structure (Forget and Weissman, 1967; Brownlee et al., 1968; Ehresmann et al., 1970). Recently, the sequence of 381 nucleotides in the RNA constituting the gene of the coat protein of a bacteriophage has been determined in a remarkable study by Jou et al. (1972). The chain contains 11 folded sequences with up to 27

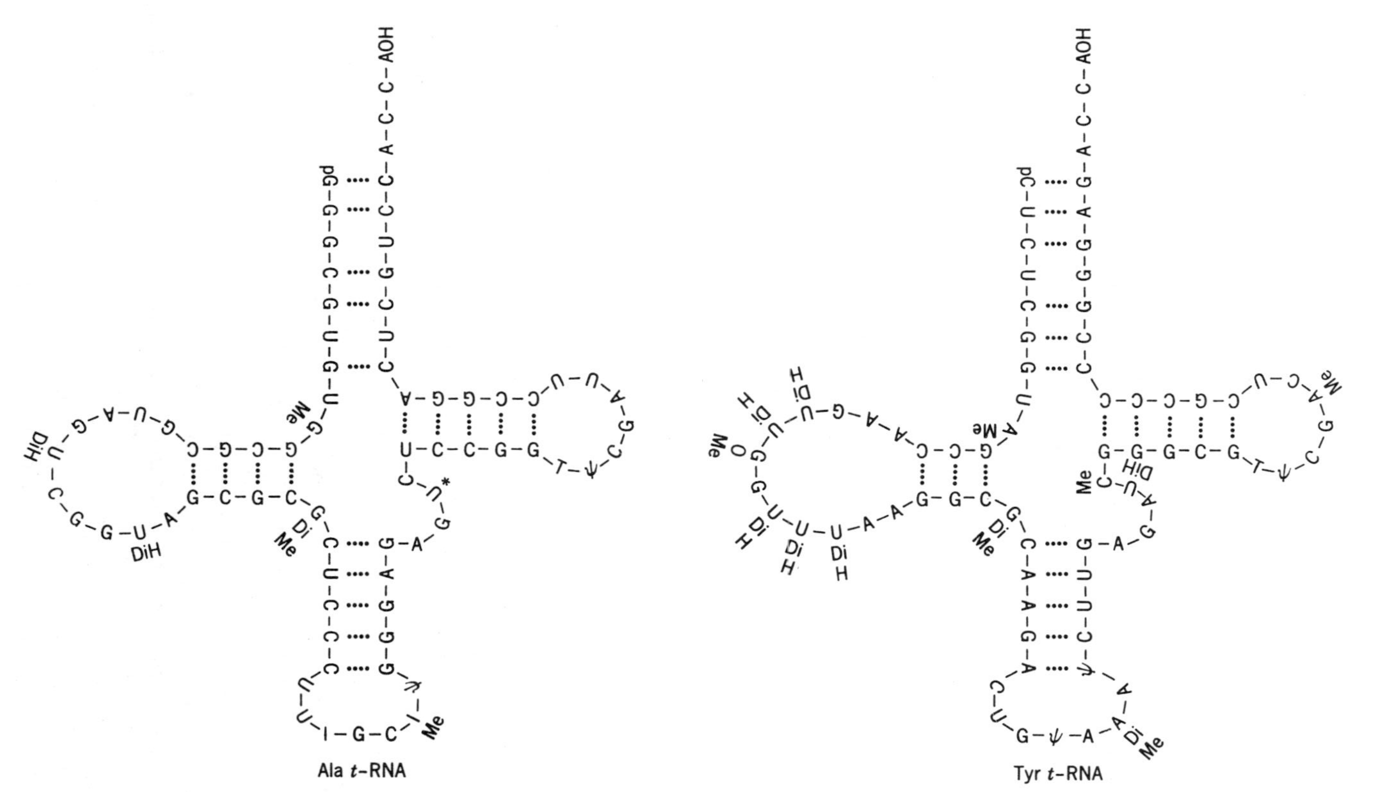

Fig. III.28. The "cloverleaf" representation of yeast transfer RNA molecules specific for alanine and tyrosine. Note that in spite of a completely different base sequence the four sequences in which the residues are connected by dashed lines contain, with a single exception, C juxtaposed to G and A juxtaposed to U (or a pseudouridine, ψ).

complementary base pairs, so that two thirds of the residues are in double-helical structures. A planar representation analogous to that of Fig. III.28 shows the double helices radiating from a central region of the molecule like the petals of a flower.

Polyribonucleotides and polydeoxyribonucleotides may be synthesized by enzymatically catalyzed processes, and in this way chains can be obtained which carry only one kind of purine or pyrimidine base. Early studies of such models have been reviewed by Felsenfeld and Miles (1967). A few of the conclusions, some of which have an obvious bearing on factors determining conformations of nucleic acids, are as follows:

(a) Spectroscopic evidence proves that adenine-uracil and guanine-cytosine pairing occurs also with the monomers, even in strongly hydrogen-bonding media.

(b) The behavior of single polynucleotide species has been studied in particular detail on polyriboadenylic acid (poly-rA) (Rich et al., 1961; Leng and Felsenfeld, 1966a; Eisenberg and Felsenfeld, 1967; Holcomb and Timasheff, 1968). At low temperature and low pH, electrostatic interactions between the phosphate and the protonated adenine residues stabilize a two-stranded helical form which exhibits a sharp melting point. At neutral pH, the single-stranded poly-rA is characterized by a very slow transition on heating, which may be interpreted as a gradual elimination of the stacking of neighboring adenine residues. As a result, the chain becomes more flexible and the "unperturbed dimensions" tend to decrease. The altered character of the chain leads to the existence of two Θ-temperatures and changes the enthalpy of solution from a negative to a positive value. This leads to an unusual temperature dependence of solubility in that the polymer first precipitates and then redissolves on heating.

(c) In the formation of heterocomplexes from two complementary polynucleotides, small structural changes may have a pronounced effect on the stability of the double helix. For instance, a 5-methyl substitution into uracil or cytosine raises the melting points of polyribonucleotide double helices depending on A—U or G—C pairing by about 20 °C (Szer and Shugar, 1966). For any given pair of complementary bases the stability of the double helix is also affected by substitution of a ribose (r) for a deoxyribose (d) residue. Thus, Chamberlin (1965) found the double helix to be much more stable for rG—rC than for dG—dC. Rather unexpectedly, the hybrid complexes dG—rC and rG—dC differed greatly in stability, with melting points of 71 and 90 °, respectively.

(d) The synthesis of oligonucleotides with defined sequences of the purines or pyrimidines makes it possible to assess the extent to which the double helix is destabilized by juxtaposition of some noncomplementary bases. For instance, the double helix formed by the antiparallel dimerization of the oligoribonucleotides A_5U_5, A_5CU_5, and A_5GU_5 melts, under a given set of conditions, at temperatures of 20, 5, and 15°C (Uhlenbeck et al., 1971).

(e) The contribution of base stacking interactions to the stability of the double helix is illustrated in a spectacular fashion by the observation that complexation of poly (rC) and poly (rU) with *monomeric* guanine and adenine, respectively, leads to helix formation (Howard et al., 1964, 1966; Huang and Ts'o, 1966). We shall return in Section VIII. A1 to a discussion of such complexes.

In some systems, triple-stranded helices may also be observed. This phenomenon has been studied in detail for polyriboadenylic acid. Conditions under which the triple-stranded and the double-stranded helices AU_2 and AU are stable have been defined by Stevens and Felsenfeld (1964) as a function of temperature and NaCl concentration and by Massoulié (1968) as a function of temperature and pH.

In view of the "twisted hairpin" form characteristic of the branches of the "cloverleaf" structure of transfer RNA (cf. Fig. III.28), it is of particular interest to study the behavior of copolynucleotides, which can fold to bring complementary sequences into juxtaposition. This type of behavior was first observed with a regularly alternating copolymer of dA and dT. Depending on temperature and ionic strength, these chains will form double-stranded helices either from two chains or from a single folded chain (Scheffler et al., 1968).

4. Polyisocyanates

Poly (*n*-alkyl isocyanates), with the following structure:

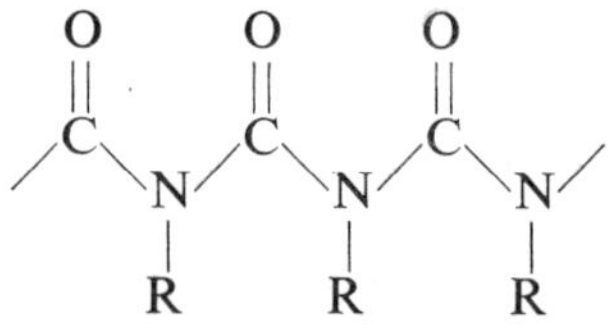

$$-\mathrm{C}(=\mathrm{O})-\mathrm{N}(\mathrm{R})-\mathrm{C}(=\mathrm{O})-\mathrm{N}(\mathrm{R})-\mathrm{C}(=\mathrm{O})-\mathrm{N}(\mathrm{R})-$$

behave in solution in a manner to be expected from rodlike particles with very small flexibility. Evidence based on dielectric behavior, hydrodynamic properties, osmometry, and light scattering, summarized by Fetters and Yu (1971) and Tsvetkov et al (1971), points to chains with a persistence length of 500–600 Å in solvents such as chloroform. X-ray crystallographic analysis of poly(*n*-butyl isocyanate) fibers by Shmueli et al. (1969) indicated that the chain has a helical conformation with a translation of 1.94 Å per monomer residue, and the behavior of the polymer in chloroform solution is consistent with this geometry. Addition of strongly acidic cosolvents increases the chain flexibility, but this transition is gradual, without any indication of the cooperativity which characterizes the breakdown of polypeptide or DNA helices (Fetters, 1972).

5. Theoretical Treatments of Helix-Coil Transitions

Transitions between ordered and disordered conformations of chain molecules are important since they concern the conditions which must be met to keep proteins and nucleic acids in the forms required for their biological functions. At the same time, the phenomenon of helix-coil transition may be regarded as a one-dimensional analogy to a melting and crystallization process, and as such it is of particular theoretical interest. Let us at first disregard the formation of multiple helices and focus our attention on transitions in isolated chains, such as are typical of polypeptides. Since the partial double-bond character of the C—N linkage precludes rotation around this bond, so that the C_α—C′—N—C_α sequence behaves as a rigid link, only two internal angles of rotation, ϕ and ψ (Fig. III.14), need be specified to fix the relative positions of three consecutive α-carbons. When the random coil is transformed into the perfectly ordered conformation, the freedom to choose the ϕ and ψ values is lost. As a result, for a chain consisting of Z amino acid residues, transformation to a helix will be opposed by a free-energy gain proportional to $Z - 2$. On the other hand, helix formation will be favored by nearest-neighbor interactions of various types. These may include intramolecular hydrogen bonding, hydrophobic bonds, and desolvation effects accompanying the transfer of side chains from the relatively exposed state in the random coil to the compact packing around the helix. Such effects will, in general, be more pronounced for residues in the interior of the helix than for those at its ends, and the contribution to the free energy of helix formation due to nearest-neighbor interactions will, therefore, be proportional to $Z - \delta$, where δ is a number taking account of the lesser stability of the ends of the helix. If $\delta > 2$ (for the α-helix $\delta = 4$ has been assumed by Schellmann, 1955), the free energy for the transition of the disordered coil to the perfect helix will become more favorable as Z is increased. However, to specify properly the conditions governing helix-coil transitions, we have to take account also of partially ordered states containing various combinations of helically wound and randomly coiled sequences.

We shall use the representation given by Zimm and Bragg (1959), to whom the reader is also referred for a bibliography of this field. Let us designate by h a monomer residue which forms part of a helical section and by c a residue which is not incorporated into the helical structure. The statistical treatment of the transition may then be carried out in terms of two characteristic parameters. The first, designated by j, is the equilibrium constant for the addition of one monomer residue from a disordered to a helical sequence, as represented schematically by

$$(\text{—h—})_n(\text{—c—})_m \rightleftarrows (\text{—h—})_{n+1}(\text{—c—})_{m-1}$$

The second parameter, designated by ζ, is the equilibrium constant for processes of the type

$$(\text{—h—})_{n+p}(\text{—c—})_m \rightleftarrows (\text{—h—})_n(\text{—c—})_m(\text{—h—})_p$$

The value of ζ will generally be much smaller than unity. It may be described as a measure of the driving force toward the nucleation of helical sequences, since to start a helical section will generally be more difficult than to add to an existing one. The model of Zimm and Bragg contains also the restriction that at least three consecutive hydrogen bonds of an α-helix have to be broken simultaneously to allow a break to be formed in a helical sequence.

The results of the statistical calculation indicate that, for a constant value of ζ, the helix-coil transition will occur over a range of j values with $j = 1$ corresponding to 50% helix content. For sufficiently long chains, the effects due to helix labilization at the chain ends will become negligible and the fraction of residues accommodated in helical sequences (f_h) will then be independent of chain length. The transition will sharpen as ζ is decreased, but an infinitely sharp transition analogous to a melting point is obtained only as ζ approaches zero. The fraction of chain residues present in the helical section of the polymer is

$$f_h = \frac{1}{2}\left[1 + \frac{j-1}{\sqrt{(1-j)^2 + 4\zeta j}}\right] \tag{III.35}$$

At the midpoint of the helix-coil transition df_h/dj attains its maximum value:

$$\left(\frac{df_h}{dj}\right)_{\max} = \frac{1}{4\sqrt{\zeta}} \tag{III.36}$$

and the average number of residues in helical or randomly coiled sections of the chain ("the cooperative length") is $1/\sqrt{\zeta}$. Since j is an equilibrium constant, $d \ln j/d\mathrm{T} = \Delta H/\boldsymbol{R}T^2$. The heat of helix-coil transition has been obtained calorimetrically for various polypeptides (Ackermann and Ruterjans, 1964; Giacometti et al., 1968), and such data may be used to obtain the temperature dependence of j and the nucleation parameter ζ from (III.36). For polypeptides, this parameter is generally of the order of 10^{-4}, corresponding to a cooperative length of the order of 100 at the midpoint of the helix-coil transition.

Whereas for long chains f_h approaches the limiting value given by (III.35), the extent of helix formation will be sensitive to Z if the chains are short. According to Zimm and Bragg, substantial helix formation requires that

$$j > 1$$

$$(j - 1)^2 j^{-Z+1} \leqslant \zeta \tag{III.37}$$

A corresponding treatment of the formation of the double-helical DNA conformation is much more complex. In analyzing helix-coil transitions in polypeptides, the contribution to the free energy, due to a monomer residue placed in a randomly coiled section, could be considered independent of the presence of helical sequences in the molecular chain. With partially ordered conformations of the DNA type, the two strands of disordered sections placed between two regions where the chains have associated to a double helix must form a closed loop. The dependence of the probability of ring closure on the size of the loop then introduces a contribution to the free energy from process such as the following:

$$\begin{array}{l} -(-h-)_n\;(-c-)_m\;(-h-)_p\;(-c-)_q\;(-h-)_r- \\ -(-h-)_n\;(-c-)_m\;(-h-)_p\;(-c-)_q\;(-h-)_r- \end{array} \rightarrow$$

$$\begin{array}{l} -(-h-)_n\;(-c-)_{m+\delta}\;(-h-)_p\;(-c-)_{q-\delta}\;(-h-)_r- \\ -(-h-)_n\;(-c-)_{m+\delta}\;(-h-)_p\;(-c-)_{q-\delta}\;(-h-)_r- \end{array}$$

The problem has been treated by Rice and Wada (1958), Gibbs and Di Marzio (1959), Zimm (1960), and Lifson and Zimm (1963). They all assumed that the probability of ring closure is inversely proportional to the 3/2 power of its contour length, as would be expected from (III.7) if the excluded volume effect can be neglected. The characteristics of the conformational transition are again independent of chain length for sufficiently long chains, and a rather sharp transition with temperature variation is predicted. However, before theoretical results may be compared to experimentally observed helix-coil transitions in DNA solutions, three factors have to be taken into account.

(a) Since DNA consists of molecules with a very high density of ionic charges, the disorganization of the double helix will lead to a pronounced decrease in electrostatic free energy. Therefore, increasing electrolyte addition, which decreases the mutual interactions of the ionic charges attached to the macromolecule, should stabilize the helical form. Experimental observations are qualitatively in agreement with this view, and a quantitative theory of the effect was proposed by Schildkraut and Lifson (1965). The situation is further complicated by the specificity of the interactions of a polynucleotide with its counterions. Krakauer and Sturtevant (1968) carried out a careful study of the heat of formation of the poly(rA)-poly(rU) double helix and found the ΔH value to be appreciably reduced when Na^+ counterions were replaced by K^+.

(b) Since A—T base pairing produces a weaker bond than G—C pairing and since the base composition may vary along the length of the chain, the melting range will be broader than predicted for chains with uniform composition. Lifson (1963) has discussed a mathematical approach to a consideration of this factor, but its application is limited at present by a lack of data concerning the sequence of the base residues in DNA.

(c) The excluded volume effect will make the ring closure probability decrease more rapidly with the size of the loop than is assumed in the above theories. This factor will tend to sharpen the helix-coil transition (Klotz, 1969).

6. The Conformation of Globular Protein Molecules

In comparison with synthetic polymers, proteins are macromolecules of very high chemical complexity, since they may contain up to 20 different amino acid residues. On the other hand, the high specificity of biosynthetic processes makes it possible for a living organism to prepare species in which the amino acid sequence is uniquely defined. The solution properties of a large class of the so-called globular proteins show that their chain molecules must be folded so as to produce compact particles containing only small amounts of solvent in their interior. The ability of the globular proteins to crystallize and the surprisingly sharp x-ray diffraction patterns characterizing these crystals prove, moreover, that the conformations of the chain backbone and of most of the side chains are defined with considerable precision.

Since the conformation of a protein is closely related to its biological function, an unusually large scientific effort has been expended in the crystallographic study of these macromolecules. Table III.1 lists some of the proteins for which molecular structures have been obtained at a high level of resolution.† A 2 Å resolution was long thought to be close to the attainable limit (North and Phillips, 1969); this not only allows the tracing of the chain backbone (as illustrated in Fig. I.4) but also provides considerable information about the location of side chains, although some of these, at the surface of the molecule, seem to have some conformational freedom. Recently,

†An excellent summary of the status of this field as of 1971 is to be found in Volume 36 of the *Cold Spring Harbor Symposia on Quantitative Biology*.

Watenpaugh et al. (1972) have shown that a refinement of x-ray structures to a 1.5 A° resolution is feasible, and this additional precision in the location of atomic positions should be most valuable in analyzing the energetics of chain folding.

The first proteins for which molecular structures were determined were myoglobin (Kendrew et al., 1960) and hemoglobin (Perutz et al., 1960). In the case of myoglobin, the protein consists of a single polypeptide chain with eight right-handed α-helix segments. The helices vary in length from 4 to 24 amino acid residues; their combined length represents approximately 78% of the polypeptide chain. Two junctions are sharp corners, while the other helical segments are separated by varying lengths of the chain with an irregular conformation. The factors determining the end of a helical section are understood in some cases (e.g., a proline residue cannot be accommodated into the α-helix), while they remain unaccounted for in others. Kendrew (1963) has discussed some striking features of the structure, which is extremely compact, with no more than a maximum of five isolated water molecules trapped in its interior. With very rare exceptions, all polar groups are found on the outside of the molecule, while the side chains of the amino acids in the interior have nonpolar character. One thus gains the impression that the cohesion of these hydrophobic residues is the most important principle in the stabilization of the protein conformation. This conclusion has been strikingly reinforced by a comparison of 18 myoglobin or hemoglobin chains from a variety of organisms, with widely varying amino acid sequences (Perutz et al., 1965). If we make the reasonable assumption (for which there is considerable evidence) that those proteins have similar conformations, we find that amino acids with nonpolar side chains in the interior of the molecular structure are invariably replaced by other residues with similar nonpolar nature.†

The concentration of most nonpolar amino acid side chains on the inside of protein molecules has proved to be a general phenomenon in all the cases for which structure determination has been obtained. On the other hand, the prominence of the α-helix as a structural feature of myoglobin and hemoglobin is now known to be exceptional. All the other proteins listed in Table III.1 have far fewer sections with the α-helical conformation, and in chymotrypsin such helices account for only 5% of the chain. In a number of the proteins, pleated sheets formed by the hydrogen bonding of extended parallel or antiparallel sections of the polypeptide chain in the β conformation become important elements of the molecular structure. Such a pleated sheet, somewhat distorted into a twisted shape, is found in the structure of carboxypep-

†The reader should realize that in the synthesis of random copolypeptides the probability of a chain folding to a compact globular structure would be negligible. Only the evolution of living organisms could have led to sequences behaving in this manner.

tidase A (Lipscomb et al., 1970) and is represented schematically in Fig. III.29.†

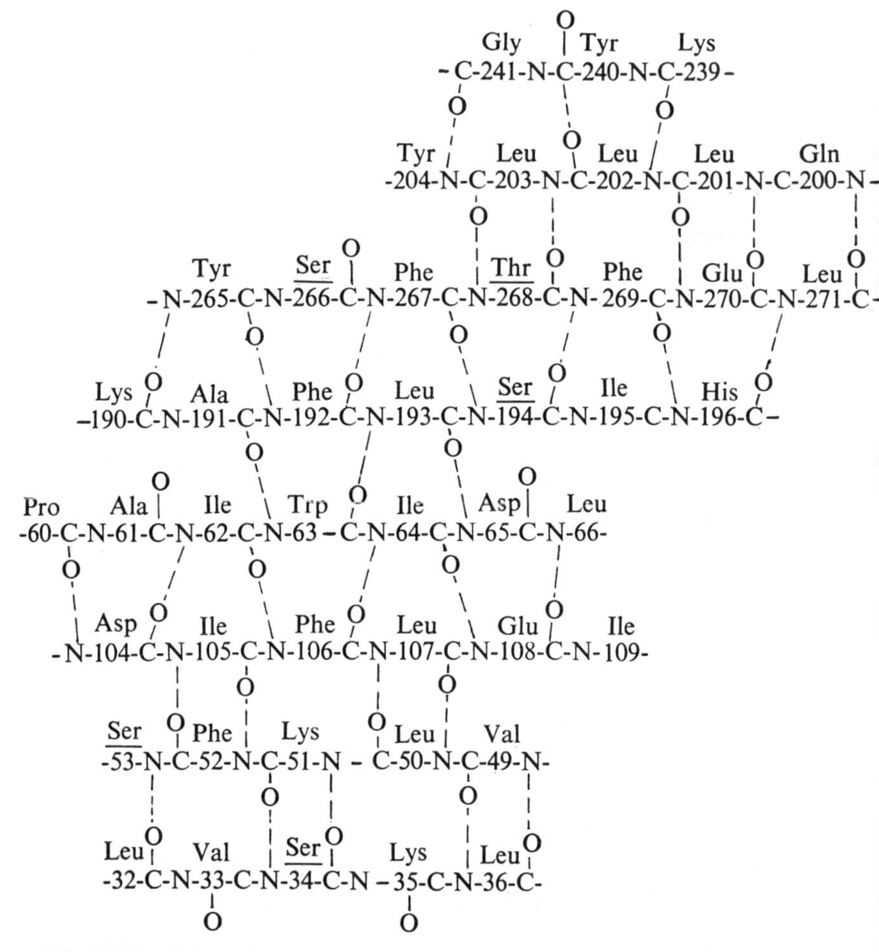

Fig. III.29. Schematic representation of the portion of the carboxypeptidase A structure which has the pleated sheet β structure. The amino acid residues are numbered starting with the amino terminal unit of the polypeptide chain. The underlined serine and threonine residues are known to favor the β conformation. Note that hydrogen-bonding between both parallel and antiparallel sections of the polypeptide chain contributes to the pleated sheet.

†Linderstrøm-Lang and Schellman (1959) suggested that the description of a protein molecule be subdivided into three steps. The "primary structure" would specify the sequence of the amino acids in the chain; the "secondary structure," the amino acid sequences in characteristic ordered conformations; and the "tertiary structure," the folding of these ordered structures into a globular particle. As crystallographic information became more

We have seen (Fig. III.15) that steric restrictions limit severely the accessible range of combinations of the internal angles of rotation ϕ and ψ in polypeptide chains. It is then of interest to find how well the conformations of the amino acid residues in a protein agree with those "allowed" by potential energy calculations. Such a comparison was first made by Liquori (1966), who found that the conformational states of amino acid residues in the nonhelical regions of myoglobin generally lie close to the five minima on the potential energy map calculated for poly (L-alanine). A similar analysis of the structures of lysozyme in terms of the map of sterically allowed conformations proposed by Ramachandran et al. (1963) was carried out by North and Phillips (1969) and is shown in Fig. III.30. Most of the residues have the conformation characteristics of the α-helix, although sequences of residues with this conformation are generally too short to develop the helical structure. In all but two cases, the residues for which the $\phi - \psi$ combination deviates significantly from the prediction of the Ramachandran plot are glycines, which are known to have considerably greater conformational freedom than the other amino acids. North and Phillips (1969) suggest that the apparent anomaly of the conformation of the remaining two residues may be the result of computational error, so that the conformation of the protein chains would be consistent with a possible structure on the basis of conformational analysis.

Our interest in the conformation of a protein molecule as derived from x-ray crystallography is conditioned by the implicit assumption that the conformation will be similar in solutions in which the protein fulfills its biological function. Evidence bearing on the relation between protein conformations in the solid state and in dilute solutions has been summarized by Rupley (1969). An impressive number of correlations of various kinds between the crystallographically determined molecular structures of proteins and the properties of these proteins in dilute solution indicate that crystallization does not lead, in general, to major changes in molecular conformations. We may cite, as typical examples, the following observations.

(a) A number of enzymes have been shown to retain their catalytic activity in the crystalline state (Doscher and Richards, 1963; Quiocho and Richards, 1964, 1966; Kallos, 1964). Since this catalytic activity implies that the conformation of the active site is preserved in great detail, we must conclude that at least the conformation of this portion of the protein molecule remains unaltered.

---

plentiful, however, it became clear that large portions of many proteins lack a repeating conformational pattern. Moreover, short sequences corresponding, for example, to one turn of a helix are hardly to be thought of as "ordered." Hence it is no longer useful to distinguish between a "secondary" and a "tertiary" protein structure.

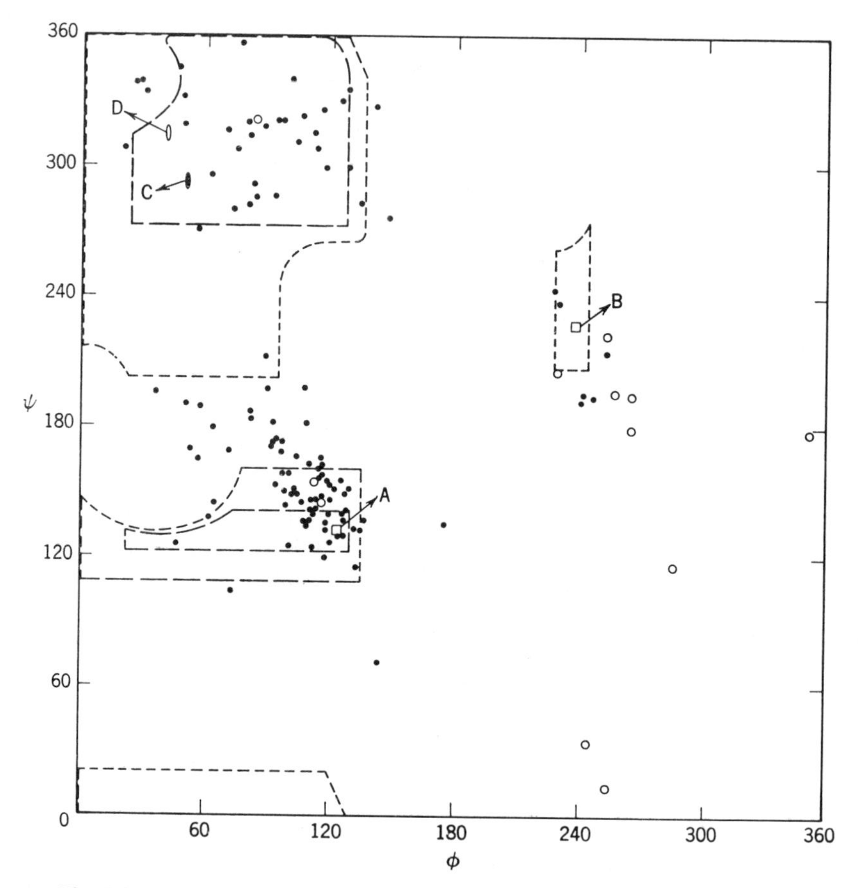

Fig. III.30. Conformations in the polypeptide chain of lysozyme. The short broken lines enclose outer limits, and the long broken lines fully allowed regions according to Ramachandran et al. (1963). (A) right-handed α-helix; (B) left-handed α-helix; (C) parallel pleated sheet; (D) antiparallel pleated sheet. Open circles denote glycine residues, which are subject to less stringent restrictions.

(b) The reactivity pattern of functional groups in the side chains of a protein molecule has been explained in terms of the molecular structure determined by crystallography. For instance, of the 12 histidine residues of sperm whale metmyoglobin 4 or 5 react very slowly with bromoacetate. The location of these residues in the sequence of the protein chain was determined, and they were found to be those which are hydrogen bonded in the interior of the myoglobin structure (Banaszak et al., 1963; Banaszak and Gurd, 1964; Gurd et al., 1968).

(c) About 77% of the polypeptide chain of sperm whale myoglobin exists in the α-helical conformation, according to crystallographic data (Kendrew et al., 1961). This is in close agreement with the helical content estimated from the optical rotatory dispersion (ORD) spectrum of myoglobin in dilute solution (Beychok and Blout, 1961; Urnes et al., 1961). Other proteins, such as lysozyme (Blake et al., 1967a,b) or papain (Drenth et al., 1968), have conformations in the crystalline state with very low contents of helical sections, again in agreement with ORD data for solutions of these proteins.

(d) Conclusive evidence as to the virtual identity of the backbone conformation of globular proteins in the crystalline state and in aqueous solution has been obtained more recently by Raman spectroscopy (see Section V.A.3).

It is not surprising that conformations of protein molecules should not be greatly perturbed when the protein crystallizes, since typical protein crystals contain 30–50% by volume of water. There are generally only a few (5 or 6) contact points between a protein molecule and its nearest neighbors, so that crystal lattice forces are small. Direct evidence that these forces are insufficient to influence greatly the protein conformation has been provided for ribonuclease, α-chymotrypsin, and lysozyme. The conformations of ribonuclease A and ribonuclease S (differing from the native enzyme in the scission of one peptide bond) have been found to be extremely similar, although the molecules are packed quite differently in their crystals (Kartha et al., 1967; Wyckoff et al., 1967). An α-chymotrypsin derivative has been shown to form crystals in which the protein is placed in two crystallographically nonequivalent environments, yet the conformations of those molecules differed only very slightly from each other (Birktoft et al., 1969). Finally, lysozyme has been shown to have conformations in tetragonal and triclinic crystals which could hardly be distingusihed at the 6 Å level of resolution (Joynson et al., 1970). However, slight differences in the conformation of proteins may be of crucial importance for their function. Also, two conformations may be in equilibrium with each other when a protein is in solution (p. 410, 415), and shortlived conformational states are involved, for instance, in enzymatic catalysis (p. 410, 448). Such insights are obviously not attainable from crystallographic studies.

Protein denaturation was classically defined as "any nonproteolytic modification of the unique structure of a native protein, giving rise to definite changes in chemical, physical, and biological properties" (Neurath et al., 1944). Changes of the state of ionization are excluded from the definition, unless they are accompanied by conformational transitions. Denaturation may be the result of heating, changes in pH, and the addition of nonpolar solvents, electrolytes such as the "solvent salts" listed in Table II.5, or certain specific denaturation reagents such as urea or guanidine salts. It may also be produced by the reductive or oxidative scission of disulfide bonds, which stabilize the native conformations of some proteins. Denaturation is typically accompanied by decreasing protein solubility. This may be easily understood since the hydrophobic bonding, which serves to stabilize the native conformation, will tend to produce intermolecular aggregation when the polypeptide chains exist in their expanded conformations. Another characteristic effect produced by denaturation is the "unmasking" of reactive groups which are hidden in the interior of the tertiary structure and become accessible to reagents when this structure is disrupted. The most useful methods for following denaturation processes include spectroscopic measurements, measurements of optical activity, and determinations of the catalytic activity of enzymes. If denaturation may be described as a transition from a "native" to a discrete "denatured" state with no intermediate forms present in significant concentrations, the fractional change in all these properties should be the same. On the other hand, if intermediate forms are relatively stable under some conditions, their physical and chemical properties may be related in different ways to those of the native and denatured states. Under these conditions, it is impossible to assess a "degree of denaturation" from the change in any one property, as illustrated by the data of Nelson and Hummel (1962), reproduced in Fig. III.31. (Note that the absorption at 287.5 nm is the same in the native form and at a point where more than 50% of the enzymic activity has been lost). With NMR spectroscopy at very high resolution, it is possible to follow changes in the resonances of various amino acid residues as a function of the concentration of a denaturant and to determine whether the fractional change from the "native" to the "denatured" state is the same for all these residues. Bradbury and King (1969, 1971) used this approach to a study of the denaturation of ribonuclease, lysozyme, and α-lactalbumin. In the case of lysozyme, for instance, the resonance due to arginine residues (which are situated on the molecular surface) changes more rapidly toward that which is characteristic of the unfolded state than the resonance due to hydrophobic residues.

The variation in the stability of the native conformation with changes in the solvent medium provides clues to the relative importance of factors determining the tertiary structure of proteins. The denaturation of aqueous protein solutions by nonpolar cosolvents is a complex process. The cohesion of nonpolar residues in an aqueous environment will stabilize the helical conformation, but may be the underlying cause of many of the "kinks" by which the helical sections of the polypeptide chain are folded into the compact structure in the native state. A decrease in the polarity of the medium may, therefore, lead at first to an increase in the content of helical conformations (Bresler, 1958; Weber and Tanford, 1959; Tanford et al. 1960) before they are eventually disrupted. The action of urea, or of guanidine salts, seems to be only partially explained by the disruption of hydrophobic bonds. Aqueous urea has been shown to be a better solvent medium for nonpolar solutes than water itself (Nozaki and Tanford, 1963), but it was also found that aqueous urea or guanidine hydrochloride exerts a specific solvating

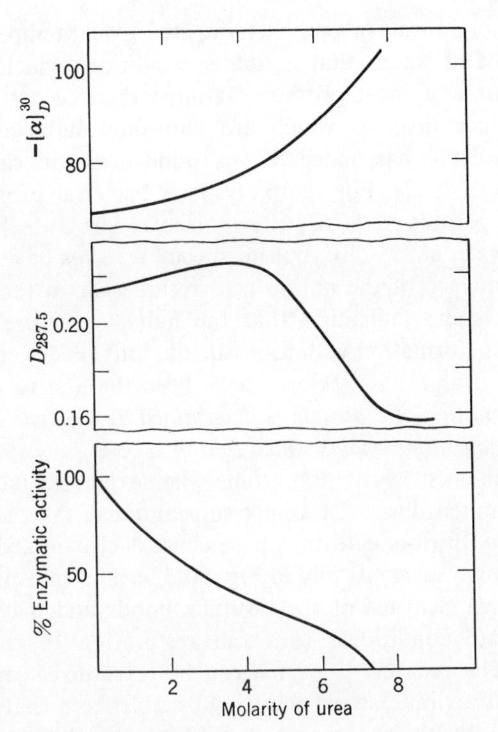

Fig. III.31. Effect of urea on the enzymatic activity, the optical density at 287.5 nm, and the optical activity at 589 nm of ribonuclease.

action on the backbone of a polypeptide chain, an effect which has characteristics quite different from the solubilization of hydrocarbon residues (Robinson and Jencks, 1963). Finally, it has been demonstrated that the stability of the native conformations of chymotrypsinogen (Brandts, 1964) and ribonuclease (Brandts and Hunt, 1967) in aqueuos media passes through a maximum at characteristic temperatures of 10 and 0°C, respectively. This is probably typical of globular proteins in general and is consistent with the assumption that denaturation involves increased exposure of nonpolar residues to the aqueous medium. We have noted previuosly (p. 49) that solution of hydrocarbons in water is characterized by large positive values of $\overline{\Delta C}_p = \partial\overline{\Delta H}_2/\partial T$, and this is also characteristic of protein denaturation. The change of the enthalpy of denaturation from a negative to a positive value as the temperature is raised requires then that the standard free energy for this conformational transition pass through a maximum.‡

‡Measurements of the heat of denaturation have recently yielded temperatures of maximum stability of native myoglobin, chymotrypsin and cytochrome (P. L. Privalov and N. N. Khechinashvili, *J. Mol. Biol.*, **86**, 665 (1974)).

A number of organisms (algae, bacteria, etc.) live at temperatures close to the boiling point of water, that is, under conditions which would lead to rapid denaturation of most proteins. It must then be assumed that such organisms produce proteins which are unusually stable against thermal denaturation, and this has, indeed, been found to be the case (Ohta et al., 1966; Amelunxen, 1967). For thermolysin, a bacterial proteolytic enzyme stable for long periods at 80 °C, the molecular structure has been determined (Matthews et al., 1972b), but no special features have been identified either in the sequence of the amino acid residues or in the folding of the chain. There is some indication that the native structure owes its high stability to chelate formation with four calcium ions (Feder et al., 1971).

Lumry and Eyring (1954) seem to have been the first to propose clearly that *"the conformation of a protein is determined by the type and sequence of amino acids in the peptide chains which fold in such a way as to minimize free energy."* This statement, written at a time when experimental evidence for it was relatively limited, has been amply substantiated. A crucial experiment was carried out with ribonuclease, whose chain is crosslinked by four disulfide bonds as shown schematically in Fig. III.32*a*. The enzyme may be denatured by reductive cleavage of the disulfide bonds and may be renatured, under appropriate conditions, with full restoration of enzymic activity (White, 1960). The renatured enzyme can be crystallized, and the crystals give an x-ray diffraction pattern indistinguishable from that of the original protein (Bello et al., 1961). This may be taken as the ultimate proof that the conformation of the native enzyme has been regained in every detail. The eight thiol groups of the reduced enzyme could combine in 105 different ways to form four disulfide bonds. Since disulfide formation precedes the recovery of catalytic power, it appears that the crosslinks form first in the "wrong" manner, but that subsequent disulfide interchange leads to the form corresponding to maximum thermodynamic stability (Epstein et al., 1962). In the time following this classical study, similar results have been reported with a variety of proteins (Anfinsen, 1967). Our current concepts of protein biosynthesis also postulate that only the linear sequence of amino acid residues has to be specified by the genetic code; the folding of the chain to a unique three-dimensional structure is accomplished spontaneously by a drift toward thermodynamic equilibrium.

Some functional proteins of living organisms are derived from protein precursors by a process in which part of the chain is eliminated. If the precursor was crosslinked by disulfide bonds, the product may consist of two or more chains held together by disulfide linkages. Examples of such proteins are α-chymotrypsin, derived from chymotrypsinogen (Matthews et al., 1967), and insulin, derived from proinsulin (Behrens and Grinnan, 1969). Figures III.32*b* and III.32*c* represent schematically the locations of the crosslinkages

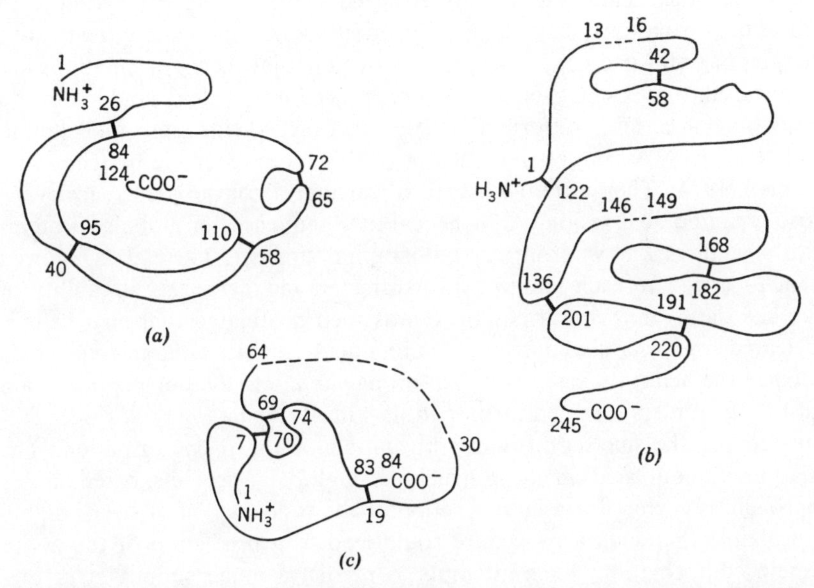

Fig. III.32. Schematic representation of the intramolecular crosslinking of globular proteins by S-S linkages in (*a*) ribonuclease, (*b*) α-chymotrypsin, and (*c*) insulin. The α-chymotrypsin and insulin are multichain proteins derived from single-chain precursors by proteolytic processes in which the dashed portions of the chain are eliminated.

and the portions of the chain eliminated in the conversion of the protein precursors. In such cases the folding of the chain apparently reflects the amino acid sequence of the precursor, so that the modified protein is in a metastable state. This has been established by showing that conditions favoring disulfide interchange rapidly inactivate both chymotrypsin and insulin (Givol et al., 1965).

Many factors make positive or negative contributions to the free energy of protein denaturation. The main driving force is the large increase in conformational entropy characterizing a transition from an ordered to a disordered conformation. Opposing this are short-range interactions of the hydrogen-bonding or hydrophobic bonding type. Coulombic interactions between ionic charges carried by the protein will, in general, make a smaller contribution, particularly in the vicinity of the isoelectric point. Since the stability of the native form depends on a relatively small difference between these opposing tendencies, a theoretical interpretation is extremely hazardous. Tanford (1970), who has discussed the problems in detail, reaches the candid and pessimistic conclusion that *"were it necessary to make a prediction in the absence of experimental knowledge, one would probably conclude that the native state should not exist."*

On the other hand, the success achieved in predicting the existence of helical conformations of synthetic polypeptides and in estimating the relative energies of right- and left-handed helices by potential energy calculations has encouraged some scientists to explore procedures which would allow a prediction of certain conformational festures of proteins from their amino acid sequences. An impressive attempt of this type was made by Chou and Fasman (1974). They carried out a statistical analysis of the occurrence of the various amino acid residues in α-helical or β sequences of globular proteins with structures known from crystallographic analysis. These data allowed them to assign to each amino acid parameters characterizing its ability to stabilize the α or β conformation. It was then postulated that an α-helix is nucleated whenever six consecutive amino acid residues contain four which stabilize the helix; the helix will then propagate along the polypeptide chain until it encounters a region incompatible with it. Analogous rules were formulated for the nucleation and propagation of the β conformation. The location of amino acids in α or β structures of globular proteins, based on this approach, is correct for a large fraction of the residues; yet, it seems highly improbable that such a procedure could lead to a prediction of the entire protein conformation. Even attempts to predict a much simpler structure, that is, that of the cyclic decapeptide gramicidin S, by potential energy calculations (Liquori et al., 1966; Vanderkooi et al., 1966), led to results which were found to be inconsistent with spectroscopic evidence (Stern et al., 1968). Moreover, a number of observations can be explained only on the assumption that the conformation of a section of the polypeptide chain in a protein is determined not only by the *local* amino acid sequence but also by interactions between residues which are distant along the contour of the chain, yet sterically close because of the folding of the polypeptide chain. Only such an effect can explain the existence of the high-energy *cis* peptide bond in carboxypeptidase A (Lipscomb et al., 1969) or the complete destruction of the native structure of a nuclease on removal of the last 23 of its 149 amino acid residues (Taniuchi and Anfinsen, 1969).

Even more striking is the observation that two nuclease fragments, both randomly coiled, associate to regenerate a structure with the optical properties of the native enzyme and with its catalytic activity (Taniuchi and Anfinsen, 1971). Prosthetic groups carried by a protein may also alter its conformation substantially (Breslow et al., 1965; Harrison and Blout, 1965; Bertland and Kalckar, 1968), and even a minor change, such as the oxidation of Fe(II) to Fe(III) in the heme group of cytochrome, results in a significant and functionally important modification of the protein structure (Takano et al., 1971).

TABLE III.1.
Protein Structures Determined at High Resolution

Protein	No. of Amino Acids	Resolution (A)	References
Myoglobin	153	2.0	Kendrew et al. (1961)
Lysozyme	129	2.0	Blake et al. (1967a)
Ribonuclease A	124	2.0	Kartha et al. (1967)
Ribonuclease S	20 + 104	3.5	Wyckoff et al. (1967)
α-Chymotrypsin	241	2.0	Sigler et al. (1968) Birktoft and Blow (1972)
Carboxypeptidase A	307	2.0	Lipscomb et al. (1969)
Subtilisin BPN'	275	2.5	Wright et al. (1969)
Papain	201	2.8	Drenth et al. (1968)
Oxyhemoglobin	$2 \times 141 + 2 \times 146$	2.8	Perutz et al. (1968)
Deoxyhemoglobin	$2 \times 141 + 2 \times 146$	3.5	Muirhead and Greer (1970)
Elastase	240	3.5	Shotton and Watson (1970)
Streptococcal nuclease	149	4.0	Arnone et al. (1969)
Lactate dehydrogenase	4×311	2.8	Adams et al. (1970)
Insulin	51	2.8	Adams et al. (1969)
Carbonic anhydrase C	258	2.0	Liljas et al. (1972)
Thermolysin	316	2.3	Matthews et al. (1972a)
Cytochrome *c*	104	2.8	Dickerson et al. (1971)
Cytochrome b_5	93	2.0	Matthews et al. (1972b)
Rubredoxin	54	1.5	Watenpaugh et al. (1972)
Concanavalin A	238	2.0	Edelman et al. (1972)
Alcohol dehydrogenase	2×374	2.9	Brändén et al. (1973)

Chapter IV

EQUILIBRIUM PROPERTIES OF DILUTE SOLUTIONS

A. THE COLLIGATIVE PROPERTIES

1. Raoult's Law as a Limiting Law

The concept of an ideal solution, in which the activities of the components are equal to their mole fractions, is useful to the physical chemist in two ways. On one hand, the ideal solution law is a reasonable first approximation to the behavior of real systems if they consist of species of similar molecular volume and if the heat of mixing is relatively small. Raoult's law [(II.17)] is then used as a frame of reference with which the properties of real systems are compared. This subject was discussed in detail in Chapter II. There is, however, another aspect of Raoult's law which is of far greater fundamental importance, namely, its quite general validity as a limiting law. We may compare the situation with the analogous case of the ideal gas law. Here we find also that the familiar expression $PV = \boldsymbol{R}T$ relating the pressure, volume, and absolute temperature of a gas is meaningful in two ways. Real gases at moderate pressures show relatively small deviations from that simple relation, and such deviations as are found in any given case may be interpreted in terms of the excluded volume of the gas molecules and the forces between them. However, *no matter how nonideal a gas may be*, the ideal gas law will always describe its behavior accurately in the limit of vanishingly small pressures.

Let us consider a solution so dilute that interactions between solute molecules are negligible. In that case the escaping tendencies of the solute molecules will be additive, and the activity of the solute will be proportional to its concentration (Henry's law). If we express concentration in terms of mole fractions, then

$$a_2 = k_H x_2 \tag{IV.1}$$

where the Henry's law constant k_H increases with decreasing solvent power of the medium. According to the Gibbs-Duhem relation, the changes in the activities of the components of a binary system are related by

$$x_1 d \ln a_1 + x_2 d \ln a_2 = 0 \tag{IV.2}$$

Combining (IV.1) with (IV.2) and using the condition $dx_2 = -dx_1$, we obtain

$$x_1 d \ln a_1 = dx_1 \tag{IV.3}$$

and since $a_1 = 1$ for $x_1 = 1$,

$$\lim_{x_1 \to 1} \left(\frac{da_1}{dx_1} \right) = 1 \tag{IV.4a}$$

$$a_1 = x_1 \tag{IV.4b}$$

We see then that Raoult's law will be valid *for the solvent* as long as Henry's law is applicable to the solute. The concentration range within which this will be the case will, of course, depend on the nature of the system. In solutions of nonelectrolytes, energetic interactions of solute molecules may be considered restricted to nearest neighbors. Then, if we have solutes of low molecular weight and if there is no pronounced clustering of the solute molecules, a system in which 1% of the volume is occupied by the solute should not deviate greatly from Henry's and Raoult's laws for the dilute and the concentrated component, respectively. For ionic solutions, in which the Coulombic forces of the charged species are exerted over much longer distances, correspondingly higher dilutions are required for approach to ideal solution behavior. Such high dilution will also be necessary to eliminate mutual interactions if the solute consists of flexible chain molecules, since the volume occupied by the molecular coil may exceed by a very large factor the volume of the "dry" molecule.

2. Relation of Solvent Activity to Measurable Quantities

In the concentration range in which Raoult's law is valid, the molecular weight M_2 of the solute in a binary system may be calculated from solvent activity by

$$a_1 = x_1 = \frac{w_1/M_1}{(w_1/M_1) + (w_2/M_2)} \tag{IV.5}$$

where w_1 and w_2 are the weights of solvent and solute. The molecular weight of the solute may, therefore, be evaluated if a convenient method is available for measuring changes of solvent activity in the concentration range in which Raoult's law is applicable. The methods which may be used for this purpose include measurements of vapor pressure, freezing point depression (cryoscopy), boiling point elevation (ebulliometry), and osmotic pressure. All of them yield, in the limit of highly dilute solutions, results which depend only on the number of solute particles per unit volume, and they are commonly referred to as the colligative properties of solutions. If the solutions contain solutes of different molecular weight, determination of the number of solute molecules at a known weight concentration leads to the number-average molecular weight $\bar{M}_n$ of the solute species.

The activity of a component of a solution is defined as the ratio of its fugacities in solution and in the standard state, respectivly. We saw in Chapter II that for a typical solvent the partial vapor pressure p_1 is a close approximation to the fugacity f_1, and since the partial vapor pressure of a solvent over a dilute solution will be very close to the vapor pressure p_1^0 of the pure solvent, we are well justified in assuming $p_1/p_1^0 = f_1/f_1^0$, so that

$$a_1 = \frac{p_1}{p_1^0} \tag{IV.6}$$

The relation between the freezing point T_m of a solution and the freezing point T_m^0 of the pure solvent may be derived simply by considering a cycle consisting of the following operations.

(a) A solution is cooled from T_m^0 to $T_m^0 - dT_m$, and δn_1 moles of the solvent are frozen out at that temperature. The solvent crystals are separated from the residual solution.

(b) The solution and the separated solvent crystals are reheated to T_m^0, and the latent heat of fusion ΔH_1^m is added per mole of solvent crystals to melt them.

(c) The molten solvent is returned to the solution in a reversible operation, yielding $-(\bar{G}_1 - \bar{G}_1^0)\,\delta n_1 = -\delta n_1\,\boldsymbol{R}T \ln a_1$ of work.

We may now relate the fraction of the heat added which was converted into work to the temperature interval of the cycle by the second law of thermodynamics:

$$\frac{-\boldsymbol{R}T\,d\ln a_1}{\Delta H_1^m} = \frac{-dT_m}{T_m^0} \tag{IV.7}$$

which leads, for dilute solutions, to the approximate result

$$-\ln a_1 = \frac{\Delta H_1^m}{\boldsymbol{R}(T_m^0)^2}(T_m^0 - T_m) \tag{IV.8}$$

These relations are valid only if no solid solutions are formed by the two components of the system. This is not a serious restriction, however, since the formation of solid solutions is a rather rare phenomenon, particularly among organic compounds.

Reasoning similar to that outlined above leads to the relation of the boiling point T_b of a solution to the boiling point T_b^0 of the pure solvent in terms of ΔH_1^v, the latent heat of vaporization of the solvent:

$$- \ln a_1 = \frac{\Delta H_1^v}{R(T_b^0)^2}(T_b - T_b^0) \qquad \text{(IV.9)}$$

provided that the volatility of the solute is negligibly small.

The osmotic pressure Π is defined as the pressure which has to be applied to a solution so as to raise the partial molar free energy of the solvent to the standard state value. Thus,

$$\bar{G}_1^0 = \bar{G}_1 + \int_0^{\Pi} \left(\frac{\partial \bar{G}_1}{\partial P}\right)_{T,x_1} dP \qquad \text{(IV.10)}$$

The variation of $(\partial \bar{G}_1/\partial P)_{T,x_1} = \bar{V}_1$ with pressure may be neglected, so that

$$\bar{G}_1 - \bar{G}_1^0 = RT \ln a_1 = -\Pi \bar{V}_1 \qquad \text{(IV.11)}$$

Since all the colligative properties essentially represent methods for measuring solvent activity, their mutual relationship is independent of the deviations from solution ideality. The choice between them is governed merely by experimental convenience.

3. The Scope of the Methods

It is instructive to compare the relative magnitudes of effects produced by the colligative properties on a typical example, that is, a benzene solution containing 1% by weight of a solute with a molecular weight of 10,000. (We shall assume that the solution behaves ideally.) The mole fraction of the solute is only 7.8×10^{-5}, and since the vapor pressure of benzene is 100 torr at 26.1 °C, the vapor pressure of the solution will be 99,9922 torr. Using the proper values for the latent heat of fusion of benzene at 5.5°C and the latent heat of vaporization at 80.2°C, we obtain a freezing point depression of 0.0051 °C and a boiling point elevation of 0.0025 °C. The osmotic pressure at 26 °C will be 25,000 dynes/cm^2. Let us then consider the relative convenience for determining such quantities experimentally.

At first sight it would appear that the extremely slight change in the vapor pressure precludes any possibility of its experimental determination. Nevertheless, an ingenious technique has been developed which allows the estimation of molecular weights up to several thousand by a method utilizing the difference in the vapor pressures of the solution and the pure solvent. In this procedure, drops of solution and solvent are held in a space filled with saturated solvent vapor. As the solvent condenses on the drop of solution, the drop is being warmed by liberation of the latent heat until a steady state is reached in which this heating is balanced by heat losses through conduction. With the use of thermistors to measure the difference in temperature of the solvent and solution drops, temperature differences of the order of 10^{-4} °C may be determined with good reproducibility. If conduction losses are negligible, the temperature difference is that predicted by the theory of boiling point elevation. Higuchi et al., (1959) have shown that it is possible to approach quite closely to this situation, and M_n values up to 20,000 have been measured by this technique (Ohama and Ozawa, 1966). Alternatively, the manometric method has been refined to the point where molecular weight determinations in the same range become possible (Tremblay et al., 1958).

The cryoscopic and ebullioscopic methods have been greatly improved in recent years by the use of thermistors for the accurate measurement of small differences of temperature. The sensitivities of the two techniques are similar and definitely exceed that of the method described above. Outstanding examples of high-precision cryoscopy are the characterizations of polystyrene (Schulz and Marzolph, 1954), and of polyethylene (Newitt and Kokle, 1966) with $\bar{M}_n$ up to 40,000. The ebullioscopic technique has been used successfully for the determination of molecular weights up to 30,000 (Glover and Stanley, 1961). The sensitivity of this method is limited not so much by the smallness of the differences of temperature which have to be measured as by various experimental difficulties, such as the foaming of boiling polymer

TABLE IV.1.
Characteristics of Cryoscopic and Ebullioscopic Solvents

Solvent	T_m^0 (°C)	T_b^0 (°C)	$(T_m^0 - T_m)/m_2$[a]	$(T_b - T_b^0)/m_2$[a]
Acetic acid	16.55		3.7	
Benzene	5.53	80.2	4.86	2.54
Butanone		79.6		2.56
Camphor	178.4		46.9	
Carbon tetrachloride		76.5		5.02
Chlorobenzene		132.0		4.06
Cyclohexane	7.09	80.9	14.8	2.79
p-Dichlorobenzene	53.2		6.98	
1,2-Dichloroethane		83.5		1.90
Dioxane	11.78		4.63	
Diphenylamine	53.2		7.68	
Ethanol		78.3		1.19
Ethyl acetate		77.1		2.68
Ethyl ether		34.6		2.10
n-Heptane		98.4		3.43
Naphthalene	80.27		7.0	
Nitrobenzene	5.65		6.8	
Phenol	40.8		7.27	
Tetrahydrofuran		63.2		2.50
Toluene		110.7		3.33
Water	0.0	100.0	1.88	0.51

[a]The symbol m_2 denotes the molality of the solute.

solutions, introducing uncertainties which do not lend themselves to a theoretical analysis. Table IV.1 lists the characteristics of some cryoscopic and ebullisocopic solvents (Bonner et al., 1958).

The determination of osmotic pressure is by far the most sensitive of the techniques by which solvent activity deviating only slightly from unity may be measured. In the usual technique a sample of the solution is separated from the pure solvent by a semipermeable membrane which will allow solvent molecules to diffuse freely but will not permit passage of the solute. Solvent will then pass into the solution until its rising level provides a hydrostatic pressure equal to Π, so that the partial molar free energy of the solvent in the solution becomes equal to that of the pure solvent. If the difference in the levels of the solvent and solution menisci, h, is expressed in centimeters, ρ is the density of the solution, and $\boldsymbol{g}$ is the gravitational acceleration of the earth (in cm/sec^2), then the osmotic pressure (in dynes/cm^2) is given by $\Pi = h\rho\boldsymbol{g}$, so that the osmotic pressure in the example given at the outset of this section corresponds to a hydrostatic head of 28 cm. Thus, a determination of solvent activity in a case which was beyond the precision attainable by measurements of vapor pressure depression and could be achieved only with great care by the use of cryoscopy or ebulliometry is very far from the limits of sensitivity attainable in osmometry. In fact, osmotic pressure measurements may be carried out with solutes whose molecular weights range up to 500,000 or even higher. The most difficult problem in the use of this technique is the selection of a suitable semipermeable membrane. This problem is less difficult in work with monodisperse natural polymers but becomes serious with synthetic polymers containing continuous distributions of molecular weights. Unless the lowest molecular weight species are removed by careful fractionation, the number-average molecular weight obtained by osmometry may reflect as much the properties of the osmotic membrane as the nature of the specimen. An extensive discussion of osmometric techniques is contained in the monograph on number-average molecular weight determination by Bonner et al. (1958). These authors consider 15,000 to be the practical lower limit of molecular weights for which the osmotic pressure method is convenient.

4. Deviations from Raoult's Law in Dilute Solutions

In discussing the nature of the ideal solution law as a limiting law (Section IV.A.1), we showed that the validity of Raoult's law for the solvent was assured if the solute particles, subject to nearest-neighbor interactions only, were not in contact with each other. This condition can be satisfied, to any degree of approximation required, by a sufficiently high dilution of the solution. As the solution becomes more concentrated, we should expect at first the appearance of binary interactions between the solute particles, and since the number of binary contacts should be proportional to c_2^2, the square of the solute concentration, the measurable intensive properties should also deviate by a term proportional to c_2^2 from the values expected for a solution containing only isolated solute molecules. At still higher concentrations, ternary interactions with a probability c_2^3 or even higher-order interactions may contribute to the properties of the system, but the interpretation of such effects is much less certain and theoretical developments are usually restricted to the interpretation of effects from binary solute interactions.

At dilutions such that Raoult's law is valid, $\ln a_1 \approx \ln (1 - x_2) \approx - x_2$ and $\bar{V}_1 \approx V_1$, so that (IV.11) becomes

$$\Pi = \left(\frac{\boldsymbol{RT}}{\bar{V}_1}\right)x_2 \tag{IV.12}$$

Since $x_2/V_1 = c_2/M_2$ (where the solute concentration c_2 is expressed in grams per milliliter), we obtain

$$\Pi = \left(\frac{\boldsymbol{RT}}{M_2}\right)c_2 \tag{IV.13}$$

At higher concentrations, where binary and higher-order interactions of solute particles have to be taken into account,

$$\Pi = \boldsymbol{RT}\left(\frac{c_2}{M_2} + \mathrm{A}_2 c_2^2 + \mathrm{A}_3 c_2^3 + \cdots\right) \tag{IV.14}$$

where A_2, A_3, etc., are the second, third, and higher osmotic virial coefficients, which will depend on the shape of the solute molecules and the forces operating between them. Analogous expressions will be obtained if we use another of the colligative properties as our experimental tool; for example, in interpreting cryoscopic data [cf. (V.8)],

$$T_m^0 - T_m = \frac{\boldsymbol{R}(T_m^0)^2 V_1}{\Delta H_1^m}\left(\frac{c_2}{M_2} + \mathrm{A}_2 c_2^2 + \mathrm{A}_3 c_2^3 + \cdots\right) \tag{IV.15}$$

where the coefficients A_2, A_3, etc., have the same values as in (IV.14). For the osmotic pressure of a solution in which solute molecules may be represented as rigid spheres much larger than the molecules of the solvent, we obtain from (II.48), after substituting $c_2 V_2/M_2$ for ϕ_2,

$$\Pi = \boldsymbol{RT}\left(\frac{c_2}{M_2} + \frac{4V_2}{M_2^2}c_2^2 + \cdots\right) \tag{IV.16}$$

Here the term $4V_2 = U_2$ represents the volume excluded, because of binary interactions, per mole of solute. It arises from the fact that the centers of two rigid spheres cannot be placed closer to each other than at a distance twice as long as their radius. Thus, any sphere excludes a volume eight times as large as that which it physically occupies. Since, however, the excluded volume due to the interaction of any pair of particles is counted twice, the volume excluded by all the particles in the system is only four times as large as this physical volume.

The relation between the second virial coefficient and the volume excluded due to binary interactions

$$A_2 = \frac{U_2}{M_2^2} \tag{IV.17}$$

is perfectly general, but the calculation of U_2 becomes much more difficult for cases other than rigid spheres. For instance, in the case of rigid ellipsoids of revolution, only certain mutual orientations are allowed if the centers of two neighboring particles are separated by a distance lying between that of the length of their major and minor axis, as illustrated in Fig. IV.1 The molar excluded volume is then given by

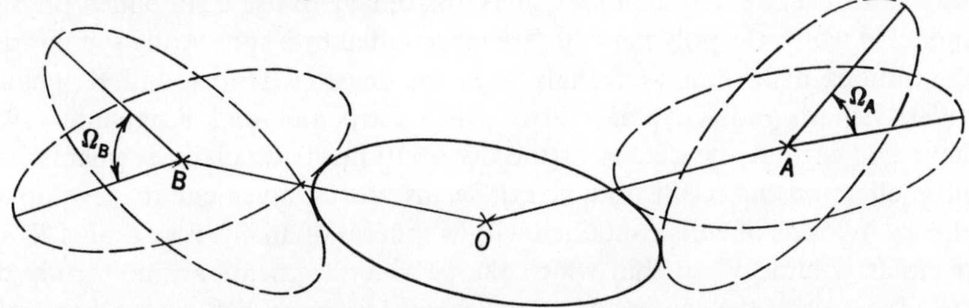

Fig. IV.1. Two-dimensional representation of the excluded volume in a system containing ellipsoidal particles. Note that for ellipses centered at A and B, at an equal distance from O, the center of a reference ellipse, the excluded angles Ω are different.

$$U_2 = \frac{N}{2}\left(\frac{1}{4\pi}\right)\iiint \Omega(x, y, z)\, dx\, dy\, dz \tag{IV.18}$$

where $\Omega(x, y, z)$ is the solid angle excluded by the presence of one ellipsoid for the orientation of the major axis of a second ellipsoid whose center of gravity is placed at x, y, z. It is intuitively reasonable that the ratio of the excluded volume to the physical volume of the ellipsoid should increase with the ratio p of the semiaxis of revolution to the equatorial radius, and Isihara (1950a) has calculated U_2/V_2 as a function of p. For very elongated rodlike particles, Zimm (1946) obtained

$$\frac{U_2}{V_2} = \mathrm{p} \qquad (\mathrm{p} >> 1) \tag{IV.19}$$

If the cross section of the rod is constant, $U_2 = \mathrm{p}V_2$ is proportional to M_2^2 so that (IV.17) gives a second virial coefficient independent of M_2. Such behavior has been observed with solutions of poly(n-butyl isocyanate) (Fetters and Yu, 1966), in which the polymer molecules retain a helical conformation (cf. Section III. C.4).

A theory of the second virial coefficient for solutions containing flexible chain molecules presents difficulties of a different order of magnitude. The region of space occupied by such macromolecules is not clearly defined, and it is possible for the molecular coils to interpenetrate one another. Such interpenetration will be resisted because of the restriction which it imposes on the number of possible chain conformations, but this factor may be opposed by a mutual attraction of the chain segments in poor solvent media, or it may be magnified by a mutual repulsion of the polymer segments if polymer—solvent interactions are favored. We may expect the polymer coils to become mutually impenetrable in the limit of very large chain length, and this limit should be reached more rapidly as the mutual repulsion of the solute molecules is increased by a favorable solvent medium.

Since the mathematical treatment of the problem in terms of the molecular chains is extremely complex, it is customary to use a simplified physical model in which the polymer coils are represented by a spherically symmetrical continuous distribution of chain segment densities (Flory and Krigbaum, 1950; Isihara and Koyama, 1956). Since Carpenter and Krigbaum (1958) have shown that the details of the distribution of the chain segments have little effect on the result obtained, it seems most convenient to develop the theory in terms of coils contained within spheres of an effective radius R_e and a molar volume V_e, within which the polymer segments are uniformly distributed. When the centers of two such spheres approach each other within a distance $2R_e y$, where $y < 1$, they will interpenetrate so as to occupy jointly a volume $(4/3)\pi R_e^3\,(y^3 - 3y + 2)/2$. If such an interpenetration is associated with an excess free energy $G^E(y)$, the excluded volume will be given by the "binary cluster integral"

$$U_2 = \frac{N}{2}\int_{y=0}^{y=1}\left\{1 - \exp\left[\frac{-G^E(y)}{kT}\right]\right\}4\pi(2R_e y)^2\,d(2R_e y)$$

$$= 4V_e\left\{1 - \int_{y=0}^{y=1}\exp\left[\frac{-G^E(y)}{kT}\right]3y^2\,dy\right\} \qquad \text{(IV.20)}$$

To evaluate the integral in (IV.20), we have to make an assumption about the form of G^E. Flory and Krigbaum (1950) suggested that it could be approximated by the Flory-Huggins theory neglecting all terms beyond that which represents binary interactions of chain segments. This leads to an excess free energy of $2kT(\frac{1}{2} - \chi)\phi_{21}^2/V_1$ per unit volume jointly occupied by the two interpenetrating spheres, where $\phi_{21} = V_2/V_e$ is the local volume fraction of polymer within the equivalent sphere. When this value is used in the estimate of $G^E(y)$, (IV.20) becomes

$$U_2 \quad = 4V_e - [1 - g(X)]$$

$$X = \frac{(\frac{1}{2}-\chi)\, V_2^2}{V_e V_1}$$

$$g(X) = \int_{y=0}^{y=1} \exp\left[-X(y^3 - 3y + 2)\right] 3y^2\, dy \qquad \text{(IV.21)}$$

The function $g(X)$ (plotted in Fig. IV.2) decays gradually with increasing X, tending to 0 as $X \to \infty$. The excluded volume to be expected in the limit of very large values of the dimensionless parameter X represents the behavior of

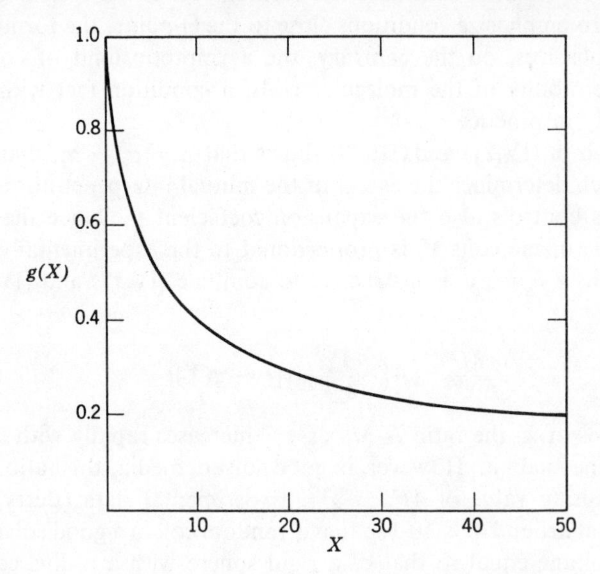

Fig. IV.2. The function $g(X)$ describing the mutual interpenetration of two polymer coils.

the macromolecules when the interpenetration of the coils has become negligible and is, therefore, independent of the form assumed for G^E. When $(\frac{1}{2} - \chi) = 0$, that is, in a Θ-solvent in which the mutual attraction of chain segments is just sufficient to cancel the effects of the reduction in the number of chain conformations resulting from the interpenetration of molecular coils, $X = 0$ and $g(X) = 1$, so that the excluded volume and the second virial coefficient vanish for chains of any length. The balance of factors which favor and oppose the mutual penetration of different sections of any one coil was shown in section III.B.5 to eliminate chain expansion due to the *intramolecular* excluded volume effect; the same conditions, in eliminating the effect of the excluded volume in *intermolecular* interactions, should assure the validity of Raoult's law in the concentration range in which only binary solute interaction need be considered. With $X \ll 1$ but finite (i.e., in a solvent medium approaching closely to Θ-conditions) the exponential term in (IV.21) may be replaced by $1 - X(y^3 - 3y + 2)$, leading to a molar excluded volume

$$\lim_{X \to 0} U_2 = \frac{(\frac{1}{2} - \chi)\, V_2^2}{V_1} \tag{IV.22}$$

which is identical with the results obtained in theories that use models with variable coil segment densities. The formulation of the results in these theories tends to emphasize conditions close to the Θ-point; the formulation in (IV.21) emphasizes, on the contrary, the asymptotic limit of complete mutual impenetrability of the molecular coils, a condition met with much more frequently in practice.

A comparison of (IV.21) and (III.29) shows that $X = \alpha_e^5 - \alpha_e^3$, that is, the parameter which determines the extent of the mutual interpenetration of the molecular coils controls also the expansion coefficient α_e. Since the molar effective volume of the coils V_e is proportional to the experimentally determinable quantity $\langle s^2 \rangle^{3/2}$, it is instructive to combine (IV.17) and (IV.21) in the form

$$\frac{A_2 M_2^2}{\langle s^2 \rangle^{3/2}} = 4\left(\frac{V_e}{\langle s^2 \rangle^{3/2}}\right)[1 - g(X)] \tag{IV.23}$$

For small values of X, the ratio $A_2 M_2^2/\langle s^2 \rangle^{3/2}$ increases rapidly with the solvent power of the medium. However, in good solvent media, this ratio should approach a limiting value of $4V_e/\langle s^2 \rangle^{3/2}$. Experimental data (Berry, 1966) place this limit at around 4×10^{24} so that a random coil in a good solvent has an excluded volume equal to that of a rigid sphere with a radius equal to $0.73 \langle s^2 \rangle^{1/2}$ of the coil. Since $\langle s^2 \rangle$ is proportional to $M_2 \alpha_e^2$ and (III.30) predicts that α_e is proportional to $M_2^{0.1}$ for very long chains in good solvent media, we should expect A_2 to decrease, under these conditions, as $M_2^{-0.2}$. This is in close agreement with observed behavior.

An analysis of the second virial coefficient in terms of the dimensionless parameter $\mathbf{z}$ [as defined in (III.33)] yields (Zimm, 1946)

$$A_2 = 4\pi^{3/2} N \left(\frac{\langle s_0^2 \rangle}{M_2}\right)^{3/2} M^{-1/2}\, \mathbf{z} F'(\mathbf{z})$$

$$F'(\mathbf{z}) = 1 - b_1 \mathbf{z} + b_2 \mathbf{z}^2 - \cdots \tag{IV.24}$$

Unfortunately, $F'(\mathbf{z})$ converges so slowly that only values for very small $\mathbf{z}$ may be estimated from theory. However, in the neighborhood of the Θ-temperature $\mathbf{z}$ should be linear in $1 - \Theta/T$, and the value of $[\partial A_2/\partial(1/T)]_{T=\Theta}$ may then be interpreted in terms of (IV.24) to estimate $\mathbf{z}$ as a function of temperature and to obtain β_0 as defined on p. 139. A plot of $A_2 M_2^{1/2}$ against $\mathbf{z}$, such as that based on Berry's (1966) data for polystyrene solutions (Fig.

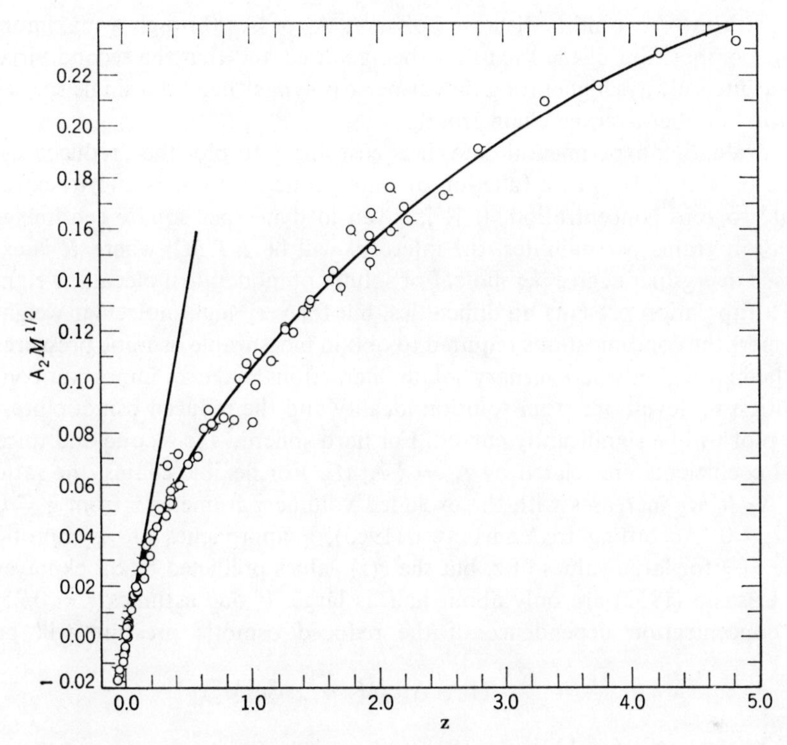

Fig. IV.3. $A_2M^{1/2}$ as a function of **z** for polystyrene in decalin; the pips distinguish samples of different molecular weights.

IV.3), may then be used to obtain $F'(\mathbf{z}) = A_2M_2^{1/2}(\mathbf{z})/\mathbf{z}\ [d\ A_2M_2^{1/2}(\mathbf{z})/d\mathbf{z}]_{\mathbf{z}=0}$.

Quite generally, the Gibbs-Helmoholtz equation (II.12) in conjunction with (IV.11) and (IV.14) yields

$$RV_1\left(\frac{\partial A_2}{\partial T} c_2^2 + \frac{\partial A_3}{\partial T} c_2^3 + \cdots\right) = \frac{\overline{\Delta H_1}}{T^2} \qquad \text{(IV.25)}$$

so that the temperature coefficient of A_2 may be used to estimate the heat of dilution in the low concentration range. Qualitatively, in a system in which the mixing of polymer and solvent is endothermic an increase in temperature will favor solvation and tend to expand the polymer coil; this in turn will increase the excluded volume and lead to a higher value of A_2.

Theoretical analysis of the second virial coefficient in systems containing polydisperse polymers is much more difficult (Flory and Krigbaum, 1950; Yamakawa and Kurata, 1960; Casassa, 1962, 1966). Of particular interest is the analysis for the case of polymers in very good solvents, where the coiled chains may be treated as impermeable spheres. Casassa (1962) concluded that for mixtures of two monodisperse polymers A_2 passes through a maximum as the composition of the mixture is being varied and that the second virial coefficient is always higher for a polydisperse polymer than for a single species with its number-average chain length.

In evaluating experimental data, it is customary to plot the "reduced osmotic pressure" Π/c_2 as a function of solute concentration c_2 and to extrapolate to zero concentration. If Π is given in dynes per square centimeter and c_2 in grams per milliliter, the intercept will be $\boldsymbol{RT}/M_2$, where $\boldsymbol{R}$ is expressed in ergs per degree per mole. For solutes of moderate molecular weight the extrapolation presents no difficulties, but for very high molecular weight polymers the concentrations required to obtain measurable osmotic pressures will lie in a region where ternary solute interactions make an important contribution to deviations from solution ideality and the reduced osmotic pressure plot will be significantly curved. For hard spheres, the second and third virial coefficients are related by $A_3 = \frac{5}{8} A_2^2 M_2$. For flexible chains, the ratio $g = A_3/A_2^2 M_2$ increases with the excluded volume parameter $\mathbf{z}$ from $g = 0$ for $\mathbf{z} = 0$. According to Yamakawa (1965), g approaches an asymptotic value of $\frac{3}{4}$ for large values of $\mathbf{z}$, but the $g(\mathbf{z})$ values predicted by Stockmayer and Casassa (1952) are only about half as large. If one assumes $g = 0.25$, the concentration dependence of the reduced osmotic pressure will be

$$\frac{\Pi}{c_2} = \frac{\boldsymbol{RT}}{M_2}(1 + A_2 c_2 M_2 + \tfrac{1}{4}A_2^2 c_2^2 M_2^2)$$

$$\left(\frac{\Pi}{c_2}\right)^{1/2} = \left(\frac{\Pi}{c_2}\right)_{c_2\to 0}^{1/2}(1 + \tfrac{1}{2}A_2 c_2 M_2) \qquad \text{(IV.26)}$$

According to (IV.26), plots of $(\Pi/c_2)^{1/2}$ against c_2 should be linear and they give generally better extrapolations than plots of Π/c_2 against c_2 (Krigbaum, 1954).

5. Osmometry with Semipermeable and with Partially Selective Membranes

The determination of osmotic pressures requires the use of a membrane which is permeable to the solvent but will not permit the passage of solute molecules. A typical experimental arrangement is indicated schematically in in Fig. IV.4. Here the membrane separates the test solution from a chamber containing pure solvent. The chemical potential gradient causes the solvent to pass into the solution, causing its level to rise in a capillary until the hydrostatic pressure equals the osmotic pressure Π—at this point, as we have seen, the activity of the solvent in the solution will be equal to its activity in the standard state, and the driving force toward solvent transfer will disappear. We should then expect the hydrostatic head to approach with time exponentially toward its equilibrium value.

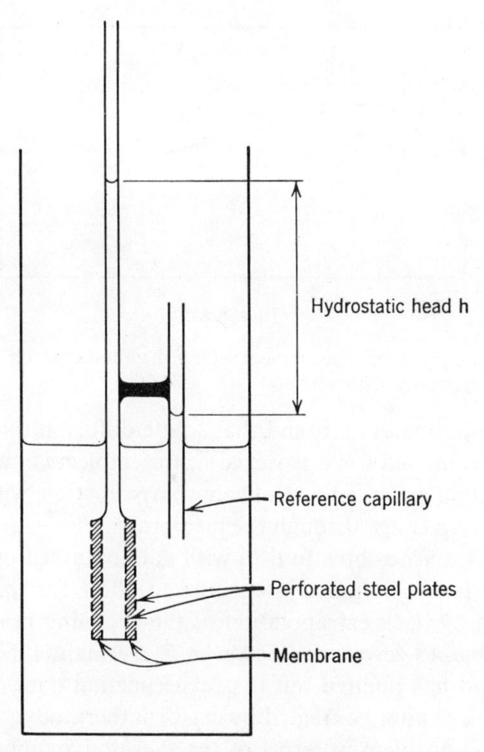

Fig. IV.4. Schematic representation of an osmometer.

An important advance in osmometry was achieved with the development of "automatic" osmometers, such as the one described by Rolfson and Coll (1964). In this instrument, any increase in the volume of the solution due to the flow of solvent through the osmotic membrane deflects a metal diaphragm. This actuates a servomechanism by which pressure is applied to reverse the solvent flow until the original level of the solution is restored. Thus, the pressure exerted on the solution builds up rapidly (within a few minutes) to the osmotic pressure without any need for appreciable volumes of fluid to pass through the membrane.

In practice, the ideal of perfect semipermeability is often difficult to attain, particularly if the sample has not been subjected to adequate fractionation, so that it contains a significant fraction of low molecular weight material, or if an attempt is made to extend osmotic pressure measurements to samples of low number-average molecular weight. Typical data illustrating this problem (Fox et al., 1962) are shown in Fig. IV.5. We see that for an unfractionated sample with $\bar{M}_n = 34{,}000$ the hydrostatic head rises to a maximum and then slowly declines because of a diffusion of the solute through the

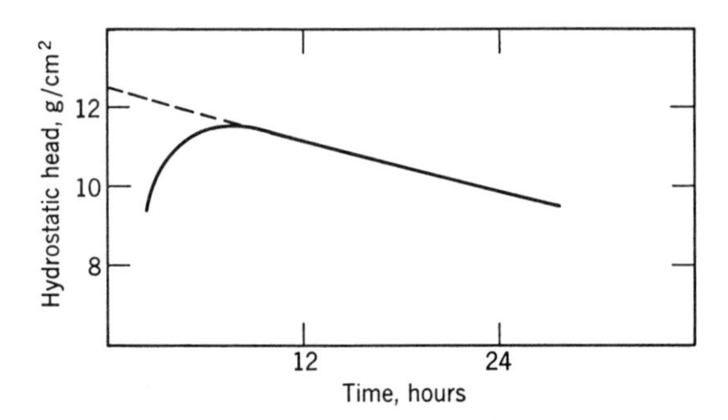

Fig. IV.5. Osmotic pressure measurement with a membrane which is partially permeable to the solute, poly(methyl methacrylate), $\bar{M}_w = 34{,}000$.

osmotic membrane. Fox et al. found that solute diffusion through the membrane could be eliminated for a material in this molecular weight range by careful fractionation, but even a relatively sharp fraction with $\bar{M}_n = 15{,}700$ showed evidence of passage through the membrane.

The question then arises how to deal with data obtained under these nonideal conditions. It is a common procedure to allow for the loss of solute through diffusion by back extrapolation of the declining branch of the plot of the pressure head to zero time. However, Staverman (1952) has criticized this procedure and has pointed out that experimental data obtained with a "leaky" membrane cannot be treated by classical thermodynamics. Theoretical analysis of the problem in terms of the thermodynamics of irreversible processes shows that the significant experimental quantity is the apparent osmotic pressure in the steady state when the volume flow due to the chemical potential gradient is counterbalanced by flow due to hydrostatic pressure. This steady-state pressure Π_{ss} is related to the osmotic pressure representing thermodynamic equilibrium Π_{eq} by

$$\frac{\Pi_{ss}}{\Pi_{eq}} = \mathscr{S}[1 - (1 - \mathscr{S})\phi_2 + \cdots] \qquad \text{(IV.27)}$$

where $\mathscr{S}$ is a "selectivity coefficient" characteristic of the membrane. This coefficient is defined so that $(1 - \mathscr{S})$ is the ratio of the concentration of the solution flowing through the membrane under the influence of a hydrostatic pressure to the concentration of the solution outside the membrane when the solutions on both sides of the membrane are of identical concentration. Thus, $\mathscr{S} = 1$ corresponds to an ideal semipermeable membrane, while $\mathscr{S} = 0$ represents a membrane with no selectivity between solvent and solute so that no osmotic effects would be observed. Talen and Staverman (1965) confirmed the theoretical analysis by an osmotic determination of the molecular weight of sucrose, using a membrane with a selectivity coefficient as low as 0.20.

Using relation (IV.27), they obtained a result within 7% of the correct value. For polymer samples the selectivity coefficient should fall off sharply with decreasing molecular weight below a certain range determined by the nature of the osmotic membrane and the extent of its swelling. For any given number average molecular weight the average selectivity coefficient will, therefore, tend to decrease with an increasing breadth of the molecular weight distribution (Elias and Männer, 1960). An application of the theory of partially selective osmotic membranes was reported by Gardon and Mason (1957), who measured osmotic pressures of solutions of lignin sulfonates in the molecular weight range of 5000. They estimated $\mathcal{S}$ by comparing the ratio of the diffusion rates of polymer and a low molecular weight solute through a sintered glass barrier (which would be expected to have $\mathcal{S} = 0$) and the membrane used in osmotic measurements. Proper treatment of the data led to essentially identical molecular weights for membranes with selectivity coefficients ranging from 0.95 to as low as 0.75. Strictly speaking, the Staverman treatment is applicable only for osmometers of very fast response. Using an automatic osmometer in which the steady state is attained within a few minutes, Coll (1967) was able to measure selectivity coefficients even for solutes with molecules no larger than those of the solvent, and he characterized a number of membranes by the dependence of the selectivity coefficient on the molecular weight of the solute.

6. Surfaces of Macromolecular Solutions

The transport of a section of a polymer chain from the bulk of a solution to an interface leads to a severe restriction of conformational freedom and a corresponding loss of entropy. Thus, polymer adsorptions will be appreciable only if the increase in conformational free energy is more than compensated by a decrease in the internal energy of the chain segments. The changes of conformational entropy and of internal energy are both proportional to the length of the chain; thus, as we increase the energetic preference of a chain segment for the interface, the transition from negligible to pronounced adsorption should occur very sharply with long chain molecules. On the other hand, if the solvent power of the medium is being continuously increased, the point at which a polymer is desorbed from a given surface should depend almost exclusively on its chemical composition and very little on its molecular weight. As a result, chromatography is most suitable for characterization of the compositional distribution of copolymers, particularly if the constituent monomer residues have very different polarities (Inagaki et al., 1968).

The theoretical and experimental work bearing on the principles governing the adsorption of flexible chain molecules at interfaces has been reviewed by Silberberg (1971), and a review dealing with the adsorption of biopolymers was published by Miller and Bach (1973). The amount of polymer adsorbed at saturation has been observed to be rather insensitive to the length of the chains. On the other hand, optical and hydrodynamic data show that the adsorbed chains extend for distances of several hundred angstroms into the bulk of the solution, that the thickness of this adsorbed layer increases with chain length and is sensitive to the polymer concentration in the bulk of the solution, and that only a fraction of the chain is in intimate contact with the interface (Stromberg et al., 1965; Rowland and Eirich, 1966; Peyser et al., 1967).

A statistical-mechanical theory of the adsorption of chain molecules at interfaces (Silberberg, 1968) uses a lattice model in which sections of polymer chains lying in the surface layer are connected by loops extending into the underlying solution. To simplify the calculation, it is assumed that the region containing these loops may be treated as a solution of uniform concentration. The calculations reproduce many of the characteristic features which have been experimentally observed, such as the dependence of the length of the loops on polymer chain length and concentration, and the fraction of the adsorbed polymer chain lying in the surface layer under various conditions.

Experimental studies bearing on the behavior of polymers at an interface have been generally concerned with the adsorption on solids. Information on the adsorption of polymers at solution-gas interfaces may be obtained from surface tension measurements. The surface energy of a solution consisting of weakly interacting components should be linear in the "surface fraction" of these components in the surface layer. In the case of chain molecules and a low molecular weight solute, this may be approximated by the volume fraction. Since the component with the lower surface energy will be concentrated in the surface layer, we should expect

$$\Delta E_s \equiv E_{s1}\phi_1 + E_{s2}\phi_2 - E_s > 0 \tag{IV.28}$$

where E_{s1} and E_{s2} are the surface energies of the solvent and the polymer, respectively, while E_s is that of the solution. Gaines (1969) measured surface tensions of polyisobutene solutions in tetralin ($E_{s1} > E_{s2}$) and in n-heptane ($E_{s1} < E_{s2}$) and confirmed inequality (IV.28) for both systems. The data can be interpreted by the Gibbs adsorption isotherm:

$$\Gamma_2 = -\left(\frac{1}{\boldsymbol{RT}}\right)\left(\frac{dE_s}{d\ln a_2}\right) \tag{IV.29}$$

where Γ_2 is the number of moles of solute adsorbed per unit area of the interface. Here a plane dividing the "surface layer" from the bulk of the solution is placed so that the excess surface concentration vanishes for the solvent (Adamson, 1960). As would be expected, the polymer is excluded from the surface (i.e., $\Gamma_2 < 0$) if its surface energy is larger than that of the solvent and the absolute value of Γ_2, positive or negative, increases with the chain length of the polymer.

In favorable cases it is possible to deposit a given molecular species in the form of a monomolecular layer on the surface of water or an aqueous solution or on an oil-water interface. The establishment of such layers requires the deposited molecules to contain polar groups to be "anchored" in the aqueous phase (hypophase), yet the solubility of these molecules in the hypophase must be negligible. If the surface coverage is small, the molecules in the surface layer will behave like a two-dimensional gas, exerting a pressure on any barrier that restricts the area over which they can spread. It can then be shown, by considerations analogous to those of the kinetic gas theory, that in the limit of low surface pressure P_S the equation of state is

$$\lim_{P_s \to 0} P_S A = n\boldsymbol{RT} \tag{IV.30}$$

where n moles of solute are spread over an area A. It is necessary to realize that complete surface coverage by a unimolecular layer represents a very small amount of material of the order of 10^{-7} g/cm^2. The validity of the ideal law (IV.30) would then require the use of extremely small samples and the measurement od correspondingly small pressures. For instance, with a surface concentration of 10^{-9}g/cm^2 of a material with a molecular weight of 10^4, the surface pressure should be about 0.0025 dyne/cm.

By analogy with the van der Waals equation of state for nonideal gases, (IV.30) may be modified by subtracting from the total area the "excluded area" due to the physical size of the adsorbed molecules. If we then define the specific area A_S as the surface area per unit weight of adsorbed substance and A_S^0 as the corresponding excluded specific area, (IV.30) must be modified to

$$P_S(A_S - A_S^0) = \frac{\boldsymbol{RT}}{M_2} \tag{IV.31}$$

so that A_S^0 is obtained as the slope of a plot of $P_S A_S$ against P_S.

The application of surface pressure measurements to the study of macromolecules has been reviewed by Crisp (1958) and by Beredjick (1963). The technique is difficult, particularly since the most meaningful data are obtained at very low sample concentrations, where the problem of excluding adventitious impurities becomes quite serious. The very small surface pressures require extremely careful experimental design. In addition, the technique for depositing the surface layer in a manner such that the individual molecules are properly spaced on the hypophase may be critical. In the case of globular proteins, surface pressures lead to molecular weights which are in very good agreement with results obtained by osmometry (Bull, 1947; Cheesman and Davies, 1954). With hemoglobin, the apparent molecular weight depends markedly on the nature of the hypophase, indicating dissociation into subunits. Insulin, on the other hand, appears to be associated in the surface layer above pH 3. Both these effects are, however, consistent with the behavior of these proteins in solution (cf. Section VIII.B.2) and do not, therefore, represent artifacts. Spreading of proteins on surfaces may also lead to their denaturation, and this will produce characteristic changes in the dependence of the surface pressure on the area (Evans et al., 1970).

With synthetic polymers, the large excluded volume of the chain molecules spread on a surface renders it impossible to make measurements at sufficient dilution to avoid extensive chain interpenetration. In the experimentally accessible region, results are independent of molecular weight (Hotta, 1954). Surface pressure measurements are, therefore, useful only in characterizing intermolecular interactions of the macromolecules (Schick, 1957). Beredjick and Ries (1967) have reported that the pressure-area isotherm is strikingly different for isotactic and syndiotactic poly (methyl methacrylate), and this result is of special interest since the solution properties of these two polymers are very similar. In the case of polyelectrolytes, the surface pressure reflects the very large repulsion between the polyions (Davies and Llopis, 1954).

7. Distribution of Macromolecules between a Solution and a Gel Phase

Gel permeation chromatography has become a powerful tool both for the study of molecular weight distribution in samples of synthetic polymers and in the analysis of solutions containing discrete species of proteins (Ackers, 1970). Usually the technique utilizes columns packed with swollen cross-linked polymer beads, and since an increasing size of the macromolecules tends to reduce the gel-solution distribution coefficient, the time of passage through the column becomes shorter as the macromolecules become larger. Benoit et al. (1966) used for the characterization of the size of the polymer chains the product of the intrinsic viscosity $[\eta]$ and the molecular weight M_2, which, as we shall see (p. 326), is proportional to the volume of a hydrodynamically equivalent sphere. Use of porous glass beads for the packing of the chromatographic column has the advantage that the pore size is independent of the nature of the solvent, and a single calibration of the time of passage as a function of $[\eta]M_2$ is valid for all solvent media (Grubisic and Benoit, 1968). The chromatographic behavior is most sensitive to the size of the polymer when the diameter of the hydrodynamically equivalent sphere is about 50% larger than the ratio of the volume and the surface of the pores (Berek et al., 1970). Strictly speaking, one should take account of the fact that $[\eta]$ characterizes polymers at infinite dilution, whereas the chromatography employs finite polymer concentration. In good solvent media, the expansion coefficient α_e tends to decrease with increasing solution concentration (p. 119), and this should be taken into account in interpreting the chromatograms in terms of a calibration curve (Rudin and Hoegy, 1972).

With globular proteins the gel-solution distribution depends on both the size and the shape of the molecule, so that no strict correlation with molecular weight is possible (Andrews, 1965). However, Fish et al. (1969) have shown that such a correlation becomes quite precise if the proteins are studied in the denatured state and with all disulfide bridges reduced, so that they behave as random coils. Since the amino acid composition varies considerably between protein species, this observation shows that such variations do not affect, to an appreciable extent, the average flexibility of the polypeptide chain in the random coil state. However, Fish et al. did not include in their study proteins with unusual amino acid composition (such as gelatin), and their behavior may well depart from the general pattern.

Casassa and Tagami (1969) have formulated a theory of the distribution of flexible chain molecules between a region whose dimensions are large compared to those of the molecule and a region which provides geometric restrictions to the shape that may be assumed by the chain. In that case an entropy loss is clearly associated with the transfer of the macromolecule to the restricted region, and the concentration of such macromolecules will be correspondingly reduced. Whether such an effect accounts for the observed behavior is hard to decide since the geometry of the cavities in porous beads available for experiment is much more complex than the simple models amenable to mathematical analysis.

B. EQUILIBRIUM CENTRIFUGATION

Although all our experiments are carried out under the influence of the terrestrial gravitational field $\boldsymbol{g}$, the effect of this field is generally neglected in considerations of the properties of systems at equilibrium. This neglect is justified in most cases since the change in the gravitational potential of a particle when placed in different locations of the system will be extremely small compared to $\boldsymbol{k}T$. The gravitational field will produce an appreciable effect either if the dimension of the system is very large in the direction normal to the surface of the earth, or if the mass of kinetic particles is sufficiently great. The first case is exemplified by the gas distribution in the atmosphere of the earth. The second case was first considered by Einstein (1906b), who pointed out that the distribution of equilibrium concentrations will be experimentally observable for particles of volume v_2 and density ρ_2 suspended in a medium with a density ρ_1 in a system of height h, if $v_2(\rho_2 - \rho_1)\boldsymbol{g}h$ is neither too large nor too small compared to $\boldsymbol{k}T$. A few years later Perrin (1908) and Westgren (1914) demonstrated that the variation with height of the equilibrium concentration may be observed with microscopic particles and may be used to evaluate Avogadro's number.

A consideration of the principles underlying the equilibrium distribution in a gravitational field led Svedberg to the realization that the method should be applicable to the determination of the molecular weight of macromolecules, if gravitational fields of the order of 10^3–10^4 g were available to the experimentalist. With the development of the ultracentrifuge in the Uppsala laboratory, such fields became accessible and by 1926 the instrument had been used for the determination of the molecular weight of hemoglobin (Svedberg, 1926; Svedberg and Fåhraeus, 1926) and of ovalbumin (Svedberg and Nichols, 1926). The popularity of the method declined gradually during the next two decades, largely because of the long times required for sufficiently close approach to equilibrium conditions. However, the importance of equilibrium centrifugation has again increased because of several factors. A number of advances in instrument design and experimental technique have greatly increased the utility of the method for high-precision measurements (Schachman, 1959, 1963). It was realized that the use of Θ-solvents permits the reliable evaluation of entire molecular weight distribution functions of polydisperse polymer samples, as against the limited characterization of such materials by other techniques in terms of molecular weight averages. Also, cell designs and techniques were developed which allow observations in liquid columns with heights of 1 mm or less, reducing the time required for a close approach to equilibrium from days to as little as an hour (Van Holde and Baldwin, 1958; Yphantis, 1960). Finally, development of the technique of density gradient centrifugation provided a badly needed method for the study of distributions of chemical composition, which has found spectacular application in the examination of biologically important macromolecules and may also be applied to investigations of synthetic polymers.

1. Methods for Determining Solute Distribution in the Ultracentrifuge Cell

The distribution of the solute in the test cell must be recorded while it is being spun at high speed in the ultracentrifuge. In Svedberg's original experiments (Svedberg, 1929; Svedberg and Fåhreus, 1926; Svedberg and Nichols 1926), this was accomplished by taking advantage of the absorption of light by the solute. The solution was therefore photographed using light of an appropriate wavelength, and the solution concentrations in various locations were estimated from the relative blackening of the photographic plate. This method is very advantageous with some important biological samples (e.g., proteins and nucleic acids) which have heavy absorption bands in the ultraviolet, so that convenient optical densities can be obtained at very high dilution (Schumaker and Schachman, 1957). Since complications arise in the interpretation of the data obtained at concentrations such that deviations from Raoult's law are appreciable, it is desirable to work at as high a dilution as possible.

A number of very sensitive methods have also been developed in which the *concentration gradient*, rather than the concentration itself, is the directly observed quantity. Among these, the so-called schlieren method has been utilized by Svensson (1939, 1940) in an optical system which has become very popular. A ray of light passing through a medium with a refractive index gradient perpendicular to the direction of light propagation is bent in the direction of increasing refractive index, n, the deflection being proportional to the magnitude of the gradient. This principle is used to distort the image of an illuminated slit into a plot of the refractive index gradient in a test cell through which the light is passed. Since the refractive index of dilute solutions is linear in solute concentration, the schlieren method yields, once the refractive index increment $(\partial n/\partial c_2)_T$ has been determined, the concentration gradient in the centrifuge cell. Refractive index gradients may also be determined by the use of interferometers (Schachman, 1959).

2. Ultracentrifuge Equilibrium in a Two-Component System

In a system subjected to a large gravitational field the gravitational potential makes an appreciable contribution to the chemical potentials of the various components. For a solute of a molecular weight M_2 placed in a centrifuge with an angular velocity ω at a distance r from the axis of rotation, the chemical potential $\bar{G}_2$ is given by

$$\bar{G}_2 = \bar{G}_2^0 + \boldsymbol{R}T \ln \gamma_2 c_2 - \frac{M_2 \omega^2 r^2}{2} \tag{IV.32}$$

Chemical equilibrium requires $\bar{G}_2$ to be invariant throughout the system, that is,

$$\frac{d\bar{G}_2}{dr} = \left(\frac{\partial \bar{G}_2}{\partial P}\right)_{c_2} \frac{dP}{dr} + \left(\frac{\partial \bar{G}_2}{\partial c_2}\right)_P \frac{dc_2}{dr} - M_2 \omega^2 r = 0 \tag{IV.33}$$

Using the relations

$$\left(\frac{\partial \bar{G}_2}{\partial P}\right)_{c_2} = \bar{V}_2 = M_2 \bar{\mathrm{v}}_2 \tag{IV.34a}$$

$$\frac{dP}{dr} = \rho \omega^2 r \tag{IV.34b}$$

where ρ is the density of the solution, substituting into (IV.33), and rearranging, we may relate the molecular weight of the solute to its concentration gradient by

$$\frac{M_2(1 - \bar{\mathrm{v}}_2 \rho)\omega^2 r c_2}{\boldsymbol{R}T} = \left[1 + c_2 \left(\frac{\partial \ln \gamma_2}{\partial c_2}\right)_P\right] \frac{dc_2}{dr} \tag{IV.35}$$

The term $(1 - \bar{v}_2\rho)$ represents the effect of the buoyancy due to the medium in which the macromolecules are suspended. If this term is positive, the molecules will sediment (i.e., travel outward from the center of rotation), while a negative value of $(1 - \bar{v}_2\rho)$ will lead to an analogous flotation in the opposite direction.

The interpretation of the data is simplest if the experiment is carried out under conditions in which Raoult's law is valid. In that case, $(\partial \ln \gamma_2/\partial c_2)_P = 0$ and a plot of $\ln c_2$ against r^2 is linear. From the slope of this plot the molecular weight of the solute may be calculated after evaluation of the partial specific volume $\bar{v}_2$. If the macromolecules are not dissolved in a Θ-solvent, the contribution of the second term in the bracket on the right-hand side of (IV.35) must be taken into account. Under conditions such that the plot of the reduced osmotic pressure against solute concentration is linear, it can be shown from the Gibbs-Duhem relation that $(\partial \ln \gamma_2/\partial c_2)_P$ will be related to the osmotic virial coefficients by

$$\left(\frac{\partial \ln \gamma_2}{\partial c_2}\right)_P = 2A_2M_2 + 3A_3M_2c_2 + \cdots \tag{IV.36}$$

so that (IV.35) may be rewritten (Mandelkern et al., 1957) as

$$\frac{(1 - \bar{v}_2\rho)\omega^2 r c_2}{dc_2/dr} = \frac{RT}{M_2}(1 + 2A_2M_2c_2 + 3A_3M_2c_2^2 + \cdots) \tag{IV.37}$$

We see then that for nonideal solutions the plot of $\ln c_2$ against r^2 will no longer be linear, but that the slope will tend to decrease with increasing values of r. The complications introduced by such solution nonideality into the interpretation of the data were discussed by Mandelkern et al. (1957), who stressed the great advantages to be derived from working in Θ-solvents or at least in systems in which deviations from ideal solution behavior are small. Data obtained in nonideal solutions may be treated either by an expression of the form (IV.37). or else by an extrapolation of the reciprocal apparent molecular weight (evaluated without regard for the deviation from solution ideality) to zero solution concentration. The latter procedure was first suggested by Wales et al. (1951), who plotted $1/M_2^{app}$ against the original solution concentration, but Williams et al. (1958) pointed out that a more reliable extrapolation could be obtained if $1/M_2^{app}$ is plotted against the arithmetic mean of the solution concentration at the top and the bottom of the cell under centrifugation equilibrium conditions.

In principle, the pressure dependence of $\bar{v}_2\rho$ and $(\partial \ln \gamma_2/\partial c_2)_P$ should be taken into account in the evaluation of experimental data. However, in practice the pressure generated with the relatively low gravitational fields used in equilibrium centrifugation leads to variations in these quantities which are small enough to be neglected.

When the solution concentration at various locations in the test cell is determined directly (i.e., by optical absorption), the use of (IV.35) and (IV.37) is straightforward. If, however, the experimental procedure measures concentration gradients, an integration is required to evaluate $c_2(r)$. The cells employed in ultracentrifugation are generally sector-shaped, so that the amount of solute in a layer between r and $r + dr$ is proportional to $c_2 r$, as is apparent from the schematic representation in Fig. IV.6. From considerations of a material balance the distribution of concentrations must then be related to the initial uniform concentration c_2^0 by

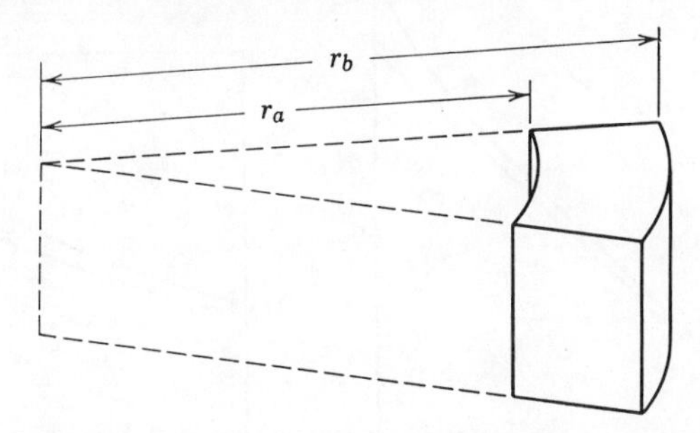

Fig. IV.6. Schematic representation of a sector-shaped ultracentrifuge cell.

$$\int_{r_a}^{r_b} r c_2 \, dr = \frac{c_2^0 (r_b^2 - r_a^2)}{2} \tag{IV.38}$$

where r_a and r_b are the distances of the top and bottom of the cell from the axis of rotation. Combining (IV.38) with (IV.35) for the case of an ideal solution for which $(\partial \ln \gamma_2 / \partial c_2)_P = 0$, we obtain for sector-shaped cells

$$\frac{RT}{M_2(1 - \nu_2 \rho)\omega^2} \int_{r_a}^{r_b} \frac{dc_2}{dr} \, dr = \frac{c_2^0 (r_b^2 - r_a^2)}{2} \tag{IV.39}$$

A number of alternative procedures by which similar evaluations may be carried out for data obtained with nonideal solutions were outlined by Van Holde and Baldwin (1958).

3. Equilibrium Centrifugation of a Polydisperse Polymer Solution

If a solution contains a number of solute species, a relation analogous to (IV.35) holds for each of them. However, if the solution is nonideal, the activity coefficient of each dissolved species will depend on the concentrations of all the other species, so that the molecular weight M_i of species i must be represented by

$$\frac{M_i(1 - \bar{v}\rho)\omega^2 r c_i}{RT} = \frac{dc_i}{dr} + c_i \sum_{k=2}^{n} \left(\frac{\partial \ln \gamma_i}{dc_k} \right)_{P, c_j \neq c_k} \frac{dc_k}{dr} \tag{IV.40}$$

In this expression the partial specific volume $\bar{v}$ is generally considered to be constant for all members of the polymer-homologous series. With certain assumptions it is possible to obtain an analytic expression for the dependence of the $\partial \ln \gamma_i / \partial c_k$ coefficients on the chain length of the interacting species (Fujita, 1962).

Characteristic plots illustrating the effects observed with a solute consisting of species with different molecular weights are illustrated in Fig. IV.7. The plot of log c_2 against r^2 is linear for an ideal solution with a single

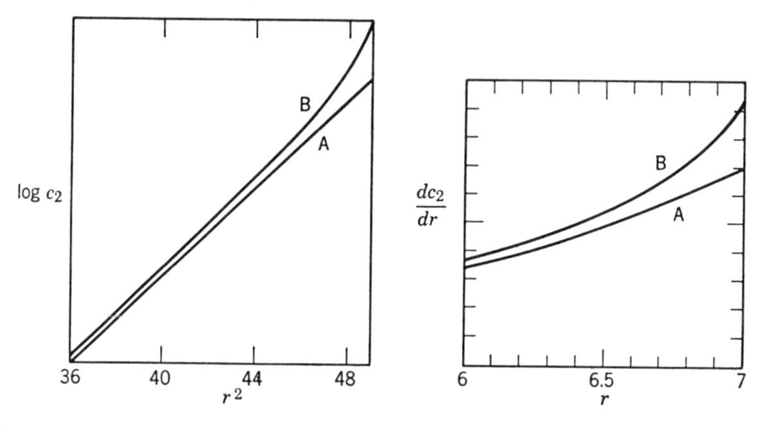

Fig. IV.7. Dependence of the equilibrium solute distribution in the ultracentrifuge on the molecular weight distribution of the solute. (A) Monodisperse solute. (B) Solute with the same weight-average molecular weight, containing 1 g of a fraction with $M = 2\bar{M}_w$ and 5 g of a fraction with $M = 0.8\bar{M}_w$. The symbol c_2 is used here for the total concentration of the polydisperse solute.

solute species, but exhibits a characteristic upward curvature for a solution containing solutes with different molecular weights.† The difference between the two kinds of systems is more pronounced in a plot of dc_2/dr against r than in a plot of c_2 against r. The method is particularly sensitive in detecting

†This criterion was historically of great importance in that it proved that solutions of native proteins contain macromolecules of a precisely defined molecular weight. In reminiscing about the early years of research with the ultracentrifuge, Svedberg (1930) wrote: "When the author, after measurements on gold sols, took up the task of studying the polydispersity of proteins, he was firmly convinced that they would turn out to be as polydisperse as the gold sol particles. . . . The calculation of the particle size of hemoglobin from concentration equilibrium yielded- quite contrary to our expectation- a system which was monodisperse within experimental error, with a molecular weight of 68,000, corresponding to four iron atoms per molecule." He further concluded that such monodispersity is typical of native proteins. The reluctance, before the advent of the ultracentrifuge, to consider proteins as consisting of well defined molecules is astonishing, since hemoglobin crystals had been observed since the middle of the nineteenth century (Edsall, 1962).

a small fraction of a high molecular weight component in the solute. When the solute contains a fraction of relatively low molecular weight, this fraction will be distributed almost uniformly under conditions such as those represented in Fig. IV.7, and therefore the data would not be suitable for determining the shape of the low molecular weight end of the molecular weight distribution function.

The concentration gradient at a distance r from the axis of rotation will be given for a polydisperse sample in a Θ-solvent by

$$\frac{dc_r}{dr} = \sum_i \frac{dc_{ir}}{dr} = \frac{(1 - \bar{v}\rho)\omega^2 r}{RT} \sum_i c_{ir} M_i \tag{IV.41}$$

and since the weight-average molecular weight at r is defined by $\bar{M}_{wr} = \sum_i c_{ir} M_i / c_r$, we may relate it to the concentration gradient by

$$\bar{M}_{wr} = \frac{RT}{(1 - \bar{v}\rho)\omega^2 r c_r} \frac{dc_r}{dr} \tag{IV.42}$$

The weight-average molecular weight of all the solute in the cell may then be obtained by a suitable integration. For a sector-shaped cell, in which the cross-sectional area is proportional to r, a material balance leads to (Williams et al., 1958)

$$\bar{M}_w = \frac{\int_{r_a}^{r_b} \bar{M}_{wr} c_r r \, dr}{\int_{r_a}^{r_b} c_r r \, dr} = \frac{2RT(c_b - c_a)}{(1 - \bar{v}\rho)\omega^2 c^0 (r_b^2 - r_a^2)} \tag{IV.43}$$

where $c_b - c_a$ may be evaluated by graphical integration of $\int_{r_a}^{r_b} (dc_r/dr)\, dr$. Van Holde and Baldwin (1958) have pointed out that for $r = \sqrt{(r_b^2 + r_a^2)/2}$ the local concentration c_r is well approximated by the original concentration c^0 and $\bar{M}_{wr} \approx \bar{M}_w$. Thus, it is possible to obtain a close estimate of the weight-average molecular weight from measurements of the concentration gradient at a single level. This procedure will tend to lead to low values of $\bar{M}_w$ if the molecular weight distribution is very broad.

In nonideal solvents, relation (IV.43) yields an apparent weight-average molecular weight $\bar{M}_w^{\text{app}}$, and the true value of $\bar{M}_w$ is obtained by extrapolation of $1/\bar{M}_w^{\text{app}}$ to zero solute concentration. The interpretation of the slope of such a plot involves assumptions about the behavior of the $\partial \ln \gamma_i / \partial c_k$ in (IV.40). According to Fujita (1969),

$$\frac{1}{\bar{M}_w^{\text{app}}} = \frac{1}{\bar{M}_w} + \frac{B(c_b - c_a)(1 + \Delta)}{2}$$

$$\Delta = \frac{\int_0^1 (\bar{M}_{wr} - \bar{M}_w)(dc^2/dx)\, dx}{\int_0^1 \bar{M}_w (dc^2/dx) dx}$$

$$x = \frac{r^2 - r_a^2}{r_b^2 - r_a^2} \tag{IV.44}$$

where $B = \ln \gamma_i / M_i c_2$ is assumed to be independent of M_i. Experimental data on polystyrene in a good solvent have confirmed that proper extrapolation of ultracentrifuge equilibrium data to zero concentration yields not only the correct value of $\bar{M}_w$ but also a good estimate of the second virial coefficient (Albright and Williams, 1967; Utiyama et al., 1969).

We noted previously that sedimentation equilibrium data obtained under conditions such that the solute concentrations at the top and the bottom of the cell are related by a relatively small factor are not suitable for defining the low molecular weight end of the molecular weight distribution function. Since this portion of the distribution function has a large effect on $\bar{M}_n$, the number-average molecular weight cannot be estimated, in general, with any assurance from equilibrium centrifugation data. However, Hermans (1963) has shown that this difficulty can be overcome if the centrifuge is operated at a speed which reduces the solute concentration at the top of the cell to a negligible value. In that case the various molecular weight averages are given by

$$\bar{M}_n = \frac{RT}{\omega^2(1 - \bar{v}\rho)} \cdot \frac{4\int_{r_a}^{r_b} (r_b^2 - r^2)(dc_r/dr)\,dr}{\int_{r_a}^{r_b} (r_b^2 - r^2)^2 (dc_r/dr)\,dr} \tag{IV.45a}$$

$$\bar{M}_w = \frac{RT}{\omega^2(1 - \bar{v}\rho)} \cdot \frac{2\int_{r_a}^{r_b} (dc_r/dr)\,dr}{\int_{r_a}^{r_b} (r_b^2 - r^2)(dc_r/dr)\,dr} \tag{IV.45b}$$

$$\bar{M}_z = \frac{RT}{\omega^2(1 - \bar{v}\rho)} \cdot \frac{(dc_r/dr)_{r=r_b}}{r_b\int_{r_a}^{r_b} (dc_r/dr)\,dr} \tag{IV.45c}$$

4. Time Required for Approach to Equilibrium

One of the main reasons for the rare use of the ultracentrifuge equilibrium method, in the years immediately following Svedberg's development of the technique, was the extremely long time required for a satisfactory approach to equilibrium conditions. The theory of the approach to equilibrium was developed by Van Holde and Baldwin (1958), who showed that for finite centrifugation times the distribution corresponds to an apparent molecular weight which is too low. The time $t(\epsilon)$ required to approach within a fraction ϵ of the true molecular weight is given by

$$t(\epsilon) = \frac{(r_b - r_a)^2}{D\pi^2 U(\alpha)} \ln\left\{\frac{\pi^2 \epsilon U^2(\alpha)}{4[1 + \cosh(1/2\alpha)]}\right\}$$

$$U(\alpha) = \frac{1 + 4\pi^2\alpha^2}{4\pi^2\alpha^2}$$

$$\alpha = \frac{2RT}{(1 - \bar{v}\rho)M_2\omega^2(r_b^2 - r_a^2)} \tag{IV.46}$$

where D is the diffusion coefficient of the sedimenting species. For most practical conditions the logarithmic term changes little with design variables, so that the time requirement of an experiment may be taken as proportional to the square of the height of the liquid column in the centrifuge cell. A shortening of the liquid column in the ultracentrifuge cell from 1 cm to 1 mm led, as expected, to a hundredfold acceleration of the approach to equilibrium (Van Holde and Baldwin, 1958; Yphantis, 1960). Use of such short cells represents, therefore, a great advantage, particularly if the experimentalist aims only at an estimate of the weight-average molecular weight. When very large species are studied, the times required to reach sedimentation equilibrium increase because of the decrease in diffusion coefficients. It is remarkable, nonetheless, that the same instrument operated at different speeds can be used for determining molecular weights of species as small as sucrose or as large as virus particles in the molecular weight range of 10^7–10^8. The ultimate limit is reached when the speed of the centrifuge is so low that terrestrial gravitation is not negligible compared to the gravitational field of the centrifuge. Stationary "gravity cells" in which the distribution of equilibrium concentration is due to the gravitational field of the earth may then be used for characterizing virus particles corresponding to molecular weights of the order of 10^9 (Weber et al., 1963).

5. Equilibrium Sedimentation in a Density Gradient

The introduction of the technique of density gradient centrifugation by Meselson et al. (1957) opened the use of the equilibrium centrifuge to a wide range of new problems, many of which cannot be explored by any other technique presently available. A review of the theoretical development and the application of the method to biochemical problems was published by Vinograd and Hearst (1962), while another review by Hermans and Ende (1963a) emphasized applications to synthetic polymers.

When a concentrated solution of a low molecular weight solute with a density very different from that of the solvent is subjected to the gravitational field of an ultracentrifuge, the density gradient corresponding to equilibrium conditions may attain a relatively large value. If the solution contains also a macromolecular solute, the macromolecules will tend to sediment or to float, depending on whether their effective buoyant density ρ_2^{eff} (a quantity which we shall consider later in more detail) is larger or smaller than the density ρ of the surrounding medium. The macromolecules will thus tend to concentrate in a band around the distance r_0 from the axis of rotation where the density of the medium equals their buoyant density. Assuming the density gradient $(d\rho/dr)_0$ to be constant across the width of the macromolecular band and the effects of deviations from solution ideality to be negligible, we obtain for the potential energy per mole of macromolecular solute due to the gravitational field

$$E_2 = \frac{(r - r_0)^2}{2} \frac{M_2(d\rho/dr)_0\omega^2 r_0}{\rho_2^{\text{eff}}} \tag{IV.47}$$

and the solute distribution around r_0 is given by the Boltzmann distribution

$$\begin{aligned} c_2(r) &= c_2(r_0) \exp\left[\frac{-E_2(r)}{\boldsymbol{RT}}\right] \\ &= c_2(r_0) \exp\left[\frac{-(r - r_0)^2}{2\sigma^2}\right] \end{aligned} \tag{IV.48a}$$

$$\sigma^2 = \frac{\boldsymbol{RT}\rho_2^{\text{eff}}}{M_2(d\rho/dr)_{r_0}\omega^2 r_0} \tag{IV.48b}$$

This relation shows that the molecular weight of the solute is, in principle, obtainable from the width of the solute band when ultracentrifugation equilibrium has been attained in a density gradient. However, in practice this technique of molecular weight determination is difficult to employ. First, the experimentalist has to be certain that the width of the solute band is not partially due to chemical heterogeneity, that is, to a range of buoyant densities. Beyond this, effects arising from a deviation from solution ideality have to be considered. Taking these complications into account, the method has yielded data on the molecular weights of DNA samples in the range of 10^6–10^8 which are in good agreement with estimates obtained by other means (Daniel, 1969; Schmid and Hearst, 1969).

The density gradient produced by the gravitational field is a consequence of effects due both to a concentration gradient and to a compression gradient. It may be represented (Hearst et al., 1961) by

$$\frac{d\rho}{dr} = \left[\frac{1}{\beta^0} - \left(\frac{\partial \ln V}{\partial P}\right)_T \rho^2\right]\omega^2 r \equiv \frac{\omega^2 r}{\beta} \tag{IV.49a}$$

$$\frac{1}{\beta^0} = \frac{(d\rho/d \ln a_3)M_3(1 - \bar{v}_3\rho)}{\boldsymbol{RT}} \tag{IV.49b}$$

where the subscript 3 refers to the cosolvent. The first term in (IV.49a) is usually larger by an order of magnitude than the term due to the compressibility of the solution, and values of the β parameter for a number of aqueous binary systems useful in concentration gradient centrifugation have been listed by Ifft et al. (1961). Typical values for β are in the range of 10^9–10^{10} cm^5/sec^2 g, so that a centrifuge with a mean value of $r = 6.5$ cm and a speed of 40,000 rpm generates density gradients between 0.01 and 0.1 g/cm^4.

In the application of density gradient ultracentrifugation to problems in biochemistry, the aim frequently is an analysis of systems containing macromolecules of identical molecular weight but different effective buoyant density Our ability to resolve bands representing different solute species will depend on the ratio γ of the distance Δr_0 between the midpoints of the bands and the standard deviation σ of the location of the macromolecules within their band. Since $\Delta r_0 = \Delta\rho_2^{\text{eff}} (d\rho/dr)$, we obtain by substitution from (IV.47) and (IV.48)

$$\gamma = \frac{\Delta r_0}{\sigma} = \Delta\rho_2^{\text{eff}} \left(\frac{M_2\beta}{RT\rho_2^{\text{eff}}} \right)^{1/2} \qquad \text{(IV.50)}$$

The resolution does not depend on the speed at which the centrifuge is operated, since Δr_0 and σ change in the same manner with the gravitational field. The resolution becomes easier with species of very high molecular weight and will be improved by an appropriate choice of a solvent system with a high β. A γ value of at least 2 is needed for satisfactory separation of solute bands.

For solutes in a binary system the effective buoyant density is related simply to the partial specific volume by $\rho_2^{\text{eff}} = 1/\bar{v}_2$. In ternary systems the situation is considerably more complex, since one of the solvent species will generally be attracted preferentially to the region occupied by the macromolecule. The effective buoyant density will then depend on the density of the solute including this solvation envelope. Hearst and Vinograd (1961) have demonstrated that the buoyant density of DNA in aqueous systems may vary within wide limits, depending on the nature and concentration of added salts, and a similar effect was observed by Hermans and Ende (1963b) with polystyrene, whose effective buoyant density was almost twice as high in methylcyclohexane-bromoform as in the benzene-bromoform system. This effect is easily understood since the heavy bromoform cosolvent will be concentrated much more effectively in the region occupied by the polymer in a medium rich in methylcyclohexane, which is a much poorer solvent for polystyrene than benzene.

The first experiment with the density gradient centrifugation method (Meselson et al., 1957) was a particularly beautiful demonstration of the value of this technique. In this experiment bacteria were grown in a culture medium rich in ^{15}N, so that the deoxyribose nucleic acid of the organisms, labeled with the heavy nitrogen isotope, had a density slightly higher than normal DNA. At a given time, the bacterial culture medium was swamped with nutrient with the natural nitrogen isotope distribution, and the change in the buoyant density of DNA isolated from successive bacterial generations was analyzed in the density gradient centrifuge. On the basis of a suggestion by

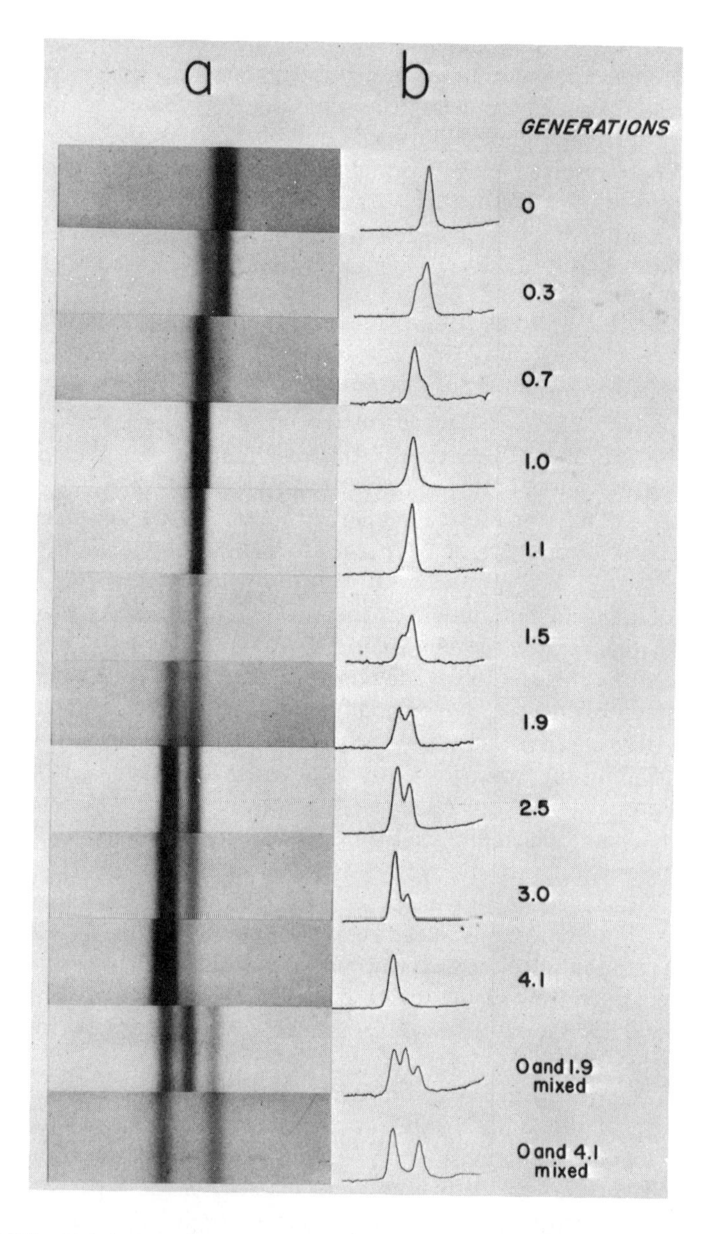

Fig. IV.8. Density gradient ultracentrifugation analysis of DNA from the bacterium *E. coli*. The bacteria were grown on a nutrient rich in ^{15}N and at time 0 a large amount of ^{15}N nutrient was added. (*a*) Ultraviolet absorption on photographs of density gradient ultracentrifuge cells. (*b*) Densitometer traces of UV absorption on photographs. Courtesy of Professor M. Meselson.

Watson and Crick (1953), it was assumed that the replication of DNA during cell division involves a separation of the two strands of the double helix (see p. 162), with each strand serving as a template for the synthesis of its complementary chain. If this mechanism is valid, the second generation of cells originating from the ^{15}N labeled organisms should contain DNA in which one of the strands of each double helix is labeled with the heavy nitrogen isotope. Succeeding cell generations would have increasing concentrations of DNA containing only the natural distribution of nitrogen isotopes, with some half-labeled DNA. At any time, only DNA molecules with three sharply defined buoyant densities, but no species of intermediate density, should be observed. This expectation was fully confirmed by the density gradient ultracentrifugation data (see Fig. IV. 8), so that the DNA replication mechanism, which had previously been of the nature of an inspired theory, could now be accepted as established fact. Since this classical experiment, the density gradient ultracentrifuge method has contributed in a variety of ways to biochemical research. We can cite here a few examples illustrating these developments. By a labeling technique similar to that described above, it was established that some ribose nucleic acid (RNA) is transferred intact through successive cell generations (Davern and Meselson, 1960). The buoyant density of DNA was found to be a linear function of the guanine-cytosine content, which is different in different microorganisms, so that the

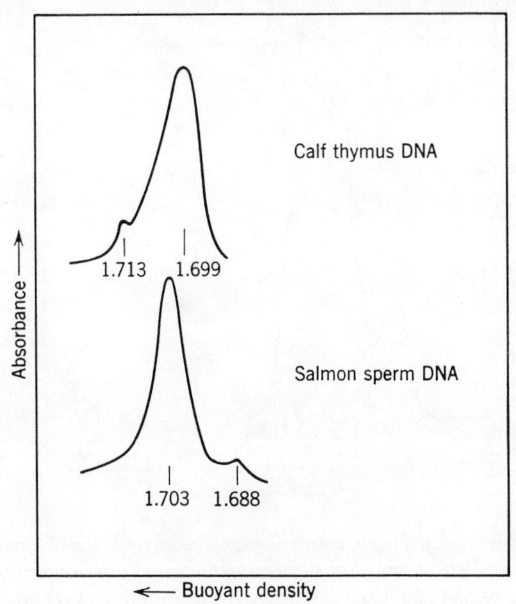

Fig. IV.9. Heterogeneity of DNA from higher organisms as revealed by density gradient centrifugation in an aqueous CsCl solution.

DNA from each species forms in the density gradient a band with a characteristic position (Rolfe and Meselson, 1959; Sueoka et al., 1959). The buoyant density was also found to change with the extent to which circular DNA is twisted into a superhelix (Gray and Vinograd, 1971). In DNA isolated from tissues of higher organisms, the density gradient ultracentrifugation technique has led to the discovery (Schildkraut et al., 1962a) of satellite bands, as exemplified in Fig. IV.9. This "satellite DNA" was later shown to have a highly repetitive nucleotide sequence, and it appears to have a structural function rather than serving as a code for protein synthesis (Yunis and Yasmineh, 1971). It was also established by Schildkraut et al. (1961, 1962b) that the separated strands of DNA originating from different organisms or viruses may recombine to hybrid double helices of intermediate density, if

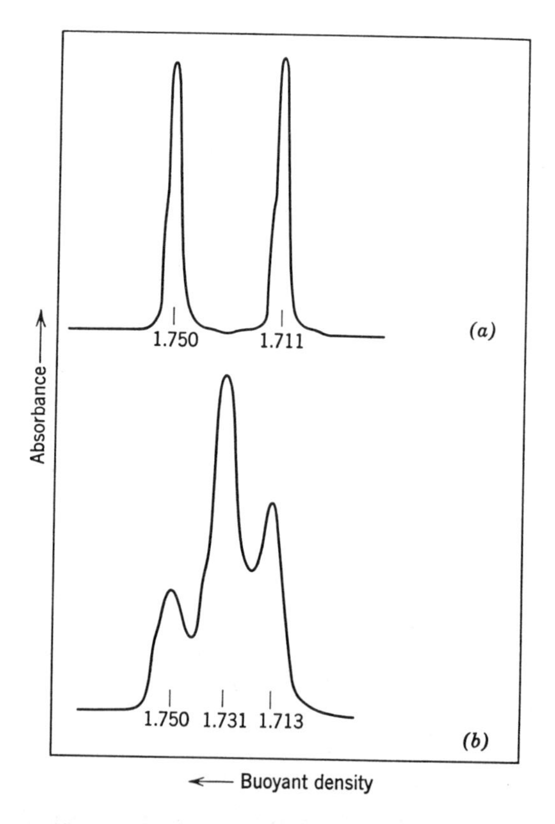

Fig. IV.10. Density gradient ultracentrifugation study of the formation of hybrid DNA. (*a*) The material with a buoyant density 1.711 came from bacteriophage T-3; that with a buoyant density of 1.750, from ^{15}N-labeled bacteriophage T-7; (*b*) After denaturation and renaturation of a mixture of the two DNA species, a new material with an intermediate density is formed.

the parent organisms or viruses were closely related. A typical result of these investigations is represented in Fig. IV.10. Another spectacular result was reported by Hall and Spiegelman (1961), who used the density gradient centrifuge to prove that the RNA synthesized by a bacterium infected by a virus forms a highly specific association complex with the DNA of the infecting virus, and by Doi and Spiegelman (1962), who demonstrated a similar specific interaction between the RNA of certain viruses and the DNA of their host cells. Some viruses, which appear to be implicated in the induction of cancer, contain RNA and an enzyme which uses this RNA as a template for DNA synthesis (Temin and Baltimore, 1972). To prove the complementariness of the two nucleic acids, the DNA is synthesized from radioactive precursors and the distribution of the radioactivity in the density gradient is determined after renaturation of a mixture of denatured viral RNA and newly formed DNA. This approach is required since the concentrations of the macromolecules are too small for detection by the usual optical techniques.

The first application of density gradient ultracentrifugation to the study of synthetic polymers was reported by Bresler et al. (1960), who used the technique for characterizing a system containing a styrene-isoprene block polymer as well as the two homopolymers. The method is of special interest for the study of copolymers, since the characterization of samples which are polydisperse with respect both to molecular weight and to chemical composition had long been a vexing problem. However, the theoretical difficulties in interpreting data obtained with a solute in which both composition and molecular weight vary continuously (Hermans and Ende, 1963a) are much greater than those encountered in biological studies dealing with well-defined chemical species. In particular, a broad polymer band in the density gradient need not indicate chemical heterogeneity, since it may be due to a low molecular weight of the solute [see (IV.48b)]. A separation of two concentration peaks will occur only if the polymer composition has a bimodal distribution and if the molecular weight is sufficiently high. On the other hand, Hermans and Ende (1963a) have pointed out that a chemically homogeneous polymer must yield a concentration distribution in the density gradient which is symmetrical around r_0, regardless of the molecular weight distribution of the solute. An unsymmetrical distribution of the solute concentration is, therefore, necessarily a qualitative indication of a variation in the buoyant density. Nakazawa and Hermans (1971) carried out a careful study of various samples of styrene-methyl methacrylate copolymers and compared the results of an analysis of the ultracentrifuge equilibrium in a density gradient with the results obtained by thin-layer chromatography, which is believed to yield reliable data on compositional distribution (Inagaki et al., 1968). If they assumed that the molecular weight distribution is independent of the chemical composition, the estimate of the breadth of the compositional distribution obtained with the ultracentrifuge was four times as large as the value indicated by the chromatographic data. Yet, in the absence of the chromatographic results, there would have been no obvious indication of a correlation between chemical composition and molecular weight. We must conclude that the density gradient ultracentrifuge does not lend itself to an easy quantitative evaluation of the chemical heterogeneity of synthetic copolymers.

Buchdahl et al. (1963a) have shown that density gradient ultracentrifugation can be used to demonstrate a small amount of microgel in a sample of linear polymer. Such material with an extremely high molecular weight will appear as a sharp spike on a plot of the polymer distribution in the density gradient cell. The location of this spike at the center of the band in which the lower molecular weight material collects proves that the material represented by the spike has a density similar to that of the bulk of the solute and renders it unlikely that the large particles come from an adventitious contamination of the system. In this manner the evidence from density gradient centrifugation is greatly superior, for instance, to that from light scattering, where the presence of large particles may as easily indicate "dust" as a very high molecular weight fraction of the material under investigation.

Data of Buchdahl et al. (1961, 1963b) have indicated that in favorable cases fractions of vinyl polymers with different stereoregularity may be resolved in a density gradient. These investigators found isotactic polystyrene to have a buoyant density in the bromoform-benzene system higher by 0.028 g/ml than that of the atactic species. This difference is much larger than the difference in the reciprocal of the partial specific volumes of isotactic and atactic polystyrene, and it seems, therefore, to indicate that preferential solvent absorption in mixed solvent media is strongly dependent on the stereoregularity of the dissolved polymer.

Chapter V

SPECTROSCOPY, OPTICAL ACTIVITY, AND THE SCATTERING OF LIGHT AND X-RAYS

No group of experimental methods is more versatile in its application to the study of macromolecular solutions than techniques utilizing the interactions of the macromolecules with radiation in various regions of the electromagnetic spectrum. The effects observed fall into a number of categories:

(a) If a quantum of radiation corresponds to an energy transition in the solute molecules which is permitted by quantum-mechanical considerations, the radiation will be absorbed. This principle is utilized in spectroscopy and will yield information typical of the region of the electromagnetic spectrum to which it is applied.

(b) The study of optical activity and circular dichroism applies only to a restricted class of high molecular weight substances, but these include the most important macromolecules of living organisms. Optical activity is highly sensitive to conformational transitions and has been invaluable in the study of helix-coil transition phenomena. A number of synthetic optically active polymers have been prepared, so as to take advantage of the peculiar insight afforded by polarimetry to further our understanding of these and other features of the behavior of dissolved chain molecules.

(c) Radiation which is not absorbed may be scattered by both solvent and solute molecules. This phenomenon, developed by Debye (1944) into a powerful tool for the investigation of polymer solutions, is studied conveniently in two regions of the electromagnetic spectrum, namely, in the x-ray region first employed by Guinier (1939) (for review see Kratky, 1960) and in the region of visible light (McIntyre and Gornick, 1964; Kerker, 1969a; Huglin, 1972). Both techniques are applicable to the determination of the weight-average molecular weight of the solute and to the evaluation of parameters which describe the thermodynamic interaction of solvent and solute. However, the wave nature of electromagnetic radiation leads also to interference phenomena, which may be used to characterize the size and shape of the scattering particles. Here it is most convenient that of the two spectral regions useful in scattering experiments the x-ray region comprises radiation with a wavelength much smaller than the overall dimensions of macromolecules, whereas visible light has wavelengths that are usually much larger than the molecules to be investigated. This difference in the relative magnitude of wavelength and particle size suggests that each of these two spectral regions is particularly well adapted to the investigation of a restricted and mutually complementary class of problems.

There is an important qualitative difference between the information obtained on the one hand by spectroscopy and polarimetry and on the other hand by the study of light scattering. The first two methods measure effects produced in relatively small regions, and the phenomena observed are generally independent of the size of the solute molecules, except in very short chains, where end-group effects may be significant, and in the special case of certain oligomers in which the molecular conformation is critically dependent on the molecular chain length. By contrast, light scattering data are eminently suited to characterization of the macromolecule as a whole. The two methods, therefore, yield complementary information.

The absorption of a quantum of energy by a solute molecule may be followed in certain cases by re-emission in a stepwise process, so that the energy lost in each step is less than the quantum absorbed and the wavelength of the emitted light is correspondingly longer than that of the absorbed radiation. This phenomenon, called "fluorescence," has some interesting applications to the study of macromolecules in solution (Steiner and Edelhoch, 1962; Oster and Nishijima, 1964). In this chapter we shall discuss only part of the applications of fluorescence studies to polymer solutions; other applications, associated with the frictional properties of the solute molecules, with the study of molecular association equilibria and the rate of reactions in polymer solution, will be discussed in Chapters VI, VIII, and IX.

All methods which depend on the interaction of the sample with electromagnetic radiation have the important advantage that the time lag between the occurrence of a change in the system under investigation and the measurement of a quantity reflecting this change is extremely short. Consider, for instance, a process in which two molecules in solution form an association complex. Such a transformation will reduce the osmotic pressure, alter the distribution of the solute in the equilibrium ultracentrifuge, and may lead to changes in the frictional properties of the solution. However, the measurement of all these changes is slow, and such techniques will not be applicable if our aim is a study of the kinetics of transformations which are accomplished in a very short time. By contrast, changes in light scattering intensity, optical activity, or absorption spectra of the system manifest themselves instantaneously, and the speed of processes which may be followed by changes in these properties, is limited only by our ability to record the time dependence of very rapid changes in the intensity of light.

A. ABSORPTION SPECTROSCOPY

The absorption of radiation by solutions containing macromolecules or low molecular weight solutes may be studied in three regions of the electromagnetic spectrum, corresponding to different modes in which the radiant energy is taken up by the system. In the region of visible and ultraviolet (UV) light the radiation causes electronic excitation. Organic molecules absorb visible light only if they contain large resonating systems, and macromolecules of this type have not been studied in solution. In some important cases however, strong absorption of visible light is due to transition metal ions forming complexes with macromolecules as, for example, in hemoglobin and other proteins containing the iron-porphyrin complex associated with the macromolecule (Lemberg and Legge, 1949). We shall not discuss the highly specialized problems involved in the spectroscopy of such materials but shall restrict ourselves to a consideration of the use of UV spectroscopy, which has wider applications in investigations of macromolecules. Spectral absorptions in the infrared (IR) and the Raman spectra arise from transitions between vibrational and rotational states. Both UV and IR spectroscopy are powerful tools for polymer analysis—we may cite, for instance, the use of UV spectra for the analysis of copolymers of styrene or vinyl pyridine with nonaromatic comonomers, or the use of IR spectroscopy for the differentiation of 1,4-*cis,* 1,4-*trans*, or 1,2-addition in polybutadiene. Such analyses depend on the assumption that contributions of monomer residues to observed optical densities are additive, and this assumption appears to represent an excellent approximation in a large number of cases. In some instances (e.g., in ethylene-propylene, acrylonitrile-methacrylonitrile, and methacrylonitrile-butadiene copolymers) the position of IR bands is sensitive to the monomer sequence, and the spectra may then be used to characterize their distribution in the chain molecule (Tosi et al., 1969; Herma et al., 1966; Schmolke et al., 1971). In any case, such studies are generally not dependent on the solubility of the sample, and they fall, therefore, outside the scope of our discussions, which will deal with studies of UV and IR spectra only insofar as they are specifically characteristic of the *dissolved* molecules. The situation is quite different with the absorption in the radio-frequency region caused by quantized transitions in the orientation of the magnetic moments of certain atomic nuclei in an external magnetic field. The resolution attainable in nuclear magnetic resonance (NMR) spectroscopy is very much higher in liquid than in solid samples, so that studies of NMR spectra of macromolecular solutions are necessary to obtain analytical information about the polymer itself which is not obtainable from solid specimens.

1. Ultraviolet Spectra

The ionization of aromatic acids or bases is generally accompanied by a pronounced change in the near-ultraviolet absorption spectrum. This phenomenon has been used particularly to study the ionization of phenolic hydroxyls in tyrosine residues of proteins. Since proteins contain a variety of ionizable groups, it is, in general, difficult to obtain a reliable interpretation of titration data in terms of the ionization of a given functional group, and the unambiguous significance of the spectral shift characterizing the ionization of the tyrosine residues is, therefore, particularly welcome. Typical data obtained by use of this method are those of Tanford et al. (1955) and of Tanford and Hauenstein (1956) with the enzyme ribonuclease. They showed that of the six tyrosyl residues in the enzyme molecule three ionize reversibly between pH 9 and pH 11.5, as would be expected if the negative charge of the protein in this pH range is taken into account. However, the ionization of the remaining three tyrosyl groups occurs only at higher pH values and is irreversible, suggesting that these groups are stabilized in their acid forms by the tertiary structure of the native protein and that their ionization accompanies irreversible denaturation. This view was reinforced when it was demonstrated (Sage and Singer, 1958) that all the tyrosyl residues titrate normally in a solvent medium in which the ribonuclease is denatured.

The technique of spectrophotometric titration may also be employed to advantage in the study of synthetic polymers. It enabled Katchalsky and Miller (1954) to establish the equilibrium between the dipolar ionic and the uncharged forms in copolymers of acrylic acid and 4-vinylpyridine:

$$-CH_2-\underset{\substack{|\\ COO^{\ominus}}}{CH}-CH_2-\underset{\substack{|\\ C_5H_4N^{\oplus}-H}}{CH}- \rightleftharpoons -CH_2-\underset{\substack{|\\ COOH}}{CH}-CH_2-\underset{\substack{|\\ C_5H_4N}}{CH}-$$

Similarly, Ladenheim et al. (1959), who studied the kinetics of nucleophilic displacements by the basic residues of poly (4-vinylpyridine), were able to use UV spectroscopy to obtain a precise estimate of the state of ionization of highly dilute solutions of the polymeric base in various buffer systems.

The near-UV spectrum of proteins is largely produced by the phenolic groups of the tyrosine and the indole groups of tryptophan residues with a much smaller contribution from phenylalanine. Since the phenolic hydroxyl of tyrosine does not ionize appreciably below pH 8, the spectrum would be expected to remain unchanged with increasing acidification of the medium. It has been found, however, that the UV absorption is pH sensitive even in acid solutions and that it varies also when changing the ionic strength of the solution, on addition of denaturation reagents, such as urea, or on partial hydrolysis of the protein (Williams and Foster, 1959; Laskowski et al., 1960). Although the change in optical density caused by these variables is not large, it can be easily observed by measuring directly in the spectrophotometer the difference in the spectra of a protein solution in a given standard state and under some other set of conditions. Figure V.l, showing the wavelength

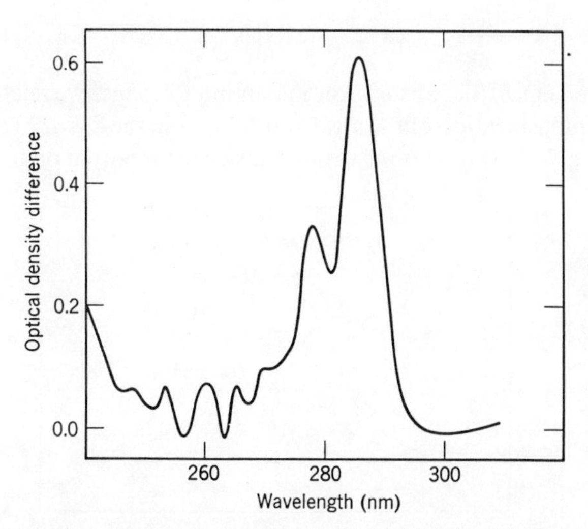

Fig. V.1. Differential spectrum of trypsin-digested insulin referred to the native protein.

dependence of the enhancement of the optical density of an insulin solution after trypsin-catalyzed hydrolysis, is typical of the effects observed. These may be interpreted as resulting from a change in the effective solvent medium of the chromophores, which are surrounded in the native protein by the paraffinic side-chains of nonpolar amino acid residues concentrated in the interior of the globular structure and exposed to direct contact with water on denaturation or hydrolysis. Yanari and Bovey (1960) showed that the near-UV absorption bands of benzene, phenol, and indole (analogous to the chromophores in phenylalanine, tyrosine, and tryptophan) move to shorter wavelength (are "blue-shifted") when transferred from isooctane to water, and the effect observed with these analogs accounts for the blue shift observed on protein denaturation. A similar spectral shift results also from the placement of an ionic charge in the neighborhood of a chromophore, as was demonstrated on suitably chosen model systems. Wetlaufer et al. (1958) studied *O*-methyltyrosinc, in which the ionizable groups are isolated by three saturated bonds from the aromatic chromophore. Nevertheless, conversion of the cationic form to the dipolar ion and to the anionic form:

$$\begin{array}{ccccc} OCH_3 & & OCH_3 & & OCH_3 \\ | & & | & & | \\ C_6H_4 & \longrightarrow & C_6H_4 & \longrightarrow & C_6H_4 \\ | & & | & & | \\ CH_2 & & CH_2 & & CH_2 \\ | & & | & & | \\ HOOC{-}CH{-}NH_3^+ & & \bar{O}OC{-}CHNH_3^+ & & \bar{O}OC{-}CHNH_2 \end{array}$$

causes a movement of the absorption maximum to longer wavelengths (red-shift) and an intensification of the absorption band in the 270–290 nm region, as shown in Fig. V.2. Similar observations have been reported on the behavior

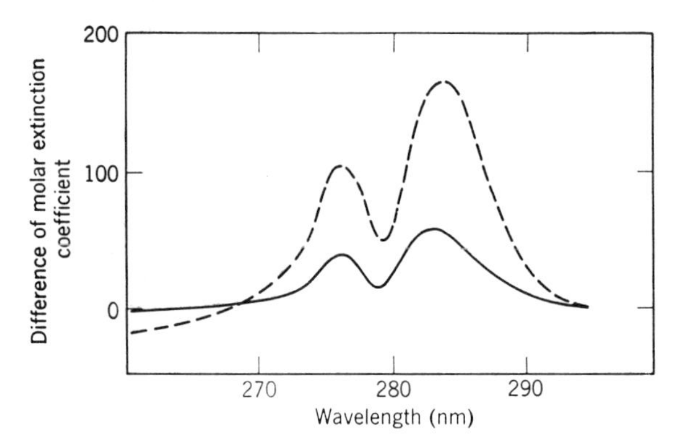

Fig. V.2. Difference spectrum of *O*-methyltyrosine at pH 5.7 (solid line) and pH 9.6 (dashed line) referred to the spectrum at pH 1.08.

of phenylalanine and tryptophan (Donovan et al., 1961). These data suggest that absorption spectra of aromatic residues are subject to perturbation caused by the electrical field of suitably placed charges. In the case of tyrosine residues, hydrogen-bonding between phenolic hydroxyl and ionized carboxyl groups may also produce a spectral shift, and a striking correlation has, in fact, been found between the degree of carboxyl ionization calculated from titration data and the difference spectrum of insulin (Leach and Scheraga, 1960).

When chromophores have a fixed mutual orientation, interactions between their transition dipoles may lead to substantial changes in absorption intensity, without a spectral shift. Tinoco et al. (1962) have discussed such phenomena with special reference to effects observed in the formation of ordered polypeptide structures. From theoretical considerations it is expected that a colinear arrangement of transition dipoles leads to hyperchromism (enhanced absorption) of the transition characterized by the longest wavelength. The additional absorption is believed to occur at the expense of absorption bands at shorter wavelength, but these are usually not experimentally accessible. The inverse effect, hypochromism (reduced absorption) of the lowest energy transition, is indicative of a parallel stacked arrangement of the transition moments. The peptide chromophore undergoes a variety of electronic transitions with different orientations of the transition moments, but the most pronounced effect accompanying conformational transitions is due to the $\pi_1 \rightarrow \pi^*$ transition at 190 nm, which exhibits hypochromism in the α-helical form and hyperchromism in the β-form (Rosenheck and Doty, 1961). These effects are illustrated in Fig. V.3 on the spectral behavior of

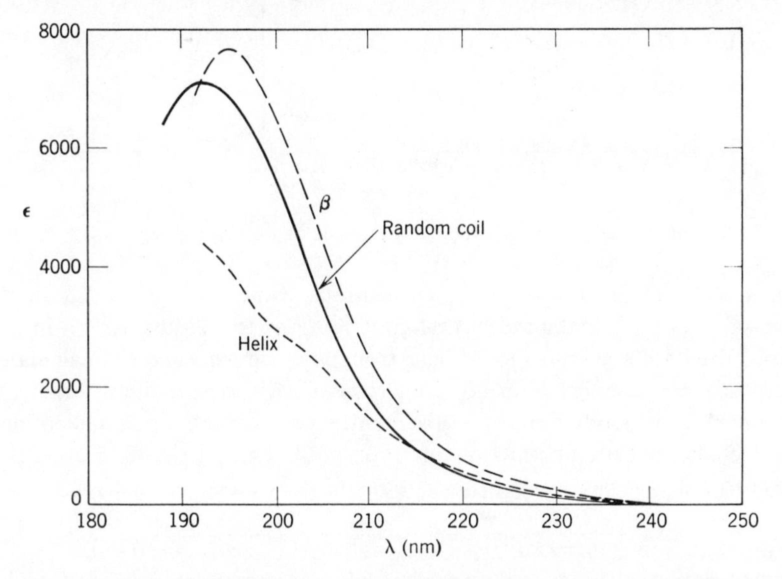

Fig. V.3. Ultraviolet absorption spectra of poly(L-lysine) in aqueous solution: random coil, pH 6.0, 25°; α-helix, PH 10.8, 25°C; β conformation, PH 10.8, 52°C.

poly (L-lysine) in the three conformational states. Since stability of the helical conformation requires a minimum length of the polypeptide chain and is favored by chain extension, spectral changes similar to those observed by Rosenheck and Doty would be expected when comparing spectra of a homologous series of oligopeptides. This effect was clearly established by Goodman and Listowsky (1962) for oligopeptides of γ-methyl-L-glutamate (Fig. V.4).

Transitions to a helical conformation lead also to a pronounced hypochromic effect in the case of nucleic acids and synthetic polynucleotides. Tinoco (1960) and Rhodes (1961) have carried out calculations of the effect to be expected from the geometry of the DNA double helix and the optical properties of the purine and pyrimidine bases on the characteristic absorp-

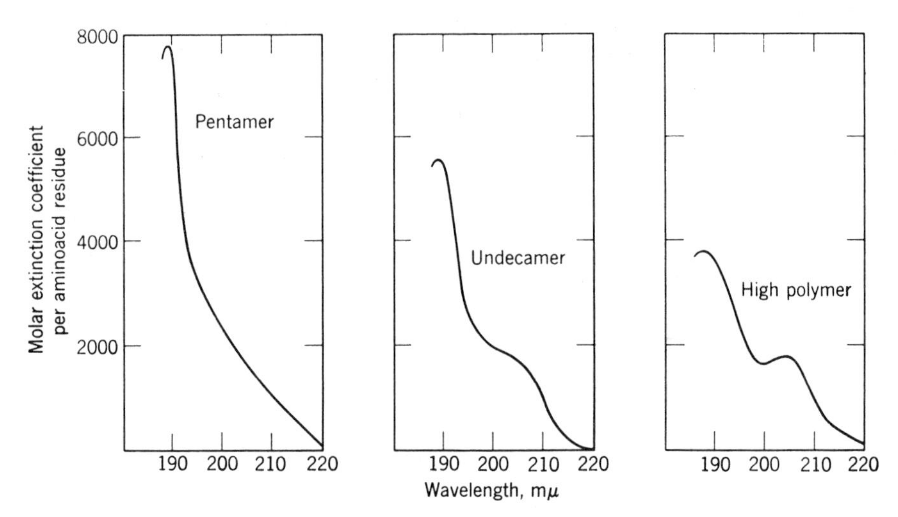

Fig. V.4. Ultraviolet spectra of γ-methyl glutamate oligomers and high polymer.

tion at 260 nm. Since the $\pi \rightarrow \pi^*$ transitions, which are polarized in the plane of the bases, make the largest contributions to the absorption in this region, the base stacking should lead to hypochromism, and the calculated magnitude of the effect is in good agreement with experimental data. The phenomenon is studied most conveniently on synthetic polynucleotides, whose chains consist of units of one type only. Thus, Felsenfeld and Rich (1957) found that the mixing of solutions of poly (adenylic acid) and poly-

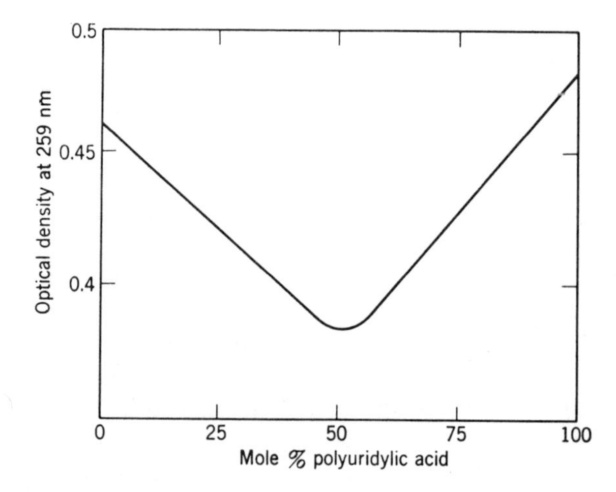

Fig. V.5. Spectrophotometric titration of poly(adenylic acid) with poly(uridylic acid): PH 7.4, 25°C, 0.1M NaCl, 0.01M glycylglycine.

(uridylic acid), which is known to lead to the formation of the double-helical structure, is accompanied by a pronounced reduction of the optical density peak at 259 nm, so that spectrophotometric titration is a convenient method for establishing the stoichiometry of the reaction (Fig. V.5). Later it was shown (Doty et al., 1959) that changes in optical density may be used to follow the thermal breakdown of the helical conformation in DNA (Fig. V.6). More detailed information may be obtained since dissociation of an

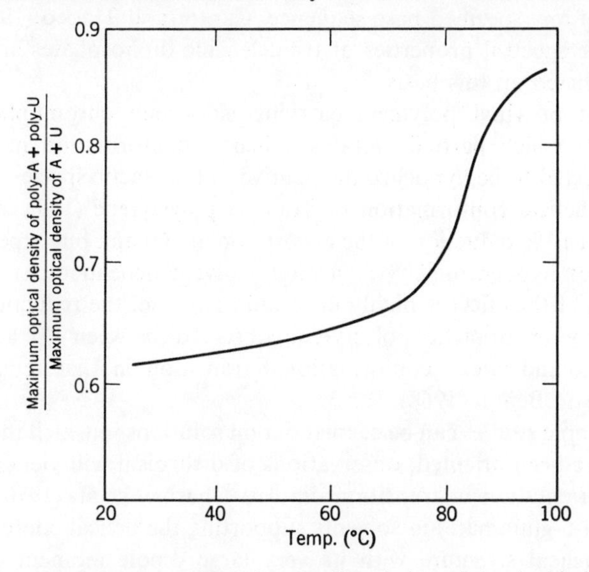

Fig. V.6. Spectroscopic evidence for the dissociation of the double helix of poly(adenylic acid) with poly(uridylic acid) at elevated temperatures.

adenine-thymine (or adenine-uracil) pair produces an increase in absorption in a different region of the spectrum than does the dissociation of a guanine-cytosine pair. Following the changes in optical density at two wavelengths may, therefore, be interpreted in terms of the sequence in which A—T and G—C pairs become separated from one another (Felsenfeld and Sandeen, 1962). This type of analysis was used by Hirschman et al. (1967) to show that DNA from λ bacteriophages consists of two regions which have very different base compositions. The technique is so sensitive that the DNA of mutants could be characterized in terms of the length of the deletion in the two regions of the chain. Another interesting application involves the study of the denaturation of *t*-RNA. The molecules of these substances may be represented, as we saw in Fig. III.28, by a "cloverleaf" structure, and the spectral data suggest that the double-helical branches melt at different temperatures, their stability decreasing with an increasing content of A—U pairs (Felsenfeld and Cantoni, 1964).

Part of the hypochromism of double-stranded DNA relative to a mixture of its constituent nucleotides is retained after denaturation. This indicates that, even in the single-stranded nucleic acid, neighboring bases have some tendency to stack up parallel to one another. If the hypochromicity of the polymer chain relative to the monomer residues is a result of nearest-neighbor interactions only, hypochromicity data on the various dinucleoside phosphates should be sufficient to predict the hypochromicity of a single-stranded nucleic acid of any specified base sequence. Cantor and Tinoco (1967) have found that the spectral properties of trinucleoside diphosphates are indeed correctly predicted on this basis.

In the case of vinyl polymers carrying side-chain chromophores, an isotactic chain which partially retains a helical conformation in solution would be expected to be hypochromic relative to the atactic species. Calculations for the helical conformation of isotactic polystyrene (Vala and Rice, 1963) lead to a 13% reduction of the absorption at 260 nm, but experimental data show that hypochromicity is strongly solvent dependent, so that the interpretation of the effect is highly uncertain. Even so, the reduction in the hypochromicity of isotactic polystyrene observed between 70 and 90°C would seem to indicate a conformational transition in that temperature range (Reiss and Benoit, 1968).

If spectroscopic studies can be carried out on solutions in which the macromolecules have been oriented, observations of dichroism will yield information on the orientation of transition dipoles. Charney et al. (1970) studied poly (γ-benzyl-L-glutamate) in solvents supporting the helical conformation, so that the helical structure with its very large dipole moment could be oriented in a strong electrical field. The electric dichroism then defined the orientation of the benzyl groups relative to the axis of the helix.

2. Infrared Spectra

Samples of vinyl polymers differing in stereoregularity frequently have significantly different IR spectra (for a review of this subject see Bawn and Ledwith, 1962). These differences are most pronounced for bulk samples of the polymers, particularly if the crystalline and amorphous materials are being compared. We have seen, however (Chapter III), that the conformation of dissolved chain molecules is governed to some extent by the nature and degree of their stereoregularity; and since the skeletal vibrational frequencies should be conformation-sensitive, some difference in the IR spectra of polymers with different tacticity should persist even in dilute solution. Kobayashi et al. (1968) and Helms and Challa (1972) identified bands which characterize tt and tg sequences, as well as longer $(\mathrm{tg})_n$ conformations, and studied the stability of helical sequences in isotactic polystyrene as a function of temperature. The data show that the helix content decreases smoothly with increasing temperature, without any suggestion of a cooperative helix-coil transtion. According to Kobayashi et al., the transfer of a mole of monomer residues from a helix to a disordered conformation is characterized by $\Delta H = 3.1$ kcal and $\Delta S = 9.9$ eu. Similar results were obtained for isotactic polypropylene, although it could not be kept in solution at low temperatures but had to be studied in the form of an isotropic gel.

In chains of low stereoregularity, IR spectroscopy may be employed for a characterization of tacticity, since in syndiotactic sequences the tt and $g^{\pm}g^{\pm}$ conformations have a very small difference in energy, while isotactic sequences require a very large addition of energy for any transition from the preferred—tgtgtg—conformation. An increasing dependence of the IR spectrum on temperature should then be indicative of an increasing content of syndiotactic sequences in a vinyl polymer chain, and this criterion appears to give correct results in the characterization of poly(vinyl chloride) (Germer et al., 1963).

The dependence of infrared spectra on conformational transitions would be expected to be most pronounced when strong energetic interactions are involved, and hence it is not surprising that the method is most valuable for the study of the hydrogen-bonded structures characteristic of proteins and synthetic polypeptides. The conformation of polypeptide chains in oriented films was determined unambiguously by crystallographic methods, and this information was correlated with the spectral properties of the specimens. In this way it was established that the so-called amide I absorption occurs at 1656 and 1650 cm^{-1} for the α-helix, the parallel β form absorbs strongly at 1630 cm^{-1} and weakly at 1645 cm^{-1}, while the antiparallel β form has a strong and a weak band at 1632 and 1685 cm^{-1}, respectively. Finally, the absorption band of the disordered polypeptide chain was placed at 1656 cm^{-1} (Ambrose and Elliott, 1951; Miyazawa and Blout, 1961). Such data were then used to identify polypeptide conformations in solution, being particularly valuable in proving the existence of dissolved β structures (Doty et al., 1954; Fasman and Blout, 1960; Goodman et al., 1962).

Infrared spectroscopy was used in a particularly elegant manner by Bird and Blout (1959), who studied the behavior of polypeptide solutions subjected to a large rate of shear. Elongated stiff molecules are partially oriented under these conditions so that their long axes lie preferentially parallel to the flow lines. If such orientation is produced in a solution in which polypeptide chains exist in the form of α-helices and if the solution is observed with polarized infrared light, the absorption due to the carbonyl stretching vibration (which is parallel to the axis of the helix) should be most intense if the plane of polarization is parallel to the flow lines. This expectation was confirmed by experiment. For a number of other absorption bands the variation in the intensity of the absorption peaks with varying orientation of the plane of polarization was also in agreement with predictions based on the geometry of the α-helix and the behavior of the α form of synthetic polypeptides in mechanically oriented films (Ambrose and Elliott, 1951). This is one of the most direct demonstrations that this model is valid for the conformation of polypeptides in suitable solvent media. As would be expected, the infrared dichroism disappears under conditions which lead to the destruction of the helix.

An entirely different problem may also be attacked with the use of infrared spectroscopy. When a polymer chain carries a small number of widely spaced, strongly interacting groups, association complexes may form involving either groups carried by the same chain or groups carried by different chains. It is then of interest to ascertain what determines the balance between the intramolecular and the intermolecular association processes. This problem may be studied by combining molecular weight measurements, which define the extent of intermolecular association, with spectroscopic data, which measure the total concentration of associated functional groups. Chang and Morawetz (1956) studied in this manner copolymers of styrene with 0.8–15.2 mole % methacrylic acid in nonpolar solvent media and found that the overwhelming majority of carboxyl groups which formed hydrogen-bonded dimers did so with carboxyls attached to the same chain. The extent of intramolecular association was independent of the polymer concentration, depending only on the local carboxyl concentration within the swollen polymer coil.

3. Raman Spectra

Vibrational modes which do not involve a change in dipole moment do not lead to infrared absorption. This excludes, for instance, from observation by infrared spectroscopy the conformation-sensitive vibrations of the backbone of vinyl polymers. In addition, a serious restriction is imposed on infrared spectroscopy by the high opacity of water over a large part of the spectrum, so that studies of aqueous solutions are usually not practicable.

A line in the Raman spectrum, on the other hand, is observed whenever the polarizability changes during a molecular vibration. Thus, a C — C stretching vibration which is IR-inactive contributes generally a strong line to the Raman spectrum. Also, no difficulty is associated with Raman spectroscopy of aqueous solutions. Nevertheless, for many years Raman spectroscopy was used rather infrequently because exposure times of many hours were required for the recording of a spectrum. In the particular case of macromolecular solutions, the disturbing effect of light scattering imposed an additional limitation on the utility of this technique.

The availability of laser sources for the excitation of Raman spectra produced a dramatic change because of both the high intensity and the monochromacity of the exciting radiation. The technique is particularly valuable for studies of protein conformation. For instance, a comparison of the spectrum of crystalline lysozyme with the spectrum of the dissolved enzyme shows that the backbone conformation remains virtually unaltered on dissolution, but there is evidence for conformational transitions in the side chains (Yu and Jo, 1973a). A similar comparison of crystalline and dissolved ribonuclease (Yu and Jo, 1973b) reveals changes in the dihedral angles of the C — S — S — C cystine crosslinks, to which the Raman spectrum is particularly sensitive. An important result was also obtained in comparing the spectra of insulin and proinsulin (Yu et al. 1972). The conformation of the hormone was shown to remain unchanged during its formation from the precursor, and the difference between the two spectra made it possible to draw conclusions concerning the conformation of the peptide chain which is eliminated during the activation of proinsulin (see Fig. III. 32).

The most interesting Raman spectroscopic study of synthetic polymers reported to date concerns the changes accompanying the titration of poly-(methacrylic acid) (Koenig et al., 1969; Lando et al, 1973). The chain expansion, which had been known to occur at a critical density of ionic charges, was shown to be accompanied by spectral changes, and the transition was found to be much sharper with the isotactic polymer.

4. NMR Spectra

Of all spectroscopic techniques used in the study of macromolecular solutions, high-resolution nuclear magnetic resonance (NMR) spectroscopy has found the most varied applications. Here we can only illustrate on characteristic examples the ways in which it is being used. For extensive treatments of this field the reader is referred to a review by Jardetzky and Wade-Jardetzky (1971), who discuss applications to protein and nucleic acid studies, and to a monograph by Bovey (1972) which covers both synthetic and biological macromolecules.

In some cases, NMR spectroscopy has been invaluable in determining the structure of a polymer. For instance, the product of the cationic polymerization of 3-methyl-1-butene, $CH_2 = CH — CH(CH_3)_2$, was expected to have structure A, which would give a very complex NMR spectrum:

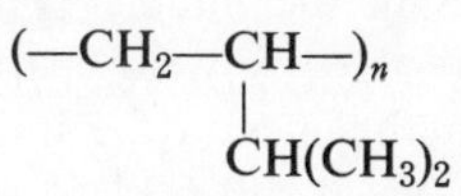

$$\underset{\text{A}}{(\text{—CH}_2\text{—}\underset{\substack{|\\ \text{CH(CH}_3)_2}}{\text{CH}}\text{—})_n} \qquad \underset{\text{B}}{(\text{—CH}_2\text{—CH}_2\text{—}\overset{\substack{\text{CH}_3\\ |}}{\underset{\substack{|\\ \text{CH}_3}}{\text{C}}}\text{—})_n}$$

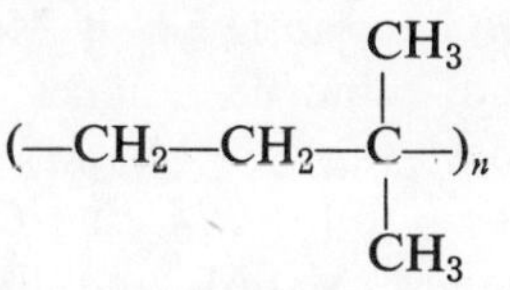

Actually, the spectrum of this material had only two sharp peaks corresponding to structure B, containing two sets of equivalent hydrogens (Kennedy et al., 1964). Similarly, NMR spectroscopy is a convenient method for differentiating between head-to-tail, head-to-head, and tail-to-tail addition in vinyl polymerization. Such an analysis has been carried out for poly-(vinylidene flouride), where the sequences C, D, E, and F:

$$-CF_2-CH_2-\underline{CF_2}-CH_2-CF_2 \qquad -CH_2-CH_2-\underline{CF_2}-CH_2CF_2-$$
$$\text{C} \qquad\qquad\qquad\qquad \text{D}$$

$$-CF_2-CH_2-\underline{CF_2}-CF_2-CH_2- \qquad -CH_2-CH_2-\underline{CF_2}-CF_2-CH_2-$$
$$\text{E} \qquad\qquad\qquad\qquad \text{F}$$

were found to be characterized by fluorine peaks 91.6, 94.8, 113.6, and 115.9 ppm upfield from the $CFCl_3$ reference (Wilson, 1963). This polymer is particularly suitable for microstructure analysis by NMR because of the high sensitivity of the chemical shift of fluorine to the chemical environment.

Nuclear magnetic resonance spectroscopy is a powerful tool in characterizing the distribution of monomer residues in copolymers. In block or graft copolymers the spectra are, of course, a superposition of those of the constituent homopolymers, but chains containing high concentrations of dissimilar nearest neighbors bear spectral evidence of such environmental factors. For instance, in styrene-methyl methacrylate copolymers, diamagnetic shielding by the aromatic rings will affect the location of the methoxyl band of the comonomer, so that comonomers lying next to 0, 1 or 2 styrene residues may be distinguished (Bovey, 1962). In copolymers of vinyl chloride and vinylidene chloride the sequence analysis concentrates on the absorption band characteristic of poly(vinylidene chloride). This appears as a singlet in the $-CCl_2-CH_2-CCl_2-$ sequence but is subject to spin-spin splitting by the α-hydrogen whenever a vinyl chloride lies next to a vinylidene chloride residue to produce a sequence $-CHCl-CH_2-CCl_2-$ (Chujo et al., 1962). Sequence analysis by NMR has led to spectacular results in the case of vinylidene chloride-isobutene copolymers. This system is particularly favorable because of the absence of spin-spin splitting and a very large difference (2.42 ppm) in the chemical shifts characterizing the methylene hydrogens of the homopolymers. Moreover, the position of the methylene peak is sensitive to the nature of the chain substituents over a considerable distance (Hellwege et al., 1966). For instance, if M_1 and M_2 are used to denote vinylidene chloride and isobutene residues, methylenes in the middle of the tetrads $(M_1)_4$, $(M_1)_3M_2$, and $M_2(M_1)_2M_2$ are characterized by well-separated peaks centered at $\tau = 6.14$, 6.34, and 6.53, and methylenes in the hexads $(M_1)_6$, $M_2(M_1)_5$, and $M_2(M_1)_4M_2$ are still distinguishable at $\tau = 6.12$, 6.14, and 6.16:

$$
\begin{array}{cccccccccccc}
 & \text{Cl} & & \text{Cl} & & \text{Cl} & & \text{Cl} & & \text{Cl} & & \text{Cl} \\
 & | & & | & & | & & | & & | & & | \\
- & \text{C} & CH_2 & \text{C} & CH_2 & \text{C} & \underline{CH_2} & \text{C} & CH_2 & \text{C} & CH_2 & \text{C}- \\
 & | & & | & & | & & | & & | & & | \\
 & \text{Cl} & & \text{Cl} & & \text{Cl} & & \text{Cl} & & \text{Cl} & & \text{Cl}
\end{array}
\qquad \tau = 6.12
$$

$$
\begin{array}{cccccccccccc}
 & CH_3 & & \text{Cl} & & \text{Cl} & & \text{Cl} & & \text{Cl} & & \text{Cl} \\
 & | & & | & & | & & | & & | & & | \\
- & \text{C} & CH_2 & \text{C} & CH_2 & \text{C} & \underline{CH_2} & \text{C} & CH_2 & \text{C} & CH_2 & \text{C}- \\
 & | & & | & & | & & | & & | & & | \\
 & CH_3 & & \text{Cl} & & \text{Cl} & & \text{Cl} & & \text{Cl} & & \text{Cl}
\end{array}
\qquad \tau = 6.14
$$

$$
\begin{array}{cccccccccccc}
 & CH_3 & & \text{Cl} & & \text{Cl} & & \text{Cl} & & \text{Cl} & & CH_3 \\
 & | & & | & & | & & | & & | & & | \\
- & \text{C} & CH_2 & \text{C} & CH_2 & \text{C} & \underline{CH_2} & \text{C} & CH_2 & \text{C} & CH_2 & \text{C}- \\
 & | & & | & & | & & | & & | & & | \\
 & CH_3 & & \text{Cl} & & \text{Cl} & & \text{Cl} & & \text{Cl} & & CH_3
\end{array}
\qquad \tau = 6.16
$$

Nuclear magnetic resonance analysis of monomer sequence distributions in copolymers is most valuable in proving the tendency of certain initiator systems to favor the production of alternating copolymers (Hirooka et al., 1968). A striking example of this application has been reported by Harwood (1971). The methine resonance in poly(β,β-dideuteriostyrene) is observed at $\tau = 7.8$, while in copolymers with perdeuteriomethacrylonitrile containing a low proportion of styrene residues this peak is shifted to $\tau = 6.88$. If an equimolar copolymer is prepared with the use of $ZnCl_2$ catalyst, only the $\tau = 6.88$ peak is observed, showing that all the styrene residues are flanked by methacrylonitrile units. However, even in such alternating copolymers more complex patterns of NMR spectra may be observable because of the influence of tacticity on the chemical shifts. For instance, Kuntz and Chamberlain (1974) studied the alternating copolymer of α-methylstyrene and acrylonitrile and observed three methyl peaks due to the following stereosequences:

$$
\begin{array}{cccccc}
 & CN & & CH_3 & & CN \\
 & | & & | & & | \\
- & \text{C} & -CH_2- & \text{C} & -CH_2- & \text{C}- \\
 & | & & | & & | \\
 & H & & C_6H_5 & & H
\end{array}
\qquad
\begin{array}{cccccc}
 & H & & CH_3 & & CN \\
 & | & & | & & | \\
- & \text{C} & -CH_2- & \text{C} & -CH_2- & \text{C}- \\
 & | & & | & & | \\
 & CN & & C_6H_5 & & H
\end{array}
$$

$$
\begin{array}{cccccc}
 & H & & CH_3 & & H \\
 & | & & | & & | \\
- & \text{C} & -CH_2- & \text{C} & -CH_2- & \text{C}- \\
 & | & & | & & | \\
 & CN & & C_6H_5 & & CN
\end{array}
$$

A similar superposition of spectra assignable to stereoisomeric sequences has been reported for alternating copolymers of maleic anhydride (Bacskai et al., 1972; Nishihara and Sakota, 1974).

At the present time, NMR spectroscopy provides the most powerful tool for the quantitative characterization of the stereoregularity of some polymers. The method was first introduced by Bovey and Tiers (1960), who showed that NMR spectra of poly(methyl methacrylate) produced under different conditions show characteristic differences (Fig. V.7). These differences appear

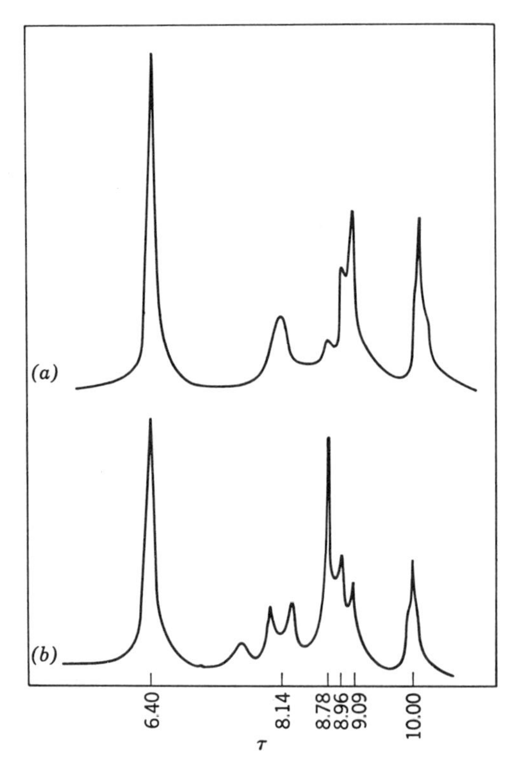

Fig. V.7. NMR spectra of poly(methyl methacrylate) with different stereoregularity. Sample (*a*) was prepared with a free-radical initiator, sample (*b*) with an anionic catalyst. The band at $\tau = 10$ is due to tetramethylsilane added as an internal standard.

in two regions of the spectra. First, the absorption peak due to the hydrogens of the α-methyl group is split into three bands, with maxima at $\tau = 8.78$, 8.96, and 9.09. Bovey and Tiers showed that these bands may be assigned to the central α-methyl groups of isotactic, heterotactic, and syndiotactic triads, respectively (see p. 101). Second, characteristic differences appear in the region corresponding to the hydrogens of the methylene groups ($\tau = 8.14$). This absorption is represented by a singlet in samples rich in syndiotactic triads, whereas polymer with a high content of isotactic sequences has this absorption represented by a quadruplet. This difference is easily understood if we consider in detail the stereochemistry of the methylene group placed between two asymmetric centers with the same or the opposite steric configuration:

Isotactic	Syndiotactic
H, H on central C; CH_3, CH_3; $COOCH_3$, $COOCH_3$	H, H on central C; CH_3, $COOCH_3$; $COOCH_3$, CH_3

In the isotactic sequence the methylene hydrogens are nonequivalent, one being in the fully extended chain closer to the ester and the other to the α-methyl groups. As a consequence, their chemical shifts are slightly different and each absorption peak is split into a doublet because of the two possible spin quantum numbers of the other methylene hydrogen. By contrast, the hydrogens of the methylene in a syndiotactic sequence are equivalent and, therefore, give rise to a single absorption peak. Thus, the NMR spectra not only distinguish between polymers of different tacticity but also allow, in this case, an assignment of the nature of stereoregularity. However, although the magnetic nonequivalence of the methylene hydrogens is an unambiguous proof of an isotactic structure in vinyl and vinylidene polymers, the failure of a spectrum to reflect such a nonequivalence is not necessarily significant. When the difference in the chemical shift of the two methylene hydrogens is small compared to the spin-spin coupling constant J, the splitting of the methylene peak will not be observable. Such a situation arises, for instance, with isotactic polystyrene when observed at 100 MHz, but when the resolution is increased by use of a 220 MHz spectrometer, the nonequivalence of the methylene hydrogens is clearly revealed (Heatley and Bovey, 1968). Sometimes the magnitude of the magnetic nonequivalence depends on the solvent medium. Thus, the splitting of the methylene peak in isotactic polyacrylonitrile was not detected with a 100 MHz instrument when the polymer was in dimethylformamide solution but was observable in dimethyl sulfoxide (Matsuzaki et al., 1968).

The increased resolution obtainable with a 220 MHz instrument makes it possible to distinguish longer stereoisomeric sequences. Here it is convenient to describe isotactic and syndiotactic dyads as having the *meso* (m) and *racemic* (r) structures, respectively. Analysis of the region of methylene absorption in a predominantly isotactic poly(methyl methacrylate) allowed Bovey (1972) to assign peaks to *mmm*, *rmm*, and *rmr* sequences, and the α-methyl peaks due to isotactic triads could be similarly subdivided as belonging to *mmmm*, *rmmm*, and *rmmr* sequences.

The interpretation of NMR spectra of polymers of vinyl derivatives is much more difficult than that of chains in which every second carbon of the backbone is doubly substituted. The splitting of absorption peaks arising from the coupling of methylene and α-hydrogens leads, in general, to very complex patterns which are not easily analyzed. For instance, in syndiotactic triads of poly(vinyl chloride) the absorption by the equivalent methylene hydrogens is split into a triplet by interaction with the two neighboring α-hydrogens. In an isotactic triad, each absorption peak of the quartet, due to the nonequivalent methylene hydrogens, is further split into a triplet by interaction with the α-hydrogens, so that the methylene absorption would be expected to lead to a system of 12 absorption bands. Characterization of the stereoregularity of the polymer may then require a combination of various approaches, such as the calculation of spectra for perfectly stereoregular polymers, the use of various model compounds, and the study of partially deuterated polymers (Bovey, 1972). Among model compounds, the isomeric 2.4.6.-trisubstituted heptanes:

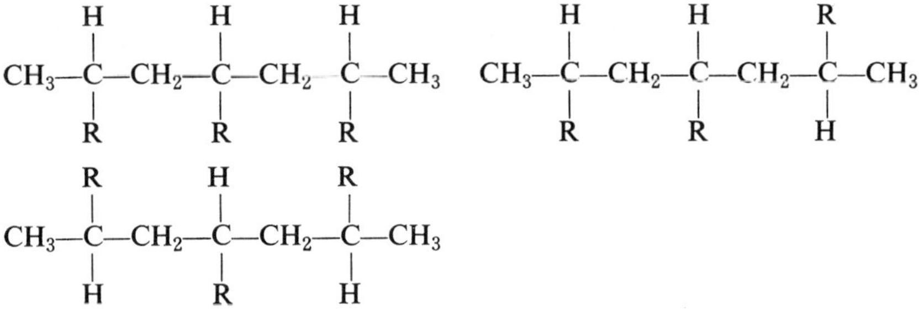

$$\begin{array}{ccccccccccccc} & & H & & & & H & & & & H & & \\ & & | & & & & | & & & & | & & \\ CH_3 & — & C & — & CH_2 & — & C & — & CH_2 & — & C & — & CH_3 \\ & & | & & & & | & & & & | & & \\ & & R & & & & R & & & & R & & \end{array} \qquad \begin{array}{ccccccccccccc} & & H & & & & H & & & & R & & \\ & & | & & & & | & & & & | & & \\ CH_3 & — & C & — & CH_2 & — & C & — & CH_2 & — & C & — & CH_3 \\ & & | & & & & | & & & & | & & \\ & & R & & & & R & & & & H & & \end{array}$$

$$\begin{array}{ccccccccccccc} & & R & & & & H & & & & R & & \\ & & | & & & & | & & & & | & & \\ CH_3 & — & C & — & CH_2 & — & C & — & CH_2 & — & C & — & CH_3 \\ & & | & & & & | & & & & | & & \\ & & H & & & & R & & & & H & & \end{array}$$

have proved particularly useful. Such models have been studied with R = Cl (Doskočilová et al., 1967), R = COOH or $COOCH_3$ (Clark, 1965), R = CN (Murano and Yamadera, 1968), and R = phenyl (Pivcová et al., 1969).

Excessively complex spectra can be greatly simplified by selective deuteration of the material under investigation. The resonance of the deuteron is very far removed from that of ^1H, so that it is not observed in the region employed for ^1H spectroscopy, and coupling is much weaker for H—D than for H—H, so that splitting of NMR peaks is effectively eliminated. For instance, polypropylene prepared from $CH_2 = CD—CD_3$ yields a sharp singlet and a quadruplet for syndiotactic and isotactic dyads, respectively (Zambelli et al. 1967a). Another application is illustrated in Fig. V.8, which compares spectra of undeuterated poly(propylene sulfone) with those of the polymer deuterated in various positions (Ivin and Navrátil, 1970). Alternatively, the technique of "spin decoupling" may be used. In this method the absorption due to one kind of hydrogen is recorded, while the sample is irradiated simultaneously with a high intensity of a radiation with a frequency corresponding to the absorption of nonequivalent neighboring hydrogen atoms.

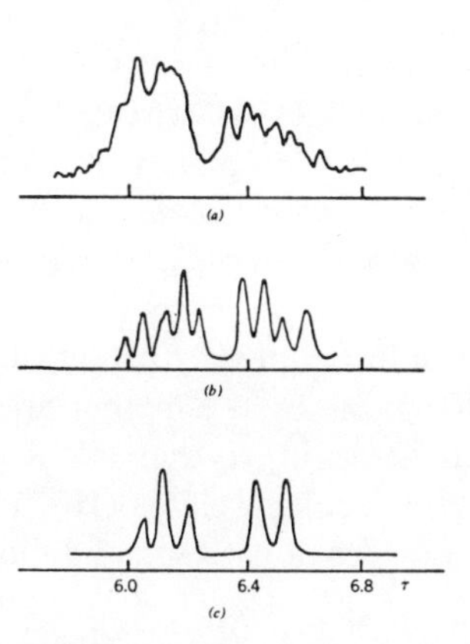

Fig. V.8. 100 MHz NMR spectra of poly(propylene sulfone) in dimethyl sulfoxide: (*a*) $[-CH_2CH(CH_3)SO_2-]_n$; (*b*) $[-CH_2CD(CH_3)SO_2-]_n$; (*c*) $[-CDHCD(CH_3)SO_2-]_n$.

This irradiation produces rapid transitions between their spin states, so that they are effectively averaged and the splitting of absorption bands is eliminated. Bovey et al. (1963) showed that this procedure leads to a great simplification in the determination of the tacticities of poly(vinyl chloride) and poly(vinyl fluoride) from their NMR spectra.

Whereas ^{12}C has no nuclear spin and is, therefore, not observed in NMR spectroscopy, the rare ^{13}C isotope (which has a natural abundance of 1.1%) has a nuclear spin of $\frac{1}{2}$ and yields useful spectra provided that multiple scan techniques are used to enhance the signal and the proton spins are decoupled to simplify the spectral patterns. The use of ^{13}C spectroscopy is particularly advantageous for the characterization of the stereoregularity of polymers (Johnson et al. 1970; Inoue and Nishioka, 1972; Inoue et al., 1972; Randall, 1974) and the sequence analysis of copolymers (Wilkes et al., 1973) because of the large chemical shifts characterizing the nucleus.

Characterization of the stereoregularity of vinyl polymers by NMR spectroscopy assumes particular importance in the study of highly stereoregular chains. We have seen (p. 116, 130) that Flory and his school believe that the chain extension of perfectly isotactic vinyl polymers should be much higher than has ever been observed, so that the flexibility suggested by experimental data has to be the result of a significant frequency of syndiotactic dyads. This interpretation has stimulated a great deal of NMR investigation on highly isotactic polypropylene samples. Heatley et al., (1969) and Ferguson (1971) concluded on the basis of their spectral data that the deviation from perfect stereoregularity is in the neighborhood of 2%, too small to explain the flexibility of isotactic chains unless the energy required to produce breaks in the helical conformation is much less than assumed by Flory et al., (1966).

In some cases, a useful increase in the resolution of NMR spectra may be obtained by employing the so-called shift reagents. These are paramagnetic complex ions which coordinate to unpaired electrons of oxygen or nitrogen atoms and produce very large displacements in the chemical shift of hydrogens in the neighborhood of the complex. Two examples of the application of this technique to the analysis of polymers will illustrate its utility. Addition of tripivaloeuropium to a solution of poly(propylene oxide) produced a large displacement of the absorption by the methyl group of the terminal units from those in the rest of the chain, so that NMR spectra could be used to estimate number-average molecular weights (Ho, 1971). When the same reagent was added to a copolymer of trioxane and dioxolane:

$$—(—OCH_2OCH_2OCH_2—)_n—(—OCH_2CH_2OCH_2—)_m$$

the absorption of ethylene hydrogens, which is a singlet in the absence of the shift reagent, was split into a triplet representing, presumably, dioxolane residues with 0, 1, or 2 trioxymethylene nearest neighbors (Fleischer and Schulz, 1972).

Although the splitting of the NMR spectral peaks which results from the coupling of α-hydrogens and methylene hydrogens in vinyl polymers increases the difficulty of interpreting the spectra, it provides a unique source of information about the conformation of the polymer chain. This is so since the coupling constant J, characterizing two vicinal hydrogens, is sensitive to the dihedral angle. It lies typically within the range of 2–4 Hz for a *gauche* H—C—C—H sequence, while the *trans* conformation yields values from 8 to 13 Hz. Since the lifetime of these conformations is short compared to the difference between the chemical shifts of α-hydrogens and methylene hydrogens, a time-average value of J is observed which may be interpreted in terms of conformational distributions if it is assumed that all the molecules exist in one of the staggered conformations. The analysis of the complex spectra is a laborious task, but computer simulations show that all the spectral features can be accounted for. Highly instructive results were obtained using the stereoisomeric 2,4,6-trisubstituted heptanes with Cl, $COOCH_3$, and phenyl substituents (Doskočilová et al., 1967, 1968; Pivcová et al., 1969). In the isotactic model compounds at room temperature about 80% of the molecules are in the tgtg conformation, with the rest in the gttg form. With the syndiotactic models, the distribution between the all-*trans* and the ttgg conformations depends strongly on the nature of the substituents, the fraction of tttt being estimated as 0.85, 0.52, and 0.38 for the chloro-, phenyl-, and carboxymethyl-substituted compounds, respectively. The temperature dependence of the NMR spectra of the trichloropheptane yielded for the tttt → ttgg transition in the syndiotactic molecule and for the tgtg → gttg transition in the isotactic molecule ΔE values of 1.5 and 0.5 kcal/mole, respectively.

Application of the same methods to polymers is more difficult because of complications arising from imperfect stereoregularity and because of reduced resolution due to a broadening of spectral peaks. Studies of Bovey et al., (1965) and Heatley and Bovey (1968) on isotactic polystyrene and of Heatley et al., (1969) on isotactic polypropylene were consistent with the expected predominance of $(tg)_n$ sequences. In the case of polystyrene, the diamagnetic shielding of backbone hydrogens by the ring currents in the phenyl substituents is strongly conformation-dependent, so that information on chain conformation is obtainable, in principle, from chemical shifts even in the absence of spin-spin splitting. Fujiwara and Flory (1970) found that α-hydrogens in atactic polystyrene, in which all other hydrogens were substituted by deuterium, were characterized by numerous peaks which could be assigned to the various tetrads. Assuming that *m* and *r* dyads were distributed at random and estimating the α-hydrogen shielding from conformational distributions obtained on the basis of potential energy calculations, Fujiwara and Flory arrived at a computed spectrum in close agreement with experiment.

The sensitivity of the NMR spectrum to the conformation of a polymer chain is important in that it provides an experimental tool to study the dependence of the conformational distribution on the solvent medium. Liu and Parsons (1969) have established in this way that the preferred conformations of the poly(ethylene oxide) chain are quite different in water and in organic solvents. This corresponds, of course, to different "unperturbed dimensions," and the NMR technique is particularly valuable in that it is not restricted to Θ-solvents for the demonstration of such variations in $\langle h_0^2 \rangle$. The use of aromatic solvents produces frequently a striking change of a polymer spectrum, since the chemical shift of the nuclei in the polymer chain will then be very sensitive to the proximity and orientation of the solvent molecules. Liu (1969) has made the surprising observation that the positions of the absorption peaks of poly(ethylene oxide) and polydimethylsiloxane in aromatic solvents shift with the chain length of the polymer. Further work would be desirable to clarify this phenomenon. Liu found that it is even possible to resolve two peaks in solutions containing polymer fractions of widely different molecular weights.

The dependence of the chemical shifts on the stereoregularity of polymers of vinyl and vinylidene derivatives is related frequently to the solvent medium. Thus, although the position of the α-methyl peak in poly(methyl methacrylate) is different for isotactic, heterotactic, and syndiotactic triads in all solvents, the methyl ester peak is resolved only in benzene solution (Ramey and Massick, 1966). Similarly, the α-methyl peaks of the various triads in poly(methacrylic acid) are resolved in dimethylformamide, but not in formamide, pyridine, or dimethyl sulfoxide (Klesper et al., 1970a). These differences in spectral behavior may be due both to the effect of the solvent on conformational distributions and to the direct effect of solvent molecules on chemical shifts of nuclei in the various stereoisomeric sequences of the polymer chain.

The breadth of the peaks in NMR spectra depends critically on the velocity of motion of the absorbing nuclei. Since the rotation of a macromolecule is a relatively slow process, the breadth of the absorption lines of nuclei attached to the backbone of randomly coiled polymer chains is determined by the rate of conformational transitions. The theory of this dependence has been developed by Ullman (1965), and we shall return in Chapter IX to a description of the manner in which NMR spectroscopy has aided our understanding of the conformational mobility of macromolecules. Here we should emphasize only that the conformational rigidity characteristic of the helical forms of polypeptides and polynucleotides reduces strikingly the resolution of the spectrum. A broadening of NMR peaks is also characteristic of globular proteins in their native states. Such effects were observed, for instance, by Bovey (1968) with poly(γ-benzyl-L-glutamate) and by McDonald and Phillips (1967) with lysozyme. A similar broadening of the NMR spectra of solutions of polyisocyanates (Goodman and Chen, 1970) is consistent with other evidence (see Section III.C.4) that these polymers assume in solution rigid helical conformations.

The NMR spectra of biological macromolecules are exceedingly complex, but some spectacular results have, nevertheless, been obtained, particularly since the 220 MHz spectrometer became available. In DNA solutions, the peak of the methyl group of the thymine residue appears in a different position, depending on whether the 5′ nearest neighbor is a purine or a pyrimidine (McDonald et al., 1967) so that the NMR spectrum yields some information on the sequence of the nucleotides. Moreover, since the doubling of this peak is observed at high temperatures, where the double-stranded helical form has completely disappeared, the stacking of the nucleic acid bases (which is responsible for the sensitivity of the chemical shift to the nature of the nearest-neighbor nucleotide) must persist to a considerable extent in the single-stranded chain.

In the NMR spectroscopy of protein solutions, differences between the spectra of the denatured and the native protein are of particular interest. The spectrum of the denatured protein may be accounted for in terms of the contributions to be expected from the α-amino acid residues, but in the native structure some chemical shifts are greatly altered as a result of various environmental influences. Thus, the aliphatic methyl groups in denatured lysozyme form a single large peak at $\tau = 9.1$,but in the native enzyme more than 10 peaks may be resolved, extending from $\tau = 8.7$ to $\tau = 12.0$ (McDonald et al., 1971). This is due to the extreme sensitivity of the chemical shift to the location of neighboring aromatic residues; and since the structure of lysozyme has been determined by x-ray crystallography, it is possible to assign the anomalous peaks to specific residues. Such assignment is greatly facilitated by the use of proteins isolated from organisms which have been grown on a medium containing some deuterated amino acids (Puttar et al., 1970).†

5. Fluorescence Spectra

The possibilities inherent in fluorescence spectroscopy as a tool for the study of dissolved macromolecules were first explored on protein solutions (Weber, 1960; Teale, 1960), and applications of this technique were reviewed by Edelman and McClure (1968) and by Stryer (1968). The value of fluorimetry as an experimental tool is a consequence of its high sensitivity to a variety of environmental effects, both short and long range. These may be classified as follows.

(a) Many fluorescing groups exhibit a dependence of the emission spectrum and of the fluorescence quantum yield on the polarity of the solvent, and extreme differences may be observed in water and in organic media.

(b) Studies with polypeptide models show that tyrosine and tryptophan fluorescence is quenched by un-ionized amine and ionized carboxyl (Rosenheck and Weber, 1964; Fasman et al., 1966b). In proteins at physiological pH, the ionized carboxyls are particularly effective in reducing fluorescence intensity.

(c) Nonradiative energy transfer may occur over a considerable distance (up to 60 Å) if the emission spectrum of one group (the donor) overlaps the absorption spectrum of a second (acceptor) fluorescing group. In that case irradiation at a wavelength at which the donor only absorbs will lead to an emission spectrum characteristic of the acceptor.

Three of the amino acids present in proteins fluoresce; the maxima of the emission spectra lie at 282 nm for phenylalanine, at 303 nm for tyrosine, and at 348 nm for tryptophan (Teale and Weber, 1957). As a consequence of nonradiative energy transfer, phenylalanine fluorescence is observed only if tyrosine and tryptophan are both absent (i.e., in gelatin), and tyrosine fluorescence requires the absence of tryptophan (e.g., in insulin). Most pro-

†The spectacular results which can be obtained with NMR spectra of protein complexes of paramagnetic ions for the mapping of the active sites of enzymes will be discussed on p. 412.

teins exhibit only tryptophan fluorescence with quantum yields ranging from 0.05 to 0.48, and while emission spectra vary considerably in native proteins, they become identical after denaturation in 8*M* urea (Teale, 1960).

Steiner et al., (1964) studied the effect of the solvent medium on the fluorescence of acetyl tryptophanamide, which may be considered a model of a typtophan residue in a protein. They found that most organic solutes tended to raise the quantum yield of fluorescence, while dipolar ions had a strong quenching effect. However, effects on the fluorescence intensity of proteins produced by cosolvents do not show even a qualitative consistency. Thus, urea, which enhances the fluorescence of acetyl tryptophanamide, may either intensify or quench the fluorescence of proteins. In the case of pepsin, the fluorescence is enhanced by urea with the native, but quenched with the denatured protein. Thermal denaturation has been shown in two cases to lead to an increase in fluorescence intensity (Gally and Edelman, 1964), although the transfer of the tryptophan residues from the interior of the native protein molecule (which is presumably rich in nonpolar residues) to a region where they are in direct contact with water would have been expected to have the opposite effect. It seems, then, fair to say that phenomena observed in protein fluorescence are rather imperfectly understood. Since the method uses the tryptophan residues as probes to explore the state of the molecule and these residues are not easily studied by other spectroscopic means, fluorescence studies should complement data obtained by other methods.

The sensitivity of fluorimetry as a tool for the study of conformational transitions was greatly enhanced when Weber and Laurence (1954) discovered dyes which fluoresce in nonpolar media up to two orders of magnitude more strongly than in water. A variety of such materials are now known (Kenner and Aboderin, 1971). When these dyes associate with protein molecules in aqueous solution, they are transferred to a region of high local concentration of hydrophobic residues, and this is revealed by a sharp increase in fluorescence. The effect is particularly pronounced with serum albumin, which is known to have an unusually large number of nonpolar groups exposed on the surface of the globular molecule. Not only protein denaturation (Weber and Laurence, 1954; Gally and Edelman, 1965) but even small conformational transitions can be monitored by these fluorescence probes. A striking example is the conversion of chymotrypsinogen to chymotrypsin, which leads to a change in the fluorescence intensity of tryptophan residues (Teale, 1960) and is followed with much higher sensitivity by the large increase in the fluorescence of an adsorbed "reporter dye" (McClure and Edelman, 1967), although the change in the structure of the dissolved molecules is extremely slight. Other examples of the use of this powerful experimental tool will be given in Chapter VIII.

The phenomenon of nonradiative energy transfer is unique in its sensitivity to the spacing r of interacting groups in the range of 10–60 Å. According to theory (Steinberg, 1971), the efficiency E of the transfer is given by

$$E = \frac{r^{-6}}{r^{-6} + R_0^{-6}} \qquad \text{(V.1)}$$
$$R_0^6 = 8.8 \times 10^{-25} Jn^{-4}K^2Q$$

where J is the integral of the overlap between the emission spectrum of the donor and the absorption spectrum of the acceptor, n is the refractive index, K^2 is a function of the mutual orientation of the interacting groups (with $K^2 = \frac{2}{3}$ when this orientation is random), and Q is the fluorescence quantum yield of the donor in the absence of the acceptor. We see then that nonradiative energy transfer may be used to "measure" the distance between donor-acceptor pairs, with E being most sensitive to r when $r = R_0$. Haugland et al., (1969) confirmed that R_0^6 is proportional to J and R_0 can, therefore, be varied by selecting donor-acceptor pairs with the proper spectral overlap. Stryer and Haugland (1967) and Gabor (1968) synthesized proline oligomers with a donor-acceptor pair at the chain ends and found that the dependence of nonradiative energy transfer on the length of the oligopeptide is correctly predicted by (V.1) assuming that the polyproline helix behaves as a rigid rod. Thus, E may be used to "measure" distances within macromolecules. Studies of nonradiative energy transfer between two chromophores of biological macromolecules or of such macromolecules modified by reaction with a properly selected chromophore were reviewed by Steinberg (1971).

A number of fluorescing aromatic compounds exhibit in concentrated solution a new emission band, which is displaced toward longer wavelengths as compared to the emission spectrum observed in dilute solutions. This phenomenon is due to the formation of an association complex between an excited molecule and a molecule in the ground state, which is referred to as an "excimer." Birks (1970) has reviewed the extensive studies in this field. By analogy to excimer emission from crystals in which the spatial disposition of the chromophores is known from crystallographic analysis, it was concluded that excimer formation requires the aromatic rings to lie face-to-face to each other at a distance of about 3.0–3.5 Å. Hirayama (1965) found that excimers could also form intramolecularly, but in a series of diphenylalkanes only compounds in which the benzene rings were separated by three carbon atoms showed excimer emission. Polystyrene and poly(1-vinylnaphthalene) are also characterized by strong excimer emission bands (Vala et al., 1965); Nishijima et al., 1970a) which were assigned, by analogy to Hirayama's observations, to interactions between nearest-neighbor chromophores on the polymer chain. This interpretation was confirmed by the finding that the intensity of the excimer band is proportional, in vinylnaphthalene copolymers, to the probability that a vinylnaphthalene residue has at least one similar nearest neighbor (Nishijima et al., 1970b). On the other hand, polyacenaphthylene also exhibits excimer fluorescence, although steric restraints make it impossible, in this case, for two nearest-neighbor aromatic chain substituents to lie face-to-face to each other, and David et al., (1972) suggested that excimers may form, in this polymer, by the interaction of next-to-nearest neighbors. However, studies on alternating acenaphthylene copolymers show that excimer formation is extremely sensitive to conformational changes in the chain backbone. With methyl methacrylate or methacrylonitrile as the comonomer, excimer emission is favored even more than in the acenaphthylene homopolymer, whereas the methyl acrylate and acrylonitrile copolymers exhibit very little excimer fluorescence (Wang and Morawetz, 1975).

Vala et al., (1965) and Nishijima et al., (1972) assumed that after excitation of an aromatic chromophore the polymer chain has to undergo a conformational transition to produce the juxtaposition of two neighboring aromatic rings required for excimer formation. If this is the case, then excimer fluorescence could be used as a tool to study conformational mobility in polymer chains. An indication that this approach might be fruitful is the striking difference in the behavior of poly(β-benzyl-L-aspartate), which has an excimer emission band, and poly(γ-benzyl-L-glutamate), which shows only the normal fluorescence. Longworth (1966) interpreted this as reflecting the difference in the stability of the helical conformations of the two polymers, assuming that local "melting" of the helix (which is easier in the polyaspartate) is a necessary prerequisite for excimer formation. On the other hand, Heisel and Laustriat (1969), who studied the dependence of the normal and excimer fluorescence of polystyrene as a function of concentration of a fluorescence quenching agent, concluded that a large fraction of the chromophore pairs from which excimers are formed must have existed in a favorable conformation *before* electronic excitation. Finally, Nishijima et al., (1970a) found that in poly(1-vinylnaphthalene) the chromophores close to the chain end are much less likely to form excimers than those in the interior of long polymer chains. This is difficult to understand, since conformational mobility should be favored by an approach to the chain end.

B. OPTICAL ACTIVITY AND CIRCULAR DICHROISM

1. General Considerations

The phenomenon of optical activity has fascinated scientists ever since its discovery in the early nineteenth century. Pasteur, who devoted many years of his life to its study (Dubos, 1950), was particularly impressed with the fact that all substances yielding optically active solutions are derived from products of the life process. He became convinced that the molecular asymmetry, which he perceived to be the fundamental cause of optical activity, must constitute a principle essential to the existence of living organisms. Although he searched in vain for that principle, he would have undoubtedly derived great pleasure from scientific discoveries of our days, which are highly suggestive of the reasons why the evolution of life utilized optically active building blocks for its most important macromolecules.

Optical activity may be observed in crystals of substances inactive in the liquid state if the crystal structure has neither planes of symmetry nor a center of symmetry. However, if a substance is to be optically active in the liquid state, where its molecules are oriented at random with respect to the direction of the light beam, the molecules themselves must lack planes and a center of symmetry. The asymmetry may be of two types, as illustrated in Fig. V.9.

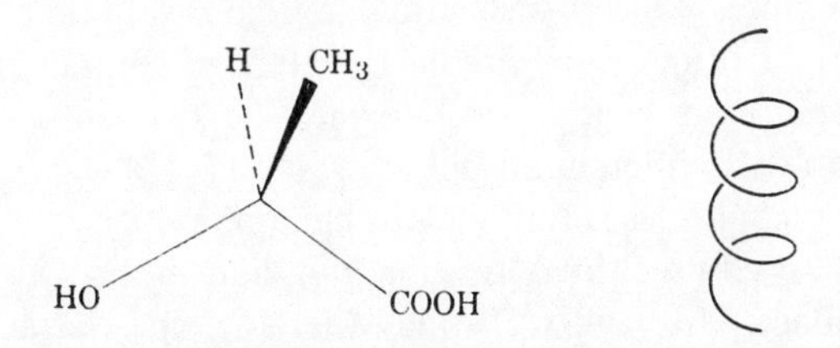

Fig. V.9. Typical examples of asymmetry.

In a substance such as lactic acid the asymmetry is due to the attachment of four different substituent groups to a carbon atom. On the other hand, in a helical coil the structure as a whole is asymmetric. It is important to stress that a system containing chain molecules all coiled into helices of the same sense would be optically active even if it contained no asymmetrically substituted atoms.

The theory of optical activity is very complex, but it is the subject of excellent reviews (Kuhn, 1958; Moscowitz, 1962) to which the reader is referred for more information. Linearly polarized light may be considered as the resultant of two coherent circularly polarized components of opposite sense of rotation. If the refractive index of the medium depends on the sense of rotation of the circularly polarized rays, the difference in their velocity will bring them out of phase as they emerge from the medium. Addition of the two components will then result in a new plane-polarized beam, but one whose plane of polarization has been rotated by an angle α_{obs} with respect to the plane of polarization of the incident beam. If the refractive indices for the two circularly polarized components are n_L and n_R and if light of a wavelength λ (*in vacuo*) has to pass through a thickness l of the liquid, the rotation of the plane of polarization, expressed in degrees, will be

$$\alpha_{obs} = \frac{180}{\pi}\left(\frac{l}{\lambda}\right)(n_L - n_R) \qquad \text{(V.2)}$$

Since l/λ is very large in any experimental arrangement, even $n_L - n_R$ of the order of 10^{-6} will produce large optical activities.

Experimental results are expressed as $[\alpha]_\lambda$, the specific activities at the wavelength λ, defined by

$$[\alpha]_\lambda = \frac{10\alpha_{obs}}{lc} \qquad \text{(V.3)}$$

where l is in centimeters and c is the concentration of the optically active substance (in g/cm³), or as a molar rotation $[M]_\lambda$:

$$[M]_\lambda = \frac{M[\alpha]_\lambda}{100} \qquad \text{(V.4)}$$

Asymmetric solutes exhibit also circular dichroism, that is, a different absorption for the left and right circularly polarized components of a plane-polarized incident light beam. As a result, the two components emerge from the solution with different amplitudes, A_l and A_d. The sum of two vectors of different magnitudes rotating with the same velocity in opposite directions traces out an ellipse; thus, the emerging light beam will be elliptically polarized. Circular dichroism (CD), may therefore, be characterized by the ellipticity $\Theta = \tan^{-1}[(A_l - A_d)/(A_l + A_\alpha)]$, which is proportional to the pathlength of the light and the solution concentration. A molar ellipticity $[\Theta]_\lambda$, defined, in analogy with molar rotation, as Θ in units of degrees per square centimeter per decimole, can be shown to be related to the extinction coefficients ε_L and ε_R for the left and right circularly polarized light by

$$[\Theta]_\lambda = 3300(\varepsilon_L - \varepsilon_R) \qquad \text{(V.5)}$$

The CD spectrum may be represented as a superposition of Gaussian absorption bands. However, since the rotational strength of a given electronic transition depends on the interaction of *pairs* of electric and magnetic moments associated with it, the relative intensities of bands in the CD spectrum are generally quite different from those in the absorption spectrum. This is illustrated in Fig. V.10 (Crabbé, 1964), where the weak absorption band centered at 352 nm has a corresponding CD band, while the much stronger absorption peak occurring at a shorter wavelength does not appear in the CD spectrum. The phenomena of optical activity and circular dichroism are closely related, and the dependence of the optical activity on the wavelength of the light—the rotatory dispersion (RD) spectrum—may, in fact, be calculated from the CD spectrum by the Kronig-Kramers relation:

$$[M]'_\nu = \frac{2\nu^2}{\pi}\int_0^\infty \frac{[\Theta]'_{\nu_1}}{\nu_1(\nu_1^2 - \nu^2)}\,d\nu_1 \qquad \text{(V.6)}$$

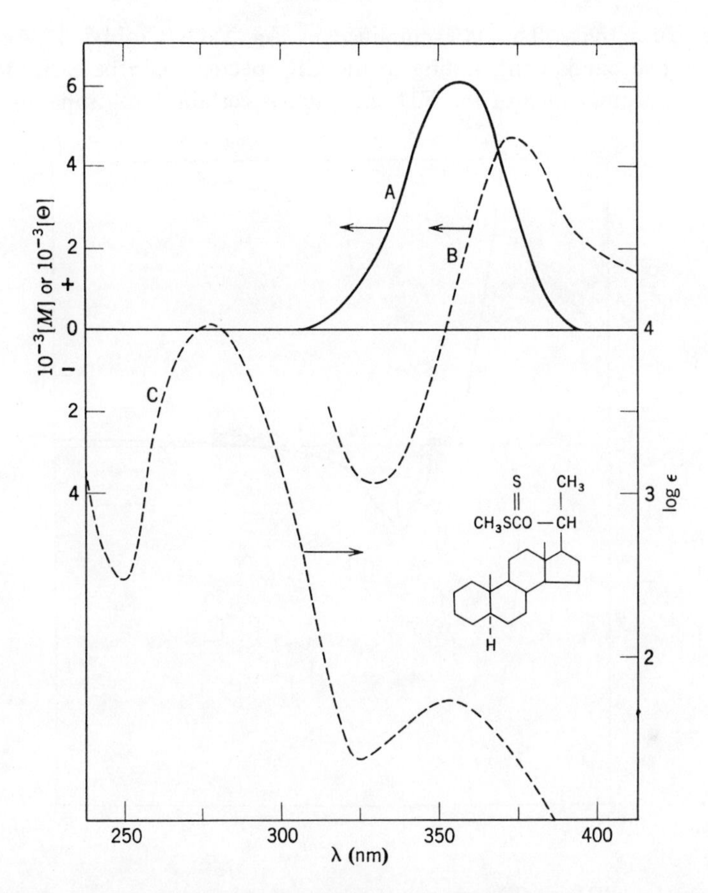

Fig. V.10. Circular dichroism (A), optical rotatory dispersion (B), and UV sbsorption spectra (C), of 20-β-methylxanthate (5α)-pregnane.

where $[M]'_\nu = 3[M]_\nu/(n^2 + 2)$ and $[\Theta]' = 3[\Theta]/(n^2 + 2)$ are "reduced" quantities corrected for the effect of the refractive index, and ν is the frequency of the light. The appearance of this transform when the CD spectrum is characterized by a single band is illustrated in Fig. V.10. We see that the optical activity exhibits a sharp maximum and minimum in the region of the CD band. The sign reversal occurs at the wavelength at which the CD band is centered, and the positive maximum occurs at the longer wavelength if the CD band is positive. This behavior is referred to as a "positive Cotton effect." Conversely, a negative CD band would correspond to a RD spectrum in which the minimum is located at the longer wavelength.

Although CD and RD spectra furnish, in principle, the same information, CD spectra are much easier to analyze if many absorption bands contribute to optical activity. This is exemplified in Fig. V.11 (Crabbé, 1964), where four or five bands contributing to the CD spectrum can be easily located, while the significance of the RD spectrum is certainly not apparent on in-

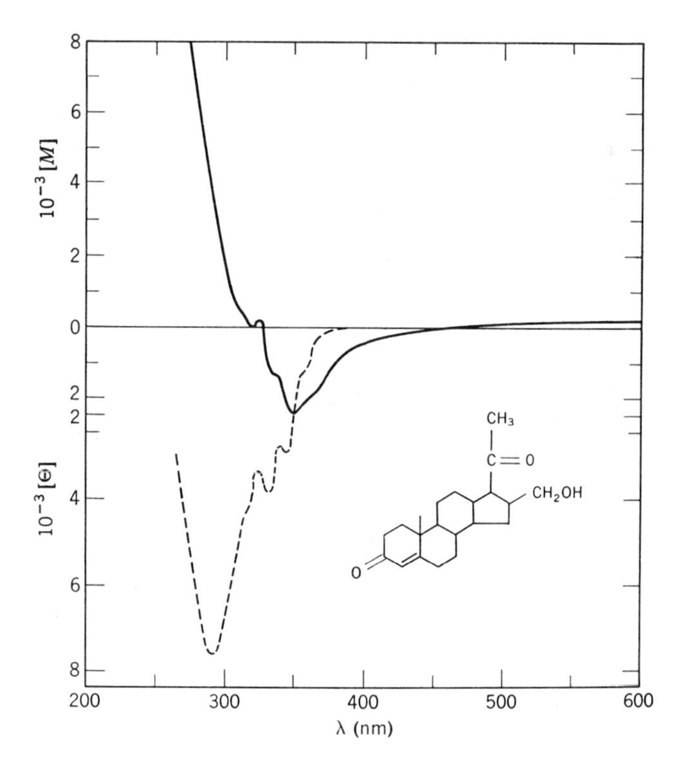

Fig. V.11. ORD and CD spectra of 16-β-hydroxymethylisoprogesterone.

spection. Note particularly the negative peak in the CD spectrum at 292 nm and the difficulty of characterizing this chromophore from the optical activity data. The ease of resolution has made the measurement of circular dichroism the method of choice since the necessary instrumentation became widely available after about 1960.

Optical activity and circular dichroism are of particular interest because of their great sensitivity to molecular conformation. If a molecule can exist in several conformations, the observed effect will be the sum of characteristic contributions made by the various conformers. This may be illustrated on 1,2-dichloropropane, which can exist in the three staggered conformations depicted in Fig. V.12. Wood et al., (1952) carried out a theoretical analysis of the optical activity of this compound and concluded that conformations in which the chlorine atoms are *trans* to each other should lead to $[\alpha] = 40°$,

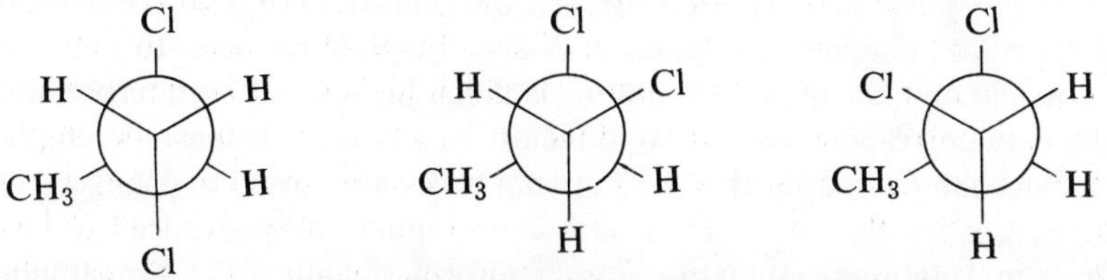

Fig. V.12. Conformations of 1,2-dichloropropane.

while the conformation in which the chlorine is *trans* to the methyl group would correspond to $[\alpha] = -40°$. The experimental data are close to the former value; from this it is concluded that there is a large preference for the conformation in which the chlorine atoms are *trans* to each other. In some cases the distribution of the conformational isomers may be strongly dependent on the nature of the solvent medium. A case of this type, whose significance is particularly easy to analyze, is *trans*-2-chloro-5-methylcyclohexanone (Djerassi et al., 1960) which can exist in two conformations with the RD spectra depicted in Fig. V.13. The conformation with the equa-

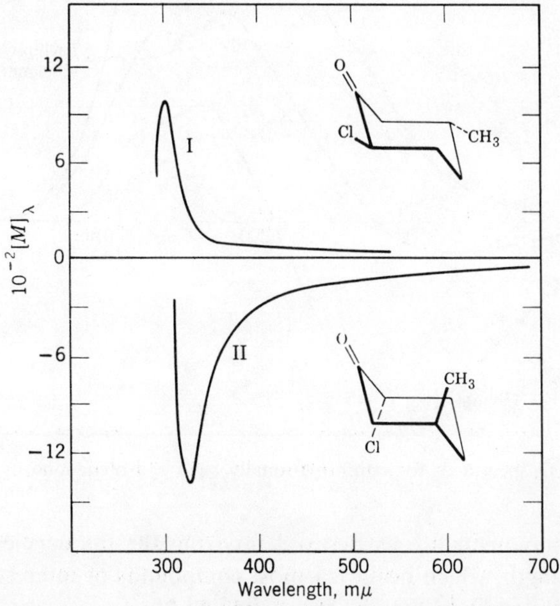

Fig. V.13. Optical activity of *trans*-2-chloro-5-methylcyclohexanone in methanol solution (I) and in octane solution (II). Also shown are the conformations believed to be present in the two solvent media.

torial substituents is stable in polar solvents, while nonpolar solvents favor the form with the axially oriented substituents. The conformational change is attended here by a change in sign of optical activity. A similar reversal in the sign of optical activity was observed by Tanford (1962) on the addition of dioxane to an aqueous solution of *N*-acetyl-L-glutamic acid. In that case, the possible number of conformations is much higher and an interpretation of the data correspondingly more difficult. In any case, it must be emphasized that changes in optical activity are not necessarily proof of changed conformation, since altered solvation of a chromophore may also lead to large changes in rotational strength. Thus, conformationally rigid norcamphor derivatives have been shown by Coulombeau and Rassat (1966) to have CD spectra which change in a striking manner with the solvent medium. An example of their data is reproduced in Fig. V.14.

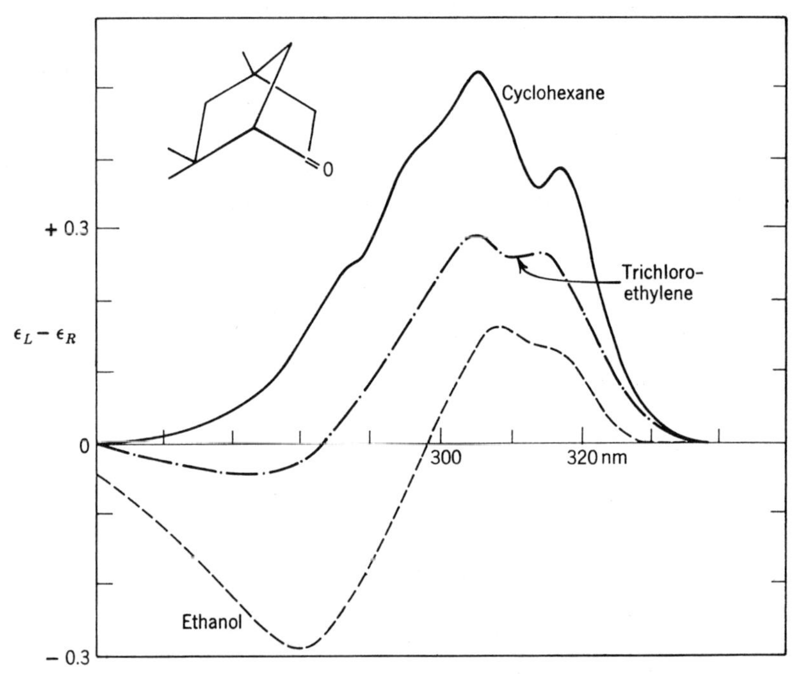

Fig. V.14. CD spectra of the conformationally rigid epiisofenchone in three solvent media.

When measurements are carried out far from the frequencies with high rotational strength, which occur for most compounds of interest in the far-ultraviolet, relation (V.6) may be approximated by

$$[M]_\lambda = \frac{n^2 + 2}{3} \frac{A'\lambda_c^2}{\lambda^2 - \lambda_c^2} \tag{V.7}$$

where n is the refractive index and λ_c is a characteristic wavelength which depends on the location and values of the CD maxima. This simple relation is generally referred to as the one-term Drude equation. Deviations from it provide, as we shall see, a powerful diagnostic tool in characterizing helical structures of macromolecular chains.

2. Synthetic Polypeptides and Proteins

We saw in Chapter III that the native forms of proteins and nucleic acids are characterized by highly specific conformations which may be disrupted relatively easily by a variety of mild treatments. These treatments cause pronounced changes in the physical and chemical behavior of the macromolecules without altering their covalent bonding. Such changes, lumped under the generic term of "denaturation," are of great interest to molecular biology, since they illuminate the forces that are responsible for holding the macromolecules in their native conformation, so crucial for their biological function.

Proteins are composed of L-amino acids and are characterized by negative $[\alpha]_D$ values. It has long been observed that this optical activity undergoes a pronounced change when the protein is denatured. In the case of globular proteins, denaturation typically leads to an increase in the negative value of $[\alpha]_D$, the specific optical activity at the wavelength of the D-line of sodium (589 nm), by an amount that may be as high as 80°. In the case of collagen, the change is in the opposite direction and is frequently very large; for example, $[\alpha]_D$ of bovine collagen changes from —350 to —146° (Harrington and von Hippel, 1961). A related phenomenon is the sharp increase in the optical activity of gelatin solutions accompanying the gelation process (Kraemer and Fanselow, 1928; Pchelin et al., 1963 a, b), which involves, as we saw in Section II. B.7, partial reconstitution of the collagen coiled-coil structure.

The observation that the optical activity of globular proteins always changes in the same direction on denaturation suggested to Cohen (1955) that all these proteins must share some conformational feature which is destroyed during the denaturation process. She suggested that helical conformations in the native samples make, because of their form asymmetry, a contribution to the optical activity which is superimposed on the contributions due to the asymmetric centers of the α-amino acid residues. If all such helices are wound in the same direction, their destruction by a denaturation reagent should always make a contribution of the same sign to the observed optical activity.

In the intervening years it has been established that helical conformations do not play as prominent a role in protein structure as was once believed and that factors other than helix-coil transitions make important contributions to the optical activity change which characterizes protein denaturation. Yet, Cohen's suggestion was most fruitful in stimulating a concentrated study of the optical activity of synthetic polypeptides which can exist in solution in either the helical or the random coil form. It was found that transitions from one form to the other are accompanied by large changes in optical activity. The original observation made by Doty and Yang (1956) was soon followed by studies employing a variety of polypeptides and solvent media. The field has been summarized in a number of excellent reviews (Blout, 1960; Urnes and Doty, 1961; Yang, 1967; Beychok, 1967; Deutsche et al., 1969). It is obvious that the interpretation of data from a synthetic polypeptide containing one or at most two types of building blocks is incomparably simpler than the analysis of phenomena observed on proteins, with their complex (and often unknown) amino acid sequences, disulfide crosslinkages, and a variety of other restraints responsible for their unique tertiary structure. We shall give two typical examples of the striking effects which may be observed with synthetic polypeptides. Figure V.15 represents the results of experiments

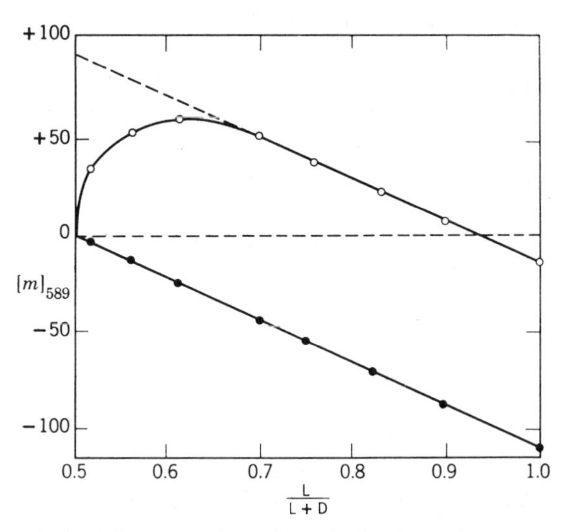

Fig. V.15. Optical activity of copolypeptides of L-leucine and D-leucine in benzene (O) and in trifluoroacetic acid (●). The dashed line extrapolates the data in benzene solution to indicate the optical activity to be expected from a racemic polymer if it could retain the helical conformation.

carried out by Downie et al. (1957), who copolymerized L- and D-leucine and determined the optical activities of the copolymers in benzene, a solvent in which the polypeptides exist in the helical conformation, and in trifluoroacetic acid, in which the polypeptides are present as random coils. The results are expressed in terms of the "mean residue rotation" $[m]_\lambda$, defined by

$$[m]_\lambda = \frac{M_0}{100} [\alpha]_\lambda \qquad \text{(V.8)}$$

where M_0 is the molecular weight of the monomer residue. For the random coils the optical activity is linear in the composition of the copolymer, indicating that the asymmetric centers make independent contributions to the effect observed. The optical activity behavior is more complex in a solvent medium supporting helix formation. Here the optical activities are more positive than for the random coils, the difference being a constant quantity for copolymers containing more than 70% of the L-isomer, but for copolymers containing 50–70% of the L-isomer this difference decreases. The linear portion of the plot of the optical activities in benzene solution may be interpreted as the range of copolymer composition in which the helical conformation is stable. Here the form asymmetry of the molecule as a whole makes a constant contribution to the optical activity, which is added to the contributions due to the asymmetric amino acid residues. If this linear portion is extrapolated to $x_L = \frac{1}{2}$, we obtain $[m]_{589} = 96°$, a value corresponding to the hypothetical racemic copolymer, which retains the helical conformation with the helices wound in the direction characteristic of chains composed of L-amino acid residues. In such copolymers the optical activity would result only from the helical conformation, since the contribution of the asymmetric centers of the L- and D-amino acid residues would be equal in magnitude and opposite in sign. We should note that a very small excess of the L-isomer leads to substantial positive values of $[m]_D$, suggesting that helices with the sense characteristic of poly (L-leucine) can incorporate a fairly large proportion of D-leucine residues. The vanishing optical activity in polypeptides with equal numbers of D- and L-amino acid residues does, of course, not prove that they contain no helical structures but only that right- and left-handed helices are equally probable. In fact, the hypochromism observed at 190 nm in solutions of racemic poly (γ-benzyl glutamate) in a helicogenic solvent points to a large content of helical structures (Masuda et al., 1969), and similar conclusions are reached on the basis of NMR spectra (Bovey et al., 1971; Paolillo et al., 1973).

Optical activity has also been used to pinpoint the critical length of a polypeptide chain at which the helical conformation may be stabilized. Specific optical activities of a series of γ-methyl-L-glutamate oligomers in dichloroacetic acid (in which no helix can form) shifted smoothly to more negative values as the chain length was being extended. In dioxane solution, however, $[\alpha]$ exhibited a discontinuity at the heptapeptide, changing precipitously from negative to large positive values (Goodman and Schmitt, 1959). The break was assumed to be associated with incipient helix formation, and this interpretation was strikingly confirmed by CD spectroscopy (Goodman et al. 1969). As may be seen in Fig. V. 16, the appearance of a spectrum with a positive and a negative band, characteristic of exciton splitting of the $\pi \rightarrow \pi^*$ transition (see below), first appears for the heptapeptide and becomes more pronounced on extension of the polypeptide chain.

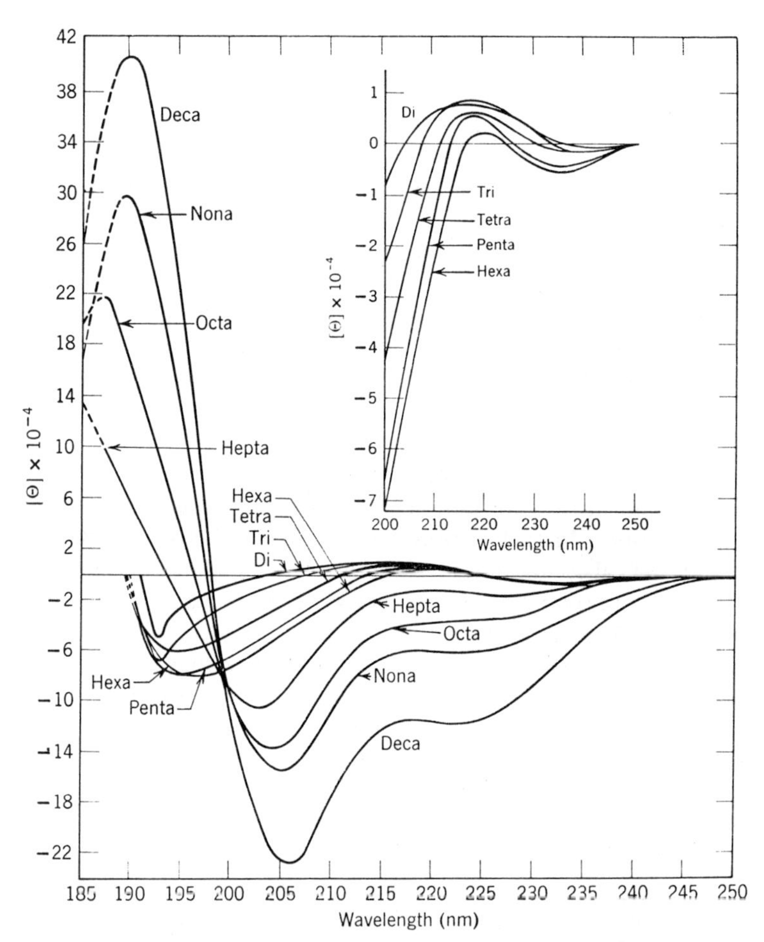

Fig. V.16. CD spectra of N-carbobenzoxy-γ-ethyl-L-glutamate oligomers in trimethyl phosphate.

The availability of synthetic polypeptides, with an optical activity behavior which can be analyzed with much more assurance than that of proteins, stimulated a number of theoreticians to formulate a mathematical description of the differences to be expected between randomly coiled and helical chain molecules. Moffitt (1956) showed that exciton splitting of the $\pi \rightarrow \pi^*$ transition in helical polypeptides should manifest itself by the appearance of two Cotton effects centered about 10 nm apart, with rotational strengths of comparable magnitude and opposite sign. In analogy with (V.7) we may then describe the wavelength dependence of the contribution of two bands centered at $\lambda_1 = \lambda_0 - \Delta$ and at $\lambda_2 = \lambda_0 + \Delta$ to the reduced residue rotation by

$$[m]'_\lambda = \frac{a_1\lambda_1^2}{\lambda^2 - \lambda_1^2} + \frac{a_2\lambda_2^2}{\lambda^2 - \lambda_2^2} \tag{V.9}$$

which yields, for the special case of $a_1\lambda_1^2 = - a_2\lambda_2^2$ and for $\Delta << \lambda_0$,

$$[m]'_\lambda \approx \frac{C}{(\lambda^2 - \lambda_0^2)^2 - \lambda_0^2(\lambda_2 - \lambda_1)^2} \tag{V.10}$$

where $C = a_1\lambda_1^2 \ (\lambda_1^2 - \lambda_2^2)$. Thus, providing that $\lambda^2 - \lambda_0^2 >> \lambda_0 \ (\lambda_2 - \lambda_1)$, an absorption band subjected to exciton splitting will be characterized by a term in the RD spectrum which is inversely proportional to $(\lambda^2 - \lambda_0^2)^2$. This is formally expressed in the relation of Moffitt and Yang (1956):

$$[m]'_\lambda = a_0\left(\frac{\lambda_0^2}{\lambda^2 - \lambda_0^2}\right) + b_0\left(\frac{\lambda_0^2}{\lambda^2 - \lambda_0^2}\right)^2 \tag{V.11}$$

which gives a good description of the RD spectra of synthetic polypeptides in the region of helix-coil transition. For poly (γ-benzyl-L-glutamate) and poly (α-L-glutamic acid), two polypeptides which have been the subject of extensive studies, $\lambda_0 = 212$ nm, b_0 has an average value of -630 in the fully helical form and is close to zero in the random coil, while a_0 varies greatly with the nature of the solvent. It is generally assumed that b_0 is linear in the fraction of the polypeptide present in the helical conformation, and this assumption is consistent with the behavior of other solution properties sensitive to the conformation of the polypeptide chain.

Moffitt's original theoretical analysis is now known to have been oversimplified, and b_0 is used more as an empirical parameter. The CD spectra of poly (α-L-glutamic acid) or of poly (L-lysine) in the disordered and α-helix forms and of poly (L-lysine) in the β form (Gratzer, 1967) are shown in Fig. V.17 Corresponding RD spectra (Greenfield et al. 1967) are represented in Fig. V.18. The "disordered" form corresponds here to the state of the polypeptide at pH values at which the side chains are highly ionized, so that the α-helical or β forms are disrupted by the mutual repulsion of the ionic charges. However, these repulsions favor a conformation quite distinct from the "random coil" conformation of uncharged polypeptides, and this difference leads to characteristic changes of the CD spectrum (Tiffany and Krimm, 1969). It may be noted that the CD spectrum of the α-helical form has not only the two exciton components of the $\pi \rightarrow \pi^*$ absorption band at 189 and 206 nm, but also a prominent band at 222 nm, ascribed to a $n \rightarrow \pi^*$ transition. (Moffitt had erroneously assumed that the contribution of the $n \rightarrow \pi^*$ transition to the RD spectrum was negligible.) The value of $\lambda_0 = 212$ nm in,

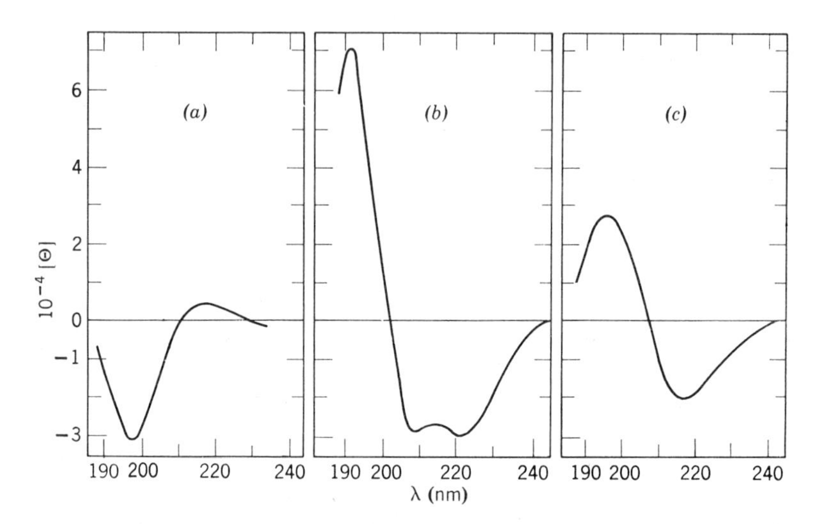

Fig. V.17. Dependence of the CD spectra on polypeptide conformation. (*a*) Randomly coiled poly (α-L-glutamic acid) or poly (L-lysine). (*b*) Poly (α, L- glutamic acid) or poly (L-lysine) in the α-helical form. (*c*) Poly (L-lysine) in the β form.

for exanple, poly (α,L-glutamic acid) results, therefore, from the location and relative intensities of all these bands. The observed RD and CD spectra correspond, of course, to a random orientation of the helices with respect to the direction of the light beam. Theoretical analysis leads to the prediction that the rotational strength of the peptide absorption bands should be reduced for helices oriented parallel to the light beam, and experiments in which helical polypeptides were oriented by an electrical field have confirmed this conclusion (Tinoco, 1959; Hoffman and Ullman, 1970). If polypeptide side chains contribute CD bands at longer wavelength (e.g., the aromatic residues of tyrosine and tryptophan), λ_0 will shift to higher values.

An alternative procedure for the estimation of θ_h, the fraction of a polypeptide chain present in the helical conformation, represents the RD spectrum of a partially helical polypeptide by the two-term Drude equation (V.9). When the optical activity is treated as a sum of contributions from randomly coiled and helical portions of the chain, θ_h is obtainable as a linear function of the coefficients a_1 and a_2 (Shechter and Blout, 1964).

Most polypeptides composed of L-amino acid residues have been found to have their rotatory dispersion spectra characterized by negative values of b_0 and negative Cotton effects. Molecular models indicate that such polypeptides should form right-handed helices more easily than left-handed ones, and Moffitt's analysis also associated a right-handed helix with a negative b_0 value. Neverthless, the correlation of the sense in which the helix is wound

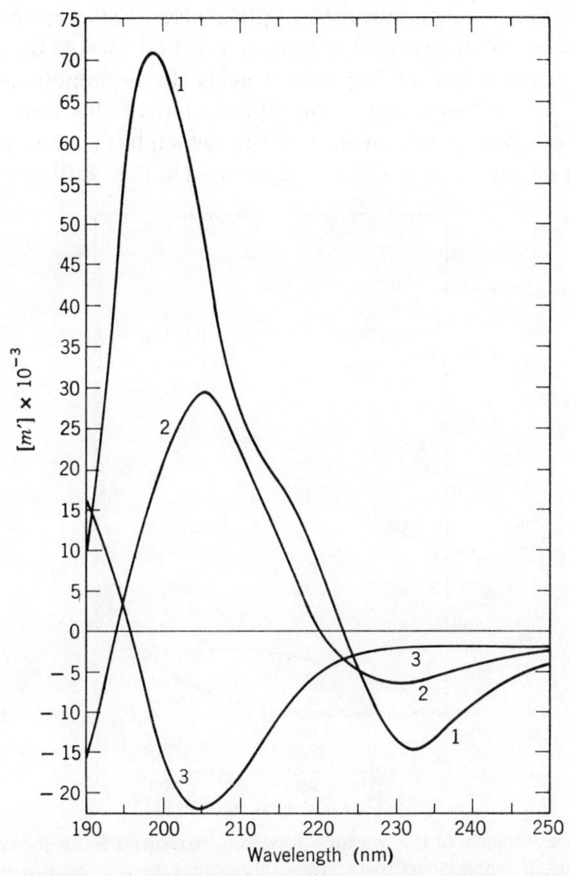

Fig. V.18. Dependence of the RD spectrum of poly(L-lysine) on chain conformation. (1) α-Helix. (2) β Conformation. (3) Random coil.

with the nature of the rotatory dispersion was definitely settled only by a study of the rotatory dispersion of myoglobin, for which the presence of right-handed α-helices (cf. p. 171) had previously been established by x-ray analysis of the protein crystal (Urnes et al., 1961; Beychok and Blout, 1961). Since the possibility of a reversal of the sense of the helix during the dissolution of the protein seems too far-fetched to be seriously considered, the negative Cotton effects in polypeptides composed of the L-isomers of amino acids, such as glutamic acid, alanine, lysine, leucine, and methionine, may be taken as characteristic of right-handed α-helices.

However, a reversal of the sign of the Cotton effect need not necessarily reflect a reversal of the sense of the helix. This can be brought about also by a change in the nature of a chromophore with an asymmetric environment.

The two cases may be distinguished by optical activity studies on copolymers, since the helix would be expected to remain intact on varying the composition of copolymers composed of two amino acids the homopolymers of which form helices of the same sense. The application of this criterion to poly-(β-benzyl-L-aspartate) (Karlson et al., 1960), which has a rotatory dispersion characterized by a positive b_0 value, is illustrated in Fig. V.19. On introducing

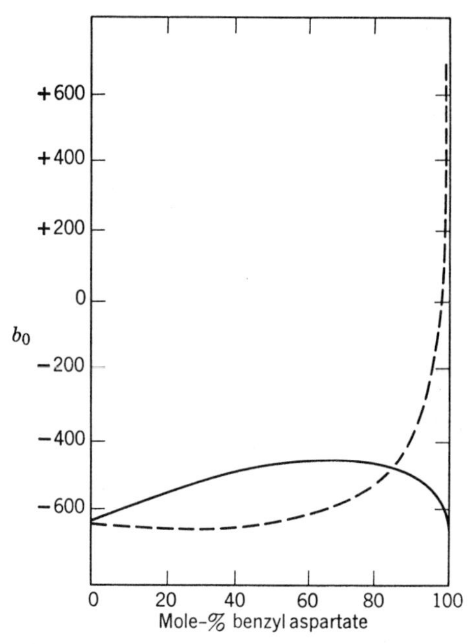

Fig. V.19. Dependence of the rotatory dispersion parameter b_0 on the composition of γ-benzyl-L-glutamate copolymers with β-benzyl-L-aspartate (—) and with β-benzyl-D-aspartate(——).

even small proportions of γ-benzyl-L-glutamate into a poly (β-benzyl-L-aspartate) chain, the b_0 undergoes an abrupt change of sign, while copolymers of γ-benzyl-L-glutamate and β-benzyl-D-aspartate have very similar b_0 values over the entire composition range. This leads to the conclusion that the positive b_0 in poly(β-benzyl-L-aspartate) is a consequence of a helical sense opposite to that of poly(γ-benzyl-L-glutamate). The application of the same criterion to poly(L-tyrosine), which is also characterized by a positive b_0 value, leads to the opposite result; that is, for copolymers of L-tyrosine and γ-benzyl-L-glutamate b_0 is linear in the polypeptide composition (Fasman, 1962a). Although this result would tend to point to a right-handed helix for poly-L-tyrosine, this conclusion must be considered uncertain, since the dielectric dispersion behavior of poly-L-tyrosine suggests a left-handed helical structure (Applequist and Mahr, 1966).

A fascinating phenomenon, which would have probably gone undetected without the availability of optical activity as an experimental tool, is the transition of dissolved poly(L-proline) from one helical conformation to another one of the opposite sense. This phenomenon, first discovered by Kurtz et al. (1956), was studied further by Steinberg et al. (1960). In glacial acetic acid the dextrorotatory form I ($[\alpha]_D = +50°$), obtained when the polymerization is carried out in ether solution, is converted to the strongly levorotatory form II ($[\alpha]_D = -540°$), while the reverse process occurs on diluting a glacial acetic acid solution of form II with large amounts of 1-propanol. The nature of these transformations was previously considered (pp. 149–150); here we should only note that poly(L-proline) offers a good example of the importance of obtaining optical activity data in the region of the Cotton effect. Poly(L-proline) II is characterized by a negative Cotton effect, which is, however, centered around 203 nm (Blout et al., 1963) and is distinct from the Cotton effect centered at 225 nm, which dominates the near-ultraviolet rotatory dispersion of the helical form of poly(γ-benzyl-L-glutamate) and similar polypeptides (Simmons et al., 1961). These two effects are known to arise from different electronic transitions, and they have opposite correlations between the sign of the Cotton effect and the sense of the helix. The RD and CD spectra of poly (L-proline) II and of collagen are similar in magnitude and location (Blout et al., 1963; Brown et al., 1969), reflecting the close relationship between the two structures. Nonethless, they are distinguishable, the collagen spectra being blue-shifted by about 10 nm.

So far, the discussion has been concerned with phenomena reflecting the properties of isolated macromolecules. If the macromolecules can form aggregates, the optical activity of the solution may be greatly affected if the aggregation alters either the molecular conformation or the solvation of the chromophore. A case of this type has been described by Jennings et al. (1968), who found a large hysteresis loop when following the optical activity of poly-(α-L-glutamic acid) at a concentration of 0.25% through a cooling and heating cycle at pH 4.4. Under these conditions the polymer exists in the α-helical conformation, and spectroscopic data suggest only a minor change in this conformation accompanying the aggregation process. Still, this aggregation results at 10°C in the very considerable change of $[\alpha]_{365}$ from -135 to $-188°$. An effect of quite a different type is observed if the *structure of the aggregate* is asymmetric. We saw in Section II. B.4 that solutions of optically active polypeptides in the helical conformation may form, at higher solution concentrations, an anisotropic liquid phase in which the rodlike helices are tilted with respect to each other by a small angle so as to form a helical structure. Such structures typically have a pitch in the range from 5 to 50 μ (Robinson et al., 1958), and this *asymmetric arrangement of the macromolecules* gives rise to extremely high optical activities. Robinson (1956) has reported rotations of several hundred degrees by a 1 mm layer of a 23% solution of poly (γ-benzyl-L-glutamate) in methylene chloride, several orders of magnitude higher than the optical activity to be expected from the isolated macromolecules.

Although the changes in the rotatory dispersion of synthetic polypeptides undergoing the helix-coil transition constitute in themselves a most fascinating subject for study, their greatest importance lies in the insight which they provide into analogous phenomena occurring in protein molecules. If a protein is fully denatured in guanidine hydrochloride solution and the disulfide bridges are reduced, the RD spectrum may be rationalized as the sum of contributions characteristic of the α-amino acid residues present in the molecule (Tanford et al., 1967). The RD spectra of a number of globular proteins in their native states have been reported (e.g. by Jirgensons, 1965, 1966), but they seem to be difficult to interpret in detail. It is questionable whether treatment of the optical activity behavior of the peptide chromophores as a sum of contributions from helical, random coil, and β structures is justified, since helical sequences may be quite short, β regions are frequently distorted, and no part of the precisely defined protein conformation may properly be described as "random" (Madison and Schellman, 1972). In addition, in proteins containing cystine crosslinkages, the asymmetric disulfide bond makes a substantial contribution to the optical activity (Tanford et al., 1967), being characterized by an intense negative Cotton effect centered at 252 nm in cystine derivatives (Coleman and Blout, 1968) and at 272 nm in insulin (Beychok, 1965). The rotatory strengths of the aromatic chromophores of L-tyrosine, L-tryptophan, and L-phenylalanine are very low, but they are greatly enhanced in some proteins, presumably because of the asymmetric environment in the native protein structure. This effect was studied by Rosenberg (1966) and by Beychok et al. (1966) on carbonic anhydrase, where absorption bands with appreciable rotatory strength extend to the neighborhood of 300 nm.

Some proteins have a very high helical content, and in their case an analysis of the RD spectrum by the Moffitt equation (V.11) yields valuable insight. This is particularly so with the muscle protein myosin (Cohen and Szent-Gyorgyi, 1957). Myosin molecules, which contain a rodlike structure with globular portions at the ends of the rod (Section III. C.2), may be split in a very specific manner by treatment with the enzyme trypsin, and one of the fractions has a rotatory dispersion characterized by $b_0 = -660°$, that is, as large as synthetic polypeptides in their fully helical form. Another protein with an unusually high helical content is myoglobin, for which Urnes et al. (1961) estimated, on the basis of the b_0 values, that 73% of the chains are present as α-helices. This estimate is in excellent agreement with the crystallographically determined structure of this protein.

Since optical activity and circular dichroism are highly sensitive to molecular conformation, they may be used to monitor even minor changes in molecular structure. A striking example is the change in the CD spectrum characterizing the conversion of chymotrypsinogen to chymotrypsin (Fasman et al., 1966*a*), shown in Fig. V.20. These spectra suggest that aromatic

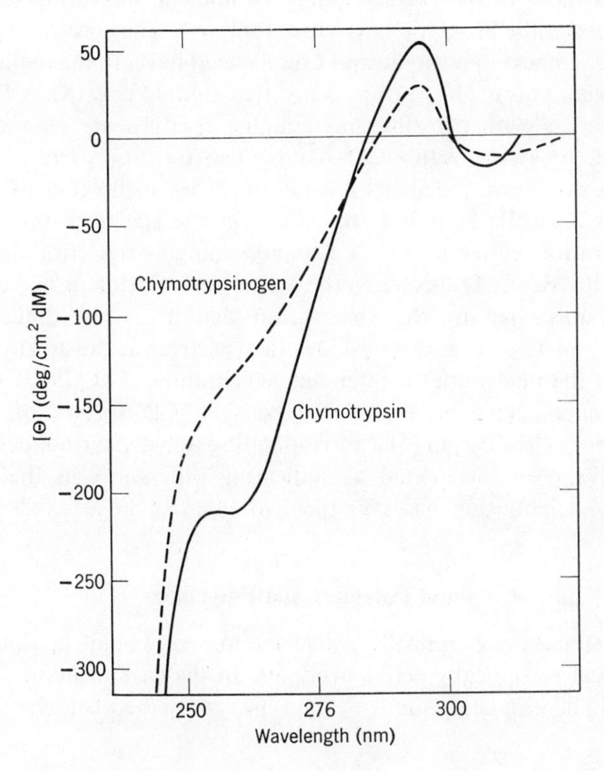

Fig.V.20. Change in the long-wavelength end of the CD spectrum characterizing the conversion of chymotrypsinogen to chymotrysin.

residues, which are responsible for the long-wavelength CD bands, are placed in regions of higher asymmetry when the active enzyme is formed from its precursor. Another interesting example of the application of CD spectroscopy concerns the mechanism of the thermal denaturation of ribonuclease (Simons et al., 1969). The spectral maximum, located at 241 nm, was found to increase with rising temperature above 15°C, while the minimum at 277 nm began to change only above 55°C. This evidence is clearly inconsistent with a two-state equilibrium between the native and the denatured enzyme and shows that at least two steps must be involved in the denaturation process.

3. Polynucleotides

Circular dichroism measurements on synthetic polynucleotides have demonstrated that this technique has some striking advantages for the study of this group of substances. Oligomers and polymers have CD spectra quite different from those of the corresponding monomers, indicating strong interactions between the aromatic base chromophores. The spectra are of two types: Short oligomers of adenylic and uridylic acid have, in the region of the aromatic chromophore absorption, a positive and a negative CD band, whereas oligomers with cytosine and guanine residues are characterized, in this region, by two positive CD bands. The transition from a single-stranded to a double-stranded helix leads to an intensification of the CD bands in poly-A, while in poly-C the shape of the spectrum undergoes a profound alteration, with one of the CD bands changing sign (Brahms, 1968). The high sensitivity of CD spectra to the relative orientation of two chromophores is demonstrated by the observation that in $3' \rightarrow 5'$-dinucleotides containing G and U or C and A residues the spectrum is drastically altered if the order of the nucleotides is interchanged (Brahms et al., 1967). Characteristic differences have been observed between CD spectra of double-helical polyribonucleotides and the corresponding polydeoxyribonucleotides, and these have been interpreted as indicating differences in the mutual orientation of neighboring bases in the two types of helices (T'so et al., 1966).

4. Vinyl Polymers and Polyethers

The polymerization of optically active monomers would, in general, be expected to lead to optically active products. In the case of homopolymers of vinyl or vinylidence compounds of the type represented below:

$$\begin{array}{c} X \\ | \\ (\text{—C—CH}_2\text{—})_n \\ |\,* \\ R_1\text{—C—}R_2 \\ | \\ R_3 \end{array}$$

the optical activity is due to the asymmetric centers in the side chains, and it disappears if these side chains are modified so as to remove the centers of asymmetry (Marvel and Overberger, 1946; Overberger and Palmer, 1956). This is easily understood since the tertiary or quaternary carbons in the chain backbone are not truly asymmetric centers (see p. 99) and therefore will not contribute to optical activity. Although isotactic chains may have a significant tendency to assume helical conformations even in solution, helical sections are equally likely to be right-handed and left-handed if the chain contains no true centers of asymmetry, so that contributions of the helices to optical activity will cancel out.

An entirely different situation arises in the polymerization of materials such as L-propylene oxide. If this monomer is polymerized under conditions leading to an isotactic product (Price and Osgan, 1956):

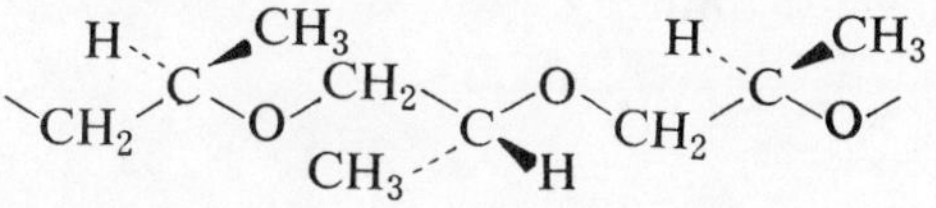

$$\sim CH_2-\overset{H\ \ CH_3}{C}-O-CH_2-\underset{CH_3\ \ H}{C}-O-CH_2-\overset{H\ \ CH_3}{C}-O\sim$$

the polymer chains contain centers of true asymmetry, since the carbons carrying the methyl substituents are preceded by methylene and followed by oxygen in the chain backbone, so that the two directions along the chain are clearly differentiated. This polymer, therefore, exhibits the expected optical activity, and fractionation procedures designed to separate the most isotactic fractions will also provide material of the highest optical activity. Price and Osgan found that the polymer is dextrorotatory in chloroform, but levorotatory in benzene solution. Yet Hirano et al., (1972) found that the coupling constants deduced from the NMR spectra are independent of the medium, so that the change in the sign of the optical activity does not appear to reflect a solvent dependence of the polymer conformation. Tsuji (1973) assumed that the reversal of optical activity is due to solvent interaction with the asymmetric carbons and supported this interpretation by the finding that poly[(S)-isopropylethylene oxide]:

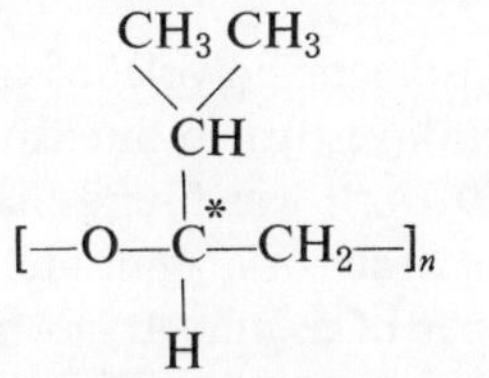

$$[-O-\overset{\overset{CH_3\ CH_3}{\diagdown\diagup}\atop{\overset{CH}{|\,*}}}{C}-CH_2-]_n \quad (\text{C bears } H \text{ below})$$

in which the chain substituents shield the asymmetric carbon from the solvent, has similar optical activities in all media.

Schuerch and his collaborators (Beredjick and Schuerch, 1958; Schmitt and Schuerch, 1960) pointed out that copolymers of vinyl compounds contain true centers of asymmetry in the chain backbone which may lead to significant optical activities. For instance, in the repeating unit of an alternating maleic anhydride copolymer:

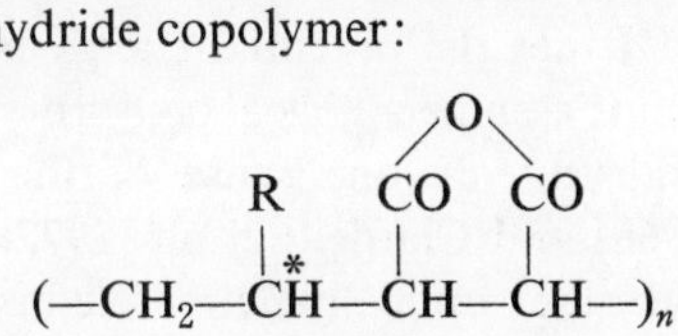

$$(-CH_2-\overset{R}{\overset{|\,*}{C}}H-\overset{CO}{\overset{|}{C}}H-\overset{CO}{\overset{|}{C}}H-)_n \quad (\text{the two CO groups joined by O})$$

the carbon atom marked with an asterisk is asymmetric, and the polymer should be optically active if it can be prepared so that one configuration predominates over the other. This result may be achieved by asymmetric induction. For example, if an optically active monomer, such as α-methylbenzyl vinyl ether, is copolymerized with maleic anhydride (Schmitt and Schuerch, 1960), the repeating unit of the alternating copolymer may exist in two diastereoisomeric forms:

$$
\begin{array}{c}
C_6H_5 \\
| \\
CH_3{-}\overset{*}{C}{-}H \\
| \\
O \\
| \\
{-}({-}CH_2{-}\overset{*}{C}{-}H \\
| \\
CH{-}CH{-}){-} \\
|\quad\;\; | \\
CO\;\; CO \\
\diagdown O \diagup
\end{array}
\qquad\qquad
\begin{array}{c}
C_6H_5 \\
| \\
CH_3{-}\overset{*}{C}{-}H \\
| \\
O \\
| \\
{-}({-}CH_2{-}\overset{*}{C}{-}CH{-}CH{-}){-} \\
|\qquad |\qquad | \\
H\quad CO\quad CO \\
\diagdown O \diagup
\end{array}
$$

The configuration of the side chain of the optically active vinyl ether influences the mode of addition of the maleic anhydride to some extent, so that one of the diastereoisomeric forms is kinetically favored. Therefore, the copolymer remains optically active after reduction of the ether side chains to hydroxyl groups.

The optical rotations of isotactic polyolefins carrying asymmetric centers in their side chains have been investigated intensively at the University of Pisa (Pino, 1965; Pino et al., 1970). It was found that the optical activities of the polymers are higher, by an order of magnitude, than those of low molecular weight analogs, if the center of asymmetry is placed α or β with respect to the chain backbone. The temperature coefficients of the optical activities are similarly enhanced in the polymers. Both these effects decrease rapidly, however, if the center of asymmetry is shifted further out in the side chain. The optical rotation increases with the stereoregularity of the polymer and is similar in solution and in the solid state. Typical values are $[m]_{589} = +161$ and $d[m]_{589}/dT = 0.36$ for poly [(S)-3-methyl-1-pentene] as against $[M]_{589} = +9.9$ and $d[M]_{589}/dT = -0.010$ for the analogous (S)-3-methylhexane.

We saw in Chapter III that the potential energy of isotactic polymers is minimized when the chains assume a helical conformation and that in chains carrying asymmetric substituents one sense of the helix is energetically favored. Pino et al. (1968) and Ciardelli et al. (1972) copolymerized (R)-3, 7-dimethyl-1-octene and other optically active olefins with a small proportion of styrene and observed a CD band in the region of absorption by the phenyl residue. This demonstration that the aromatic ring is in an asymmetric environment provides an elegant proof that the polymer retains, even in solution, a preference for one sense of the helical winding. Birshtein and Luisi (1964) and Allegra et al. (1966) reached the conclusion that this preferred sense of the helix is responsible for the high optical activities observed. Pino (1965) and Pino et al. (1970) used the empirical method of Brewster (1959) to calculate optical activities which could be compared with experimental data. This method first sets down rules for "allowed" and "forbidden" conformations and then suggests procedures for assigning optical activities to the various allowed forms. The first part is obviously a crude approximation, since different conformations are in fact characterized by different statistical weights arising from their different energies. Hence calculations by this method do not reveal the temperature dependence of optical activity, one of the important experimental parameters. In addition, Brewster's rule is incorrect in forbidding structures such as the *gauche* form of 2,3-dimethylbutane (see Fig. III.13), which has, in fact, a statistical weight very similar to that of the *anti* conformation (Szasz and Sheppard, 1949). To avoid this difficulty, Abe (1968) estimated the conformational distribution of optically active polyolefins from their conformational energies, using the statistical methods developed by Flory and his school. On the basis of Brewster's optical activity parameters, he obtained for polyolefins with asymmetric centers in the α or β position $[m]$ values close to those observed for the isotactic polymer, whatever the tacticity of the chain for which the computation was carried out. Thus, the helical forms of isotactic polymers would, presumably, not contribute significantly to the optical activity.

It has long been known that compounds with a high degree of conformational rigidity tend to have high optical activities. For instance, Kauzmann et al. (1940) pointed out that cyclic sugars have optical activities which are generally higher by an order of magnitude than those of their open chain analogs. This would be expected on statistical grounds, since different conformations may make positive or negative contributions to $[\alpha]$, as was illustrated in our discussion of 1,2-dichloropropane on p. 246, and the probability that these contributions will cancel one another will increase with the number of accessible conformations. There is no doubt that olefins with branch points in their side chains are highly rigid structures, but it is hard to reconcile Abe's conclusions with extensive evidence that enhanced optical activity parallels increasing stereoregularity. It is also significant that plots of optical activity against the optical purity of the monomer are concave toward the abscissa (Pino et al., 1967). This could be interpreted as signifying that the helix does make an important contribution to the optical activity of the polymer and that a strong preference for one sense of the helix is attained at a rather low degree of steric purity (in analogy with the behavior of the copolypeptides represented in Fig. V.15). Unfortunately, the Cotton effect of polyolefins is too far in the ultraviolet for its experimental observation with presently available instruments.

Studies of the optical activity of ionizable polymers carrying asymmetric substituents are of interest since a change in the degree of ionization may produce large changes in the conformation of the chain backbone. Typical data obtained with poly(*N*-methacrylyl-L-glutamic acid) and its analog, *N*-pivalyl-L-glutamic acid (Kulkarni and Morawetz, 1961):

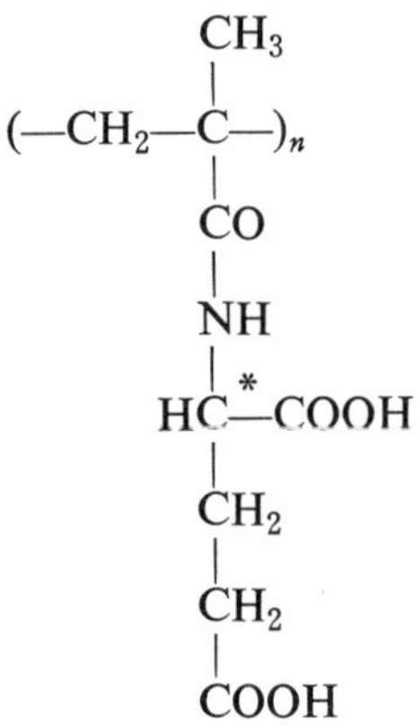

$$\begin{array}{c} CH_3 \\ | \\ (-CH_2-C-)_n \\ | \\ CO \\ | \\ NH \\ | \\ H\overset{*}{C}-COOH \\ | \\ CH_2 \\ | \\ CH_2 \\ | \\ COOH \end{array} \qquad \begin{array}{c} CH_3 \\ | \\ CH_3-C-CH_3 \\ | \\ CO \\ | \\ NH \\ | \\ H\overset{*}{C}-COOH \\ | \\ CH_2 \\ | \\ CH_2 \\ | \\ COOH \end{array}$$

Poly(*N*-methacrylyl-L-glutamic acid) 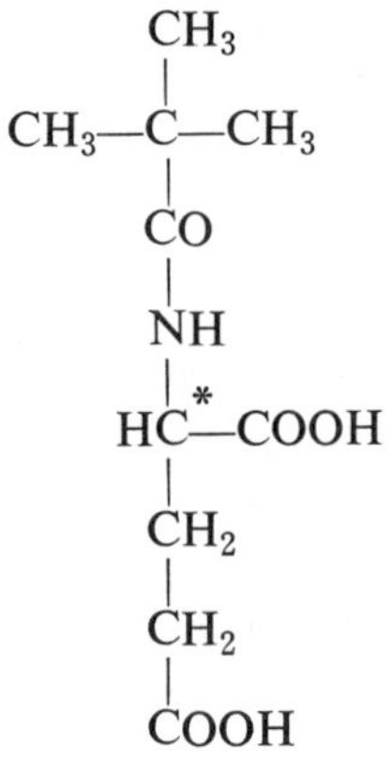 *N*-pivalyl-L-glutamic acid

and represented in Fig. V.21 show a striking difference between the behavior of the polymer and that of its analog. The large negative values of $[\alpha]$ observed with the partially ionized polymer seem to be a consequence of changes in the side chain conformations as the polymer backbone assumes a more expanded form. Such changes could be produced either by interactions of side chains with the backbone or by interactions of side chains with each other. The optical activity depends significantly on the conditions under which the polymer was prepared, indicating a sensitivity of the optical activity to small changes in stereoregularity. A copolymer containing 15% of D-monomer units behaves in a manner similar to that observed with the homopolymer of the L-monomer, confirming the assumption that the complex dependence of optical activity on the degree of ionization is not caused by a helix-coil transition. We are led to the conclusion that, although optical activity studies are a sensitive tool for revealing qualitatively the existence of conformational transitions, the multiplicity of possible effects may preclude a detailed interpretation of data.

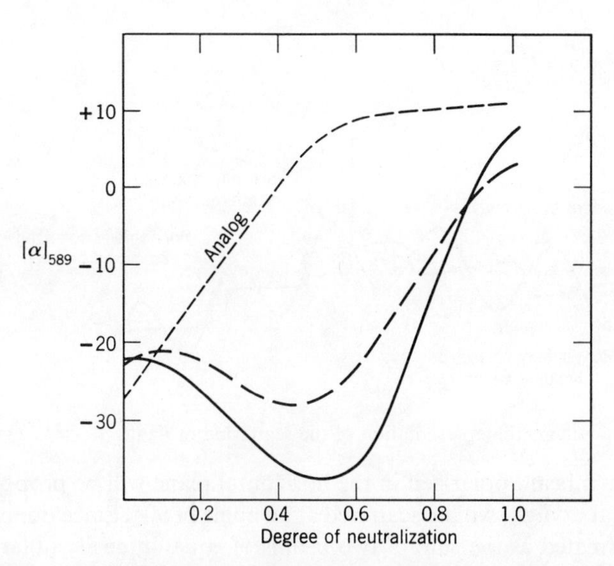

Fig. V.21. Dependence of the optical activity of poly(*N*-methacrylyl-L-glutamic acid) on its degree of neutralization. (—)Polymer prepared in water; (- - -)polymer prepared in dioxane. Results for the analog, *N*-pivalyl-L-glutamic acid, are also shown.

C. LIGHT SCATTERING

1. Scattering by Gases

Many of the fundamental concepts in the interpretation of light scattering were formulated by Lord Rayleigh (1871, 1881), who concerned himself with the scattering by gases, in which the scattering molecules are placed at random at large distances from each other and make independent contributions to the effects observed. Scattering occurs because the oscillating electrical field of the incident light wave induces an oscillating dipole in molecules lying in its path, which will then radiate light in all directions. It is assumed that only a very small fraction of the incident light beam is scattered, so that the effect of multiple scattering may be neglected. The dependence of the scattered light intensity i_θ on the angle θ, subtended by the directly transmitted and the scattered light ray, is derived most easily if the radiating molecules are electrically isotropic and small compared to the wavelength λ of the light employed. In the schematic representation in Fig. V.22 the scattering from the horizontal incident beam AO is observed in various horizontal direction OC. If the incident light is polarized in the plane AOV, the scattered light will also be polarized in a vertical plane (e.g., COV) and its intensity will be independent of θ. On the other hand, the intensity of light

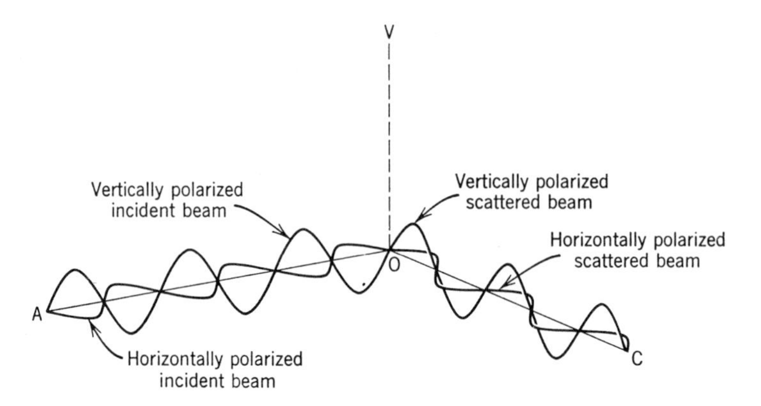

Fig. V.22. Schematic representation of the scattering of light.

scattered from a beam polarized in the horizontal plane will be proportional to $\cos^2\theta$, so that no light will be scattered at an angle of 90°. Since unpolarized light may be treated as the sum of two beams of equal intensity polarized in planes perpendicular to each other, the intensity of the light scattered from an unpolarized beam will be proportional to $(1 + \cos^2\theta)$ and the scattered light will be fully polarized if observed at a scattering angle of 90°. All the scattered light intensities will be inversely proportional to the square of the distance l from the scattering source O.

The amplitude of the oscillating dipole induced in a molecule by the oscillating field of the incident light is proportional to the polarizability α of the molecule. However, when the scattered ray comes from a large number of molecules spaced at random, the electrical vector due to any one of them may be positive or negative with equal probability at any given time and location. The summation of such vectors is then analogous to the "random walk" problem discussed on p. 120, so that it will result in a value in which the amplitude of the vibration characterizing the ray scattered by a single molecule is multiplied by the square root of their number. The intensity of the radiated energy, which is proportional to the square of the amplitude of the ray, is then proportional to the square of the molecular polarizability but the first power of the number of radiating molecules. According to the Rayleigh theory, if I_0 is the intensity of the incident light, the scattered beam intensity i_θ due to a unit volume of the scattering medium containing N^* molecules of polarizability α is given by

$$\mathscr{R}_\theta \equiv \frac{i_\theta l^2}{I_0} = \frac{8\pi^4 N^* \alpha^2}{\lambda^4}(1 + \cos^2\theta) \tag{V.12}$$

where $\mathscr{R}_\theta$, commonly referred to as the Rayleigh ratio, is defined so that the scattering may be characterized without reference to the distance l at which the light is observed. Using the Clausius-Mosotti relation between the molecular polarizability α nad the dielectric constant $\mathscr{D}$:

$$\alpha = \frac{3}{4\pi N^*}\frac{\mathscr{D} - 1}{\mathscr{D} + 2} \tag{V.13}$$

we obtain, after substituting $N^* = \rho \boldsymbol{N}/M$, where ρ is the density and M the molecular weight, for a gas in which $\mathscr{D} \approx 1$,

$$\mathscr{R}_\theta = \frac{\pi^2}{2}\frac{M(\mathscr{D} - 1)^2}{\lambda^4 \boldsymbol{N}\rho}(1 + \cos^2\theta) \tag{V.14}$$

For a series of substances, such as the homologous series of paraffins, for which the dielectric constant would be expected to remain constant at constant gas densities, the scattering will then be proportional to the molecular weight. This is so since vibrating dipoles in different parts of any one molecule are in phase with one another, in contrast to the random shift in phase between dipole vibrations in different molecules. When the molecular weight of the scattering molecules is known, relation (V.14) may be used to estimate Avogadro's number $\boldsymbol{N}$.

When the scattering molecules are not isotropic, the magnitude and direction of the oscillating dipole of the scattering molecule will depend on its orientation and will not necessarily lie in the plane of polarization of the incident beam. As a consequence, light scattered at a right angle from an unpolarized beam is no longer polarized in a vertical plane but has a horizontally polarized component H_u, whose ratio to the vertically polarized component V_u is given by (Rayleigh, 1918)

$$\rho_u \equiv \frac{H_u}{V_u} = \frac{2(\alpha_1^2 + \alpha_2^2 + \alpha_3^2 - \alpha_1\alpha_2 - \alpha_1\alpha_3 - \alpha_2\alpha_3)}{4(\alpha_1^2 + \alpha_2^2 + \alpha_3^2 + \alpha_1\alpha_2 + \alpha_1\alpha_3 + \alpha_2\alpha_3)} \tag{V.15}$$

where α_1, α_2, and α_3 are the three principal polarizabilities of the scattering molecules. According to (V.15), ρ_u attains a maximum value of 0.5 for long rod-shaped particles in which one of the polarizabilities is much larger than the other two. Depolarization measurements on light scattered from gases lead to ρ_u values ranging from 0 to 0.125. It has also been shown (cf. Kerker, 1969c) that the intensity of light scattered at 90° from anisotropic molecules is increased over the value predicted by (V.14) by the "Cabannes factor" $(6 + 6\rho_u) / (6 - 7\rho_u)$.

2. Scattering from Solutions of Molecules Much Smaller than the Wavelength of Light

When the scattering molecules are not distributed at random, destructive interference will tend to reduce the intensity of the scattered radiation. This factor becomes important in liquids with their partially ordered structures.

It could be dealt with, in principle, by calculating explicitly the interference corresponding to a known radial distribution of molecules in a given liquid; but the procedure would be extremely laborious, and it is more convenient (Debye, 1944) to approach the problem from fluctuation theory, as first suggested by Einstein (1910).

Let us consider any intensive thermodynamic property g such as density or concentration in a system of one or several components. Although the system is macroscopically uniform, with a bulk-average value g_0 of the property with which we are concerned, the local value of this property fluctuates on a scale small compared to the wavelength of the light used. The probability distribution of such fluctuations, subject to the restriction of a constant volume and constant energy of the system, will be given by

$$\mathscr{W}(g)\,dg = \frac{\exp[-F(g)/\boldsymbol{k}T]dg}{\int_0^\infty \exp[-F(g)\boldsymbol{k}T]\,dg} \tag{V.16}$$

where $F(g)$ is the Helmholtz free energy associated with g. For systems close to equilibrium, $F(g)$ may be expanded in a Taylor series around g_0:

$$F(g) = F(g_0) + \left(\frac{dF}{dg}\right)_{g=g_0}(g - g_0) + \frac{1}{2}\left(\frac{d^2F}{dg^2}\right)_{g=g_0}(g - g_0)^2 + \cdots \tag{V.17}$$

where the first derivative must vanish, since it is evaluated at the position of equilibrium. Neglecting terms beyond the second power of $(g - g_0)$, we can write

$$\mathscr{W}(g)\,dg = \frac{\exp[-a(g - g_0)^2]\,dg}{\int_{-\infty}^{\infty}\exp[-a(g - g_0)^2]\,dg}$$

$$a = \frac{1}{2\boldsymbol{k}T}\left(\frac{d^2F}{dg^2}\right)_{g=g_0} \tag{V.18}$$

and, on evaluating the integral, we obtain

$$\mathscr{W}(g)\,dg = \sqrt{a/\pi}\,\exp[-a(g - g_0)^2]\,dg \tag{V.19}$$

A local fluctuation in any given property will, in general, lead to a corresponding fluctuation in the dielectric constant $\mathscr{D}$. In turn, such a fluctuation will lead, according to the Rayleigh theory, to

$$\mathscr{R}_\theta = \frac{\pi^2}{2\lambda^4}\langle\Delta\mathscr{D}^2\rangle\,\delta V(1 + \cos^2\theta) \tag{V.20}$$

where $\langle\Delta\mathscr{D}^2\rangle$ is the mean square deviation of the local dielectric constant, within a volume element δV, from the bulk average value of $\mathscr{D}$. Using the distribution function (V.19), we obtain

$$\begin{aligned}\langle \Delta\mathscr{D}^2\rangle &= \left(\frac{\partial\mathscr{D}}{\partial g}\right)^2 \langle \Delta g^2\rangle \\ &= \left(\frac{\partial\mathscr{D}}{\partial g}\right)^2 \int_{-\infty}^{+\infty} (g - g_0)^2 \sqrt{a/\pi}\, \exp\,[-\, a(g - g_0)^2]\, dg \\ &= \frac{(\partial\mathscr{D}/\partial g)^2 \boldsymbol{k}T}{(d^2F/dg^2)_{g=g_0}} \end{aligned} \tag{V.21}$$

We may now apply this result to two cases of special interest. In a pure liquid, fluctuations in the local value of the dielectric constant may be caused, in principle, by both density and temperature fluctuations, but the effect of temperature fluctuations is generally negligible. It can be shown from thermodynamics that

$$\left(\frac{\partial^2 F}{\partial \rho^2}\right)_T = \frac{V}{\rho^2}\left(\frac{1}{\beta} - 2P\right) \tag{V.22}$$

where β is the isothermal compressibility. We may neglect the second term on the right and obtain, by substitution into (V.20) and (V.21),

$$\mathscr{R}_\theta = \frac{\pi^2}{2\lambda^4}\, \boldsymbol{k}T\beta\left[\rho_0\left(\frac{\partial\mathscr{D}}{\partial\rho}\right)_T\right]^2 (1 + \cos^2\theta) \tag{V.23}$$

For light of $\lambda = 546$ nm, the $\mathscr{R}_{90}$ value (i.e., Rayleigh ratios for scattering at an angle of 90°) for water is 0.86×10^{-6}; it is 4.56×10^{-6} for cyclohexane, 5.38×10^{-6} for carbon tetrachloride, 4.18×10^{-6} for butanone, 15.8×10^{-6} for benzene, and 83.9×10^{-6} for carbon disulfide (Kerker, 1969b).

If, on the other hand, we deal with a solution, the scattering resulting from density fluctuations and that caused by local fluctuations in the concentration of the solute may be considered to be additive. It may be shown that

$$\left(\frac{\partial^2 F}{\partial c_2^2}\right)_T = -\left(\frac{\delta V}{\bar{V}_1 c_2}\right)\left(\frac{\partial \bar{F}_1}{\partial c_2}\right)_T \tag{V.24}$$

and with the osmotic pressure $\Pi = -(\bar{F}_1 - \bar{F}_1^0)/\bar{V}_1$, we obtain, by substitution into (V.20) and (V.21), for the contribution $\Delta\mathscr{R}_\theta$ to the Rayleigh ratio resulting from fluctuations in solute concentration

$$\Delta\mathscr{R}_\theta = \frac{\pi^2(\partial\mathscr{D}/\partial c_2)_T^2}{2\lambda^4(\partial\Pi/\partial c_2)_T}\, \boldsymbol{k}Tc_2(1 + \cos^2\theta) \tag{V.25}$$

The dielectric constant may be equated to the square of the refractive index n, and for dilute solutions we may approximate

$$\left(\frac{\partial\mathscr{D}}{\partial c_2}\right)_T^2 = 4\text{n}^2\left(\frac{\partial \text{n}}{\partial c_2}\right)_T^2 \approx 4\text{n}_0^2\left(\frac{\partial \text{n}}{\partial c_2}\right)_T^2 \tag{V.26}$$

where n_0 is the refractive index of the solvent. The Rayleigh ratio increment is then given by

$$\Delta\mathscr{R}_\theta = \frac{K_\theta \boldsymbol{R}Tc_2}{(\partial\Pi/\partial c_2)_T}$$

$$K_\theta = \frac{2\pi^2 n_0^2(\partial n/\delta c_2)_T^2}{\lambda^4 \boldsymbol{N}}(1 + \cos^2\theta) \tag{V.27}$$

Finally, expressing the osmotic pressure as in (IV.14), we obtain

$$\left(\frac{\partial\Pi}{\partial c_2}\right)_T = \boldsymbol{R}T\left(\frac{1}{M_2} + 2A_2c_2 + 3A_3c_2^2 + \cdots\right) \tag{V.28}$$

so that

$$\frac{K_\theta c_2}{\Delta\mathscr{R}_\theta} = \frac{1}{M_2} + 2A_2c_2 + 3A_3c_2^2 + \cdots \tag{V.29}$$

and a plot of $K_\theta c_2/\Delta\mathscr{R}_\theta$ against c_2 will yield the reciprocal molecular weight of the solute as the intercept. Let us now consider the qualitative significance of this result. The local fluctuations in the concentration of the solute reflect an equilibrium in which the buildup of concentration gradients, due to random molecular motion, is opposed by osmotic pressure gradients, which tend to restore the uniformity of the system. Since the osmotic forces for any given concentration gradient become smaller with an increase in the size (and decrease in number) of the solute molecules, the fluctuations in concentration will increase with an increasing molecular weight of the solute. Because of this intimate relationship between osmotic and fluctuation phenomena, the concentration dependence of light scattering may be used to evaluate the osmotic virial coefficients A_2 and A_3, whose significance was discussed in Chapter IV.

The result for a single solute species may be extended to a polydisperse polymer sample, in which the refractive index increment $(\partial n/\partial c_i)_T$ is identical for all molecular weight fractions. In the limit of high dilution

$$\lim_{c_i\to 0}\left(\frac{K_\theta c_i}{\Delta\mathscr{R}_{\theta i}}\right) = \frac{1}{M_i}$$

$$\Delta\mathscr{R}_\theta = \sum_i \Delta\mathscr{R}_{\theta i} = K_\theta \sum c_i M_i \tag{V.30}$$

so that

$$\lim_{c_2\to 0}\left(\frac{K_\theta c_2}{\Delta\mathscr{R}_\theta}\right) = \frac{\sum_i c_i}{\sum_i c_i M_i} = \frac{1}{\bar{M}_w} \tag{V.31}$$

where $\boldsymbol{c}_2$ stands for the sum of concentrations of all species in a polymer-homologous solute. Theoretical analysis predicts also that with polydisperse polymers the A_2 value derived from light scattering data may differ from that characterizing the osmotic behavior (Flory and Krigbaum, 1950; Yamakawa and Kurata, 1960; Casassa, 1966). For mixtures of two monodisperse poly-(α-methylstyrene) samples, A_2 was found to pass through a maximum as the composition of the system was varied (Kato et al., 1968), an effect which had been expected from theory.

The range of molecular weights conveniently investigated by the light scattering technique extends from about 20,000 to several millions. The lower limit corresponds to the molecular weight of the solute, for which the excess in the light scattering intensity is small compared to the scattering intensity of the pure solvent. The upper limit is brought about by the experimental difficulty of separating satisfactorily very high molecular weight solutes from adventitious extraneous impurities ("dust"), which are inevitably introduced into the system, and by instrumental limitations, which usually do not allow meaningful measurements to be made at small scattering angles. With various methods employed to surmount these problems, the range of molecular weights accessible to measurement by the light scattering method may be extended at least by an order of magnitude.

3. Estimation of the Size and Shape of Macromolecules from the Angular Dependence of the Scattered Light Intensity

When the dimensions of the scattering particles exceed about one-twentieth of the wavelength of the light used, the intensity of the scattered light will be reduced by a measurable amount as a consequence of destructive interference. A schematic representation of light scattered by two point scatterers is given in Fig. V.23. The displacement of the scattering elements from one another is characterized by a vector **r**, which is pictured for simplicity as lying in the plane defined by the incident and the scattered light

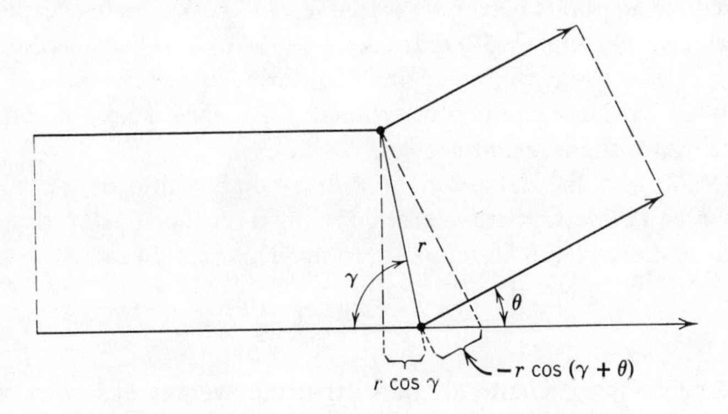

Fig. V.23. Two-dimensional representation of the difference in the path length of rays scattered by a particle with two scattering centers.

beam. If **r** subtends an angle γ with the incident beam and the scattered light is observed at an angle θ from the directly transmitted beam, the difference in the path lengths of two parallel rays scattered by the two point scatterers is $r[\cos \gamma - \cos (\gamma + \theta)]$, so that the two scattered rays will be out of phase by an angle $2\pi r[\cos \gamma - \cos (\gamma + \theta)]/\lambda'$, where λ' is the wavelength of the light in the scattering medium. This phase angle will increase with increasing θ, leading to a corresponding increase in the destructive interference and a decrease in the intensity of the scattered light. To carry out a quantitative evaluation of this effect for a particle which may be represented as an assembly of N scatterers, the destructive interference must be averaged over all possible relative orientations of the assembly to the incident beam. Such a calculation leads to

$$\mathscr{P}(\theta) = \frac{1}{N^2} \sum_i \sum_j \frac{\sin \mathrm{k} r_{ij}}{\mathrm{k} r_{ij}}$$
$$\mathrm{k} = \frac{4\pi \sin (\theta/2)}{\lambda'} \qquad \text{(V.32)}$$

where $\mathscr{P}(\theta)$ is the factor by which the beam scattered at an angle θ is attenuated because of destructive interference. If the scattering particle is deformable so that the distances r_{ij} between the scattering points are variable, the double summation in (V.32) has to be averaged over time. Although relation (V.32) was originally derived for the angular dependence of x-ray scattering from gases (Debye, 1915), it is equally applicable to the light scattering by dissolved macromolecules provided that the difference in the refractive indices of the solution and the solute is small (Debye, 1947). In practice, this condition does not represent a serious limitation to the study of macromolecular solutions. For scattering points very close to one another, so that $\sin \mathrm{k} r_{ij}/\mathrm{k} r_{ij} \approx 1$, the quantity $\mathscr{P}(\theta)$ will approach unity. On the other hand, when the distance between scattering points becomes very large, $\sin \mathrm{k} r_{ij}/\mathrm{k} r_{ij} \approx 0$, except for the terms in which $i = j$, and (V.32) reduces then to $\mathscr{P}(\theta) = 1/N$. This result again shows that in dilute systems and at small scattering angles the intensity of the light scattered by the solute is proportional, for a given solution concentration, to the size of the solute molecules.

From (V.27) and the definition of $\mathscr{P}(\theta)$ a combination of the effects of polarization and of destructive interference gives, for the angular dependence of the light scattered from an unpolarized incident beam,

$$\Delta\mathscr{R}_\theta = \frac{\Delta\mathscr{R}_0 \mathscr{P}(\theta)(1 + \cos^2\theta)}{2} \qquad \text{(V.33)}$$

If the solution is sufficiently dilute so that the average distances between solute molecules are large compared to the wavelength of light, the form of $\mathscr{P}(\theta)$ may be considered to reflect solely the effects of destructive interference of light scattered from different points of an individual solute molecule. We may expand (V.32) into a power series to yield

$$\mathscr{P}(\theta) = \frac{1}{N^2}\left(N^2 - \frac{1}{3!}\sum_i \sum_j \mathrm{k}^2 r_{ij}^2 + \frac{1}{5!}\sum_i \sum_j \mathrm{k}^4 r_{ij}^4 - \cdots\right) \quad \text{(V.34)}$$

The radius of gyration of any assembly of identical particles is related to the r_{ij} values by

$$\langle s^2 \rangle = \frac{1}{2N^2}\sum_i \sum_j r_{ij}^2 \quad \text{(V.35)}$$

so that $\mathscr{P}(\theta)$ assumes the form

$$\begin{aligned}\mathscr{P}(\theta) &= 1 - \tfrac{1}{3}\,\mathrm{k}^2\langle s^2\rangle + \cdots \\ &= 1 - \tfrac{1}{3}\left(\frac{4\pi}{\lambda'}\right)^2 \langle s^2\rangle \sin^2\left(\frac{\theta}{2}\right) + \cdots \end{aligned} \quad \text{(V.36)}$$

We see then that the dependence of the scattered light intensity on the scattering angle, in the limit of small θ, yields an unambiguous measure of the mean radius of gyration of the scattering particle.† As the kr_{ij} terms assume larger values, the shape of $\mathscr{P}(\theta)$ becomes dependent on the higher moments of the distribution of the scattering elements in the solute particle. Calculations of $\mathscr{P}(\theta)$ have been carried out for spheres of uniform density (Rayleigh, 1911), rigid thin rods (Neugebauer, 1942; Zimm et al., 1945), Gaussian chains (Debye, 1947), and polydisperse Gaussian chains with a normal distribution of chain lengths (Zimm, 1948).

These calculations yield the following expressions in terms of the dimensionless parameter, $\mathrm{u} = \mathrm{k}^2\langle s^2\rangle$.

Spheres of radius R:

$$\mathscr{P}(\theta) = \left[\frac{3}{\mathrm{x}^3}(\sin \mathrm{x} - \mathrm{x}\cos \mathrm{x})\right]^2$$

$$\mathrm{x}^2 = \mathrm{k}^2 R^2 = \frac{5\mathrm{u}}{3} \quad \text{(V.37)}$$

Rods of length L:

$$\mathscr{P}(\theta) = \frac{1}{\mathrm{x}}\int_0^{2\mathrm{x}} \frac{\sin \mathrm{y}}{\mathrm{y}}\,d\mathrm{y} - \left(\frac{\sin \mathrm{x}}{\mathrm{x}}\right)^2$$

$$\mathrm{x}^2 = \frac{\mathrm{k}^2 L^2}{4} = 3\mathrm{u} \quad \text{(V.38)}$$

†This conclusion is strictly valid only for optically isotropic particles. In the case of scattering by rigid rods, the relation between $\mathscr{P}(\theta)$ and $\langle s^2\rangle$ has been shown to depend on the optical anisotropy of the scattering particle (Horn et al,. 1951; Horn, 1955). For Gaussian chains, the effect of the anisotropy of the chain segment is generally negligible (Horn, 1955; Benoit and Weill, 1957).

Monodisperse Gaussian coils:

$$\mathscr{P}(\theta) = \frac{2}{\mathrm{u}^2}(e^{-\mathrm{u}} + \mathrm{u} - 1) \quad \text{(V.39)}$$

Polydisperse Gaussian coils:

$$\mathscr{P}(\theta) = 1 - \frac{\langle \mathrm{u} \rangle_w}{2 + \langle \mathrm{u} \rangle_w} \quad \text{(V.40)}$$

where $\langle \mathrm{u} \rangle_w$ corresponds to the weight-average chain length.

In principle it should be possible to eliminate the effects of destructive interference by measuring the scattered light intensity at sufficiently small scattering angles θ. In practice, however, conventional light scattering photometers are restricted to the angular range $30° \leqslant \theta \leqslant 150°$ because of experimental difficulties to be discussed later. At such scattering angles the effect of destructive interference on the intensity of the light scattered by the macromolecules is significant and has to be taken into account to obtain correct molecular weights. This is done by modifying (V.29) for solutions of a single macromolecular species to give

$$\lim_{c_2 \to 0} \left(\frac{K_\theta c_2}{\Delta \mathscr{R}_\theta} \right) = \frac{1}{M_2 \mathscr{P}(\theta)} \quad \text{(V.41)}$$

A general method for treating experimental light scattering data so as to eliminate the effects of destructive interference and obtain the correct values for the molecular weight of the solute was outlined by Zimm (1948). In this procedure data are obtained for a number of solute concentrations, each at a number of scattering angles θ. The ratio $K_\theta c_2 / \Delta \mathscr{R}_\theta$ is then plotted as a function of $\sin^2(\theta/2) + \mathrm{q}c_2$, where q is an arbitrarily selected constant. Experimental points obtained at any given scattering angle may then be extrapolated to $c_2 = 0$. From (V.41) we know that $K_\theta c_2 / \Delta \mathscr{R}_\theta$ is proportional to $1/\mathscr{P}(\theta)$, and this must be linear in $\sin^2(\theta/2)$ for sufficiently small angles (V.36), so that it should be possible to extrapolate data obtained at different scattering angles to $\theta = 0$. By a double-extrapolation procedure it is possible, therefore, to eliminate both effects due to deviations from solution ideality and effects due to the destructive interference of the scattered light and obtain a value of $\bar{M}_w$ independent of any assumptions. A typical example of such a "Zimm plot" is shown in Fig. V.24. It also follows from (V.41) and (V.36) that

$$\lim_{c_2 \to 0} \left(\frac{K_\theta c_2}{\Delta \mathscr{R}_\theta} \right) = \frac{1}{\bar{M}_w} \left[1 + \frac{16\pi^2}{3(\lambda')^2} \langle s^2 \rangle \sin^2\left(\frac{\theta}{2}\right) \cdots \right] \quad \text{(V.42)}$$

so that the mean square radius of gyration may be calculated from the initial slope and the intercept of the $c_2 = 0$ line of the Zimm plot.

When the solution contains a polydisperse sample of macromolecules,

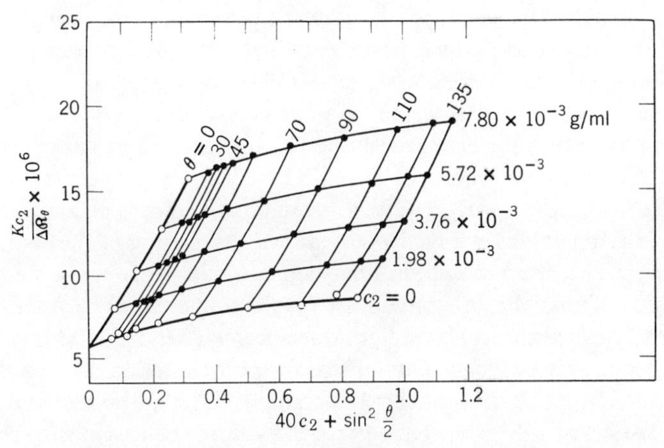

Fig. V.24. Zimm plot of light scattering data. Polystyrene, $\bar{M}_w = 170{,}000$, in dioxane solution.

the $\mathscr{P}(\theta)$ values corresponding to the individual species will be averaged to an experimentally observed $\bar{\mathscr{P}}(\theta)$, given by

$$\frac{\sum \Delta\mathscr{R}_i(\theta)}{K_\theta \sum c_i} = \frac{\sum \bar{M}_i c_i \mathscr{P}_i(\theta)}{\sum c_i}$$
$$= \bar{M}_w \bar{\mathscr{P}}(\theta) \tag{V.43}$$

which may be expressed in terms of the weight fraction w_i of the individual species as

$$\bar{\mathscr{P}}(\theta) = \frac{\sum M_i w_i \mathscr{P}_i(\theta)}{\sum M_i w_i}$$
$$= \frac{\sum N_i M_i^2 \mathscr{P}_i(\theta)}{\sum N_i M_i^2} \tag{V.44}$$

so that $\bar{\mathscr{P}}(\theta)$ is a z-average quantity. It follows then that the mean square radius of gyration obtained from a Zimm plot is $\langle s^2 \rangle_z$, involving the z-averaging of the sizes of the individual molecules in the sample.

For extremely large particles, plots of $1/\mathscr{P}(\theta)$ against $\sin^2(\theta/2)$ may be curved, down to rather small scattering angles. Thus, data from a conventional light scattering apparatus (in which readings are confined to the angular range 30–150°) cannot be reliably extrapolated to $\theta = 0$. This problem is particularly serious with DNA solutions in which the molecular weight of the solute is of the order of 10^7 or higher. In this case, the molecular weight will be seriously underestimated unless special efforts are made to measure light scattering intensities down to θ values as low as 10° (Harpst et al., 1968; Krasna, 1970). This requires not only special instrumental design (to ensure that none of the energy of the very intense direct beam affects the sensor measuring the intensity of the weak scattered beam) but also unusual care to eliminate dust particles, which scatter strongly at small scattering angles. With such improvements, Slagowski et al. (1971) were able to characterize a polystyrene with a molecular weight as high as 5×10^7.

The experimentally determined shape of $\mathscr{P}(\theta)$ at higher scattering angles may lead to useful conclusions of a qualitative nature. Inspection of (V.34) and (V.35) shows that, for any given mean square radius of gyration, $\mathscr{P}(\theta)$ at large scattering angles will be decreased by a sharpening of the distribution of r_{ij} values. The effect has been analyzed quantitatively for a number of special cases. Thus, the magnitude of $1/\mathscr{P}(\theta)$ is increased by a decreasing polydispersity of the molecular weight distribution (Benoit, 1953; Goldstein, 1953), by decreasing polydispersity of size for particles of equal weight (Rice, 1955), by increasing chain branching (Benoit, 1953), and by increased chain flexibility (Peterlin, 1953). Since the excluded volume effect tends to lead to a sharpening of the maximum in the radial distribution function of chain segment densities (see p. 137), we should also expect in the case of flexible chains in good solvent media that $1/\mathscr{P}(\theta)$ would rise at large scattering angles above the values predicted by (V.39) for Gaussian coils (Hyde et al., 1958). It is clear from this enumeration that the general case, where all these parameters might be variable, does not lend itself to a reliable interpretation. On the other hand, these general principles may find application in special cases, where there is reasonable certainty that only one of these properties is being varied. When dealing with very large solute particles, the asymptotic behavior of $\mathscr{P}(\theta)$ may be useful as an additional source of information. Two cases are of particular importance. For polydisperse Gaussian coils Benoit (1953) derived, for the asymptotic behavior of the $c_2 = 0$ line in a Zimm plot (the "Zimm envelope"),

$$\lim_{u\to\infty}\left(\frac{K_\theta c_2}{\Delta\mathscr{R}_\theta}\right) = \frac{1}{2\bar{M}_n} + \frac{8\pi^2}{(\lambda')^2}\frac{\langle s^2\rangle_w \sin^2(\theta/2)}{\bar{M}_w} \tag{V.45}$$

so that the intercept of the asymptote yields the number-average molecular weight, while its slope defines the weight-average radius of gyration. For chains with a normal distribution, so that $\bar{M}_w = 2\bar{M}_n$, the intercept of the asymptote coincides with that of the initial tangent, that is, the envelope of the Zimm plot will be linear over the entire course. An upward or downward curvature corresponds, therefore, to a narrower and broader distribution, respectively, than that given by the normal molecular weight distribution function. For the case of a system containing thin rods of constant cross section and variable length L, Casassa (1955) obtained

$$\lim_{u\to\infty}\left(\frac{K_\theta c_2}{\Delta\mathscr{R}_\theta}\right) = \frac{2\sum M_i w_i/L_i^2}{\pi^2(\sum M_i w_i/L_i)^2} + \frac{4\sin(\theta/2)}{\lambda'\sum M_i w_i/L_i} \tag{V.46}$$

so that a plot of $K_\theta c_2/\Delta\mathscr{R}_\theta$ against $\sin(\theta/2)$ yields limiting slopes proportional to the particle length per unit weight (i.e., the cross-sectional area of rods of known density). This result may be used advantageously to distinguish between end-to-end and side-by-side aggregation of rigid elongated macromolecules.

The collection of data required for a Zimm plot and their evaluation are fairly time consuming, and a less rigorous but much more rapid procedure for the estimation of molecular weights and the sizes of dissolved macromolecules is often desirable. Since the factor $(1 + \cos^2\theta)$, which takes into account the variation in the scattered light intensity caused by the polarization of the scattered light, has the same value for any two angles symmetrically distributed around 90°, the ratio of the light intensities observed at two such angles will reflect the ratio of the $\mathscr{P}(\theta)$ values. It is customary to make measurements at scattering angles of $\theta = 45°$ and $\theta = 135°$ and to express the results as the dissymmetry $z^* = \mathscr{P}(45°)/\mathscr{P}(135°)$. Substituting $\sin^2(22.5°) = (2 - \sqrt{2})/4$ and $\sin^2(67.5°) = (2 + \sqrt{2})/4$ into (V.36), we obtain

$$
\begin{aligned}
z^* &= 1 + \frac{8\pi^2\sqrt{2}\,\langle s^2\rangle}{3(\lambda')^2} + \cdots \\
&= 1 + \frac{37.2\langle s^2\rangle}{(\lambda')^2} + \cdots
\end{aligned}
\tag{V.47}
$$

As long as the radius of gyration is amall compared to the wavelength of the light in the scattering medium, z^* will be a reliable measure of $\langle s^2\rangle^{1/2}/\lambda'$. For larger particles, the terms in the higher moments of the distribution of scattering elements become important, and the relation between z^* and $\langle s^2\rangle^{1/2}/\lambda'$ will depend on particle shape. This is illustrated by values calculated by Doty and Steiner (1950) and plotted in Fig. V.25, which show that for $z^* > 1.7$ a given dissymmetry corresponds to significantly different radii of gyration for spherical, rodlike, and random coil particles.

The dissymmetry z^* may be used also to correct the scattered light intensity observed at 90° for the effects of destructive interference. From (V.36) we have

$$
\begin{aligned}
\frac{1}{\mathscr{P}(90°)} &= 1 + \frac{(8\pi^2/3)\langle s^2\rangle}{(\lambda')^2} - \cdots \\
&= 1 + \frac{26.3\langle s^2\rangle}{(\lambda')^2} - \cdots
\end{aligned}
\tag{V.48}
$$

and combining this with relation (V.47) gives

$$
\frac{1}{\mathscr{P}(90°)} = 1 + \frac{z^* - 1}{\sqrt{2}} + \cdots \tag{V.49}
$$

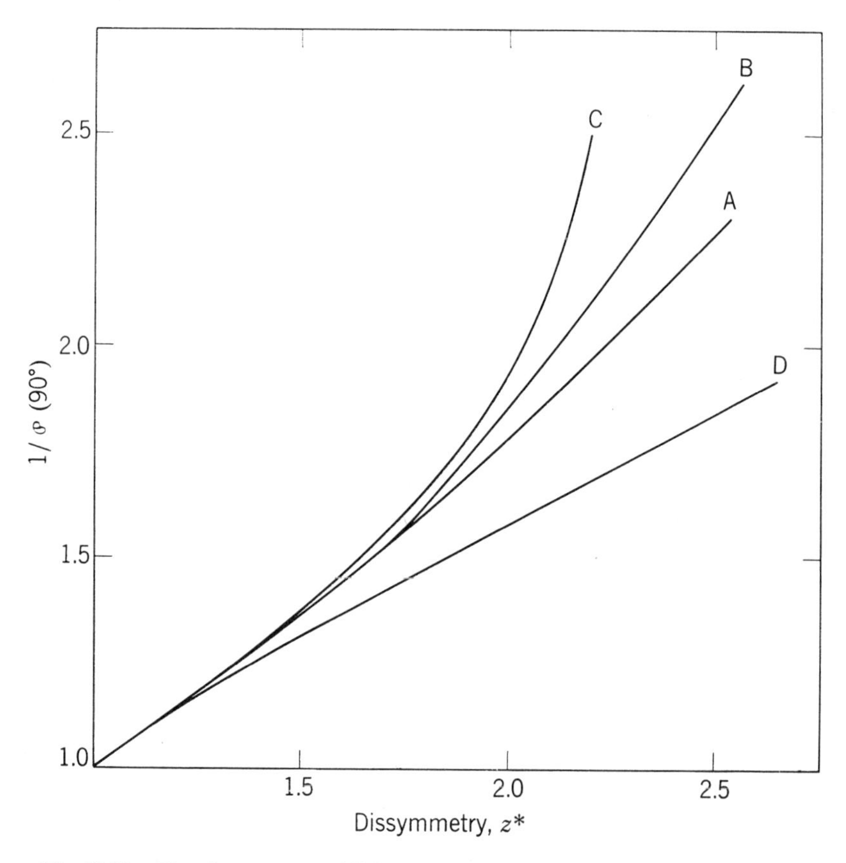

Fig. V.25. The dissymmetry of light scattering as a function of the ratio of the root-mean-square radius of gyration of scattering particles to the wavelength of light. (A) Monodisperse coils. (B) Coils with a normal chain length distribution ($\langle s^2 \rangle = \langle s^2 \rangle_z$). (C) Rods. (D) Spheres.

Again we shall find that the relation between $1/\mathscr{P}(90°)$ and z^* will depend on the particle shape as the dimension of the particles exceeds a certain size. The correlation plotted from values calculated by Doty and Steiner (Fig. V.26) shows that estimates of $1/\mathscr{P}(90°)$ will require information about the shape of the scattering molecules for $z^* > 1.4$.

In studies of synthetic chain molecules, the ratio $\langle s^2 \rangle^{1/2}/\lambda$ will rarely exceed 0.1, corresponding to a root-mean-square radius of gyration between 300 and 400 Å. In this range the correlation between $\langle s^2 \rangle^{1/2}/\lambda'$, the dissymmetry z^*, and the function $1/\mathscr{P}(90°)$ differs only slightly for monodisperse coils, polydisperse coils, and rods, and one might conclude that the analysis based on the dissymmetry measurement is quite adequate both for the mole-

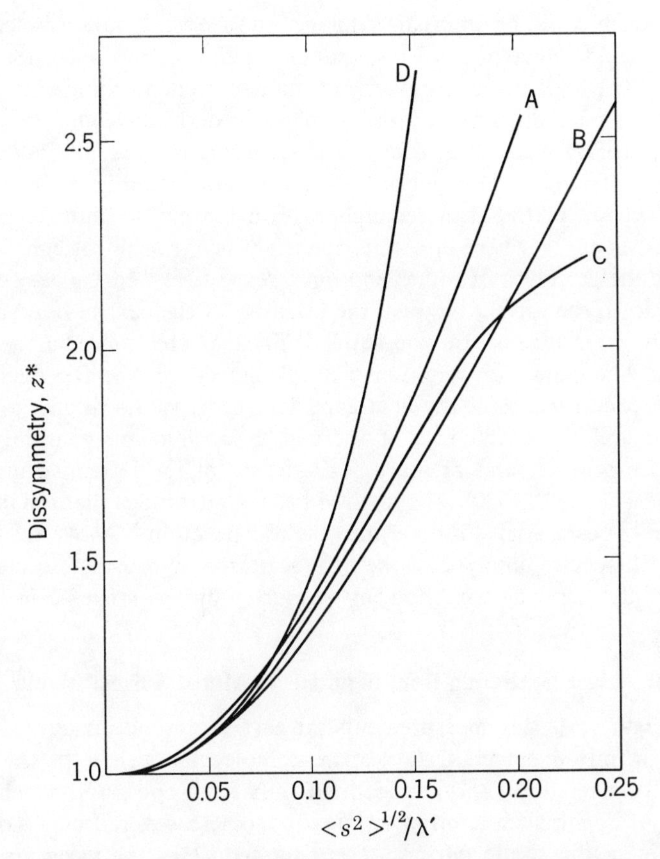

Fig. V.26. Dependence of $1/\rho$ (90°) on the dissymmetry z^*. (A) Monodisperse coils. (B) Coils with a normal chain length distribution. (C) Rods. (D) Spheres.

cular weight determination and for the estimation of the radius of gyration. Nevertheless, in practice it is desirable, even in such cases, to carry out the double-extrapolation procedure, since insufficient care in the removal of dust will lead to a characteristic increase of scattering intensities at small angles, which will be clearly revealed in a Zimm plot.

The availability of lasers as sources of very intense monochromatic radiation for light scattering studies may be employed to advantage in several ways. Kaye et al. (1971) used a laser source to circumvent the problem of scattering from dust particles in the determination of very high molecular weights. They found that with a very intense incident beam the volume of the solution from which the light scattering was measured could be reduced to 2.5×10^{-5} ml. This small volume was free of dust particles most of the time; whenever such a particle was present, the scattered light intensity registered a "spike" which could be ignored. With this arrangement, data were collected over the angular range of 1.5–7°, so that extrapolation to $\theta = 0$ presented no problems. The high monochromacity of the laser light also makes it possible to study very small differences in the wavelengths of the incident and scattered radiation. Brillouin (1922) predicted that scattering of light on inhomogeneities resulting from the propagation of sound waves should lead to side bands in the spectrum of the scattered light with a frequency shift $\Delta\nu$ equal to $\pm 2\nu_0(\upsilon/\mathbf{c}) \sin(\theta/2)$, where ν_0 is the frequency of the incident light, while υ and $\mathbf{c}$ are the velocities of sound and light, respectively. In the case of polymer solutions, the ratio J between the intensity of the central band and the sum of the intensities of the side bands is linear in the molecular weight of the solute. The molecular weight may be calculated from the dependence of J on solute concentration if the heat capacities at constant volume and constant pressure, the coefficients of thermal expansion and compressibility, and the temperature and pressure dependences of the refractive index are known (Miller et al., 1970). The method has the advantage that the intensity of the direct beam does not enter into the computation.† As we shall see in Chapter VI, spectral analysis of the light scattered by macromolecular solutions may also serve as a method for measuring diffusion coefficients of the solute.

4. Light Scattering from Solutions in Mixed Solvent Media

When light scattering measurements are carried out with macromolecular solutions in mixed solvents, the apparent molecular weight of the solute deduced from relation (V.41) varies frequently from one solvent medium to another. Although such a variation may, in some cases, reflect a molecular association of the solute particles (see Chapter VIII), the variation in the apparent molecular weights is encountered in many cases where such association is extremely unlikely. The phenomenon was first analyzed by Ewart et al. (1946), who showed that it may be fully accounted for by optical effects caused by the concentration of the thermodynamically better solvent in the domains occupied by the macromolecules. We may illustrate the principle by a case of a polymer with a refractive index n_2 dissolved in two solvents with refractive indices n_1 and n_3, where $n_3 > n_2 > n_1$ and n_2 is larger than the refractive index of the mixed solvents. Let us assumes, for simplicity, that the mutual interaction of the components does not affect their refractivity, so that the observed n is a linear function of the composition of the system. We shall then obtain the correct molecular weight from light scattering data only if

†In the conventional method, determination of the ratio of incident and scattered light intensities presents experimental difficulties, since I_0 and I differ by many orders of magnitude. Indirect methods, such as the estimation of I_0 from the intensity of light scattered by benzene and its theoretical Rayleigh ratio, may introduce some uncertainty into the results obtained.

the ratio of the concentrations of the two solvents in the domain occupied by the polymer is the same as in the system as a whole. Only in this case will the macroscopically determined dn/dc_2 value represent correctly the difference in refractive index experienced by a light ray on passing from the mixed solvent medium into the region occupied by the macromolecule. If the macromolecule adsorbs component 3 preferentially, the effective refractive index of the polymer domains will be increased, while that of the intervening regions will decline. This will result in an increased light scattering intensity and a higher apparent molecular weight. Conversely, if the macromolecules bind component 1, the difference in the refractive indices of the polymer domains and of the intervening regions will be reduced, resulting in a decrease of the apparent molecular weight. A variable apparent molecular weight may be used, therefore, to evaluate the relative affinities of polymers for the components of a mixed solvent system. According to Stockmayer (1950), the apparent molecular weight M_2^{app} is related, in a three-component system, to the true value M_2 by

$$\frac{M_2^{app}}{M_2} \approx \frac{1 - (a_{23}/a_{33})(\Gamma_3/\Gamma_2)^2}{1 - a_{23}^2/a_{33}a_{22}}$$

$$\Gamma_i = \left(\frac{\partial n}{\partial c_i}\right)_{T,P,m_j \neq i}$$

$$a_{ij} = \left(\frac{\partial \bar{G}_i}{\partial m_j}\right)_{T,P,m_i \neq j} \tag{V.50}$$

where m_j are the quantities of the various components, and the index 2 refers to the polymeric component. An alternative formulation which has a more easily visualized significance (Strazielle and Benoit, 1961) is

$$X_3 = \frac{[(M_2^{app}/M_2)^{1/2} - 1]\Gamma_2}{\Gamma_3} \tag{V.51}$$

where X_3 gives the weight of component 3 bound to a unit weight of the polymer.

The experimentalist is frequently not interested in preferential solvation phenomena but is concerned merely with obtaining the correct molecular weight from light scattering data on polymers which must be studied in mixed solvent media. This is particularly the case with polyelectrolytes, which as we shall see in Chapter VII, are most easily studied in solutions containing added salt. Fortunately, Casassa and Eisenberg (1960) proved that the complications produced by preferential solvation in mixed solvent systems are eliminated if the excess light scattering and the refractive index increment are expressed with reference to the solvent mixture with which the polymer solution is in dialysis equilibrium, rather than the solvent mixture used in the preparation of the polymer solution. With this modified procedure light scattering data of multicomponent systems may be treated just as data obtained from polymer solutions in a single solvent, yielding the correct value for the molecular weight.

5. Light Scattering from Solutions of Polymers with a Variable Chemical Composition

So far we have considered only samples consisting of macromolecules which might differ in molecular weight, but not in chemical composition. It is legitimate to assume that the refractive index increment dn/dc_2 is identical for all molecular weight fractions, and this assumption was implied in deriving (V.31). If we deal with chemically heterogeneous samples, in which both the refractive index and the molecular weight of the individual molecular species will vary, the situation will be considerably more complex. Note, in particular, one characteristic difference in the behavior of chemically homogeneous and chemically heterogeneous samples: for homogeneous samples the excess light scattering intensity vanishes if $dn/dc_2 = 0$, while with a heterogeneous sample it is impossible to match the refractive index of the solvent simultaneously with that of all the dissolved species. Thus, a finite excess light scattering intensity will be observed even if the refractive index increment, averaged over all the solute species, is equal to zero. If we do not allow for the chemical heterogeneity of the sample, we will arrive at molecular weights which will be, in general, too high and will tend toward infinity as the average refractive index of the polymer approaches that of the solvent medium.

The complications arising from this problem were first pointed out by Stockmayer et al. (1955), who considered, in particular, light scattering from a copolymer sample of variable composition containing units A and B. If the average refractive index increment of the sample is linear in the weight fraction x_B of B units, the apparent molecular weight M_w^{app} is related to the true weight-average molecular weight $\bar{M}_w$ by

$$M_w^{app} = \bar{M}_w + 2\left(\frac{\Gamma_B - \Gamma_A}{\Gamma}\right)\langle M\Delta x\rangle + \left(\frac{\Gamma_B - \Gamma_A}{\Gamma}\right)^2 \langle M(\Delta x)^2\rangle \quad \text{(V.52)}$$

where Γ_A and Γ_B are refractive index increments of poly-A and poly-B, Γ is the refractive index increment of the copolymer, Δx and $(\Delta x)^2$ denote the deviations and the square deviations from the mean composition, and $\langle\ \rangle$ stand for average quantities. The second term on the right vanishes if the mean chemical composition is the same for all molecular weight fractions. In principle, relation (V.52) allows us to obtain not only the weight-average molecular weight, but also a measure of the chemical heterogeneity by evaluating light scattering data in three solvent media (Bushuk and Benoit, 1958; Krause, 1961; Leng and Benoit, 1962a).‡ The statistical variation in

‡Lamprecht et al. (*Makromol. Chem.*, *148*, 285 (1971)) have described experimental problems in the application of this method.

the composition of copolymer chains prepared under identical conditions (see p. 24) is generally too small to lead to the complications discussed above. However, unless copolymerization is stopped at very low conversion, it will be dangerous to disregard the possibility of chemical heterogeneity.

Even if a copolymer is monodisperse with respect to its chemical compoistion, some complications will arise in connection with the interpretation of the angular dependence of its light scattering intensity. The theory for this case has been developed by Benoit and Wippler (1960), who showed that the apparent mean square radius of gyration obtained by applying (V.36) to solutions of such copolymers may be interpreted by

$$\begin{aligned} \langle s^2 \rangle^{\text{app}} &= \alpha^2 \langle s^2 \rangle_{\text{A}} + \beta^2 \langle s^2 \rangle_{\text{B}} + \alpha\beta[\langle s^2 \rangle_{\text{A}} + \langle s^2 \rangle_{\text{B}} + \langle r_g^2 \rangle] \\ \alpha &= \frac{\Gamma_{\text{A}}(1 - x_{\text{B}})}{\langle \Gamma \rangle} \\ \beta &= \frac{\Gamma_{\text{B}} x_{\text{B}}}{\langle \Gamma \rangle} \end{aligned} \tag{V.53}$$

where x_{B} is the weight fraction of B units in the copolymer, $\langle s^2 \rangle_{\text{A}}$ and $\langle s^2 \rangle_{\text{B}}$ are the mean square radii of gyration of the A and B units in the copolymer, while $\langle r_g^2 \rangle$ is the mean square distance between their centers of gravity. If Γ_{A} and Γ_{B} have opposite signs, $\alpha\beta$ may be negative and $\langle s^2 \rangle^{\text{app}}$ may also become negative, particularly with a block copolymer containing two blocks, so that $\langle r_g^2 \rangle$ is relatively large (Leng et al., 1963).

If the refractive index of the solvent matches that of one or the other monomer units of a copolymer [i.e., α or β vanishes in (V.53)], then the angular dependence of the scattered light reflects only the spatial distribution of the units with a nonzero refractive index increment. Such a measurement has been used in an interesting manner by Leng and Benoit (1962b), who studied a block copolymer containing a central sequence of styrene residues flanked by two poly (methyl methacrylate) blocks. If this block copolymer is dissolved in a medium which matches the refractive index of poly (methyl methacrylate), only the central section will contribute to the light scattering. From the angular dependence of the scattered light intensity it was shown that the polystyrene block is more expanded than a homopolymer of the same length. This was attributed to the excluded volume effect operating between the two "invisible" poly (methyl methacrylate) blocks.

D. X-RAY SCATTERING

The theory outlined in Section V. C. applies also to scattering in the x-ray region of the electromagnetic spectrum. However, whereas the wavelength λ' of visible light usually employed in light scattering ranges, in different solvents, between 3000 and 4000 Å, the most commonly used x-rays (those of Cu Kα radiation) have $\lambda = 1.54$ Å. We employ then, in the case of visible light, a wavelength much longer than the dimensions of the scattering macromolecules, while the wavelength of the x-rays is much shorter than even the most compact macromolecules which we may investigate. This quantitative difference allows us to obtain, from x-ray scattering, information about the distribution of scattering centers over distances much shorter than those studied by the angular distribution of the scattering intensity of visible light.

In the limit of zero scattering angle, the amplitude of the x-rays scattered by an atom is proportional to its number of electrons. The quantity which is analogous to the refractive index increment in the scattering of visible light is then, in the case of x-ray scattering, the "excess electron density" $\Delta\rho_e$ of the solute, defined by

$$\Delta\rho_e = \frac{e_2}{M_2} - \bar{v}_2\rho_e^0 \tag{V.54}$$

where e_2 is the number of electrons carried by a solute molecule, $\bar{v}_2$ is the partial specific volume of the solute, and ρ_e^0 is the number of moles of electrons per unit volume of the solvent medium. The molecular weight of the solute may then be obtained, in principle, from values of the Rayleigh scattering ratio extrapolated to $\theta = 0$, using for $\lambda = 1.54$ Å the relation (Kratky and Kreutz, 1960; Kratky, 1962a,b)

$$\Delta\mathscr{R}_0 = 0.048 M_2 c_2 (\Delta\rho_e)^2 \tag{V.55}$$

In practice, such determinations are difficult, since they require an absolute determination of the Rayleigh ratio, that is, comparison of the very high intensity of the incident beam with the very weak intensity of the scattered x-rays. Although these difficulties can be overcome (Kratky and Kreutz, 1960; Luzzati et al., 1961c), the technique is not particularly advantageous for molecular weight determinations.

The main value of x-ray scattering studies lies in the possibilities inherent in the interpretation of the angular dependence of light scattering intensities. Relations (V.32) and (V.36) show that the factor by which the scattered light is attenuated as a result of the destructive interference of rays scattered from two points of a scattering particle separated by a distance r_{ij} is a function of $(r_{ij}/\lambda') \sin(\theta/2)$. Thus, the same relative attenuation which is observed with $\lambda' = 3000$ Å over an angular range of 0–90° is compressed for x-rays with $\lambda = 1.54$ Å to an angular range of about 1 minute of arc. This means that we should have to make measurements with x-rays at exceedingly small angles to characterize the size of macromolecules whose expansion is studied conveniently by the angular distribution of scattered visible light. On the other hand, the angular distribution of scattered x-ray intensities will vary appreciably over a narrow angular range for scattering particles with dimensions of the order of 10–50 Å, much too small to lead to any measurable destructive interference effects in the scattering of visible light. This is the range of dimensions characteristic of many globular proteins, and their radii of gyration may be estimated from the relative scattered x-ray intensities over a range of scattering angles. For small θ values, relation (V.36) may be approximated by

$$\mathscr{P}(\theta) = \exp\left[-\left(\frac{4\pi^2}{3\lambda^2}\right)\langle s^2\rangle\theta^2\right] \tag{V.56}$$

and the square radius of gyration may be obtained, therefore, from the slope of a plot of the logarithm of the x-ray intensity against the square of the scattering angle. (Note that the "radius of gyration" obtained from x-ray scattering is defined in terms of the distribution of excess electron densities, rather than the distribution of mass.) Early measurements of this type were carried out by Kratky (1948) and Kratky et al. (1955). Later, similar studies using improved methods were reported by Kratky and Kreutz (1960) and by Luzzati et al. (1961d), who were able to observe by the x-ray scattering method the expansion of bovine serum albumin molecules at pH 3.6, which had previously been inferred from the hydrodynamic behavior of its solutions.

For rodlike particles, $\mathscr{P}(\theta)$ depends on the length of the particles only at very small scattering angles. It soon reaches its asymptotic behavior, where $\mathscr{P}(\theta)$ depends only on the distribution of scattering electrons around the axis of the rod. This asymptote has the form

$$\ln[\theta\mathscr{P}(\theta)] = \ln[\theta\mathscr{P}(\theta)]_0 - \frac{2\pi^2}{\lambda^2}\langle s_a^2\rangle\theta^2 \tag{V.57}$$

where $\langle s_a^2\rangle$ is the mean square radius of gyration of the excess electron density around the axis of the rod (Kratky and Kreutz, 1960; Kratky, 1962b). The intercept of this asymptote may be used, provided the absolute value of the Rayleigh ratio is known, to calculate the mass per unit length of the rod m_L (Luzzati, 1961; Kratky, 1962b) from

$$m_L = \frac{27.3(\theta\Delta\mathscr{R}_\theta)_0}{c_2\Delta\rho_e} \tag{V.58}$$

This type of analysis has been used by Luzzati et al. (1961b) to show that m_L for DNA in solution conforms to the Watson-Crick double-helical model. Similar results were obtained by Timasheff et al. (1961) for RNA solutions, in agreement with the concept that the single-stranded chain has a structure resembling that of a twisted hairpin.

Interpretation of data from flexible chain molecules is much more complex. The problem was analyzed by Kratky and Porod (1949), who pointed out that different properties will determine the distribution of scattering intensities in different ranges of scattering angles. According to their analysis, we may distinguish five characteristic regions:

(a) At very small angles the scattering envelope will depend on the overall dimensions of the macromolecules and $\mathscr{P}(\theta)$ will have the shape of an error curve.

(b) At somewhat larger angles the scattering intensity will be that characterized by a Gaussian distribution of scattering particles, with $\mathscr{P}(\theta)$ proportional to $1/\theta^2$.

(c) When the scattering angles are of the same order of magnitude as the ratio of the wavelength of the x-rays to the persistence length of the chain (defined on p. 125), the intensity of the scattered x-rays is largely determined by interference effects between rays originating in short sections of the chain. Such sections are considered to be essentially rodlike and to have $\mathscr{P}(\theta)$ proportional to $1/\theta$.

(d) At still larger angles the shape of $\mathscr{P}(\theta)$ may be determined by the spacing of nearest-neighbor scattering centers, particularly if the chain carries heavily scattering atoms of a high atomic number.

(e) Finally, we may reach a region where the scattering intensity reflects the distribution of electrons in the individual atoms contained in the system.

One of the difficulties encountered in the interpretation of x-ray scattering data is due to the overlap of these various effects. Nevertheless, in flexible chains much longer than the persistence length, the transition between regions a, b, and c should be clearly marked. A plot of $\theta^2\mathscr{P}(\theta)$ against θ, re-

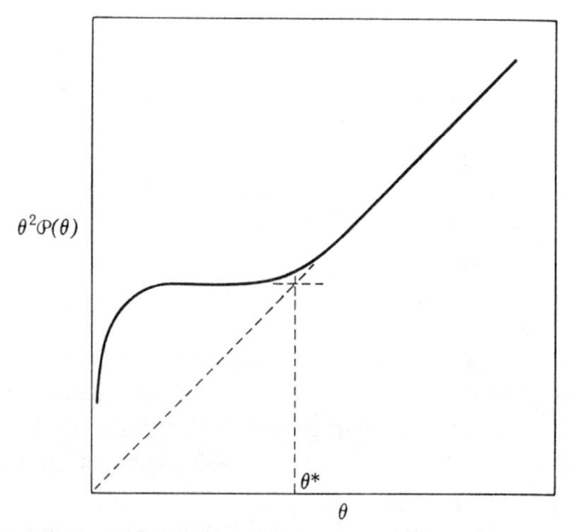

Fig. V.27. Schematic representation of the angular dependence of x-ray scattering intensity to be expected in solutions of flexible chain molecules.

presented schematically in Fig. V.27, will then have two linear branches intersecting at a point corresponding to θ^* which is related to the persistence length a (Porod, 1953; Peterlin, 1960; Heine et al., 1961) by the relation

$$a = \frac{\lambda}{2\pi\theta^*} \tag{V.59}$$

The method has been used by Heine *et al* (1961) to characterize the persistence length of cellulose nitrate. Its application to highly flexible polymer chains, such as those of Hevea rubber (Kratky and Sand, 1960) or poly-(methyl methacrylate) (Kirste and Kratky, 1962) is however questionable. Here the persistence length is of the order of only 5–6 Å, so that the chain cannot be considered, even locally, to be rodlike. The scattering pattern at large angles reflects then the preferred chain conformations and this leads to characteristic differences in the scattering patterns of isotactic and syndiotactic poly (methyl methacrylate) (Kirste and Wunderlich, 1964), although their chain flexibilities are quite similar.

As stated above, chain polymers carrying heavily scattering atoms should have $\mathscr{P}(\theta)$ at large scattering angles controlled by the nearest-neighbor spacing of these heavy atoms. The feasibility of determining such distances in model compounds of low molecular weight has been demonstrated by Kratky and Worthmann (1947) on rigid molecules such as *p*, *p'*-diiododiphenyl or *p*, *p'*-diiododiphenylmethane. Extension of the method to flexible chains with heavy atoms at the two ends should yield, in principle, a probability distribution of end-to-end distances. This approach applied to α, ω-diiodoalkanes (Holleman, 1960; Brady et al., 1967) yields results consistent with end-to-end distances estimated from conformational analyses of paraffinic chains. The study of short-chain analogs of polymers has the advantage that long-range interactions manifested in the excluded volume effect are absent, so that the dependence of the end-to-end distance on the solvent medium may be interpreted in terms of the local conformational preferences, which are related to the "unperturbed" polymer dimensions. Li et al. (1972) used for this purpose model compounds of the type I—(—CH_2CH_2O—)$_n$—CH_2CH_2I and found that these molecules are considerably less expanded in ethanol than in xylene solution. On the basis of this result, it was concluded that a corresponding difference exists between the "unperturbed dimensions" of poly (ethylene oxide) in these two media. (We pointed out in Section V.A.4 that spectroscopic observations also indicate a solvent dependence of the preferred conformation of this polymer.) Studies on high polymers carrying heavy atom substituents on all monomer residues were reported from Brady's laboratory. No difference was found in the scattering patterns of isotactic and atactic poly (*p*-iodostyrene) (Brady and Salovey, 1964). Presumably, the distance between nearest-neighbor iodines is here poorly defined. On the other hand, with poly (3,5-dibromo-L-tyrosine) the helical form is sufficiently rigid to exhibit a scattering pattern quite distinct from that of the monomer. A change in this pattern is observed in the same range of solvent composition in which optical activity changes indicate a helix-coil transition of the polypeptide (Brady and Salovey, 1967).

In all of the preceding discussion, we concerned ourselves only with the diffuse scattering from isotropic solutions. However, we have seen (Section II.B.4) that anisotropic phases may form in solutions containing stiff rodlike solutes or in solutions containing more than about 40% of block polymers and a solvent interacting preferentially with one of the blocks. In such a case, sharp x-ray diffraction photographs, characteristic of liquid crystals, may be obtained. Observations on systems of the first type were first recorded by Bernal and Fankuchen (1941), who studied solutions of tobacco mosaic virus and found that the diffraction pattern of the anisotropic phase corresponds to a hexagonal packing of the rodlike virus particles. The spacing between nearest neighbors was inversely proportional to the square root of the concentration of the anisotropic phase. Luzzati et al. (1961a) carried out similar studies on solutions of poly(γ-benzyl-L-glutamate) in solvents favoring the helical conformation of the polypeptide. Again, the macromolecular particles were arranged in a hexagonal array, but in this case the spacing of the molecules remained constant over a wide range of concentrations. The liquid crystals tend to align themselves with the long dimension of the solute particles parallel to the wall of thin capillaries (Bernal and Fankuchen, 1941). Even better orientation is attainable when anisotropic solutions are subjected to a shear gradient. With poly(γ-benzyl-L-glutamate) oriented in this manner, the x-ray diffraction pattern resembles that of a typical fiber, and the repeat period, parallel to the axis of the helix, is very close to that observed in solid specimens (Parry and Elliott, 1955).

Extensive investigations of concentrated solutions of block copolymers were carried out by Skoulios and his collaborators. In an account of the earlier work (Skoulios et al., 1964) involving block copolymers of the type ABA they emphasized the analogy between the behavior of block copolymers and soaps. In both cases, the preferential interaction of the solvent with the end of the solute molecule may lead to three types of structures, in which the solute is contained (a) in lamellar regions parallel to one another, (b) in hexagonally packed cylinders, or (c) in spheres packed in a cubic array. The first two structures produce an anisotropic phase. A detailed study of systems containing chains with a polystyrene and a poly (2-vinylpyridine) or a poly(4-vinylpyridine) block showed how the nature of the packing depends on the concentration and the composition of the copolymers (Grosius et al., 1969, 1970a). Increasing dilution of a system favors a transition from lamellar to cylindrical and to spherical structures. The same order of phase stabilities was observed with solvents solvating preferentially the polystyrene blocks as the length of the polyvinylpyridine block was reduced. This result is quite in accord with expectation, since spherical particles with the poorly solvated polyvinylpyridine blocks in the centers of the spheres will allow the greatest degree of dilution of the polystyrene blocks. It is surprising that even block copolymers of 2-vinylpyridine and 4-vinylpyridine are sufficiently different at the two ends of the molecular chain to produce ordered structures of this type (Grosius et al., 1970b).

Chapter VI

FRICTIONAL PROPERTIES OF DISSOLVED MACROMOLECULES

When a driving force acts on a particle suspended in a viscous medium, the particle will accelerate until the driving force is balanced by the frictional forces due to its motion. Phenomena based on this general principle may be used in three ways to obtain information about the nature of the suspended particle: we may measure (a) the steady-state velocity of linear translation of a particle subjected to a driving force, (b) the velocity with which the particle would rotate under the influence of a force couple, or (c) the increase in viscosity resulting from a suspension of the particle in a viscous fluid. (An excellent discussion of the viscosimetric method has been presented by Frisch and Simha, 1956.)

Each of these methods will yield a parameter which depends in a characteristic manner on the size and the shape of the body moving through the fluid. If we know that this body is a rigid sphere, a single parameter will suffice to characterize its size, and this may then be obtained from a single measurement. If we are dealing with an ellipsoid of revolution, two parameters will be required, specifying its volume and its axial ratio. Fortunately, the three methods mentioned above differ in their relative sensitivities to these two parameters, so that an ellipsoidal particle may be characterized by a combination of any two of them. However, a particle with a more complicated shape may require many more parameters for its detailed description, and such a description is not attainable, therefore, on the basis of frictional measurements. We must then content ourselves with the statement that the unknown particle behaves in a given experiment (or set of experiments) in the manner expected of a model, such as a sphere of a given size or an ellipsoid of revolution with certain specified dimensions. These are referred to as "hydrodynamically equivalent spheres" or "hydrodynamically equivalent ellipsoids." It should be understood that a comparison of an unknown body to these hydrodynamic equivalents in no way implies that the body is in reality a sphere or an ellipsoid.

A number of important special cases of the motion of rigid bodies through viscous media have been subjected to hydrodynamic analysis. Such treatments invariably assume that the fluid may be represented as a structureless continuum. This is a valid assumption when we deal with macroscopic particles with dimensions many orders of magnitude larger than the molecular dimensions of the fluid through which they move. However, it is by no means obvious that the hydrodynamic results derived for such macroscopic systems retain their validity when the particle in whose motion we are interested is more nearly comparable in size to the molecules of the surrounding fluid. Nevertheless, a great deal of experimental evidence indicates that a treatment of the frictional properties of dissolved macromolecules in terms of theories valid for macroscopic models leads to reasonable conclusions. Had this not been the case, the theoretical development of this subject would have been immensely more difficult.

Like colligative properties and light scattering, the frictional properties of solutions at finite concentrations may be interpreted in terms of the intrinsic particle contributions, to which terms taking account of particle interactions must be added. Only the behavior of the isolated macromolecules lends itself to fairly safe theoretical interpretation. Since macromolecular interactions can frequently not be neglected even in the most dilute solutions on which measurements are possible, a suitable extrapolation procedure for data obtained over a range of concentrations is required to estimate the behavior of the macromolecules at infinite dilution.

A. LINEAR TRANSLATION

1. The Frictional Coefficient of Rigid Particles

As pointed out above, all theories dealing with frictional properties of dissolved macromolecules assume that laws derived for macroscopic particles moving in a structureless fluid are applicable. It is also assumed that viscous forces are much larger than inertial forces, so that the flow of the liquid is laminar rather than turbulent. In Fig. VI.1 we represent schema-

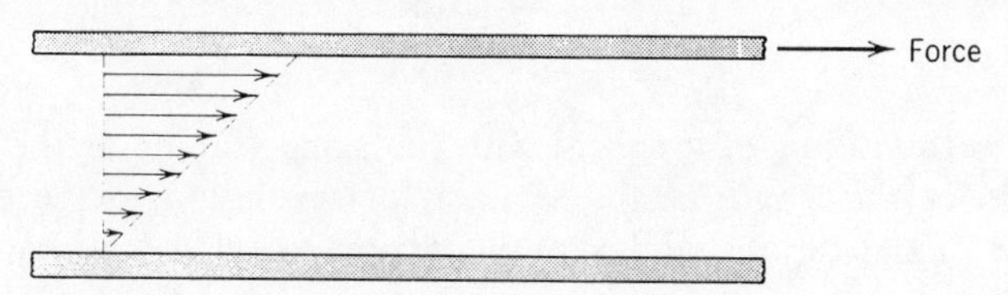

Fig. VI.1. Velocity distribution in viscous fluid between two parallel plates (laminar flow).

tically two parallel plates separated by a distance x with the intervening space filled by a viscous fluid. If one of the plates is to be moved relative to the other, a force has to be applied to overcome the resistance caused by fluid friction.

We may consider that a layer of molecules of the liquid adheres to each of the surfaces of the moving plates. A layer of the liquid, part of the way between the plates, is subjected to the frictional pull of the two neighboring layers, which exert forces in opposite directions. Since the frictional force is a function of the velocity gradient q, a given layer will experience no net force if $dq/dx = 0$. For many simple fluids the velocity gradient is proportional to the applied force, and such fluids are called "Newtonian."

For a spherical particle moving slowly through a Newtonian fluid the viscous force F resisting the motion is proportional to the velocity of the particle:

$$\mathrm{F} = \mathrm{f}u \qquad \text{(VI.1)}$$

The proportionality constant f, commonly called the "frictional coefficient," denotes the force in dynes required to maintain a particle velocity of 1 cm/sec. For spheres moving through a fluid far from the walls of the container,

$$\mathrm{f}_{\mathrm{sphere}} = 6\pi\eta_0 R_s \qquad \text{(VI.2)}$$

where η_0 is the viscosity of the fluid and R_s the radius of the sphere. A detailed discussion of this result, first derived by Stokes, is given by Lamb (1945). Experimental studies have shown that Stokes's law is strictly valid for velocities of the spherical particle below $\eta_0/2R_s\rho$, where ρ is the density of the fluid(Schlichting, 1960). Thus, the validity of Stokes's law extends, for particles of colloidal dimensions, far beyond the highest velocities with which we shall be concerned.

If the moving particle is asymmetric (i.e., an ellipsoid of revolution), the solution of the hydrodynamic problem becomes much more difficult, since the motion of the particles may now lead to their orientation. At low rates of shear, where Brownian motion can effectively randomize the particle orientations, the frictional coefficient of ellipsoids of revolution may be expressed as

$$\mathrm{f}_{\mathrm{ell}} = \frac{6\pi\eta_0 R_s'}{\varkappa(\mathrm{p})} \qquad \text{(VI.3)}$$

where R_s' is the radius of a sphere with the same volume as the ellipsoid, p the ratio of the length of the semiaxis of revolution to the equatorial radius (the "axial ratio"), and $\varkappa(\mathrm{p})$ a function calculated by Perrin (1936) as

$$\varkappa(\mathrm{p}) = \frac{\mathrm{p}^{1/3}}{\sqrt{\mathrm{p}^2 - 1}} \ln\left(\mathrm{p} + \sqrt{\mathrm{p}^2 - 1}\right) \qquad (\mathrm{p} > 1) \qquad \text{(VI.4a)}$$

$$\varkappa(\mathrm{p}) = \frac{\mathrm{p}^{1/3}}{\sqrt{1 - \mathrm{p}^2}} \tan^{-1}\left(\frac{\sqrt{1 - \mathrm{p}^2}}{\mathrm{p}}\right) \qquad (\mathrm{p} < 1) \qquad \text{(VI.4b)}$$

It is common practice to approximate the behavior of long rodlike particles by a similar prolate ellipsoid. It is then more convenient to express f_{ell} in terms of $\boldsymbol{a}_1$, the semiaxis of revolution, rather than R_s'. Since the volume of the ellipsoid is $\frac{4}{3}\pi \boldsymbol{a}_1^3/p^2$, we have $R_s' = \boldsymbol{a}_1/p^{2/3}$, and from (VI.3) and (VI.4a)

$$f_{ell} = \frac{6\pi\eta \boldsymbol{a}_1 \sqrt{p^2 - 1}}{p \ln (p + \sqrt{p^2 - 1})} \tag{VI.5}$$

For very long particles of given equatorial radius, f_{ell} then becomes proportional to $\boldsymbol{a}_1/\ln (2\boldsymbol{a}_1)$; this is the same conclusion as reached by Riseman and Kirkwood (1950) in their analysis of the behavior of thin rods.

2. Frictional Coefficients of Chain Molecules

In treating the frictional properties of chain molecules, it is customary to represent them by a "pearl necklace" model in which rigid beads are connected by infinitely thin linkages. Kuhn and Kuhn (1943) and Debye and Bueche (1948) pointed out that two limiting situations may be envisaged. In the first case, the beads are relatively far from one another, so that the disturbances of the flow caused by the individual beads may be considered not to interact. This model is generally referred to as the "free-draining coil." If such a coil is forced to move through a viscous fluid, each unit will be subjected to a frictional resistance independent of the presence of other, similar units, and the effective frictional coefficient of the coil as a whole will be proportional to the number of units composing it. In the second limiting case the interactions between the flow disturbances are so large that the solvent is effectively trapped within the coil, which may be treated as a rigid "hydrodynamically equivalent sphere." The radius of this equivalent sphere, R_t, controlling the frictional resistance to linear translation, will be proportional to some characteristic dimension of the coil, such as the root-mean-square radius of gyration $\langle s^2 \rangle^{1/2}$. We have then, in analogy to (VI.2),

$$f_{coil} = 6\pi\eta_0 R_t = 6\pi\eta_0 C \langle s^2 \rangle^{1/2} \tag{VI.6}$$

where C is a characteristic constant. This result predicts a much lower resistance to translation of the coil than the free-draining model. In a Θ-solvent, where $\langle s^2 \rangle$ is proportional to chain length, (VI.6) predicts that the frictional coefficient of chain molecules should increase as the square root of their molecular weight. In better solvents, where the molecular coil is more expanded, the frictional coefficient is correspondingly higher.

The criteria for the applicability of these limiting cases were discussed by Debye and Bueche for a simplified model in which Z beads, each with a frictional coefficient f_0, are uniformly distributed in a spherical volume v_e. The resistance offered by the beads to the flow of the fluid will cause the fluid velocity to decay exponentially with the distance below the surface of the sphere. The characteristic depth L_f at which the flow velocity has been reduced to e^{-1} of its value at the surface is given by

$$L_f = \left(\frac{\eta_0 \mathrm{v}_e}{Z\mathrm{f}_0}\right)^{1/2} \tag{VI.7}$$

The free-draining model is then applicable if the radius of the equivalent sphere is much smaller than L_f, while $R_t \gg L_f$ corresponds to a situation in which most of the sphere is shielded from any flow outside, so that it may be thought of as a single unit with the entrapped liquid.

The most difficult problem concerns the intermediate case. The model proposed by Debye and Bueche leads to a result depending on the "shielding factor" R_t/L_f and gives, for the frictional coefficient in the limit of very small and very large shielding factors,

$$\begin{aligned} \mathrm{f}_{\text{coil}} &= Z\mathrm{f}_0 \left[1 - \frac{4}{15}\left(\frac{R_t}{L_f}\right)^2 + \cdots\right] \\ &= Z\mathrm{f}_0 \left(1 - \frac{1}{5\pi}\frac{Z\mathrm{f}_0}{\eta_0 R_t} + \cdots\right) \qquad (R_t/L_f \ll 1) \end{aligned} \tag{VI.8}$$

$$\begin{aligned} \mathrm{f}_{\text{coil}} &= 6\pi\eta_0 R_t \left(1 - \frac{L_f}{R_t} - \cdots\right) \\ &= 6\pi\eta_0 R_t \left(1 - 2\sqrt{\pi\eta_0 R_t/3Z\mathrm{f}_0} - \cdots\right) \qquad (R_t/L_f \gg 1) \end{aligned} \tag{VI.9}$$

A completely different approach to the problem of the frictional resistance to translation offered by chain molecules was employed by Kirkwood and Riseman (1948). They used the pearl necklace model with perfectly flexible joints for the chain and random flight statistics for the distribution of the chain segments, without consideration of the excluded volume effect. Then, taking account of the interactions of the flow disturbances produced by the individual beads, they arrived at

$$\begin{aligned} \mathrm{f}_{\text{coil}} &= \frac{Z\mathrm{f}_0}{1 + \Xi} \\ \Xi &= \frac{4}{9\pi^{3/2}}\frac{Z\mathrm{f}_0}{\eta_0 \langle s^2 \rangle^{1/2}} \end{aligned} \tag{VI.10}$$

The similarity of this result to that obtained by Debye and Bueche is impressive when we consider the pronounced difference in the models on which the two theories are based. In the limit of $\Xi \gg 1$, (VI.10) reduces to the form of (VI.6) for an impermeable coil, with

$$\frac{R_t}{\langle s^2 \rangle^{1/2}} = \frac{3\pi^{1/2}}{8} \tag{VI.11}$$

The dependence of the frictional coefficient on the length of flexible chain molecules and on the solvent power of the medium was carefully studied on poly(methyl methacrylate) (Lütje and Meyerhoff, 1963) and on poly(α-methylstyrene) (Cowie and Bywater, 1970). They used diffusion and ultracentrifuge sedimentation, to be discussed in the following sections, for the calculation of f_{coil}. In Θ-solvents, the frictional coefficients were found to be proportional to the square root of the molecular weight, as predicted by (VI.6) for impermeable coils with $\langle s^2 \rangle$ proportional to M_2. However, this relation held even for chains which were clearly too short to fit the description of "impermeable coils." The constancy of the $f_{coil}/M_2^{1/2}$ ratio in this range must, therefore, be accounted for by mutual compensation of two effects; partial drainage of the coil, and a drift of the $\langle s^2 \rangle / M_2$ ratio below the asymptotic limit characteristic of very long chains. In the dependence of f_{coil} on the solvent medium, three regions of molecular weights may be distinguished. For very short chains, the theoretical analysis in terms of a structureless medium breaks down so that f_{coil} is not proportional to the bulk viscosity of the solvent. In the intermediate molecular weight range, the discontinuous nature of the solvent is no longer important and the excluded volume effect is still negligible; thus, f_{coil}/η_0 is the same in good and bad solvents. Finally, for a polymer with $M_2 > 10^4$, the expansion of the molecular coils in a good solvent was found to lead to a decrease in the $R_t/\langle s^2 \rangle^{1/2}$ ratio. According to Lütje and Meyerhoff, this ratio is reduced by about 20% in a good solvent medium, while Cowie and Bywater fitted their data by making $R_t/\langle s^2 \rangle^{1/2}$ proportional to $\alpha_e^{-0.4}$. A decrease in the ratio of the hydrodynamic radius to the radius of gyration is the result of a deviation from a Gaussian distribution of chain segment densities, since the excluded volume effect is most pronounced in the center of the molecular coil. Kurata and Yamakawa (1958) predicted that for impenetrable coils $R_t/\langle s^2 \rangle^{1/2}$ should be proportional to $\alpha_e^{-0.35}$, in good agreement with experimental data. Ptitsyn and Eizner (1960) predicted a smaller dependence of $R_t/\langle s^2 \rangle^{1/2}$ on polymer solvation; their theory seems to have underestimated the effect.

3. Diffusion

The diffusion of solute molecules in a concentration gradient may be thought of as a motion of the molecular particles in response to a driving force provided by the gradient of chemical potential. The force acting on a molecule of the ith species is then

$$\mathrm{F} = \frac{1}{N}\left(\frac{\partial \bar{G}_i}{\partial x}\right)$$

$$= kT\left(\frac{\partial \ln c_i}{\partial x}\right)\left(1 + \frac{\partial \ln \gamma_i}{\partial \ln c_i}\right) \qquad \text{(VI.12)}$$

If the solution is sufficiently dilute, so that Henry's law applies to the solute, $\partial \ln \gamma_i / \partial \ln c_i = 0$ and the velocity of particles with a frictional coefficient f becomes

$$u = \frac{\mathrm{F}}{\mathrm{f}} = \frac{\boldsymbol{k}T}{\mathrm{f}} \left(\frac{1}{c_i}\right)\left(\frac{\partial c_i}{\partial x}\right) \tag{VI.13}$$

Experimental data on diffusion are conventionally described in terms of Fick's first law, according to which the rate of transport per unit cross-sectional area is proportional to the concentration gradient, the proportionality constant being the diffusion coefficient D. Since the rate of transport also equals the product of solute concentration and flow velocity,

$$uc_i = D\left(\frac{\partial c_i}{\partial x}\right) \tag{VI.14}$$

Comparing (VI.13) and (VI.14), we see that the diffusion coefficient is related to the frictional coefficient of the diffusing particle by

$$D = \frac{\boldsymbol{k}T}{\mathrm{f}} \tag{VI.15}$$

This relation was first suggested by Einstein (1905), who showed that frictional coefficients of even small molecules are adequately approximated by the values predicted by hydrodynamic theory. This assumption has been shown to be surprisingly close to the real behavior of diffusing molecules. When small, spherical molecules move in a medium of similar particles (i.e., in the self-diffusion of a liquid such as argon or mercury), the frictional coefficient calculated from (VI.15) is generally smaller than the value predicted by Stokes's law (VI.2). Even in such extreme cases, however, the frictional coefficient deviates from the predicted value by less than a factor of 2 (Corbett and Wang, 1956).

Alternatively, the diffusion coefficient may be derived from the theory of Brownian motion of an isolated particle. From our discussion in Section III.B.2, it follows that in a three-dimensional random flight the mean square displacement $\langle \Delta x^2 \rangle$ is proportional to the length of the path, and it can be shown (Einstein, 1905; Chandrasekhar, 1943) that

$$\langle \Delta x^2 \rangle = 2Dt \tag{VI.16}$$

Although the mean particle velocity, given by Boltzmann's distribution law as $(8\boldsymbol{R}T/\pi M)^{1/2}$, may be extremely high, the mean free path is so short that the particle diffusion is extremely slow. For instance, the mean velocity of benzene molecules at 25°C is 285 m/sec, but the diffusion coefficient of benzene in heptane ($\eta_0 = 3.86 \times 10^{-3}$ poise) is only 2.47×10^{-5} cm^2/sec. This diffusion coefficient may then be interpreted either by (III.9) and (VI.16), assigning a free path of 0.17 Å to the benzene molecules, or else in terms of (VI.2) and (VI.15), assigning to the equivalent sphere a radius of 2.3Å.

Diffusion coefficients of flexible chain molecules reflect the dependence of the frictional coefficient on chain length and the solvent power of the medium, as discussed in the preceding section. Results by Lütje and Meyerhoff (1963) on poly(methyl methacrylate) yield $R_t / \langle s^2 \rangle^{1/2} = 0.82$ in Θ-solvents, only 20% above the value predicted by (VI.11). In such media the diffusion coefficient may be used to estimate a characteristic dimension of the polymer coil; Tsvetkov and Klenin (1958) suggest $\langle s^2 \rangle^{1/2} = 1.1 \times 10^{-7} T/\eta_0 D$.

Fick's law is valid only in the limit of highly dilute systems, in which the diffusing particles are too distant from each other to interact. As we increase the concentration of the diffusing species, the apparent diffusion coefficient will, in general, not remain constant. Its variation will be caused both by a thermodynamic and by a hydrodynamic factor. First, we may recall that the derivation of Fick's law in (VI.12)–(VI.15) presupposed the validity of Henry's law. At higher concentrations, where deviations from ideal solution behavior may occur, the driving force due to the chemical potential gradient is no longer adequately represented by $\boldsymbol{k}T\,\partial \ln c_i/\partial x$. In particular, if the solution exhibits a negative deviation from solution ideality, as it will with polymers in any medium better than a Θ-solvent, $\partial \ln \gamma_i/\partial \ln c_i$ will be positive, and this factor will tend to increase the diffusion coefficient as the solution concentrations is increased. Second, the simple derivation given above assumes that the hydrodynamic resistance to the motion of a particle is independent of the presence of other, similar particles. Again, this assumption is reasonable only if the diffusing particles are distant from one another. At higher concentrations the hydrodynamic disturbances produced by their motion will interact, and this may be expressed as a slow drift in the effective frictional coefficient. Since f will, in general, increase with concentration, this factor will tend to reduce the diffusion coefficient and should thus act in a direction opposite to the thermodynamic causes for a drift in D.

An excellent study of the concentration dependence of the diffusion coefficient has been reported by Tsvetkov and Klenin (1958, 1959). They found that in polystyrene, poly(methyl methacrylate), and poly(*p-tert*-butylphenyl methacrylate) solutions the diffusion coefficient of the polymer is an S-shaped function of concentration in good solvent media. However, when the diffusion coefficient was studied in a Θ-solvent, it was found to remain constant up to a concentration of 1 g/100 ml. It was concluded therefore, that the variation of the frictional coefficient with concentration is too small to be detected in the concentration range used. This result is difficult to understand, since the molecular weight of the polymer used by these authors was extremely high (4.6×10^6), and considerable interpenetration of the molecular coils would be expected at the higher solution concentrations employed. Results obtained by Cantow (1959) with polystyrene in cyclohexanone also showed D to become independent of solute concentration in dilute solutions at the Θ-temperature. Nevertheless, we should be cautious in assuming that such behavior will be encountered in all cases.

The experimental determination of the diffusion coefficient is carried out with highest precision in a cell in which a sharp boundary is initially produced between a solution with a polymer concentration c_2^0 and the solvent. As the solute diffuses, the concentration of the solute at the location of the original boundary remains fixed at $c_2^0/2$, and the concentration gradient at a distance x and a time t is given by

$$\frac{dc_2}{dx} = \frac{c_2^0}{2\sqrt{\pi D t}} \exp\left(-\frac{x^2}{4Dt}\right) \tag{VI.17}$$

We see then that an experimental procedure which records directly the concentration gradient as a function of distance (cf. Section IV.B.1) should yield a Gaussian curve, provided that the solution contains a single solute and the diffusion coefficient is independent of concentration. If the system contains several solutes with different diffusion coefficients, the variation of the concentration gradient will correspond to the sum of Gaussian curves with different standard deviations, so that the trace will appear broadened at its base. The magnitude of this effect may be expressed quantitatively as the ratio of the third and the second moments of the dc_2/dx curve. If the diffusion coefficient changes with concentration, the dc_2/dx curve will lose its symmetry. In the case illustrated in Fig. VI.2 for the diffusion of polystyrene in

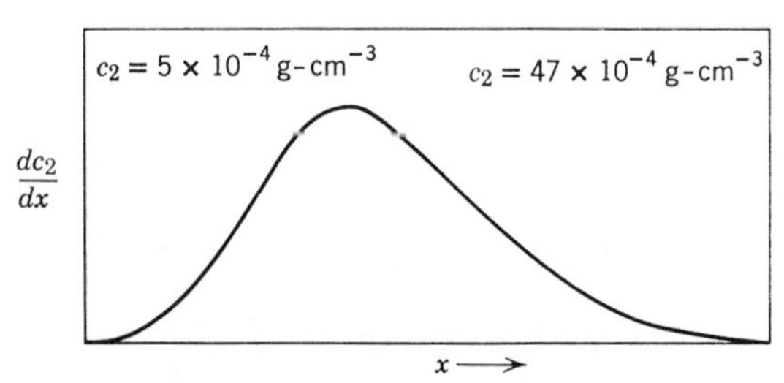

Fig. VI.2. Diffusion of polystyrene ($\bar{M}_w = 10^5$, $c_2 = 4.2 \times 10^{-3}$ g/cm^3) from a more concentrated to a more dilute carbon tetrachloride solution. The skewness of the dc_2/dx plot indicates the concentration dependence of the diffusion coefficient.

carbon tetrachloride (Tsvetkov and Klenin, 1958), the curve is much flatter on the side of the original boundary on which the solution is more concentrated, that is, D is here an increasing function of solution concentration.

Diffusion of solute molecules at a fixed concentration of the system can be followed by tagging part of the solute with a radioactive tracer. Two spectroscopic techniques have also been developed in which the need for a tracer is eliminated. The NMR "spin echo" method (Carr and Purcell, 1954) has been employed by Tanner et al., (1971) to measure self-diffusion coefficients in solutions of poly(ethylene oxide) over a broad range of concentrations. The second method utilizes the spectral distribution of the light scattered from solutions irradiated with a laser source. We pointed out previously (p. 280) that the scattered light contains three spectral bands: a band centered at the wavelength of the exciting radiation and two side bands arising from scattering on inhomogeneities produced by propagating sound waves. If we now concentrate our attention on the central band, we find that its finite width arises from Doppler shifts resulting from the interaction of the incident light with moving scattering particles. For monodisperse rigid particles of moderate size, the half-width $\Delta\nu$ of the spectrum of the scattered light is related to the diffusion coefficient by

$$D = \frac{\pi\Delta\nu}{\mathrm{k}^2} \tag{VI.18}$$

where k has the significance given in (V.32). If the particles become very large, their rotation will affect the shape of the spectrum of the scattered beam, and with very long flexible chains complications arise because of intramolecular motions of chain segments. This method for measuring diffusion coefficients has been used by Dubin et al., (1967) for rigid particles such as DNA and tobacco mosaic virus and by Reed and Frederick (1971) for flexible chains. The technique requires sophisticated instrumentation to measure extremely small spectral shifts. For instance, with an exciting radiation having a frequency of 4.737×10^{14} Hz, a solution of polystyrene with $\bar{M}_w = 179{,}000$, and a scattering angle of 30°, the half-width of the spectrum of the scattered light was only $\Delta\nu = 460$ Hz. If the test solution contains a polydisperse polymer, the ratio of the diffusion coefficient calculated from (VI.18) to the diffusion coefficient characterizing chains of weight-average length will decrease with increasing polydispersity (Frederick et al., 1971).

4. Sedimentation Velocity

A driving force for the linear translation of a dissolved macromolecule may be provided by a gravitational field. As we saw in the discussion of the equilibrium distribution in Section IV.B, the gravitational field of the earth is too weak to affect appreciably the distribution of macromolecules, unless we deal with species having a molecular weight higher than about 10^7. We must, therefore, utilize the large gravitational fields attainable in ultracentrifuges before the motion of macromolecules from the original uniform distribution toward a new equilibrium distribution becomes experimentally observable.

Consider a solute with a molecular mass $m_2 = M_2/\boldsymbol{N}$, a partial specific volume $\bar{v}_2$ in a solution of density ρ placed at a distance r from the axis of rotation of a centrifuge spinning with an angular velocity ω. The force acting on the particle will then be

$$\mathrm{F} = m_2 r\omega^2(1 - \bar{v}_2\rho) \tag{VI.19}$$

where the second term in the parentheses represents the effects of buoyancy. Equating this driving force with the frictional resistance of the medium, we obtain for the steady-state velocity in a unit gravitational field

$$\begin{aligned} \boldsymbol{s} &\equiv \frac{dr/dt}{r\omega^2} = \frac{d\ln r/dt}{\omega^2} \\ &= \frac{M_2(1 - \bar{v}_2\rho)}{\boldsymbol{N}\mathrm{f}} \end{aligned} \tag{VI.20}$$

The sedimentation constant $\boldsymbol{s}$ is customarily specified in units of 10^{-13} sec, called Svedberg units. It may be seen that $\boldsymbol{s}$ does not define the molecular weight of the solute unless the frictional coefficient is known. If the particle is a hard sphere with a specific volume $\bar{v}_2$, we obtain, after use of the Stokes value for the frictional coefficient,

$$\boldsymbol{s} = \frac{1 - \bar{v}_2\rho}{3\eta_0} \frac{M_2^{2/3}}{(6\pi^2\bar{v}_2)^{1/3}\boldsymbol{N}^{2/3}} \qquad \text{(spheres)} \tag{VI.21}$$

For ellipsoids of revolution, $\boldsymbol{s}$ will be decreased for any given particle weight because of the increase in the frictional coefficient. In the case of rodlike particles of a given cross section and a length L, f is proportional to $L/\log L$ (Riseman and Kirwood, 1950), while the gravitational driving force is proportional to L, so that the sedimentation velocity should increase only as $\log L$. For flexible chain molecules in a Θ-solvent with the molecular coil impermeable to solvent flow, the frictional coefficient is proportional to $\langle s^2\rangle^{1/2}$ and thus also to $M_2^{1/2}$ (cf. p. 122); the sedimentation constant will then be proportional to the square root of the molecular weight. In better solvent media $\langle s^2\rangle$ may be represented over a limited range as proportional to $M_2^{1+\varepsilon}$, where ε is a measure of the excluded volume effect, and $\boldsymbol{s}$ will then be proportional to $M_2^{(1-\varepsilon)/2}$. A theory for stiff chains with contour lengths corresponding to a small number of statistical chain elements was developed by Hearst and Stockmayer (1962) on the basis of the Kratky-Porod "wormlike chain" model and the hydrodynamic treatment of Kirkwood and Riseman (1948). For this case, a plot of $\boldsymbol{s}$ in a Θ-medium against $M_2^{1/2}$ has a nonzero intercept and an asymptotic slope of $0.12(1 - \bar{v}_2\rho)\,(M_L/b_s)^{1/2}/\eta_0$, where M_L is the molecular weight per unit length of the chain, so that the sedimentation data allow us to estimate the length b_s of the statistical chain element. In non-Θ-media, the corresponding analysis requires the substitution of M_L/b_s by

$M_L^{1+\varepsilon}/b_s^{1-\varepsilon}$ (Hearst et al., 1968). In the presence of 0.2*M* NaCl, to minimize electrostatic interactions between DNA particles, Hearst et al., found **s** proportional to $M_2^{0.44}$ in the range $22 \times 10^6 < M_2 < 96 \times 10^6$. Nicolaieff and Litzler (1969) measured the chain length in DNA preparations by electron microscopy and found on that basis that **s** increased as $M_2^{0.3}$ for $2 \times 10^5 < M_2 < 1.6 \times 10^6$.

Ultracentrifuge sedimentation played a crucial part in the discovery of circular DNA (Fiers and Sinsheimer, 1962). The DNA from a bacteriophage was found to have two components sedimenting at different rates; short exposure to an enzyme which leads to a scission of the DNA chain produced a conversion of the slower to the faster sedimenting component. Since the molecular weights of the two components were shown to be identical, it was concluded that the higher **s** value characterizes a circular structure, while the more slowly sedimenting species is a linear chain derived from it. Qualitatively, it would be expected that the hydrodynamic radius of a randomly coiled chain would be reduced if the chain ends are joined to each other to from a ring. A quantitative theory of the effect is due to Bloomfield and Zimm (1966). Much smaller changes in **s** produced by conformational transitions in biological macromolecules may be detected by "differential sedimentation," in which the optical system compares directly the distributions of the solute in two sample cells in the ultracentrifuge (Gerhart and Schachman, 1968; Schumaker and Adams, 1968). This method can measure differences as small as 0.016 **s**.

So far our discussion has been concerned with a single driving force acting on the macromolecules, namely, that provided by the gravitational field. If this were, in fact, the only cause for the motion of the solute, all molecules would sediment at identical rates and an infinitely sharp boundary would from between the sedimenting solution and the pure solvent. This is, of course, not the case, since diffusional flow will be superimposed on sedimentation. As a result, the sedimentation of a single solute species will form a boundary which will broaden with increasing sedimentation time in the manner described by (VI.17). This broadening poses no problems in the interpretation of results in which a single sedimenting species is present, since the sedimentation velocity is correctly represented by the rate at which the location of the concentration gradient maximum moves along the test cell. The situation is more complicated when the system contains solutes with a continuous distribution of sedimentation coefficients; in this case the effects of diffusion have to be eliminated before the experimental data can be interpreted. Since the distance covered by a particle subjected to a gravitational force is proportional to time, whereas the distance through which a particle moves because of random Brownian motion is proportional to the square root of time, the relative importance of diffusive flow decreases with increasing duration of the sedimentation experiment. We may, therefore, eliminate the effect of diffusion by extrapolating the apparent distribution of sedimentation coefficients to infinite time (Williams et al., 1952).

A second complication arises because of the concentration dependence of the frictional coefficient. For a suspension with a volume fraction ϕ of rigid spheres, the Stokes value of f in (VI.2) has to be multiplied by a factor $1 + (55/8)\phi$. Thus, the concentration dependence of $\boldsymbol{s}$ is of the form

$$\frac{1}{\boldsymbol{s}} = \frac{1}{\boldsymbol{s}_0}(1 + k_s c_2) \tag{VI.22}$$

where k_s depends on the volume of the hydrodynamically equivalent spheres per unit weight of the macromolecules (Wales and Van Holde, 1954). In the sedimentation of a polydisperse sample, the more slowly sedimenting compounds find themselves in regions of lower solution concentration, and the ratio of the sedimentation velocities of light and heavy molecules is, therefore, reduced. As a result, the apparent distribution of sedimentation constants broadens with a decreasing concentration of the solution. Johnston and Ogston (1946) treated the case of a heavy component with a sedimentation coefficient $\boldsymbol{s}_3$ and a light component whose sedimentation velocities are $\boldsymbol{s}_2$ and $\boldsymbol{s}_2'$ and whose concentrations are c_2 and c_2' in front of and behind the sedimentation boundary of component 3, respectively. The concentration of component 2 behind the sedimentation boundary of component 3 will then increase per unit gravitational field at the rate $c_2(\boldsymbol{s}_3 - \boldsymbol{s}_2) - c_2'(\boldsymbol{s}_3 - \boldsymbol{s}_2')$, and for the steady state c_2' and c_2 will be related by

$$\frac{c_2'}{c_2} = \frac{\boldsymbol{s}_3 - \boldsymbol{s}_2}{\boldsymbol{s}_3 - \boldsymbol{s}_2'} \tag{VI.23}$$

Since $\boldsymbol{s}_2 < \boldsymbol{s}_2'$, we obtain $c_2' > c_2$, and this will lead to an underestimation of the concentration of the heavy component (Fig. VI.3). Trautman et al., (1954) formulated a quantitative theory of the effect, taking account of the variation in the cross section of the sector-shaped cells and the dependence of the gravitational field on r. When the analysis is applied to flexible chain molecules (rather than globular proteins), an additional complication is introduced since the coefficient k_s in (VI.22) becomes very sensitive to the chain length (Fujimoto and Nagasawa, 1967).

All these difficulties may be circumvented by a procedure called "band centrifugation" (Vinograd et al., 1963; Vinograd and Brunner, 1966). In this technique a shallow layer of the solution to be analyzed is carefully placed on a heavier binary solution before centrifugation. Each sedimenting component will then form a band of its own, and since the macromolecular species are separated from each other, interaction effects will be eliminated. To

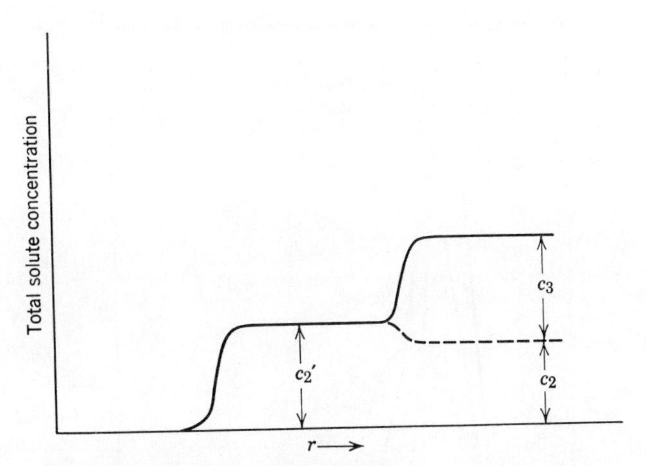

Fig. VI.3. The Johnston-Ogston effect.

use this method successfully, it is necessary to employ a system which will yield $d\rho/dr > 0$ at all points so that no complications arise from convection.

If the distribution function of sedimentation constants is to be obtained in a system in which the frictional coefficient is concentration dependent, the apparent distributions obtained at a number of finite solute concentrations have to be extrapolated to $c_2 = 0$. A procedure for accomplishing this result was described by Williams and Saunders (1954). Figure VI. 4, taken from their work, gives a striking illustration of the gradual broadening of the apparent distribution of sedimentation constants as the polymer solution is diluted. Work by Cantow (1959) on polystyrene solutions in cyclohexane suggests that the concentration dependence of the sedimentation constant disappears at the Θ-point. If this phenomenon could be generalized (cf. Tsvetkov and Klenin, 1959), the interpretation of sedimentation data would be greatly simplified. However, there is no theoretical reason why the frictional coefficient should be independent of concentration in Θ-media, and Scholtan and Marzolph (1962) found the sedimentation constant of polyacrylonitrile to remain concentration dependent even under Θ conditions. In any case the use of Θ-solvents is desirable, since it appears that the resolution attainable in the sedimentation is maximized in Θ-media (McCormick, 1959a).

An unusual concentration-dependent phenomenon has been observed on centrifuging solutions of very high molecular weight monodisperse DNA (Rosenbloom and Schumaker, 1967). In this case, any DNA above a critical concentration that depends on the centrifugal field behaves like a precipitate, while the DNA that remains in solution sediments normally. It appears that the formation of the precipitating fraction reflects aggregation resulting from

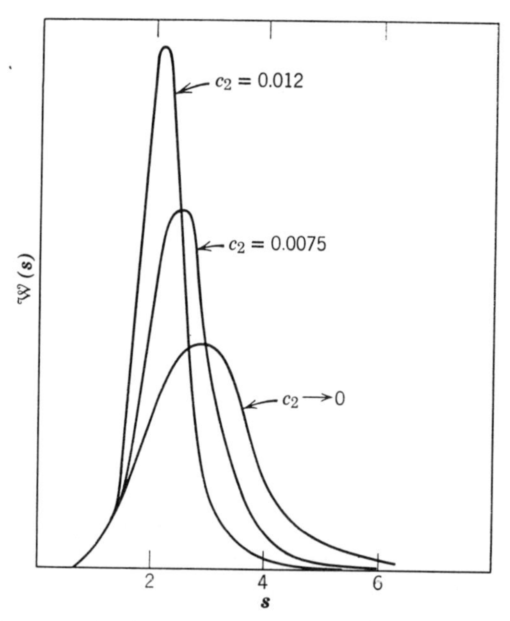

Fig. VI.4. Apparent distribution function of sedimentation constants at two concentrations of dextran and the distribution function obtained by extrapolating experimental data to $c_2 = 0$.

chain entanglements, which are favored by an increasing collision rate between sedimenting macromolecules. Plots of the reciprocal of the critical concentration against the square of the rotor speed are linear with a slope proportional to $M_2^{3/2}$.

If the distribution of the molecular weights of a polymer sample is to be determined, we may combine (VI.20) with (VI.15) to obtain

$$M_2 = \frac{RT}{1 - \bar{v}_2\rho} \frac{s^0}{D^0} \qquad \text{(VI.21)}$$

where the superscript 0 refers to quantities extrapolated to zero concentration. When all the corrections described above are properly carried out, an analysis of sedimentation data becomes the most powerful method for the determination of molecular weight distributions. It is particularly valuable in characterizing samples prepared under conditions which should yield very narrow molecular weight distributions (Cantow, 1959: McCormick, 1959a, b) and has been shown to be superior to analysis by gel permeation chromatography (Noda and Nagasawa, 1970).

The above discussion implies that sedimentation data cannot be interpreted in terms of molecular weights unless additional information relating to the frictional coefficient of the macromolecule is available. However, Archibald (1947) showed that it is possible to obtain unambiguous molecular weights from sedimentation velocity data alone. From (VI.14) and (VI.20) the velocity of the molecules due to both the gravitational field and diffusion is

$$\frac{dr}{dt} = \mathbf{s}r\omega^2 - \frac{D\, d \ln c}{dr} \qquad \text{(VI.25)}$$

and since dr/dt must vanish at the bottom and the top of the cell ($r = r_a$ and $r = r_b$), we obtain

$$\frac{\omega^2 \mathbf{s}}{D} = \frac{1}{r_a c_2}\left(\frac{dc_2}{dr}\right)_{r=r_a} = \frac{1}{r_b c_2}\left(\frac{dc_2}{dr}\right)_{r=r_b} \qquad \text{(VI.26)}$$

Provided that the solution is sufficiently dilute so that $\mathbf{s}/D$ may be approximated by $\mathbf{s}^0/D^0$, we obtain by substitution from (VI.24)

$$\frac{1}{r_a c_2}\left(\frac{dc_2}{dr}\right)_{r=r_a} = \frac{1}{r_b c_2}\left(\frac{dc_2}{dr}\right)_{r=r_b} = \frac{M_2\omega^2(1 - \bar{v}_2\rho)}{\mathbf{R}T} \qquad \text{(VI.27)}$$

Thus, the concentration gradient at the top and the bottom of the cell defines the molecular weight of the sedimenting macromolecules. For optical reasons it is not possible to measure dc_2/dr close to the meniscus or close to the bottom of the cell, but $(1/rc_2)(dc_2/dr)$ can be measured at various values of r, and Paetkau (1967)) has outlined procedures for the extrapolation of these data to $r = r_a$ and $r = r_b$.

For a detailed treatment of the theoretical foundation of sedimentation analysis the reader is referred to a monograph by Fujita (1962).

5. Mobility of Small Molecules and Ions in Polymer Solutions

Since the translational frictional coefficient is proportional to the viscosity of the medium, both the coefficient of diffusion and the electrophoretic mobility are expected to vary as $1/\eta$. Neverthleess, the mobility of small particles in a given solvent is affected very little by the presence of macromolecules, although the viscosity of the macromolecular solution may be much higher than the viscosity of the pure solvent. In a dramatic demonstration of this effect, carried out more than a century ago, Graham (1862) showed that the rate of diffusion of salts through a gelatin gel could not be distinguished from the diffusion rate in pure water. Taft and Malm (1939) found later that the presence of gelatin has a small effect on ionic mobilities, but no change in these mobilities was detected when the gelatin solution set to a gel.

These results are remarkable, since gels which cannot flow in response to a shearing stress must be assigned, by definition, an infinite viscosity. It is clear, however, that the significance of "viscosity" is different if we consider the macroscopic flow of a system and if we consider the passage of a small molecular particle through the same medium. This leads to the distinction between "macroscopic viscosity," describing the flow properties of the system as a whole, and "microscopic viscosity," which characterizes the resistance to the motion of a molecular particle and will, in general, depend on the dimensions of that particle. Although the viscosity of solutions of chain molecules depends on their length, the reduction in the diffusion rate of a small particle, resulting from collisions with segments of a polymer chain, will not depend on the length of the chain to which these segments are attached. Even the formation of a three-dimensional network by these chains will make little difference to the obstacle presented to small diffusing species, although macroscopic flow will then be impossible.

Nishijima and Oster (1956) found that the diffusion coefficient of sucrose in water is independent of added poly(*N*-vinylpyrrolidone) below a critical concentration of the polymer and approaches a lower constant value at higher polymer concentrations. The concentration at which the microscopic viscosity decreases most rapidly is reduced when the molecular weight of the polymer is increased; apparently it corresponds to the concentration at which the mutual interpenetration of polymer chains becomes important. At polymer concentrations above 10% the diffusion rate of the sucrose was entirely independent of the chain length of the polymer, as would be expected on the basis of the considerations outlined above.

B. ROTARY DIFFUSION

1. Rotary Diffusion Coefficient of Rigid Particles

We saw in our discussion of linear diffusion that Brownian motion tends to transport solute molecules from regions of high local concentration to regions of higher dilution. In the absence of any other driving forces, such diffusion eventually results in a complete equalization of the local composition of the system. In the presence of another force, such as gravitation, the system approaches an equilibrium in which gravitational and diffusive transport balance one another.

An analogous situation arises if we consider a system containing asymmetric particles and characterize it by the distribution of the particle orientations. If we consider the behavior of solids of revolution, we may describe their orientation by specifying the distribution function $\mathscr{W}(\varphi)$, where $\mathscr{W}(\varphi)d\varphi$ is the number of particles whose axes of revolution subtend angles between φ and $\varphi + d\varphi$ with a given direction of the coordinate system. If the system is completely disordered, $\mathscr{W}(\varphi) = \text{const}$; on the other hand, if $\mathscr{W}(\varphi)$ is variable, there will be a statistical probability for the particle orientations to pass from more occupied to less occupied φ values. We have then, in analogy with Fick's first law for linear translation (VI.14), a diffusive rotation given by

$$\frac{d\varphi}{dt} = -\left[\frac{D_r}{\mathscr{W}(\varphi)}\right]\frac{d\mathscr{W}(\varphi)}{d\varphi} \tag{VI.28}$$

where D_r is the rotational diffusion coefficient. Again in analogy with (VI.16), D_r may be related to the mean square change in the angle of particle orientation, $\langle\Delta\varphi^2\rangle$, by

$$\langle\Delta\varphi^2\rangle = 2D_r t \tag{VI.29}$$

while (VI.15) is replaced by

$$D_r = \frac{kT}{\mathrm{f}_r} \tag{VI.30}$$

where the rotational frictional coefficient f_r is numerically equal to the force-couple in dyne-centimeters required to impart to the particle a rotational velocity of 1 rad/sec around an axis perpendicular to the axis of revolution. For spheres of radius R_s Stokes (1880) obtained

$$\mathrm{f}_r = 8\pi\eta_0 R_s^3 \qquad \text{(spheres)} \tag{VI.31}$$

while for ellipsoids of revolution with a semiaxis of revolution $\boldsymbol{a}_1$, an equatorial axis $\boldsymbol{a}_2$, and an axial ratio $\mathrm{p} = \boldsymbol{a}_1/\boldsymbol{a}_2$, the result is (Gans, 1928; Perrin, 1934)

$$\mathrm{f}_r = \frac{8\pi\eta_0\boldsymbol{a}_1\boldsymbol{a}_2^2}{\boldsymbol{y}(\mathrm{p})}$$

$$\boldsymbol{y}(\mathrm{p}) = \frac{3}{2}\frac{\mathrm{p}^2}{\mathrm{p}^4-1}\left[1 + \frac{2\mathrm{p}^2-1}{\mathrm{p}\sqrt{\mathrm{p}^2-1}}\ln\left(\mathrm{p}+\sqrt{\mathrm{p}^2-1}\right)\right] \qquad (\mathrm{p}>1)$$

$$\boldsymbol{y}(\mathrm{p}) = \frac{3}{2}\frac{\mathrm{p}^2}{1-\mathrm{p}^4}\left[1 + \frac{1-2\mathrm{p}^2}{2\mathrm{p}\sqrt{1-\mathrm{p}^2}}\tan^{-1}\left(\frac{\sqrt{1-\mathrm{p}^2}}{\mathrm{p}}\right)\right] \qquad (\mathrm{p}<1) \tag{VI.32}$$

The resistance to rotation increases sharply with an increasing elongation of ellipsoids of revolution. For ellipsoids with $\mathrm{p} > 5$, we may use in good approximation

$$\mathrm{f}_r = \frac{8\pi\eta_0\boldsymbol{a}_1^3}{3\ln\mathrm{p} + 0.57} \tag{VI.33}$$

and Gans (1928) obtained a similar expression for long, thin rods. It is then apparent that for sufficiently elongated particles the resistance to rotation at right angles to the long axis increases approximately as the cube of the long dimension and is almost independent of the thickness of the particle. For p < 1 (i.e., for disk-shaped objects), the rotational frictional coefficient is rather insensitive to the axial ratio.

2. Rotary Diffusion of Chain Molecules

In treating the rotary diffusion of flexible polymer chains we may again employ the pearl necklace model composed of chain elements with a translational frictional coefficient f_0, at a distance r_i from the axis of rotation. Each element contributes then $f_0 r_i^2$ to the rotational frictional coefficient of the chain, and since r_i^2 must be, on the average, two-thirds of the square distance from the center of gravity, $f_r = \frac{2}{3} Z f_0 \langle s^2 \rangle$. For flexible chains in Θ-solvents $\langle s^2 \rangle$ is proportional to Z, and f_r in free-draining coils is, therefore, proportional to Z^2.

In the previous discussion of the frictional resistance to translation of flexible chain molecules we noted that real chains are far from free draining and that their behavior approaches, in practice, more closely to that of an impermeable spherical body. The general problem for a pearl necklace chain with partial permeability to the fluid in which it moves was treated by Riseman and Kirkwood (1949). Their result may be expressed in analogy to (VI.31) as

$$f_r = 8\pi\eta_0 R_r^3 F(\Xi) \qquad \text{(VI.34)}$$

where R_r is the radius of a sphere which would encounter the same frictional resistance to rotation as a flexible coil in the limit of complete impermeability. The parameter Ξ, defined in (VI.10), is equal to the ratio of the frictional coefficients of the sum of the chain elements and the impermeable coil, respectively. As $\Xi \to \infty$ and the coil becomes impermeable, $F(\Xi) \to 1$; according to Riseman and Kirkwood, a chain molecule with a root-mean-square radius of gyration $\langle s^2 \rangle^{1/2}$ will then behave like a sphere whose radius is $R_r = 0.89 \langle s^2 \rangle^{1/2}$. The sphere which is hydrodynamically equivalent to the coil with respect to rotation is somewhat larger than the sphere which is equivalent to the coil with respect to linear translation [cf. (VI.11)].

A theoretical treatment of wormlike chains (Hearst, 1963) leads to the conclusion that a plot of $\eta_0 M_2^2/f_r$ against $M_2^{1/2}$ has a slope of $0.72(M_L/b_s)^{3/2}$. Such a plot may, therefore, be used to estimate the length of the statistical chain element.

3. Flow Birefringence

The classical method for evaluation of the rotational diffusion constant employs the anisotropy induced in a solution when a shear gradient leads to a partial orientation of asymmetric particles. This anisotropy is observed most conveniently as an optical birefringence induced by a velocity gradient.

Theoretical and experimental studies of this effect were reviewed by Jerrard (1959), Tsvetkov (1964), and Janeschitz-Kriegl (1969).

The theory of flow birefringence may be divided into two parts. First we have to concern ourselves with the hydrodynamical problem of characterizing the probability distribution of the orientations of particles placed in a velocity gradient of a viscous fluid. Once this problem has been solved, we have to define the optical consequences of this particle orientation.

Any particle placed in a velocity gradient will be subject to a torque that will lead to the rotation of the particle. The situation is represented schematically in Fig. VI.5, where the x direction is the direction of the flow lines and

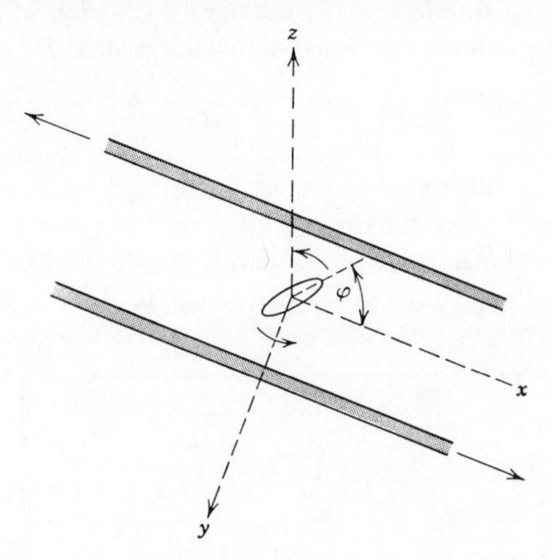

Fig. VI.5. Ellipsoidal particle suspended in a viscous liquid with a velocity gradient.

the velocity gradient increases in the y direction, while the particle is being observed in the z direction perpendicular to the x—y plane. If the particle is spherical, its rotational speed will be uniform, but for ellipsoidal particles the torque due to the viscous drag will depend on particle orientation. If the axis of symmetry of the ellipsoid lies in the x—y plane, the driving force will be largest with this axis in the y direction and smallest with the long axis parallel to the flow lines, so that the rotational velocity will change periodically during each revolution. The probability $\mathscr{W}(\varphi)$ of finding the particle in any given orientation, where φ is the angle subtended by the symmetry axis of the ellipsoid with the flow lines, will be inversely proportional to the rotational speed, so that

$$\frac{\mathscr{W}(\varphi)d\varphi}{dt} = \text{const} \qquad \text{(VI.35)}$$

If the particles are small, Brownian motion will counteract the hydrodynamic orientation of the particles. We may then express $d\varphi/dt$ as the sum of ω_h, the angular velocity due to hydrodynamic forces, and the velocity produced by rotational diffusion, related to the rotational diffusion constant by (VI.28). This leads to†

$$\mathscr{W}(\varphi)\omega_h(\varphi) - \frac{D_r d\mathscr{W}(\varphi)}{d\varphi} = \text{const} \tag{VI.36}$$

Equation (VI.36) has been solved by Boeder (1932) for ellipsoids whose symmetry axes lie in the x—y plane. Since $\omega_h(\varphi)$ must be proportional to the velocity gradient in the solution, $\mathscr{W}(\varphi)$ becomes a function of the ratio of the velocity gradient q and the rotational diffusion constant D_r:

$$\mathscr{W}(\varphi) = \mathrm{f}(\boldsymbol{\alpha}), \qquad \boldsymbol{\alpha} = \frac{q}{D_r} \tag{VI.37}$$

Boeder's results are illustrated in Fig. VI.6. For low values of $\boldsymbol{\alpha}$ the function $\mathscr{W}(\varphi)$ is rather flat but tends to have a maximum at $\varphi = \pi/4$. As $\boldsymbol{\alpha}$ increases, the particles are oriented more nearly parallel to the flow lines, as shown by the decreasing value in $\varphi_{\max}$ and the sharpening of the distribution function $\mathscr{W}(\varphi)$. In reality the long axis of the ellipsoids is of course not restrained to lie

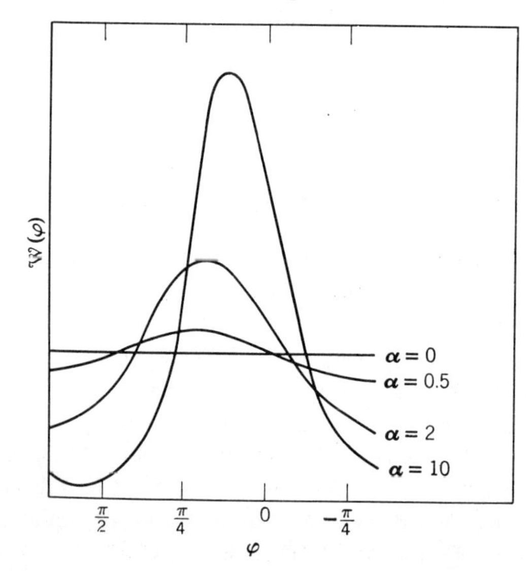

Fig. VI.6. Orientation of ellipsoidal particles as a function of the ratio of the velocity gradient to the rotatory diffusion coefficient.

†In this treatment it is assumed that inertial forces, which would tend to smooth out the periodic changes in the angular velocity of the rotating particles, are negligible.

in the x—y plane and its orientation will have to be specified by two angles: the angle φ between the $\boldsymbol{a}_1$—z plane and the x—z plane, and the angle ϑ between $\boldsymbol{a}_1$ and z. This complicates the problem of deriving the steady-state probability distribution function $\mathscr{W}(\varphi, \vartheta)$ of particle orientations, but a solution has been obtained (Peterlin, 1938).

We must now consider the second part of the problem, namely, the optical consequences of partial orientation of the solute particles. The experimental arrangement is shown schematically in Fig. VI.7. The test solution is contained in the annulus between two cylindrical surfaces, one stationary and one rotating. A plane-polarized beam of light enters the solution parallel to the cylindrical axis of the apparatus (the z axis) and is observed through a crossed analyzer. If the solution is isotropic, no light will pass through the analyzer, but if it contains an anisotropic particle, this particle will be visible unless its optical axis is parallel to the plane of polarization of the polarizer or the analyzer. If we have a solution of ellipsoidal particles and if the φ angle as defined above is the same for all the particles, Fig. VI.7 shows that the con-

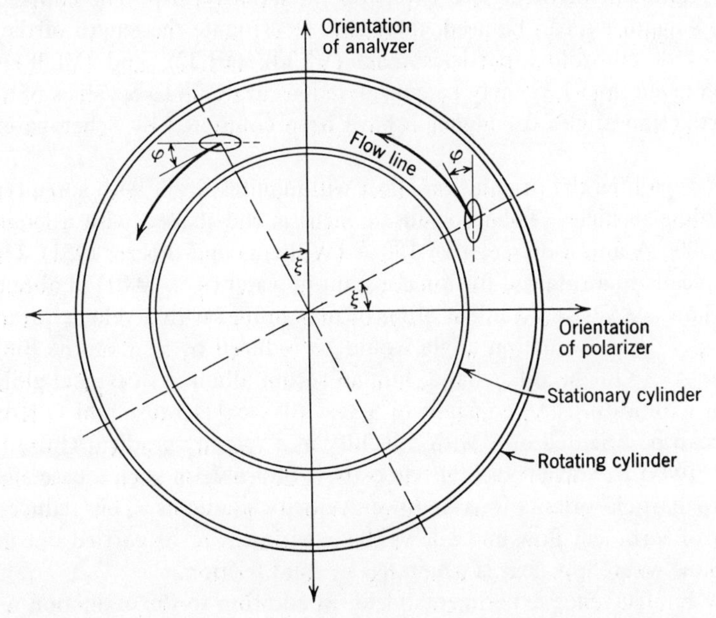

Fig. VI.7. View of doubly refractive medium in the gap between a stationary and a rotating cylinder.

dition for light extinction will be met for all particles whose centers lie in a plane rotated around the z axis from the plane of polarization of the polarizer or the analyzer by an angle $\xi = \varphi$. We shall then observe light transmission except for a dark cross ("the cross of isocline"). For a solution in which the orientation angles φ are variable, the condition of light extinction requires that the electrical vectors of light passing through the anisotropic particles have components parallel to the orientation of the analyzer which add up to zero.

Peterlin and Stuart (1939) showed that for rigid ellipsoids of revolution the extinction angle ξ should be given by

$$\xi = \frac{\pi}{4} - \frac{\boldsymbol{\alpha}}{12}\left\{1 - \frac{\boldsymbol{\alpha}^2}{108}\left[1 - \frac{24}{35}\left(\frac{\mathrm{p}^2 - 1}{\mathrm{p}^2 + 1}\right)^2\right] + \cdots\right\} \qquad \text{(VI.38)}$$

We see then that the extinction angle approaches to 45° as the velocity gradient approaches zero. The decay of ξ with increasing velocity gradient depends at first only on $\boldsymbol{\alpha} = q/D_r$, so that the rotational diffusion constant may be derived from

$$\left(\frac{d\xi}{dq}\right)_{q\to 0} = -\frac{1}{12D_r} \qquad \text{(VI.39)}$$

At higher values of $\boldsymbol{\alpha}$, the dependence of the extinction angle on the velocity gradient involves also explicitly the axial ratio p. The course of a plot of ξ against $\boldsymbol{\alpha}$ can be used, therefore, to estimate the length of the two semiaxes of ellipsoidal particles from (VI.30), (VI.32), and (VI.38). The relation given in (VI.38) may be considered accurate up to $\boldsymbol{\alpha}$ values of about 1.5; extinction angles for higher $\boldsymbol{\alpha}$ have been computed by Scheraga et al., (1951).

Let us see how the orientation effect will manifest itself with some typical rigid solute particles. Tobacco mosaic virus is rod-shaped with a length of about 3000 Å and a diameter of 150 Å (Williams and Steere, 1951). Use of (VI.33) leads to a rotary diffusion constant in water ($\eta_0 = 0.01$) of about 400 sec^{-1}. This is a very convenient order of magnitude; with a velocity gradient of 10^3 sec^{-1}, the extinction angle would be reduced by 9° from its limiting value fo 45°. On the other hand, human serum albumin, a typical globular protein with a diffusion constant of 8.3×10^5 sec^{-1} (Krause and O'Konski, 1959), can be oriented only with difficulty in a velocity gradient (Edsall and Foster, 1948). A solvent of high viscosity is desirable in such a case since it leads to particle orientations at lower velocity gradients. This reduces the danger of turbulent flow and allows the experiment to be carried out under conditions where less heat is generated by fluid friction.

Flow birefringence experiments yield, in addition to the extinction angle, the magnitude of the birefringence $\Delta\mathrm{n}$. This quantity depends both on the extent of orientation of the asymmetric solute particles and on the optical anisotropy $g_1 - g_2$. At a low degree of particle orientation Peterlin and Stuart (1939) give, for systems containing a volume fraction ϕ_2 of solute,

$$\frac{\Delta\mathrm{n}}{\phi_2} = \frac{(2\pi/15)(g_1 - g_2)}{\mathrm{n}}\,\mathrm{f}(\mathrm{p}, \boldsymbol{\alpha}) \qquad \text{(VI.40)}$$

The optical anisotropy $g_1 - g_2$ depends on the refractive index of the solvent, the axial ratio of the ellipsoidal solute, and the refractive indices n_1 and n_2 parallel and perpendicular, respectively, to the symmetry axis of the ellipsoids. It contains two contributions; the first is due to the intrinsic anisotropy of the particle if $n_1 \neq n_2$, while the second ("the form anisotropy") results from the orientation of an asymmetric particle in a medium of different refractive index and does not vanish even if $n_1 = n_2$. If the dimensions of the particle are known from the dependence of the extinction angle on the velocity gradient, (VI.40) provides one relation between n_1 and n_2. The second relation is obtained from measurements of the average refractive index of the solute in bulk, which must be $(n_1 + 2n_2)/3$. We may then solve for the principal refractive indices of the solute particle.

The interpretation of flow birefringence of solutions of flexible chain molecules is much more involved. The frictional forces due to the streaming fluid will not only tend to rotate the coil, but will also lead to a periodic change in its shape as a consequence of the alternating tension and compression indicated in Fig. VI.8. This change of shape will be counteracted by the

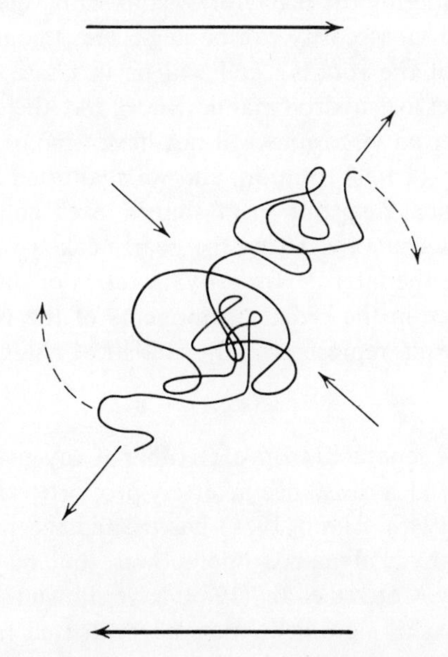

Fig. VI.8. Tensile and compressive stresses on a flexible coil in a viscous fluid with a velocity gradient. The heavy lines indicate the flow lines in the fluid, the light solid lines the stresses on the chain molecule, the dashed lines the direction in which the molecule rotates.

rubber-like elasticity of the coils, which resist changes to less probable chain conformations (Kuhn and Kuhn, 1943). If we place one chain end at the origin of the coordinate system, while the other lies on one of the axes, we may use the probability distribution function (III.6) to obtain for the elastic retractive force F_{el} (on the assumption that the internal energy of the chain is independent of the chain end separation h)

$$\begin{aligned} F_{el} &= \left(\frac{\partial G}{\partial h}\right)_{T,P} \\ &= -T\left(\frac{\partial S}{\partial h}\right)_{T,P} = -\boldsymbol{k}T\left[\frac{\partial \ln \mathscr{W}(h)}{\partial h}\right]_{T,P} \\ &= 3\left(\frac{\boldsymbol{k}T}{h}\right)\left(\frac{h^2}{\langle h_0^2\rangle}\right) \end{aligned} \tag{VI.41}$$

where $\langle h_0^2\rangle$ refers to the mean square end-to-end displacement of the unperturbed chain. However, Kuhn and Kuhn (1946) pointed out also that, for any given rate at which the chain is to be stretched or compressed, the potential energy barriers that have to be overcome in conformational transitions require an additional force, which may be characterized by an internal viscosity η_i of the chain. The significance of this parameter is best illustrated by considering the behavior of coils with very low and with very high η_i. If the internal viscosity can be neglected, the alternating expansion and compression of the rotating coil will be in phase with the alternating tensile and compressive hydrodynamic forces. At the other extreme, coils with very high internal viscosities will not have time to change their shapes appreciably during a single rotation, and we shall then have an assembly of coils with a statistical distribution of shapes, each coil rotating essentially as if it were a rigid structure. Using the pearl necklace model for the chain, Cerf (1957) defined the internal viscosity, in terms of the force f_i required to produce a difference in the radial components of the velocity vectors $\mathbf{u}_r$ of two consecutive beads representing the statistical chain elements, as

$$f_i = \eta_i(\mathbf{u}_{r,j+1} - \mathbf{u}_{r,j}) \tag{VI.42}$$

If the chain ends are separated from each other at any given velocity, the internal viscosity will offer a resistance inversely proportional to the chain length (Kuhn and Kuhn, 1946). Zimm (1956) has treated the pearl necklace model taking account of hydrodynamic interactions, but he neglected the effect of internal viscosity. Cottrell et al., (1969) have shown that the distortion of polymer coils in a shear gradient may be evaluated from light scattering data, and their results prove the importance of internal viscosity. We shall consider its signiflnance in more detail in connection with the viscoelastic behavior of highly dilute polymer solutions.

Peterlin (1963) has pointed out that the orientation of flexible chain molecules in a shear gradient is proportional to the shear stress $q\eta_0$, whereas the deformation of the molecular coil depends on $(q\eta_0)^2$. The orientational effect is then dominant at low shear stresses where the extinction angle is governed by $\boldsymbol{\alpha} = q/D_r$, just as with rigid particles. For impermeable coils, D_r is given by (VI.30) and (VI.32) in terms of the dimensions of the hydrodynamically equivalent ellipsoid, so that for low shear stresses the initial slope of a plot of ξ against q should depend on $\langle s^2\rangle^{3/2}\eta_0/\boldsymbol{k}T$. We shall see in Section VI.C that the intrinsic viscosity $[\eta]$, a very easily measured quantity, is proportional to $\boldsymbol{N}\langle s^2\rangle^{3/2}/M_2$. Thus, one would expect

$$\left(\frac{\partial\xi}{\partial q}\right)_0 = \frac{\text{const}\,[\eta]\eta_0 M_2}{\boldsymbol{R}T}\left(\frac{n_0^2+2}{3}\right)^2 \qquad \text{(VI.43)}$$

However, the implication that $(\partial\xi/\partial q)_0$ is proportional to η_0 breaks down in viscous media because of the effect of the internal viscosity of the chain. Data obtained with DNA (Leray, 1957) and with poly(methyl methacrylate) (Tsvetkov and Budtov, 1964) show that plots of $(\partial\xi/\partial q)_0$ against η_0 have an asymptote with a positive intercept on the ordinate, and theoretical analysis has demonstrated that this intercept is proportional to the internal viscosity of the chain (Cerf, 1959; Peterlin, 1973).

The magnitude of the flow birefringence of solutions of chain molecules depends, as with rigid particles, on both the intrinsic anisotropy and the form anisotropy of the macromolecule. Random flight chains have an intrinsic anisotropy if the chain segments are anisotropic, since the segments have a statistical preference for pointing in the direction of the chain end displacement. According to Kuhn and Grün (1942), the optical anisotropy of flexible chains in a solvent with an index of refraction n_0 is given by

$$g_2 - g_1 = \frac{3}{5}\frac{(g_2^s - g_1^s)\langle h^2\rangle}{\langle h_0^2\rangle}\cdot\left(\frac{n_0^2+2}{3}\right)^2 \qquad \text{(VI.44)}$$

where $g_2^s - g_1^s$ is the optical anisotropy of the statistical chain segment. However, if the refractive index of the solvent is different from that of the polymer, additional contributions to $g_2 - g_1$ arise from optical interactions between chain segments (Tsevtkov, 1964). Long-range interactions will make a contribution which depends on the asymmetry of the coil, while short-range interactions reflect the flexibility of the chain. These various contributions to $g_2 - g_1$ may have opposity signs and since they depend in different ways on shear stresses, we may then observe a reversal of the sign of Δn in experiments carried out over a range of velocity gradients (Tsvetkov, 1957). The relative contributions of intrinsic and form anisotropy determine also the dependence of Δn on the chain length of the polymer (Tsvetkov, 1964). If only the intrinsic anisotropy is important, $\Delta n/[\eta]$ (where Δn is measured at a fixed low polymer concentration) is independent of M_2; if form anisotropy makes an appreciable contribution to the birefringence, $\Delta n/[\eta]$ increases linearly with $M_2/[\eta]$.

Tsvetkov et al. (1973) have carried out a very interesting study of the solution behavior of polymers with long side chains, such as the following substance:

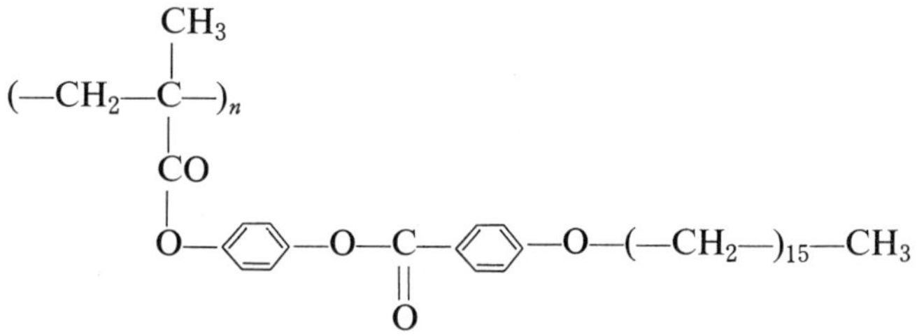

$$(-CH_2-\underset{\underset{\underset{\underset{O-C_6H_4-O-\underset{\underset{O}{\|}}{C}-C_6H_4-O-(-CH_2-)_{15}-CH_3}{|}}{CO}}{|}}{\overset{\overset{CH_3}{|}}{C}}-)_n$$

This polymer was found to have an unusually large optical anisotropy, interpreted as indicative of a packing of the side chains into liquid-crystal-like aggregates.

In view of the insensitivity of many solution properties to the stereoregularity of chain molecules, it is particularly interesting that flow birefringence measurement may be an efficient tool for the characterization of tacticity (Tsvetkov et al., 1961). Thus, the segmental anisotropy $g_2^s - g_1^s$ (in units of 10^{-25} cm^3) is -146 and -224 for atactic and isotactic polystyrene in bromoform, and $+2$ and $+25$ for atactic and isotactic poly (methyl methacrylate) in benzene (Tsvetkov, 1962).

Phenomena similar to flow birefringence may be observed with respect to the anisotropy of other physical properties of polymer solutions subjected to a shear gradient. Flow dichroism data allow not only the evaluation of the rotary diffusion constant, but also a determination of the orientation of transition dipoles responsible for the dichroic absorption bands. Use of this technique with double-helical nucleic acids yielded results in agreement with the Watson-Crick structure (Wada, 1964; Wada et al., 1971). Hartmann and Jaenicke (1956) described the measurement of the flow-induced anisotropy of the dielectric constant. A velocity gradient may also induce an anisotropy of electrolytic conductance, as we shall see in the next chapter in discussing solutions of polyelectrolytes.

4. Disorientation of Oriented Solutions

When polarizable molecules are placed in an electrical field of intensity X, the induced dipole is proportional to X and the potential energy is proportional to X^2, provided that effects caused by permanent dipoles are negligible compared to polarization effects. If the molecule can be treated as an ellipsoid of revolution with maximum polarizability in the direction of the symmetry axis, the distribution function of the angle φ between the symmetry axes of the particles and the electrical field is given by

$$\mathscr{W}(\varphi)\, d\varphi = \tfrac{1}{2} \sin \varphi \exp\left[\frac{-E(\varphi,X)}{kT}\right] d\varphi \qquad \text{(VI.45)}$$

The partial orientation of the molecules will produce a birefringence given by

$$\frac{\Delta \mathrm{n}}{\mathrm{n}} = \mathrm{K}\phi_2 X^2 \qquad \text{(VI.46)}$$

where ϕ_2 is the volume fraction of the solute which is oriented by the field, and the Kerr constant K is a function of the polarization anisotropy of the solute particles. If the field is cut off, Brownian motion will disorient the particles so that the mean value of cos φ will decay according to

$$\langle \cos \varphi \rangle(t) = \langle \cos \varphi \rangle(0) \exp\left(-\frac{t}{\tau_r}\right) \qquad \text{(VI.47)}$$

This leads to a decay of birefringence (or any other anisotropic property such as, for instance, the dichroism induced by the electrical field) proportional to $\langle \cos \varphi \rangle\ (t)$ with the relaxation time τ_r related to D_r by

$$\tau_r = \frac{1}{6D_r} \qquad \text{(VI.48)}$$

The relaxation of electrical birefringence was used by Benoit (1951a) to measure the rotational diffusion constant of tobacco mosaic virus particles. Experimental improvements allowed Krause and O'Konski (1959, 1963) to apply the method to particles with a much higher rotational diffusion constant, such as globular proteins whose longest dimension is of the order of 100 Å.

If the solution exposed to the electrical field contains flexible macromolecules, the orientation involves chain segments rather than the solute particle as a whole.‡ The electrically induced anisotropy is then no longer proportional to X^2 but becomes linear in X for sufficiently long chains, and the dependence of the anisotropy on X can be used for the quantitative characterization of chain rigidity (Milstien and Charney, 1969). The decay of the anisotropy will then generally reflect a distribution function of relaxation times, and this distribution will be independent of chain length for chain molecules which are very much longer than their persistence length. In the case of the slightly flexible DNA double helix, the decay of electrical birefringence can be interpreted in terms of the rotary diffusion of the macromolecule for relatively

‡However, solutions of polymers in which side chains form liquid-crystal-like aggregates (see p. 316) may have their Kerr constants increased by several orders of magnitude (Tsvetkov et al., *Vysokomol. Soedin.*, **A 15**, 2270, 2570 (1973)).

low molecular weight samples (Benoit, 1951b), but with the very long DNA isolated from bacteriophages the distribution of relaxation times remains unchanged after shear degradation of the nucleic acid (Golub, 1964). For a listing of a large number of rigid or flexible macromolecules which have been studied by electrical birefringence the reader is referred to Yoshioka and O'Konski (1968).

Polarizable particles will also be oriented in an alternating field. If the frequency ν is low compared to the rotational diffusion constant, the orientation will increase and decrease in phase with the field; at the other extreme, with $\nu \gg D_r$, the orientation will be constant, corresponding to the time-average value of X^2. If ν and D_r have comparable magnitudes, the particle orientation will lag behind the variation in the alternating field, with the phase angle φ related to the rotational diffusion constant by

$$\tan \varphi = \frac{2\pi\nu}{3D_r} \tag{VI.49}$$

This principle has been employed by Benoit (1952) for the determination of D_r in aqueous solutions of tobacco mosaic virus.

5. Depolarization of Fluorescence

When a fluorescent group is attached rigidly to a globular protein molecule and excited with polarized light, the fluorescence at right angles to the plane of polarization of the incident beam will be partially depolarized to an extent which will depend on how far the molecule has rotated during the lifetime τ_e of the excited state. The use of this principle in the study of rotational diffusion of globular protein molecules was introduced by Weber (1952), and work in this area has been reviewed by Steiner and Edelhoch (1962) and by Chen (1967). If we characterize polarization by $\mathbf{P} \equiv (I_{\parallel} - I_{\perp}) / (I_{\parallel} + I_{\perp})$, where $I_{\parallel}$ and $I_{\perp}$ refer to intensities of components of the fluorescence beam polarized parallel and perpendicular, respectively, to the direction of polarization of the exciting beam, it can be shown (Perrin, 1929; Weber, 1952) that

$$\frac{\mathbf{P}^{-1} - \frac{1}{3}}{\mathbf{P}_0^{-1} - \frac{1}{3}} = 1 + \frac{3\tau_e}{\tau_{rh}} \tag{VI.50}$$

Here $\mathbf{P}_0$ refers to the polarization of the fluorescent beam when the fluorescent molecule cannot rotate, and τ_{rh} is the harmonic mean relaxation time, given by $\tau_{rh} = 2\tau_r\tau_r'/(\tau_r + \tau_r')$, where τ_r and τ_r' are relaxation times for rotation around the equatorial axis and the symmetry axis, respectively. The value of $\mathbf{P}_0$ is obtained by measuring the depolarization of fluorescence in media of increasing viscosity and carrying out a linear extrapolation of $\mathbf{P}^{-1}$ against η_0^{-1} to $\eta_0^{-1} = 0$ (Perrin, 1929). For spheres, $\tau_r = \tau_r' = \tau_{rh}$, so that (VI.30), (VI.31), and (VI.48) give τ_{rh} proportional to the volume of the particle. For very elongated ellipsoids of revolution, the hydrodynamic resistance is much less for rotation around the symmetry axis than for rotation around the equatorial axis, so that $\tau_r \gg \tau_r'$ and $\tau_{rh} \approx 2\tau_r'$; for particles of equal volume, τ_{rh} will increase only very slowly with an increasing axial ratio p (Weber, 1953). This is in sharp contrast to the high sensitivity of flow birefringence or the relaxation of electrical birefringence to increasing particle elongation, since in those techniques only rotations around the equatorial axis are observed.

The range of relaxation times which may be studied by measurement of the depolarization of fluorescence is determined by the lifetimes of the excited species of the fluorescent dyes employed. Typical values are $\tau_e = 1.2 \times 10^{-8}$ sec for 1-dimethylaminonaphthalene-5-sulfonyl derivatives and 5×10^{-9} sec for fluorescein derivatives (Steiner and Edelhoch, 1962). It is then possible to measure even rotational diffusion constants of the order of 10^8 sec^{-1}, which cannot be studied by flow birefringence. This range of D_r values is important, since it covers the smaller globular protein molecules. Depolarization of fluorescence can be measured with equal ease whether the solute molecule is asymmetric or spherical. Thus, we may study rotational diffusion even with particles which cannot be oriented.

A complication may arise if nonradiative energy transfer (as discussed on p. 241) occurs between chromophores carried by the macromolecule. Such energy transfer will generally reduce the polarization of the emitted light and hence will lead to an increase in the apparent rotary diffusion constant calculated from the depolarization of fluorescence. Weber and Anderson (1969) have formulated a theory concerning this effect and reported on a system in which it was observed.

In a number of cases the τ_{rh} values calculated for conjugates of globular proteins have been found to be lower than those to be expected if the protein could be represented by a rigid sphere (Chen, 1967). Such a result can be understood only if the protein molecule is, in fact, flexible so that the attached dye can rotate faster than the molecule as a whole. The amount of information which may be extracted from experimental data is greatly increased if the polarization of fluorescence is followed after a flash irradiation of a test solution over the lifetime of the excited state of the dye, and the experimental difficulties connected with such fast measurements have been successfully overcome (Wahl, 1966). In that case, any deviation from a simple exponential decay of the polarization of fluorescence indicates that more than one mode of rotational diffusion is operative. We shall discuss in Chapter VIII an application of this technique to the structure of an antibody.

When a fluorescent dye is attached to a flexible chain molcculc, its rotational diffusion clearly does not involve rotation of the entire macromolecule but is merely indicative of the local rigidity of the chain. Frey et al. (1964) and Nishijima et al. (1967) have studied the depolarization of fluorescence in solutions of polymers carrying a fluorescent dye at the chain end. The dependence of the relaxation time τ_r, calculated from such data, on the polymer chain length (Fig. VI.9) shows that τ_r approaches its limiting value much more slowly than might have been expected.

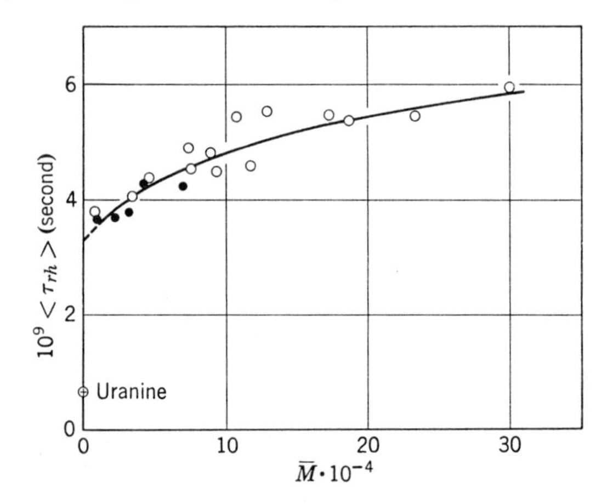

Fig. VI.9. Mean rotational relaxation times of uranine residues attached to the end of polyacrylamide chains. The open and closed circles indicate different modes of bonding of the chromophore to the polymer. The rotational relaxation time of the free dye is also shown.

Analysis of the depolarization of fluorescence involving polymers with dyes rigidly attached to the chain backbone requires a knowledge of the orientation of the transition moment relative to the polymer chain (Gottlieb and Wahl, 1963). A study of poly(methyl methacrylate) with anthracene residues incorporated into the chain backbone (Anufrieva et al., 1970) and of styrene copolymers with 9-*p*-vinylphenyl-10-phenylanthracene (Biddle and Nordström, 1970) yielded surprisingly small temperature coefficients of τ_r in any one solvent medium. These corresponded to activation energies of 2 and 3.4 kcal/mole, respectively, values which seem too low to characterize energy barriers to conformational transitions. Biddle and Nordström found also that at a fixed temperature the rotational relaxation time increases less rapidly than in proportion to the viscosity of the solvent, and this observation is of interest in connection with a recent theory of the interal viscosity of polymer chains (see p, 340).

The depolarization of fluorescence of dyes attached to the carboxyls of poly(acrylic acid) or poly(methacrylic acid) is much greater for the isotactic species, indicating a sensitivity of the dynamic flexibility to the stereoisomerism of the chain (Valeur et al., 1971). This property is also sensitive to the conformational state of the chain, as is strikingly illustrated by the behavior of poly(methacrylic acid) in water-dioxane mixtures, where τ_r decreases ninefold in the ionization range in which the chain undergoes a cooperative expansion (Anufrieva et al., 1969). A very large decrease in the depolarization of fluorescence is, of course, observed when the fluorescing residue is attached to a chain which changes from a flexible coil to a rigid helical conformation, and depolarization data provide a sensitive index of such transitions (Gill and Omenn, 1965).

C. SOLUTION VISCOSITY

1. Viscosity of Solutions of Rigid Molecules

When a Newtonian fluid of viscosity η_0 is subjected to a uniform velocity gradient q, the rate at which mechanical energy is transformed into heat by fluid friction per unit volume will be given by

$$J = \eta_0 q^2 \tag{VI.51}$$

If we now suspend rigid spheres in the fluid, the flow lines will be perturbed as indicated in Fig. VI.10. The viscous drag will induce the spheres to rotate,

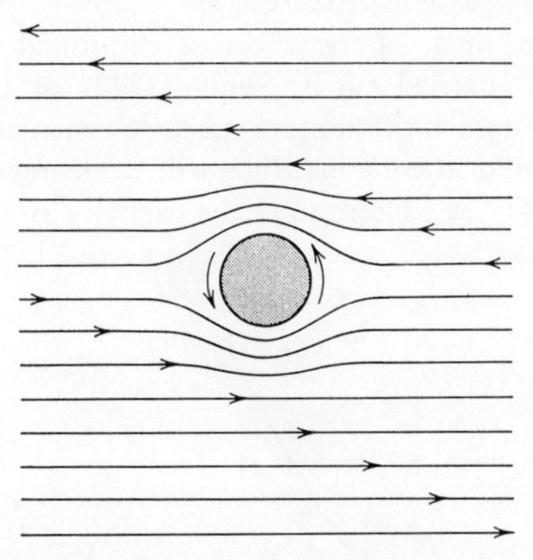

Fig. VI.10. Flow lines around a sphere suspended in a viscous fluid with a velocity gradient.

and it can be shown that the steady-state angular velocity ω is half of the velocity gradient. The flow perturbations caused by the spheres will lead to an increase in the rate of energy dissipation, which may be formally related to the viscosity η of the suspension by

$$J + \Delta J = \eta q^2 \qquad \frac{\Delta J}{J} = \frac{\eta - \eta_0}{\eta_0} \tag{VI.52}$$

The calculation of $\Delta J/J$ for a suspension of rigid spheres with interparticle distances very large compared to the particle diameters was carried out by Einstein (1906a, 1911). His result was

$$\eta_{sp} \equiv \frac{\eta - \eta_0}{\eta_0} = \tfrac{5}{2}\,\phi_2 \tag{VI.53}$$

where η_{sp} is called the "specific viscosity", and ϕ_2 is the volume fraction occupied by the spheres. It is striking that the specific viscosity depends only on the total volume occupied by the spheres and *not on the size of the individual spheres.*

The case of asymmetric particles is more complex. As was pointed out in our discussion of flow birefringence, the orientation of ellipsoidal particles will be kept random by Brownian motion at low velocity gradients; the extent of orientation will be a function of $\boldsymbol{\alpha} = q/D_r$. Since the frictional dissipation of energy due to the presence of the particles must depend on their orientation with respect to the flow lines, η_{sp} would also be expected to depend on $\boldsymbol{\alpha}$. In general, the asymmetric particles will orient themselves at high $\boldsymbol{\alpha}$ values so as to offer the least resistance to the streaming fluid, and η_{sp} must, therefore decrease as $\boldsymbol{\alpha}$ increases.

A quantitative treatment of the effect of ellipsoidal solute particles on solution viscosity was carried out by Simha (1940) for the case of $q \ll D_r$, that is, for a random distribution of particle orientations. The specific viscosity obtained under these conditions will be denoted by η_{sp}^0. The ratio η_{sp}^0/ϕ_2 depends then on the axial ratio of the particles as

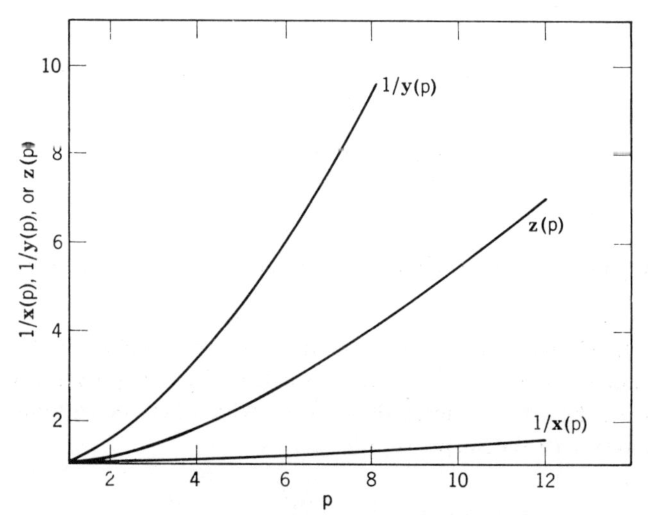

Fig. VI.11. The functions $1/x(\mathrm{p})$, $1/y(\mathrm{p})$, and $z(\mathrm{p})$ characterizing the translational frictional coefficient, the rotational frictional coefficient, and the specific viscosity of ellipsoids of revolution.

$$\frac{\eta_{sp}^0}{\phi_2} = \tfrac{5}{2}\, \boldsymbol{z}(\mathrm{p}) \tag{VI.54}$$

and Fig. VI.11 compares the function $\boldsymbol{z}(\mathrm{p})$ with $1/\boldsymbol{x}(\mathrm{p})$ and $1/\boldsymbol{y}(\mathrm{p})$ describing the dependence of the translational and rotational frictional coefficients, respectively, on the axial ratio of ellipsoidal particles. It shows that solution viscosity is much more sensitive to particle asymmetry than linear diffusion, but not as sensitive as rotary diffusion. For large values of p, Simha gives the approximate relation

$$\frac{\eta_{sp}^0}{\phi_2} = \frac{14}{15} + \frac{\mathrm{p}^2}{5}\left[(3 \ln 2\mathrm{p} - \tfrac{9}{2})^{-1} + (\ln 2\mathrm{p} - \tfrac{1}{2})^{-1}\right] \tag{VI.55}$$

A theory of the dependence of η_{sp}/ϕ_2 on the parameter $\boldsymbol{\alpha}$, which controls the orientation of asymmetric particles, was formulated by Kuhn and Kuhn (1945a). The dependence of η_{sp} on the velocity gradient is generally of the form

$$\frac{\eta_{sp}}{\eta_{sp}^0} = 1 - s_1(\mathrm{p})\boldsymbol{\alpha}^2 + s_2(\mathrm{p})\boldsymbol{\alpha}^4 - \cdots \tag{VI.56}$$

where the coefficients $s_1(\mathrm{p})$, $s_2(\mathrm{p})$, and so on, vanish for spherical particles and increase with the axial ratio p. In the limit of large p values, $s_1 = \frac{1}{32}$, $s_2 = 0.0017$. Scheraga (1955) has computed η_{sp}/η_{sp}^0 as a function of p up to large values of $\boldsymbol{\alpha}$; some of his results are shown in Fig. VI.12. As (VI.56) indicates, the decay of η_{sp} with an increase in the velocity gradient is a function of the

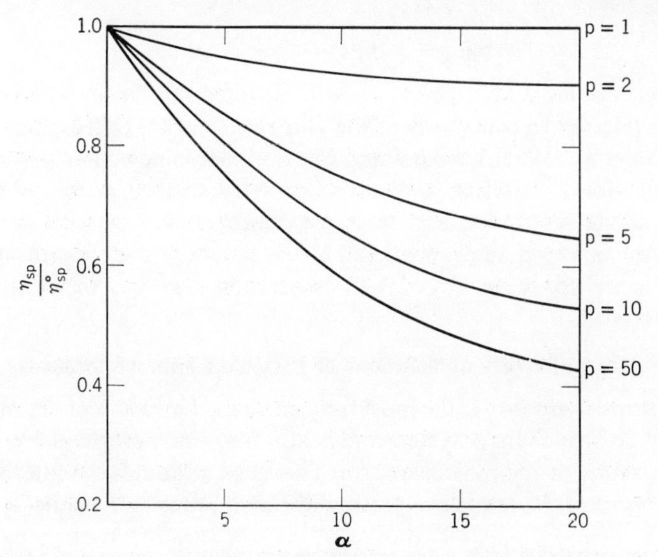

Fig. VI.12. Relative decrease of the specific viscosity of solutions of ellipsoidal particles with an increasing ratio of the velocity gradient to the rotary diffusion coefficient.

rotational diffusion constant and thus a measure of the asymmetry of the particles. This is a more reliable indication of their axial ratio than η^0_{sp}, since solvation of the macromolecules may increase their effective hydrodynamic volume, so that an interpretation of η^0_{sp} based on a ϕ_2 value corresponding to the volume of the "dry" solute may involve an appreciable error.

So far, we have considered only systems so dilute that interactions between the disturbances of the flow pattern due to the various solute particles could be neglected. In this limit, η_{sp} is proportional to the number of solute particles. As we pass to more concentrated systems where interactions of the suspended particles are no longer negligible, we have to express η_{sp} as a power series in ϕ_2. For spherical particles,

$$\eta_{sp} = \tfrac{5}{2}\phi_2 + a_2\phi_2^2 + a_3\phi_2^3 + \cdots \tag{VI.57}$$

Frisch and Simha (1956) reviewed theoretical estimates of a_2, which range from 2.5 to 14.1, and the subject was discussed later by Peterson and Fixman (1963), who arrived at $a_2 = 4.32$. Experiments on suspensions of glass beads in a medium of the same density yielded $a_2 = 7.17$ (Vand, 1948).

As noted before, the effective hydrodynamic volume of a mole of solute particles (V_e) may be appreciably different from the molar volume of the solute in bulk. Expressing the solute concentration c_2 in grams per cubic centimeter, we have $\phi_2 = (V_e/M_2)c_2$ and (VI.57) may be rewritten as

$$\begin{aligned}\eta_{sp} &= 2.5\left(\frac{V_e}{M_2}\right)c_2 + 12.6\left(\frac{V_e}{M_2}\right)^2 c_2^2 \\ &= [\eta]c_2 + k'[\eta]^2c_2^2 \end{aligned}\tag{VI.58}$$

where $[\eta]$, the limit of η_{sp}/c_2 as $c_2 \to 0$, is called the "intrinsic viscosity,"† and k' is referred to commonly as the Huggins constant (cf.Huggins, 1942d). Weissberg et al. (1951) have pointed out that macromolecular particles may form short-lived doublets, with an asymmetry smaller or larger than the isolated solute molecules, and since the concentration of such doublets in dilute systems would be proportional to the square of the concentration, the hydrodynamic consequences of such association may lead to an increase or reduction of k'.

2. Viscosity of Solutions of Flexible Chain Molecules

The intrinsic viscosity is the most frequently used measure of the molecular weight of flexible chain polymers and is also frequently employed to estimate the dimensions of the molecular coil. This is so although the interpretation of $[\eta]$ in terms of M_2 requires a prior calibration procedure against a primary

†Here we are expressing $[\eta]$ in cubic centimeters per gram for the sake of consistency of units used throughout the book. It is common to express the solute concentration in grams per 100 ml, and in this case $[\eta]$ is given in deciliters per gram.

method (usually osmometry or light scattering) and although the estimation of coil dimensions from $[\eta]$ is based on theories which cannot compare in rigor to the relation derived between $\langle s^2 \rangle$ and the angular dependence of the scattered light intensity. The popularity of the viscosimetric method is due to the ease with which experimental data of high precision may be obtained. This advantage may frequently outweigh the theoretical uncertainties, so that coil expansion may be estimated more reliably from intrinsic viscosities than from procedures which are sounder theoretically but experimentally more difficult.

We discussed the behavior of a chain molecule, suspended in a viscous fluid and subjected to a velocity gradient, in our previous considerations of flow birefringence. All the factors mentioned in that context retain their importance in the theory of intrinsic viscosity. We thus have to consider the effect of the extent to which the coil is permeated by the streaming fluid, the extent to which the coil is being periodically deformed during its rotation in the velocity gradient, and the consequences of orientation of coils whose overall shape deviates from spherical symmetry.

Kuhn and Kuhn (1945b) carried out a searching analysis of the intrinsic viscosity to be expected from a free-draining coil. We saw in section VI.B.2 that the rotational frictional coefficient for a rigid free-draining coil is proportional to the product of the number of chain elements and the mean square radius of gyration. Thus, the energy dissipated in fluid friction per unit weight of the solute will be proportional to $\langle s^2 \rangle$; and if the geometry of the coil can be described by the random flight model (which makes $\langle s^2 \rangle$ proportional to the number of chain links), then $[\eta]$ should be proportional to the length of the chain. Since the coil is not spherically symmetrical but has the overall shape of a slightly elongated ellipsoid, Kuhn and Kuhn concluded that coils with very high internal viscosities should be oriented to some extent toward the flow lines, leading to a decrease of $[\eta]$ with increasing q, as described in the preceding section for rigid ellipsoids of revolution. On the other hand, they predicted that coils with zero internal viscosity, which are expanded and compressed in phase with the alternating hydrodynamic stresses during each revolution of the coil, should have intrinsic viscosities independent of the velocity gradient.

The effects of hydrodynamic interactions between the elements of a Gaussian chain were taken into account by Kirkwood and Riseman (1948) in their theory of intrinsic viscosity. Their result may be written in the form

$$[\eta] = \frac{10}{3}\pi R_\eta^3 \left(\frac{N}{M_2}\right) F'(\Xi) \qquad \text{(VI.59)}$$

where $\frac{4}{3}\pi R_\eta^3$ is the volume of a rigid sphere which has the same intrinsic viscosity as an impermeable Gaussian coil and Ξ is the parameter, defined in

(VI.10), describing the permeability of the coil. For $\Xi \to \infty$, $F'(\Xi) \to 1$ if $R_\eta = 0.94 \langle s^2 \rangle^{1/2}$. Later it was found that the Kirkwood—Riseman treatment contained errors which led to an overestimate of R_η. According to Auer and Gardner (1955), $R_\eta = 0.87 \langle s^2 \rangle^{1/2}$, and an almost identical result was obtained by Zimm (1956) by a different method. Comparison of this result with values quoted previously for the translational and rotational diffusion coefficients shows that a rigid sphere which is hydrodynamically equivalent to an impermeable coil will have slightly different radii, depending on the property being considered.

Flory (1949, 1953c) has pointed out that most polymer chains with an appreciable molecular weight approximate the behavior of impermeable coils, and this leads to a great simplification in the interpretation of intrinsic viscosity. Substituting for the polymer coil a hydrodynamically equivalent sphere with a molar volume V_e, we obtain

$$[\eta] = \frac{5}{2}\frac{V_e}{M_2} = \frac{\Phi' \langle s^2 \rangle^{3/2}}{M_2} \qquad \text{(VI.60)}$$

TABLE VI.1.
Intrinsic Viscosities of Polymers in Θ-Media

Polymer	Θ-Medium	$[\eta]/M_2^{1/2}$ (cm³ mole$^{1/2}$/g$^{3/2}$)
Amylose	0.33*M* Aqueous KCl, 25°C	0.115
Cellulose tricaproate	Dimethylformamide, 41°C	0.245
Cellulose tricaprylate	Dimethylformamide, 140°C	0.113
Poly(trans-1,4-butadiene)	*n*-Propyl acetate, 60°C	0.232
Poly(*cis*-1,4-butadiene)	*n*-Propyl ketone, 14.5°C	0.119
Poly(acrylic acid)	Dioxane, 30°C	0.076
Polyacrylonitrile	18% Methanol-82% dimethylformamide, 20°C[a]	0.325
Poly(*n*-butly methacrylate)	isopropanol, 23.7°C	0.037
Poly(dimethylsiloxane)	Butanone, 20°C	0.081
Poly(ethylene oxide)	0.45*M* Aqueous K_2SO_4, 35°C	0.130
Poly(ethyl methacrylate)	Isopropanol, 37°C	0.047
Polyisobutene	Benzene, 24°C	0.107
Polymetaphosphate	0.415*M* Aqueous NaBr, 25°C	0.049
Poly(methacrylic acid)	0.002*M* Aqueous HCl, 30°C	0.066
Polymethacrylonitrile	Dimethylformamide, 20°C	0.220
Poly(methyl methacrylate)	50% Butanone-50% isopropanol, 25°C[a]	0.059
Poly(α-methylstyrene)	79.4% Benzene-20.6% methanol, 30°C[a]	0.077
Polystyrene	Cyclohexane, 34°C	0.082
Poly(vinyl acetate)	Ethyl-*n*-butyl Ketone, 29°C[a]	0.093

[a]Compositions of mixed solvents are given by volume.

where $\Phi' = \frac{10}{3}\pi N\,(R_\eta/\langle s^2\rangle^{1/2})^3$ should be a universal constant independent of the nature of the macromolecule (provided only that the molecular chain is sufficiently flexible) and independent of the solvent medium.† When the result of the calculations of Auer and Gardner (1955) or Zimm (1956) is used for the ratio $R_\eta/\langle s^2\rangle^{1/2}$, the constant Φ' should have the value of 4.2×10^{24}.

Relation (VI.60) leads to a number of interesting consequences. In a Θ-solvent, in which the shape of the chain is described by the random flight model, $\langle s^2\rangle$ is proportional to M_2, so that the intrinsic viscosity should be proportional to $M_2^{1/2}$. This prediction has been amply verified, and Table VI.1 is based on the listing of $[\eta]/M_2^{1/2}$ values compiled by Kurata and Stockmayer (1963) for a variety of polymers in Θ-media. We have seen above that $[\eta]$ would be proportional to the first power of M_2 if the coil were free draining; the experimental results constitute, therefore, a striking confirmation of Flory's assumption that flexible polymer coils may be considered impermeable. In solvent media better than Θ-solvents, the theory of Flory (1949) predicts that the linear expansion factor α_e increases for any polymer-homologous series with chain length. Thus, the exponent γ in the empirical equation (Mark and Tobolsky, 1950)

$$[\eta] = K'M_2^{\gamma} \tag{VI.61}$$

should be larger than 0.5. Since Flory predicts $\alpha_e^5 - \alpha_e^3$ to be proportional to $M_2^{1/2}$, we obtain in the limit of $\alpha_e^5 \gg \alpha_e^3$ a proportionality of α_e and $M_2^{0.1}$. With $\langle s^2\rangle^{3/2}$ increasing as $(M_2^{1/2}\alpha_e)^3$, this would lead to the prediction that $[\eta]$ is proportional, in very good solvent media, to $M_2^{0.8}$. The theory of Kurata et al., (1960) leads to somewhat different results. Here α_e may become, in powerful solvent media, proportional to $M_2^{1/6}$. The theory assumes, however, that the molecular coil expands only in the direction parallel to the chain end displacement, so that the chain expansion is not isotropic and the hydrodynamic properties of the coil should be represented by an equivalent ellipsoid of revolution, whose axial ratio increases with the solvent power of the medium.

Careful measurements by Krigbaum and Carpenter (1955), in which the intrinsic viscosity was compared with the radius of gyration derived from light scattering, have shown that (VI.60) is not strictly valid. The experimentally observed deviation may be expressed in two alternative ways. In the first treatment we change the functional dependence of $[\eta]$ on α_e; Krigbaum

†It is customary in the literature to quote the quantity Φ, defined by $[\eta] = \Phi\langle h^2\rangle^{3/2}/M_2$ with $[\eta]$ in deciliters per gram, as originally used by Flory. The formulation given in. (VI.60) has the advantage that $\langle s^2\rangle$ is the experimentally measured quantity. If the chains are Gaussian, $\langle h^2\rangle = 6\langle s^2\rangle$ and $\Phi' = 100 \times 6^{3/2}\,\Phi$

and Carpenter found that their data were consistent with $[\eta]$ proportional to $\alpha_e^{2.2}$. Alternatively, $[\eta]$ may be treated as proportional to α_e^3 with the factor Φ' varying with the solvent power of the medium (Ptitsyn and Eizner, 1959). The deviation from Flory's relation (VI.60) may be traced to the approximation involved in the assumption that all linear dimensions of a flexible coil change by the same factor when it is transferred from one solvent medium to another. This assumption cannot be rigorously correct, however, since the excluded volume effect will be most pronounced in the center of the coil, where the chain segment density is largest. As a result, the ratio of the radius of the hydrodynamically equivalent sphere to the radius of gyration of the coil will tend to decrease with an increase in the excluded volume effect. Kurata et al. (1959) carried out a theoretical treatment of this effect and predicted $[\eta]$ to be proportional to $\alpha_e^{2.43}$. Ptitsyn and Eizner (1959), in a different approach to the problem, concluded that a proper consideration of the excluded volume effect requires the factor Φ' in (VI.60) to be a function of γ in relation (VI.61), so that Φ' should decrease from a value of 4.2×10^{24} for Θ-solvents to 2.5×10^{24} for very strong solvent media. The Ptitsyn-Eizner range of predicted Φ' values may be compared with $\Phi' = 3.1 \times 10^{24}$, given by Flory (1953c) as an average value found in a variety of systems. We may conclude that, although the various refinements seem well justified both theoretically and experimentally, Flory's original formulation, as given by (VI.60), represents a very close approximation to the solution viscosity behavior of flexible chain polymers.

If we assume that the sphere which is hydrodynamically equivalent to the polymer coil has the same volume as the equivalent sphere which governs the excluded volume reflected in the second virial coefficient, we can combine (VI.60) with (IV.17) and (IV.21) to obtain

$$\frac{A_2 M_2}{[\eta]} = \tfrac{8}{5}[1 - g(X)] \qquad \text{(VI.62)}$$

In good solvents and for sufficiently long polymer chains, $g(X) \to 0$ and $A_2M_2/[\eta]$ should have the same value for all polymer-solvent systems. Figure VI.13 illustrates this behavior, but $A_2M_2/[\eta]$ levels off at a value somewhat lower than that indicated by (VI.62). Since $g(X)$, which characterizes the mutual interpenetration of polymer coils, depends on the excluded volume effect, which also causes chain expansion beyond the unperturbed dimensions, the $A_2M_2/[\eta]$ ratio is generally believed to be a function of the expansion coefficient α_e. If this is indeed the case, it is possible to estimate unperturbed chain dimensions from measurements performed in non-Θ-solvents. Such an indirect approach is required in the study of some crystalline polymers which dissolve only in good solvents or in the case of chains which undergo helix-coil transition before the Θ condition is reached. The relation between $A_2M_2/[\eta]$ and α_e is, according to Yamamoto et al., (1971) the same

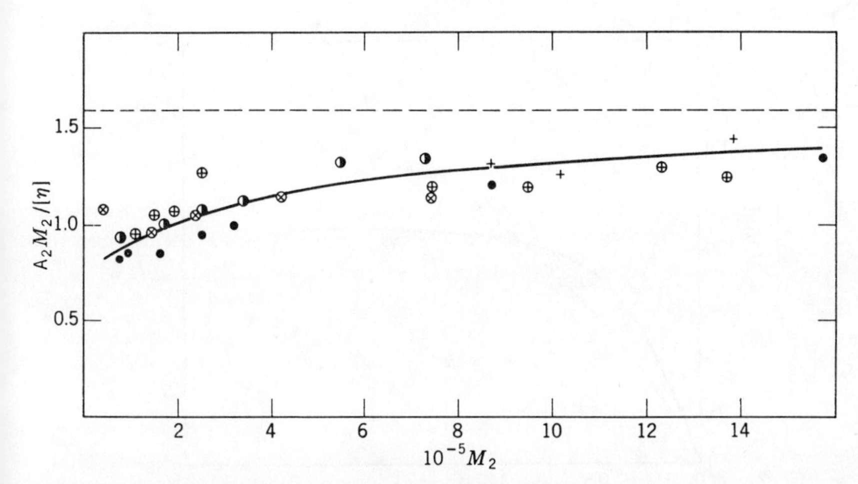

Fig. VI.13. Relation between the second virial coefficient and the intrinsic viscosity for chain molecules in good solvent media. (●) Polystyrene in toluene, (+) poly(vinyl acetate) in butanone, (◑) polyisobutene in cyclohexane, (⊕) poly(methyl methacrylate) in acetone, (⊖) poly(methyl methacrylate) in butanone. The dashed line gives the theoretical value, assuming that the thermodynamic excluded volume is equal to the volume of the hydrodynamically equivalent sphere. Data from Orofino and Flory (1957) and Casassa and Stockmayer (1962).

for polystyrene, poly(p-methylstyrene), and polychloroprene, and we may assume that this relation, as shown in Fig. VI.14, has general validity. Also indicated in Fig. VI.14 is the relation predicted theoretically by Orofino and Flory (1957):

$$\frac{A_2M_2}{[\eta]} = 1.88 \ln [1 + 0.855 (\alpha_e^2 - 1)] \qquad \text{(VI.63)}$$

It may be seen that for low values of $A_2M_2/[\eta]$ the estimate of α_e from (VI.63) is too high. . For high values of α_e, the $A_2M_2/[\eta]$ ratio reaches a limiting value [in conformity with (VI.62) but contrary to the form of (VI.63)], so that the expansion coefficient can be estimated by this approach only if its value is relatively small.

An alternative method, in which only viscosities and molecular weights are used to estimate unperturbed dimensions, is based on the relation (Stockmayer and Fixman, 1963)

$$\frac{[\eta]}{M_2^{1/2}} = K_\Theta + 0.036\Phi' BM_2^{1/2}$$

$$B = \beta\left(\frac{Z}{C_\infty M_2}\right)^2$$

$$= \left(\frac{V_2}{M_2}\right)^2 \frac{1 - 2\chi}{NV_1} \qquad \text{(VI.64)}$$

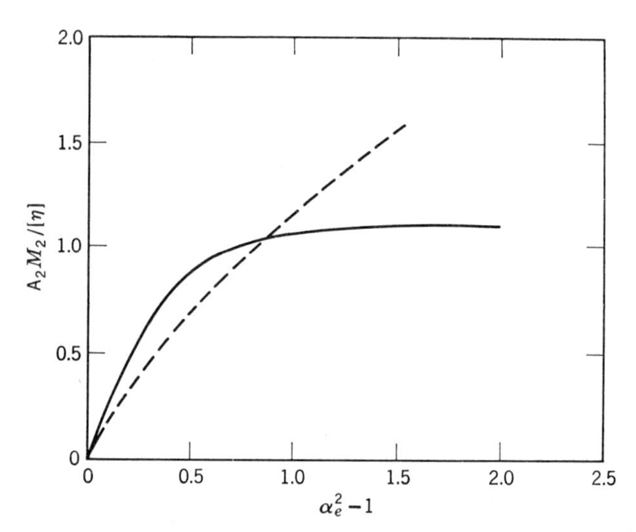

Fig. VI.14. Relation between $A_2M_2/[\eta]$ and the expansion coefficient α_e.(——) Experimental data for polystyrene, poly(p-methylstyrene), and polychloroprene; (----) relation predicted by Orofine and Flory (1957).

Thus, a plot of $[\eta]/M^{1/2}$ against $M^{1/2}$ yields K_Θ, the value of $[\eta]/M^{1/2}$ in Θ-media, as the intercept and allows either β, the excluded volume of the statistical polymer segment, or the Flory-Huggins interaction constant χ to be obtained from the slope. It should be noted that (VI.64) predicts $[\eta]$ to become proportional to the first power of M_2 for very long chains in good solvents, in contrast with the prediction based on Flory's theory of chain expansion, which makes $[\eta]$ asymptotically proportional to $M_2^{0.8}$. Experimental data by Yamamoto et al. (1971) show, in fact, that Stockmayer-Fixman plots tend to curve downward if data are obtained for polymers with molecular weights above 10^6, but for the molecular weight range usually available to the experimentalist the linear behavior is generally obeyed. The validity of this method rests on the assumption that $\langle s^2 \rangle / M_2$ is constant and that the polymer coils are impermeable over the molecular weight range in which $[\eta]$ values are obtained. With very stiff chains, such as cellulose derivatives, these conditions are attained only for extremely high molecular weights (Banks et al. 1970), and the Stockmayer-Fixman extrapolation cannot be used for such systems. Dondos and Benoit (1971) have used this extrapolation to demonstrate the dependence of unperturbed chain dimensions of poly(methyl methacrylate), poly(2-vinylpyridine), and poly(vinyl acetate) on the solvent medium, and although the method may not be rigorous, the substantial variation in K_Θ values must be considered significant.

For relatively short chains, the excluded volume effect becomes negligible, and we should then expect the intrinsic viscosity to be independent of the solvent. Such an effect has been observed for polystyrene (Fig. VI.15) and

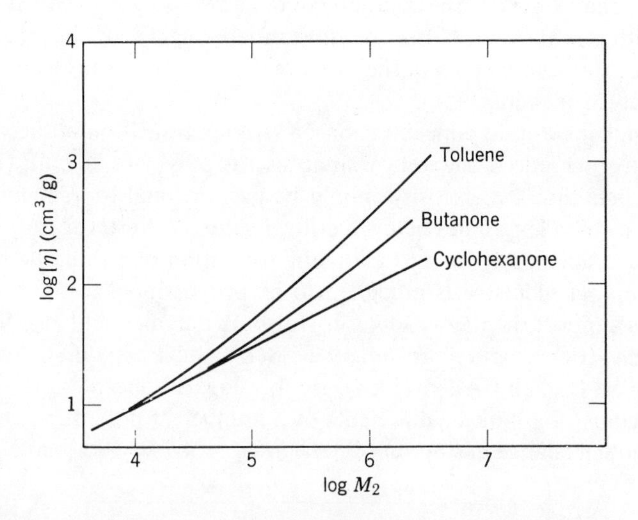

Fig. VI.15. Intrinsic viscosity-molecular weight relation for polystyrene in media of varying solvent power at 34.5°C (Okada etal., 1963).

for a number of other polymers (Bianchi and Peterlin, 1968). However, the constancy of $[\eta]/M_2^{1/2}$ extends to rather low molecular weights where the polymer molecule cannot possibly behave as an impermeable coil. For such short chains the $\langle s_0^2 \rangle / M_2$ ratio has not yet attained its asymptotic value, and the apparent validity of Flory's theory in this range seems to be due to a compensation of errors.‡

When branched molecules are compared with linear chains of the same molecular weight, the $[\eta]/\langle s^2 \rangle^{3/2}$ ratio is found to be distinctly higher for the branched species. If (VI.60) could be applied to both linear and branched chains, $[\eta]$, for any molecular weight, would be proportional to the 3/2 power of the parameter g^*, defined by (III.23). However, a theory formulated by Zimm and Kilb (1959) concluded that $[\eta]$ should vary only as $(g^*)^{1/2}$, and experimental data on monodisperse branched polystyrenes (Morton et al., 1962) bore out this prediction.

‡With a silicone ladder polymer, the $\log[\eta]$—$\log M_2$ plot curves *down word.* This is expected for chains of small bulk but low flexibility, where coils remain partially permeable up to fairly high chain lengths (Tsvetkov et al., *Vysokomol. Soedin.*, **A 15**, 400 (1973)).

The viscosity of solutions of flexible chain polymers at finite concentrations may be represented by (VI.58), where the Huggins constant k' decreases with increasing polymer solvation from about 0.7 in Θ-media to a limiting value of about 0.3 (Sakai, 1968). If the solution contains two polymers which interact strongly with one another, k' may be substantially higher for the mixed polymer than for the components of the mixture. Figure VI.16 shows an example of this effect (Morawetz, 1954).

At still higher solution concentrations, a striking transition in the viscosity behavior is observed. A theoretical analysis for polymers in bulk (Bueche, 1962a) predicts that the viscosity should be proportional to their molecular weight if the chain length lies below a critical value. If, however, the polymer molecules are sufficiently long to allow the formation of an infinite network by chain entanglements, η is predicted to be proportional to $M^{3.5}$. Experimental data support the theory, not only for bulk polymers but also for polymer solutions (Porter and Johnson, 1966; Berry and Fox, 1968). According to a study by Fox and Allen (1964), doubly logarithmic plots of solution viscosity against the molecular weights of a number of polymers exhibit the expected sudden change in slope at $(\langle s_0^2 \rangle / V_2) Z \phi_2 \approx 4.7 \times 10^{-15}$ mole/cm.†

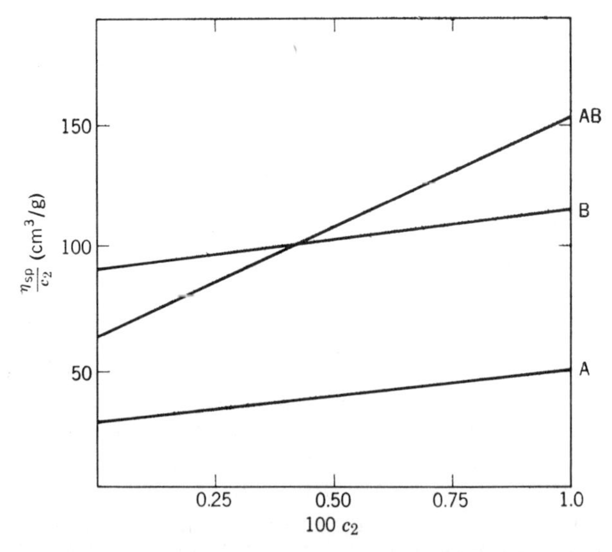

Fig. VI.16. Effect of molecular interactions on the specific viscosity of polymer solutions. Butanone solutions of a methyl methacrylate copolymer with 4.9 mole % methacrylic acid (A); a methyl methacrylate copolymer with 5.8 mole % dimethylaminoethyl methacrylate (B); and a mixture of equal weights of the two copolymers (AB).

†A very interesting effect has been described for the concentration dependence of the viscosity of solutions of rod-like particles, *i.e.*, polypeptides in helicogenic solvents, at low rates of shear. In this case, the viscosity increases with increasing solution concentration

Solution viscosities of high molecular weight flexible chain polymers depend appreciably on the rate of shear q. Since the effect observed cannot depend on the sign of q, this rate of sheer dependence must generally be of the form $\eta_{sp}/\eta_{sp}^0 = 1 - a_1 q^2 + a_2 q^4 - \cdots$. The coefficients a_1, a_2, and so forth, increase with chain length and solution concentration. With very long polymer chains, it is necessary to make measurements at extremely low rates of shear to obtain $(\eta_{sp}/\eta_{sp}^0) - 1$ proportional to q^2 and thus allow a reliable extrapolation to η_{sp}^0. This is particularly the case with high molecular weight DNA (Gill and Thompson, 1967; Lin, 1970), where the chains are very much longer than the persistence length of the double-helical structure, so that the molecule behaves as a random coil. For instance, Gill and Thompson found that a viral DNA with an intrinsic viscosity of 3.2×10^4 cm^3/g exhibited this limiting behavior only for $q < 0.2$ sec^{-1}. The deviation from Newtonian behavior exhibited by polymer solutions of finite concentration is a complicated phenomenon which includes effects due to the entanglements of the macromolecular coils. It is of particular theoretical interest that the rate of shear dependence of the contribution of the polymer to the solution viscosity does not vanish in the limit of infinite dilution. The effect of the velocity gradient on $[\eta]$ is measurable for flexible chain polymers only if the molecular weight is about 10^5, and it becomes pronounced for $M_2 > 10^6$. It is generally agreed that the non-Newtonian behavior of $[\eta]$ is due to various hydrodynamic interactions between chain segments in the polymer coil disturbed by the shear stresses and to the finite internal viscosity of the chain. The ratio of the intrinsic viscosities at finite and zero shear, $[\eta]_q/[\eta]_0$, is customarily represented as a function of the dimensionless parameter $\beta = M_2[\eta]_0\eta_0 q/\boldsymbol{R}T$, which we have already encountered in (VI.43) as governing chain orientation in a shear gradient. With increasing β the intrinsic viscosity first decreases; however, with very long chains and in viscous solvents the trend is eventually reversed, and $[\eta]_q/[\eta]_0$ may level off at values above unity (Burow et al., 1965; Wolff, 1969). It has been established that the sensitivity of $[\eta]_q/[\eta]_0$ to β increases sharply because of the excluded volume effect in good solvent media (Noda et al., 1968; Wolff, 1969). The importance of the internal viscosity has been demonstrated in two ways: (a) the decrease in the intrinsic viscosity is observed at lower β values if the length of the chain is reduced, increasing its resistance to the separation of the chain ends at any given rate (Noda et al., 1968); (b) The non-Newtonian behavior of the intrinsic viscosity in a series of Θ-solvents is reduced with rising temperature, thus increasing the rate at which the energy barriers to conformational transitions are being overcome (Carpenter and Hsieh, 1972).

---

only up to the point where an anisotropic phase is formed (cf. Section II.B.4) but drops precipitously as the concentration is increased further. This is due to the ease with which the anisotropic phase is oriented, so that the fluidity of this phase is extremely high compared to that of the isotropic solution (Hermans, 1962).

The theory of intrinsic viscosity has played a crucial part in the discovery of the helix-coil transition in synthetic polypeptides (Doty et al., 1956). When the intrinsic viscosity of poly (γ-benzyl-L-glutamate) was measured in dichloroacetic acid solution, log $[\eta]$ − log M_2 plots had a slope of 0.87, close to the value predicted by the Flory theory for flexible chains in strong solvent media. On the other hand, solutions of these polymers in dimethylformamide gave log $[\eta]$ − log M_2 plots with a much steeper slope (about 1.7), which could not possibly be accounted for by random coils, but was in close agreement with Simha's predictions [cf. (VI.55)] for ellipsoids of constant cross section and increasing length. Later it was shown by Yang (1959) that the helical form of the polypeptide in dimethylformamide and the random coil form in dichloroacetic acid could also be clearly differentiated by studies of the rate of shear dependence of the intrinsic viscosity. This dependence is relatively small for random coils of moderate molecular weight, but is very pronounced for the helical form, which behaves like a rigid rod and is, therefore, very easily oriented in a velocity gradient. Yang suggests that a comparison of the observed variation of $[\eta]$ with q and the theoretically predicted dependence may even be interpreted in terms of the distribution of lengths of the rodlike particles.

Intrinsic viscosity data have also served to reveal another type of transition. Dondos (1969) studied a graft copolymer over a range of temperatures and found an abrupt reversal in the sign of $d[\eta]/dT$ (Figure VI.17). It seems that at a critical temperature the entropic driving force becomes sufficient to overcome the unfavorable contact energy between the dissimilar monomer re-

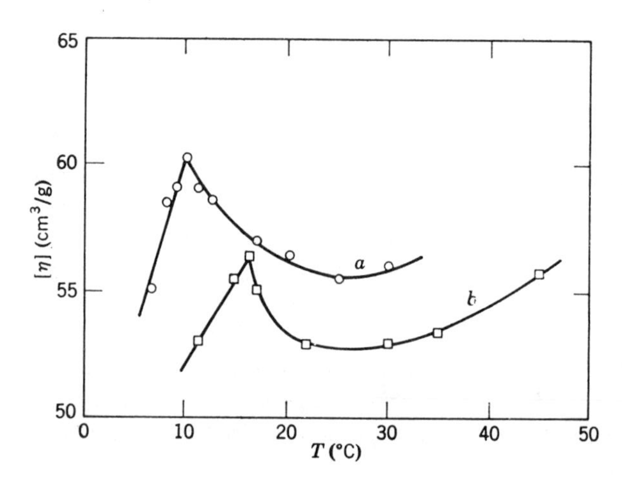

Fig. VI.17. Intrinsic viscosity-temperature relation for graft copolymer with poly-(3,3-diphenylpentene-1) backbone and poly(methyl methacrylate) branches in (*a*) benzene and (*b*) tetrahydrofuran solution.

sidues in the backbone and the side chains of the polymer, so that these portions of the macromolecule, which are segregated at lower temperatures, can now mix freely with each other. A similar phenomenon is observed with block copolymers when either the temperature or the solvent composition is changed (Girolamo and Urwin, 1971)

3. Interpretation of Intrinsic Viscosity in Conjunction with Other Solution Properties

As pointed out in the introduction to this chapter, satisfactory characterization of the state of a macromolecule in solution requires the specification of several parameters. Such an evaluation will then frequently call for a combination of information obtained with the use of several experimental methods. Let us first consider the situation with rigid globular particles (Scheraga and Mandelkern, 1953). If we are content with a description of such macromolecules as ellipsoids of revolution (and a more detailed description is beyond the reach of the study of solution properties), we shall require the specification of the molecular weight M_2, the molar volume of the hydrodynamically equivalent ellipsoids V_e, and the axial ratio p. Assuming that M_2 is known from osmotic, light scattering, or ultracentrifuge equilibrium measurements, we may obtain V_e and p from the frictional coefficient and the intrinsic viscosity in the limit of low-velocity gradients by using (VI.3) and (VI.60), rewritten in the forms

$$\mathrm{f} = \frac{6\pi\eta_0 R_s'}{\boldsymbol{x}(\mathrm{p})} = \frac{\sqrt[3]{162\pi^2 V_e/\boldsymbol{N}\eta_0}}{\boldsymbol{x}(\mathrm{p})}$$

$$[\eta] = \frac{5}{2}\left(\frac{V_e}{M_2}\right)\boldsymbol{z}(\mathrm{p}) \qquad \text{(VI.65)}$$

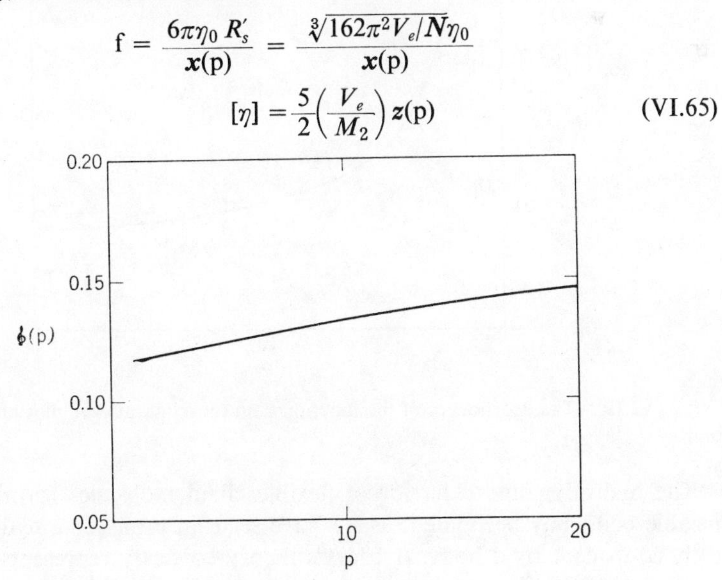

Fig. VI.18. The dependence of the function ɸ(p) on the axial ratio of ellipsoids of revolution.

when the sedimentation constant (extrapolated to zero concentration) is used as a measure of the frictional coefficient, a combination of (VI.65) and (VI.20) leads to

$$\phi(\mathrm{p}) \equiv \frac{\mathbf{s}^{\circ}[\eta]^{1/3}\eta_0 \mathbf{N}^{2/3}}{M_2^{1/3}(1 - \bar{v}_2\rho)} = 0.116\mathbf{x}(\mathrm{p})[\mathbf{z}(\mathrm{p})]^{1/3} \tag{VI.66}$$

The function $\phi(\mathrm{p})$ is given in Fig. VI.18. In a similar manner we may combine (VI.20), (VI.32), and (VI.65) to obtain the axial ratio from a combination of measurements of the rotational diffusion coefficient and the intrinsic viscosity by the use of

$$\xi(\mathrm{p}) \equiv \frac{12}{5}\left(\frac{\eta_0 D_r [\eta] M_2}{\mathbf{RT}}\right) = \mathbf{y}(\mathrm{p})\mathbf{z}(\mathrm{p}) \tag{VI.67}$$

The function $\xi(\mathrm{p})$ is plotted in Fig. VI.19, and we may note that it is much more sensitive to the asymmetry of the particle than $\phi(\mathrm{p})$.

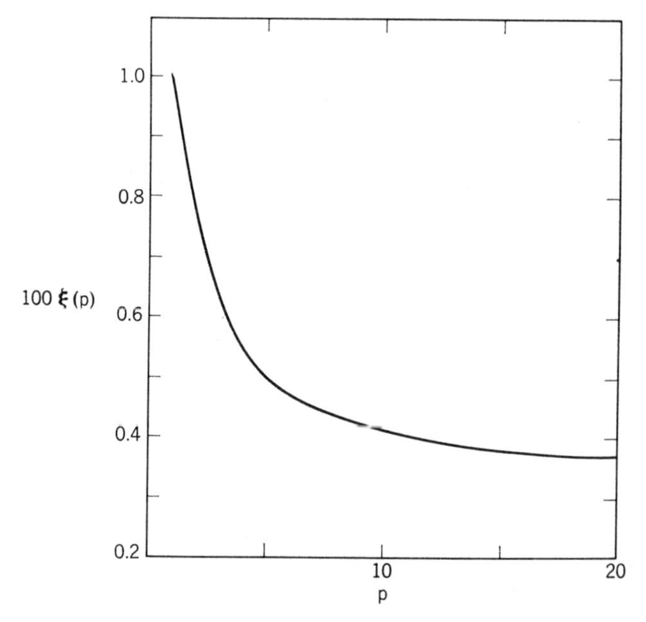

Fig. VI.19. The dependence of the function ξ on the axial ratio of ellipsoids of revolution.

The hydrodynamic behavior of flexible chain molecules forming impermeable coils may be treated, as we have seen, in terms of a hydrodynamically equivalent rigid body. If Flory's theory correctly represents the chain expansion, the coil may be considered spherical, so that only two parameters have to be assigned: the molecular weight, and the radius of the hydrodynamically equivalent sphere.† As Mandelkern and Flory (1952) have suggested, the sedimentation coefficient and the intrinsic viscosity for impermeable coils, given by

$$\mathbf{s}^0 \frac{M_2(1 - \bar{v}_2\rho)}{N\mathrm{f}} = \frac{M_2(1 - \bar{v}_2\rho)}{6\pi\eta_0 R_t N} \tag{VI.20}$$

$$[\eta] = \frac{10}{3}\left(\frac{\pi R_\eta^3 N}{M_2}\right) \tag{VI.60}$$

may then be combined to give

$$\boldsymbol{\varepsilon} = \frac{\mathbf{s}^0 [\eta]^{1/3}\eta_0 M_2^{-2/3}}{1 - \bar{v}_2\rho} = 1.7 \times 10^{-17} \frac{R_\eta}{R_t} \tag{VI.68}$$

where R_η/R_t is the ratio of the radii of spheres which are equivalent to the molecular coil in intrinsic viscosity and in hydrodynamic resistance to linear translation, respectively. The data of Mandelkern and Flory (1952), Mandelkern et al. (1952) and Mandelkern and Fox (1953) may be represented by a clustering of R_η/R_t around a value of 1.15. Sakato and Kurata (1970) have substantiated the predictions of the Mandelkern-Flory theory on highly monodisperse poly (α-methylstyrene) and have shown that the same value is obtained for $\boldsymbol{\varepsilon}$ in a Θ-solvent and in strongly solvating media. This agrees with the prediction of Ptitsyn and Eizner that, in spite of the variation of Φ' in (VI.60), $R_\eta/\langle s^2\rangle^{1/2}$ and $R_t/\langle s^2\rangle^{1/2}$ change in a very similar manner with the solvent power of the medium. With R_η/R_t known, (VI.68) may be used to estimate the molecular weights of flexible chain polymers from the sedimentation coefficient and the intrinsic viscosity.

Tsvetkov and Klenin (1958) have reported data on the solution behavior of flexible chain polymers to test the relation between the molecular weight, the intrinsic viscosity, and the frictional coefficient as derived from diffusion measurements in the limit of infinite dilution. These quantities should be related [cf. (VI.6), (VI.15), and (VI.60)] by

$$\frac{\eta_0 D^\circ([\eta]M_2)^{1/3}}{T} = \boldsymbol{k}\left(\frac{5N}{324\pi^2}\right)^{1/3}\frac{R_\eta}{R_t} = 1.4 \times 10^{-9}\frac{R_\eta}{R_t} \tag{VI.69}$$

The data of Tsvetkov and Klenin lead to R_η/R_t ratios in the range of 1.1–1.3, in remarkably good agreement with theoretical predictions.

All the above derivations depend on the assumption that the macromolecular particle can be characterized by two parameters, for example, the vol-

†If the theory of Kurata et al. (1960) describes correctly the excluded volume effect, the overall shape of the coil would become increasingly elongated with increasing solvation, and the coil would have to be described by an ellipsoid with a variable axial ratio. The development which follows would then be invalid.

ume and the axial ratio. However, this assumption is not necessarily valid. For instance, the parameter plotted in Fig. VI.18 should be at a minimum for spherical particles, increasing for both prolate and oblate ellipsoids (i.e., for $p > 1$ and $p < 1$, respectively). Yet solution properties of serum albumin lead to a ϕ *smaller* than the value predicted for a sphere by the Scheraga-Mandelkern treatment, and it must be concluded that the ellipsoidal model is inadequate for a description of the hydrodynamic properties of this protein (Tanford and Buzzell, 1956)

4. Viscoelasticity of Dilute Solutions

When solutions of macromolecules are subjected to oscillating stresses of high frequency, their characterization as fluids with a viscosity η is no longer adequate since their response to the applied stress contains also an elastic component. For an extensive discussion of the theory and the experimental study of the viscoelasticity of such solutions the reader is referred to Ferry's (1970a) monograph.

If the shearing strain in the x-y plane is $\gamma_{xy} = \gamma_{xy}^0 \sin(\omega t)$, where $2\pi\omega$ is the frequency of oscillations, the shearing stress σ_{xy} has an elastic component in phase with γ_{xy} and a viscous component proportional to $d\gamma_{xy}/dt$, so that

$$\sigma_{xy} = \gamma_{xy}^0 [G' \sin(\omega t) + G'' \cos(\omega t)] \tag{VI.70}$$

where the shear storage modules G' and the shear loss modulus G'' are functions of ω. Alternatively, the solution may be characterized by a complex viscosity

$$\eta^* = \eta' - i\eta'' \tag{VI.71}$$

where $\eta' = G''/\omega$ and $\eta'' = G'/\omega$. The solvent exhibits no elasticity over the frequency range employed for the study of these solutions, so that G' and η'' are due entirely to the presence of the macromolecules. On the other hand, the shear loss modulus contains a contribution $\eta_0\omega$ from the solvent, so that the quantities characterizing the solute are $G'' - \eta_0\omega$ and $\eta' - \eta_0$.

Let us first consider the case of rigid elongated particles. We saw in Section VI.B that in steady flow the asymmetric particles are oriented relative to the flow lines and that in the absence of a shear stress the disappearance of this orientation is characterized by a rotary relaxation time τ_r, defined by (VI.47) and (VI.48). Thus, if we have an oscillating shear at a frequency such that $\omega\tau_r \ll 1$, the particle orientation at any time will correspond to the instantaneous shearing stress, while for $\omega\tau \gg 1$ the particle orientation will be unable to follow the oscillations of σ_{xy}. On multiplying G' and $G'' - \eta_0\omega$ by $5M_2/3c_2\boldsymbol{R}T$, we obtain "reduced moduli" G_R' and $(G'' - \eta_0\omega)_R$, which are functions of $\omega\tau_r$ for any given model of the solute particle. For highly elongated ellipsoids Ferry (1970a) obtained

$$G'_R = \frac{\omega^2\tau_r^2}{1 + \omega^2\tau_r^2} \quad \text{(VI.72)}$$

$$(G'' - \eta_0\omega)_R = \omega\tau_r\,[(1 + \omega^2\tau_r^2)^{-1} + \tfrac{2}{5}] \quad \text{(VI.73)}$$

and a similar expression was obtained by Ullman (1969) for rigid rods. For $\omega\tau \ll 1$, G'_R becomes negligible compared to $(G'' - \eta_0\omega)_R$, that is, elastic effects due to the solute tend to disappear. At the other extreme, $G'_R \to 1$ as $\omega\tau_r \gg 1$. The contribution of the solute to the real part of the complex viscosity, $\eta' - \eta_0$, is reduced at very high frequencies according to (VI.73) by a factor of $\frac{2}{7}$ as compared to the low-frequency (i.e., the steady-flow) value. Experimental results obtained with solutions of paramyosin, which has a rigid rodlike structure, are in good agreement with these theoretical predictions (Allis and Ferry, 1965).

For a corresponding theory applicable to flexible chain molecules, Rouse (1953) proposed that the chain be represented by a series of beads connected by elastic springs. The motion of these beads then corresponds to series of vibrational modes in analogy to the behavior of a vibrating string. These vibrational modes are associated with relaxation times, with the longest, τ_1, characterizing a vibration in which the two halves of the chain move in opposite directions. Rouse's theory neglected hydrodynamic interactions between the elements of the chain and assumed Kuhn's statistics to apply to the mean distances between chain segments. The treatment proposed by Zimm (1956) takes account of the hydrodynamic interactions between chain elements, and a modification by Tschoegl (1964) introduces the consequences of the excluded volume effect on the distribution of chain segments. Both Rouse's and Zimm's theories predict G'_R proportional to ω^2 and $(G'' - \eta_0\omega)_R$ proportional to ω for $\omega\tau_1 \ll 1$. For high frequencies, the predictions of the two theories diverge; Rouse predicts $G'_R = (G'' - \eta_0\omega)_R$ proportional to $\omega^{1/2}$, while Zimm's result has both moduli proportional to the 2/3 power of the frequency with the loss modulus larger by a constant factor. Both treatments lead then to the conclusion that the contribution of the polymer to the real part of the complex viscosity, $\eta' - \eta_0 = G''/\omega - \eta_0$, should vanish at very high frequencies. However, Peterlin (1967) has shown that this conclusion is valid only if the internal viscosity of the chain is neglected and that the ratio of $\eta' - \eta_0$, observed for $\omega\tau_1 \gg 1$, to $\eta - \eta_0$ under steady-flow conditions is a function of the ratio of the internal viscosity η_i and the frictional coefficient f of the statistical chain element. Thus, the study of the viscoelasticity of dilute polymer solutions has become a valuable tool for estimating the internal viscosity of flexible chain molecules.

Since effects arising from the internal viscosity of a molecular chain arise from the necessity to surmount potential energy barriers when the chain undergoes conformational transitions, it has long been assumed that η_i depends only on the height of these barriers and that it is independent of the nature of the solvent. However, interpretations of experimental data in terms of Peterlin's theory have shown that in viscous media η_i becomes proportional to the viscosity of the solvent (Massa et al., 1971; Osaki and Schrag, 1971). This result has led Peterlin (1972) to advance a new theory of the origin of η_i according to which it is made up of two contributions. The first, determined by the height of the energy barriers to be overcome in conformational transitions, is dominant in media of low viscosity, so that η_i is independent of η_0 under these conditions. The second contribution to η_i arises from the frictional dissipation of energy accompanying a conformational transition and is, therefore, proportional to η_0. Peterlin estimates that this effect becomes dominant with energy barriers of 1.5–3.5 kcal/mole when the solvent viscosity lies in the range of 0.05–2 poise.

Instrumental advances have now made it possible to study the viscoelastic behavior of polymer solutions at sufficient dilution to permit a reliable extrapolation of G'_R/c and $(G'' - \eta_0\omega)_R/c$ to their limiting infinite dilution values $[G']_R$ and $[G'']_R$ (Johnson et al., 1970). The concentration dependence of the reduced moduli in the range of incipient interpenetration of the polymer coil has been shown to exhibit characteristic differences between linear and "star-shaped" molecules, as predicted by theory (Osaki et al., 1972).

In solutions of extremely long chain molecules, particularly those containing undergraded DNA, the elastic effects may be so pronounced that relaxation can be followed conveniently when the stress to which the solution has been exposed is suddenly eliminated. The Cartesian-diver rotating cylinder viscometer (Gill and Thompson, 1967), in which the torque is provided by a rotating magnetic field, is particularly suited for such studies, and the stress relaxation, indicated by a reversal in the rotation of the rotating cylinder when the torque to which it is subjected is suddenly eliminated, is dominated by the longest relaxation time τ_1 of the chain molecule. Applying the theories of Rouse (1953) and Zimm (1956), it is then possible to use τ_1 to calculate the molecular weight of the macromolecule (Chapman et al., 1969; Klotz and Zimm, 1972). This method is particularly valuable because of the difficulties encountered with other techniques in the measurement of molecular weights above about 10^7.

5. Drag Reduction in Turbulent Flow

All our discussions thus far were concerned with phenomena observed in laminar flow. For fluids flowing through tubes with a circular cross section, the transition from laminar to turbulent flow is governed by a dimensionless parameter referred to as the Reynolds number Re, defined by $\mathrm{Re} = d\rho\bar{u}/\eta$, where d is the diameter of the tube, ρ is the density, and $\bar{u}$ the mean flow velocity. The breakdown of laminar flow occurs generally in the range of $2000 < \mathrm{Re} < 5000$. Although, as we have seen, the presence of dissolved macromolecules necessarily leads to an increase in the viscosity observed in laminar flow, Toms (1949) has made the surprising discovery that the addition of minute amounts of high molecular weight polymers (a few parts per million) leads to a *reduction* in the pressure gradient required to maintain a given flow rate for high values of the Reynolds number. Since turbulent flow does not lend itself to the rigorous theoretical analysis which can be carried out for laminar flow, it is not surprising that our understanding of the "Toms phenomenon" should be relatively limited. Outstanding studies of the drag reduction observed in aqueous solutions of poly(ethylene oxide) of various chain lengths were reported by Virk et al. (1967) and by Patterson and Abernathy (1970).

The Toms phenomenon is observed only in turbulent flow. The presence of the macromolecular chains does not seem to affect the Reynolds number at which laminar flow breaks down, and drag reduction does not set in as soon as the flow becomes turbulent but commences at a critical value of the shear stress at the wall of the tube. The concentration of polymer required to attain a given drag reduction decreases with an increasing chain length of the macromolecules; however, as the polymer concentration is being increased, the effect approaches a saturation value which is independent of the molecular weight of the solute. Virk et al. have observed drag reductions by as much as 80%; and since their data suggest that the largest factor by which the pressure gradient may be reduced at a constant flow rate by polymer addition is proportional to $\mathrm{Re}^{-0.30}$, the effect increases apparently without limit with increasing flow rates through any given tube. Qualitatively, the Toms phenomenon seems to be caused by a tendency of long-chain molecules to oppose an intensification of vortices in the boundary layer of the turbulent fluid adjoining a solid surface. Extensive discussions of the experimental evidence and various approaches to its theoretical interpretation are contained in reviews by Gadd (1971) and Lumley (1973) and a paper by Kohn (1973). Apart from its theoretical interest, the effect may be of practical importance in reducing the energy requirement for the pumping of fluids and in increasing the speed attainable by ships.

Results obtained with polyelectrolytes are particularly valuable, since the expansion of the macromolecular chain may be varied, in this case, within wide limits. Kim (1973) found that the drag reduction efficiency of poly(acrylic acid) increases sharply with rising pH, leveling off above pH 6;

addition of NaCl produces a precipitous drop in drag reduction. This shows that drag reduction is enhanced by an expansion of the chain molecules responsible for the effect.

D. SHEAR DEGRADATION

When solutions of very long chain molecules are subjected to shearing stresses in viscous flow, a gradual degradation of the macromolecules is observed. This phenomenon imposes a serious limitation on experimental studies of the extremely long DNA molecules isolated from biological sources (Hershey and Burgi, 1960; Levinthal and Davison, 1961). The sensitivity to shear degradation increases with increasing chain length, and the effect is, therefore, much less pronounced with the synthetic polymers usually encountered, since these are generally much shorter than the DNA of bacteriophages or living cells. Nevertheless, high-speed stirring of polystyrene or polyisobutene solutions leads to appreciable degradation, and the process is due to a breakage of covalent bonds in the chain backbone, as can be demonstrated by the reaction of the free radicals with iodine added to the system (Johnson and Price, 1960). Even at the relatively low shear rates encountered in the flow through the capillary of a viscosimeter, degradation may be followed kinetically by passing the polymer solution through the capillary for a large number of times (Moore and Parts, 1968). We may note that shear degradation impairs the potential utility of high polymers for drag reduction, as discussed in the preceding section of this chapter.

Harrington and Zimm (1965) carried out an extensive study of the shear degradation of polystyrene and DNA in an effort to establish the principles governing the process. They found that the breakage of molecules exceeding a limiting length may be described as a first-order reaction if the stress F to which the molecular chain is exposed lies above a critical value F_c. The contribution of the polymeric solute to the shearing stress is $q(\eta - \eta_0)$, and the number of chains which cross a unit area in the shear plane is Nc_2z/M_2, where, z is the chain extension in the direction perpendicular to the shear plane, so that $F = qM_2(\eta - \eta_0)/Nc_2z$. For very dilute polymer solutions $(\eta - \eta_0)/c_2 \approx \eta_0[\eta]$, which is proportional, in a Θ-solvent, to $M_2^{1/2}$. In such a medium, characteristic linear dimensions of the molecular coil are also proportional to $M_2^{1/2}$, and although this may not be strictly true for z when the chain is distorted by a strong shear gradient, one may assume that $(\eta - \eta_0)/c_2z$ is not very sensitive to the chain length of a polymer. Thus F should be approximately proportional to $qM_2\eta_0$. The stress is highest in the middle of the chain, so that shear degradation of monodisperse DNA may lead to half-molecules with a very narrow distribution of lengths (Hershey and Burgi, 1961). Harrington and Zimm estimated $F_c = 3.5 \times 10^{-6}$ and 4×10^{-7} dyne for polystyrene in good and poor solvents and 2×10^{-5} dyne for DNA at an ionic strength of 0.1. These values are about two orders of magnitude below the force required to break a covalent bond in the chain backbone. It is hard to understand why the breaking of polystyrene chains should require a larger force when solvation of the polymer is increased.

The effect of temperature on shear degradation was studied by Harrington and Zimm, who found that the rate of breakage of polystyrene chains was approximately constant between 15 and 45 °C. They rejected, therefore, the possibility that chain scission takes place by a thermal mechanism, with the stress on the chain lowering the activation energy required for the process.

Chapter VII

POLYELECTROLYTES

The division which exists in classical chemistry between uncharged moleculs and eletrolytes composed of ions has its counterpart in the field of high polymers. By "polyelectrolytes" we refer to substances containing macromolecules carrying a large number of ionic charges—the polyions—with the small counterions which render the system electroneutral. A typical example of such a substance, which has been a frequent subject of investigation, is the sodium salt of poly(acrylic acid):

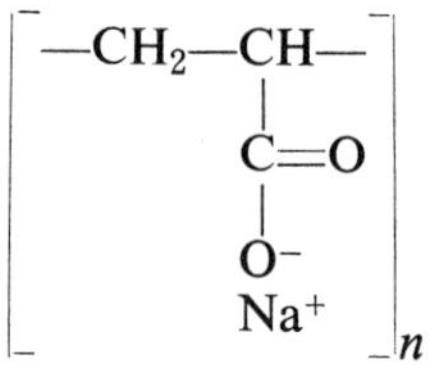

$$\left[\begin{array}{c} -CH_2-\underset{\underset{\underset{\displaystyle \overset{|}{O^-}\;Na^+}{|}}{\displaystyle C{=}O}}{CH}- \end{array}\right]_n$$

Polyelectrolytes are usually studied in aqueous solutions. The properties of such solutions are strongly affected by the addition of simple salts, such as sodium chloride. This addition increases the counterion concentration and introduces byions into the system. that is, small ions carrying charges of the same sign as the charge of the polyion.

Some typical examples of ionizable polymers with flexible molecular chains are represented in Table VII.1 to indicate the wide variety of materials available for study. Much of the early work was carried out on poly(acrylic acid), poly(methacrylic acid), and acidic polysaccharides such as alginic acid and polygalacturonic acid (cf. Fig.III.3). In these materials the ionizable groups are weakly acidic, so that the charge density along the molecular chain can be varied at will by changing the degree of neutralization. Strongly ionizing polymers are available in poly(vinylsulfonic acid) and poly(styrenesulfonic acid). In polyphophates we have an inorganic polyanion suitable for polyelectrolyte studies. Another interesting class of materials is available in hydrolyzed copolymers of maleic anhydride. It is known that maleic anhydride adds with difficulty to a growing chain terminating with a maleic anhy-

TABLE VII.1.
Typical Examples of Ionizable Chain Molecules

Polymer	Structure
Poly(acrylic acid)	$(-CH_2-CH(COOH)-)_n$
Poly(methacrylic acid)	$(-CH_2-C(CH_3)(COOH)-)_n$
Poly(vinylsulfonic acid)	$(-CH_2-CH(SO_3H)-)_n$
Poly(styrenesulfonic acid)	$(-CH_2-CH(C_6H_4SO_3H)-)_n$
Styrene-maleic acid copolymer	$(-CH_2-CH(C_6H_5)-CH(COOH)-CH(COOH)-)_n$
Vinyl methyl ether-maleic acid copolymer	$(-CH_2-CH(OCH_3)-CH(COOH)-CH(COOH)-)_n$
Acrylic acid-maleic acid copolymer	$(CH_2-CH(COOH)-CH(COOH)-CH(COOH)-)_n$
Poly(metaphosphoric acid)	$(-O-P(=O)(OH)-)_n$
Poly(vinyl amine)	$(-CH_2-CH(NH_2)-)_n$
Poly(ethylene imine)	$(-CH_2-CH_2-N(H)-)_n$
Poly(4-vinylpyridine)	$(-CH_2-CH(C_5H_4N)-)_n$
Poly(4-vinyl-*N*-dodecyl pyridinium)	$(-CH_2-CH(C_5H_4N^+C_{12}H_{25})-)_n$

dride unit, so that copolymerization leads to chains in which the maleic anhydride residues are separated from one another by one or more comonomers. Hydrolysis then leads to macromolecules containing closely spaced pairs of carboxyls separated from other such pairs by at least one comonomer unit. Such materials have been called "polydibasic acids." In the case of the acrylic acid-maleic acid copolymer, triplets of backbone carbons carrying carboxyl groups are separated from each other by a methylene residue. Cationic polymers (for a review see Hoover, 1970) may have weakly basic groups as in poly(ethylene imine) or poly(4-vinylpyridine), but the latter is easily converted into a strong base by quaternization. If the quaternization is carried out with a long-chain alkyl halide, the hydrophobic bond between the paraffinic residues will result in a stabilization of compact conformations in analogy to the formation of soap micelles. Such materials have, therefore, been described as "polysoaps." Amphoteric polyelectrolytes may be obtained by copolymerization of suitable acidic and basic monomers, for instance, methacrylic acid and vinyl pyridine. Panzik and Mulvaney (1972) have described a method by which polymers with a regularly alternating sequence of acidic and basic side chains may be prepared. We should also bear in mind that polymers which are weak Lewis bases may acquire polyelectrolyte character in strongly acidic media, although the polymer chains would not be charged in other solvents. Thus, polyamides have been observed to behave like polyelectrolytes in formic acid solutions (Schaefgen and Trivisonno, 1952; Saunders, 1962, 1964).

In addition to polyelectrolytes with flexible chain polyions, which may assume a broad range of conformations, we have to concern ourselves with charged macromolecules existing in the specific conformations discussed in Section. III.C. Proteins carry a variety of ionizable groups in the side chains of the amino acid residues: the carboxyls of glutamic and aspartic acid, the imidazole of histidine, the amino groups of lysine, the phenolic groups of tyrosine, the thiols of cysteine, and the guanidino residues of arginine. They may, therefore, carry a net positive or negative charge, depending on their state of ionization. The tertiary structure of globular proteins is usually sufficiently stable to allow the buildup of a considerable net charge density before denaturation sets in. Finally, we have to consider the behavior of substances such as poly(α-L-glutamic acid) and DNA, where a helix-coil transition is intimately related to the density of ionic charges along the molecular chain.

There is no question that the study of polyelectrolytes has been greatly stimulated by the fact that proteins and nucleic acids, materials indispensable to the existence of life as we know it, fall into this category. It was first pointed out by Staudinger (1932) that some characteristic properties of solutions of proteins and nucleic acids may be a consequence of their high ionic charges.

Since the complex chemistry of these materials introduces uncertainties into the interpretation of experimental data, he proposed that appropriate studies be carried out on synthetic polyelectrolytes containing a single type of repeating unit along the molecular chain. With our present knowledge of the specific conformations of globular proteins and DNA, the validity of the analogy between these materials and synthetic polyelectrolytes is much more limited than might once have appeared, but Staudinger's suggestion was nevertheless valuable in directing attention to an interesting field of investigation.

From one point of view, the study of polyelectrolytes may be regarded as an extension of the studies on simple electrolytes. If the polyion has a well-defined shape, as in globular proteins, the main distinction is due to the high charge of the polyion and the high values which may be assumed by the electrostatic potential. With flexible chain polyions, the fixed charges may be separated from each other, to some extent, as the chain expands when the system is diluted. Nevertheless, a limit is set on the attainable separation of the fixed charges by their attachment to the chain backbone and the rubber-like elasticity of the molecular chain. Therefore, effects produced by the interaction of ionic charges will not vanish in the limit of infinite dilution—

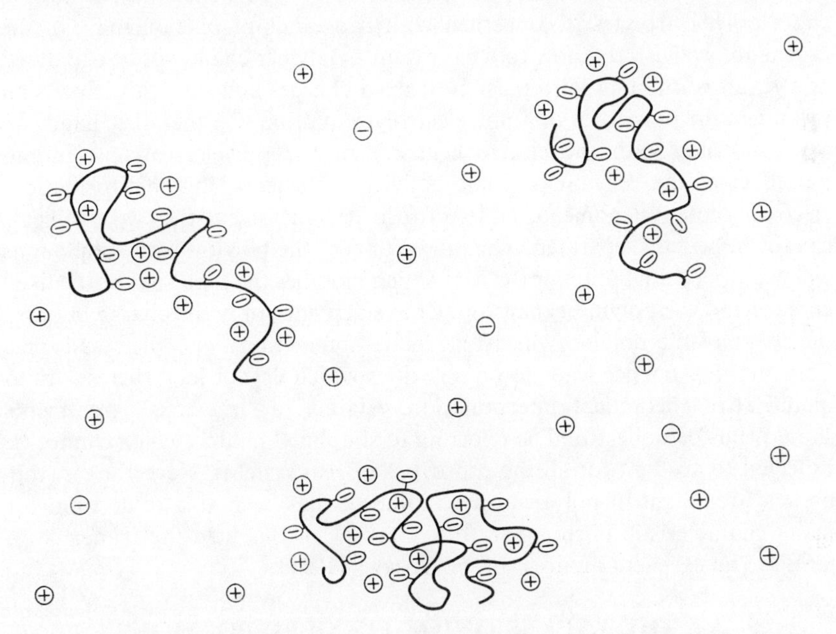

Fig. VII.1. Schematic representation of the ion distribution in dilute solutions containing a salt of a polymeric acid composed of flexible chain molecules and a small concentration of added uni-univalent electrolyte.

as they do in solutions of simple electrolytes. On the contrary, a highly dilute solution of flexible chain polyelectrolytes must be viewed as consisting of small regions in which polyions with a high density of "fixed" charges create high local electrostatic potentials, so that counterions are powerfully attracted, while the intervening spaces have very low ion concentrations (Fig. VII.1). The electrochemical behavior of polyelectrolyte solutions also strongly reflects the large difference in the charge of the polyion and the charge of the small ions. As a result of this asymmetry, the concept of ionic strength, which is so useful in characterizing electrostatic interactions in simple salt solutions, may be grossly misleading when applied to systems containing polyelectrolytes. The properties of such systems are generally much more sensitive to the charge of the counterions than to that of byions.

Whenever an uncharged polymer chain is converted by titration or by chemical modification to a chain carrying a large number of ionized groups, the mutual repulsion of fixed charges may lead to chain expansions which are far beyond the range attainable by the transfer of uncharged polymers from poor to good solvent media. But the ionic charges attached to the polymer chain do not affect only the conformation of the macromolecule. They also create a high local charge density which must affect strongly the properties of simple ions present in the solution. Thus, the study of flexible chain polyelectrolytes is concerned with two kinds of phenomena. On the one hand, we wish to know to what extent a polymer chain will be expanded as a result of the mutual repulsion of fixed charges and how this expansion will affect the properties of a polyelectrolyte solution. On the other hand, we are concerned with the electrochemistry of the polyelectrolyte solution, the effect of the polyion on ionic activity coefficients, ion-pair formation, electrophoretic phenomena, and so forth. It is clear that the two problems cannot be strictly separated. The interaction of the polyion with simple ions will lead to a distribution of the ions which modifies the repulsion of the fixed charges and the polyion expansion. Conversely, not only the charge but also the shape of the polyion will determine its interactions with the small ions. This interdependence leads to complexities which do not lend themselves to quantitative theoretical interpretation. Claims of "rigorous" treatments should thus be understood as referring to simplified models which cannot be expected to account for all the properties of real systems. Nevertheless, theoretical treatment of polyelectrolyte solutions has been able to account for many characteristic phenomena. For a review of this field the reader is referred to an excellent monograph by Oosawa (1971).

A. THE EXPANSION OF FLEXIBLE POLYIONS

In our discussions of the shape of flexible chain molecules in Section III.B we took as our first point of departure the random flight model, in which the chain links are represented as mathematical lines of zero volume and energetic interactions between chain segments are neglected. This treatment led to an expression for the probability distribution function of chain end displacements $\mathscr{W}(h)$, given by (III.7). The model was then refined to take into account both spatial interference and energetic interactions between chain segments, which will expand the chain in good solvent media beyond the dimensions predicted by the random flight treatment. In the theories of this expansion it is generally assumed that only nearest-neighbor interactions contribute to the excess free energy, attending the mixing of chain segments and solvent molecules. This assumption is amply justified if the polymer is uncharged. We are then concerned merely with the question of how the number of contacts between polymer segments depends on the extension of the molecular chain.

When ionized functional groups are attached to the macromolecules, the forces between the charges are very much larger than the dispersion forces or the dipole-dipole interactions between uncharged groups. Moreover, these forces then act over relatively long distances. In formulating a theory of the expansion of such charged chains, it is convenient to think of a polyion with its counterions as if it were being introduced into the system in two consecutive steps:

(a) The polyion and the neutralizing counterions are first added under hypothetical conditions which do not allow them to interact electrostatically with one another.

(b) The charges are then allowed to interact with one another while the chain end displacement h of the polyion is held constant. The counterions are allowed to attain their equilibrium distribution in the field of the polyion. The free energy characterizing this step is referred to as the excess electrostatic free energy $G_{el}^E(h)$.

If a polyelectrolyte is added to a solution which contains simple salt, it is customarily assumed that the change in the excess electrostatic free energy of the salt may be neglected. We shall see later that this assumption appears to be well justified by experimental evidence at very low salt concentration, although, with higher concentrations of simple electrolyte, significant deviations from this behavior may be encountered.

We may now obtain the probability distribution of end-to-end displacements of the charged chains by multiplying the a priori probability $\mathscr{W}_0(h)$ with the Boltzmann exponential involving the excess electrostatic free energy:

$$\mathscr{W}(h)\,dh = \mathscr{W}_0(h) \exp\left[\frac{-G_{el}^E(h)}{kT}\right] dh \qquad \text{(VII.1)}$$

The problem to be solved consists then in finding the proper form for $G_{el}^E(h)$ and its dependence on variables of interest, such as the chain length of the polyion, the density of ionized groups along the chain, and the concentration of simple electrolyte. The new distribution of h may be characterized by its most probable value h^*, which corresponds to $d\mathscr{W}(h)/dh = 0$:

$$\left[\frac{d \ln \mathscr{W}_0(h)}{dh}\right]_{h=h*} - \frac{1}{kT}\left[\frac{dG_{\text{el}}^E(h)}{dh}\right]_{h=h*} = 0 \qquad \text{(VII.2)}$$

Alternatively, the mean square chain end displacement $\langle h^2 \rangle$ of the charged chain may be evaluated from

$$\langle h^2 \rangle = \frac{\int h^2 \mathscr{W}(h)\, dh}{\int \mathscr{W}(h)\, dh} \qquad \text{(VII.3)}$$

For the random flight chain, for which $\mathscr{W}_0(h)$ is given by (III.7), we have $(h^*)^2 = \frac{2}{3}\langle h^2 \rangle$; but as the chain is being extended, the distribution function of h becomes sharper and $(h^*)^2/\langle h^2 \rangle$ tends toward unity. However, in the theoretical treatments which follow, we shall assume that the chain is far from its full extension, so that $(h^*)^2/\langle h^2 \rangle$ maintains the value characteristic of Gaussian chains.

1. Theories of Polyion Expansion

The first attempt to treat the problem of the expansion of a charged polymer chain was made by Kuhn et al. (1948) and by Katchalsky et al. (1950). They assumed that in dilute solutions the counterions escape from the polyion domain, so that they do not influence the mutual repulsion of the fixed charges carried by the chain molecule. In their model the polyion charge Q is divided into $Z_s + 1$ equal parts, q, which are placed at the chain ends and the junction points of the statistical chain elements. The electrostatic free energy is then obtained by summing contributions from all pairs of charges and averaging the result over all properly weighted chain conformations consistent with a chain end displacement h.

We shall see later that the model underlying the theory of Kuhn et al. is unrealistic, since a significant fraction of counterions will always remain in the polyion domain, even under conditions of extreme dilution. The charges of the counterions will shield the ionic charges carried by the macromolecular chain ("the fixed charges"), and the expansion of the chain backbone, resulting from the mutual repulsion of the fixed charges, will be correspondingly reduced. To allow for this effect, Katchalsky and Lifson (1956) assumed that each fixed charge builds up in its vicinity an ion atmosphere similar to one which would exist in a simple salt solution with an ionic strength corresponding to the concentrations of the mobile ions in the polyelectrolyte solution. The potential at a distance r from a fixed charge is assumed to be given by the Debye-Hückel limiting law as $q \exp(-\kappa r)/\mathscr{D}r$, with κ, the reciprocal of the equivalent thickness of the ion atmosphere, given by

$$\kappa = \sqrt{4\pi \boldsymbol{e}^2 \boldsymbol{N} \sum m_i \nu_i^2 / 1000 \mathscr{D} \boldsymbol{k} T} \tag{VII.4}$$

where $\boldsymbol{e}$ is the electronic charge, m_i are the molarities of ions with a valence ν_i, and $\mathscr{D}$ is the dielectric constant. Using these screened potentials, we obtain

$$G_{\mathrm{el}}^E(h) = \left(\sum_i \sum_{j>i} \frac{q_i q_j}{\mathscr{D} r_{ij}} \right)_h \tag{VII.5}$$

where r_{ij} is the distance between the ith and the jth charges. For long chains, $Z_s + 1 \approx Z_s$ and $Z_s q \approx Q$, and since the summation in (VII.5) contains $Z_s^2/2$ terms and $q_i q_j/\mathscr{D}$ is constant,

$$G_{\mathrm{el}}^E(h) = \frac{Q^2}{2\mathscr{D}} \left\langle \frac{\exp(-\kappa r_{ij})}{r_{ij}} \right\rangle_h \tag{VII.6}$$

Here the expression in the pointed brackets is the properly weighted average of $\exp(\kappa r_{ij})/r_{ij}$ over all pairs of interacting charges and all shapes of the equivalent chain consistent with a chain end separation h. According to calculations of Katchalsky and Lifson, this leads to

$$G_{\mathrm{el}}^E(h) = \frac{Q^2}{\mathscr{D} h} \ln \left[1 + \frac{6h}{\kappa \langle h_0^2 \rangle} \right] \tag{VII.7}$$

The theory assumes that the dimensions of the macromolecular chain are large compared to the thickness of the ion atmosphere ($h\kappa \gg 1$). In practice this is an unimportant restriction, requiring concentration of uni-univalent electrolyte of 0.002 M and 0.001M if the chain end separation of the polyion is 100 and 400 Å, respectively. The expansion of the polyion predicted by this theory is then given by

$$\alpha_e^2 - 1 = \frac{Q^2}{2\mathscr{D} \boldsymbol{k} T h^*} \left[\ln\left(1 + \frac{6h^*}{\kappa \langle h_0^2 \rangle}\right) - \frac{6h^*/\kappa\langle h_0^2 \rangle}{1 + 6h^*/\kappa \langle h_0^2 \rangle} \right] \tag{VII.8}$$

If we use, as an example, a polyacrylate with a polymerization degree of 1200, assume that three monomer units form a statistical chain element, and compute the chain extension in a solution containing 0.01M uni-univalent salt on the basis of relation (VII.8), we find that h^* increases fivefold with only 5% of the carboxyl groups ionized. It is certain that this result greatly exaggerates the expansion of polyions. The error is due chiefly to the manner in which Katchalsky and Lifson took account of the different shapes which a chain can assume for any given chain end separation. They used the distribution function applicable to uncharged chains, whereas in actuality shapes which are more compact (and correspond, therefore, to a higher electrostatic energy) will occur much less frequently in polyions than in analogous uncharged chains. As a result, the Katchalsky-Lifson treatment tends to overestimate $G_{el}^E\ (h)$, and this error should be largest for the smallest values of h, so that the calculated driving force toward chain expansion is too high.

The elimination of this error from the Katchaslky-Lifson theory would meet with formidable mathematical difficulties, and this led Harris and Rice (1954) and Rice and Harris (1954) to the use of a different model for the estimation of polyion expansions. They considered an equivalent chain with the ionic charges carried by the monomer units concentrated at the midpoints of each statistical chain element. The extension of the chain depends then on the probability distribution of the angles γ subtended by two adjoining chain elements. Treating the angle γ in the same manner as a bond angle in Kuhn's theory (cf. Section III.B.3), we obtain for the dependence of h on the angle γ

$$\frac{\langle h^2 \rangle}{\langle h_0^2 \rangle} = \frac{1 - \langle \cos \gamma \rangle}{1 + \langle \cos \gamma \rangle} \tag{VII.9}$$

As a first approximation, $\langle \cos \gamma \rangle$ is calculated considering only $G_{j,j+1}^E$, the potential energy resulting from the repulsion of two charges q carried by neighboring elements of the statistical chain. For any assumed dependence of $G_{j,j+1}^E$ on charge separation, $\langle \cos \gamma \rangle$ may be rigorously obtained from

$$\langle \cos \gamma \rangle = \frac{\int \cos \gamma \exp [- G_{j,j+1}^E(\gamma)/\boldsymbol{k}T]\, d\Omega}{\int \exp [- G_{j,j+1}^E(\gamma)/\boldsymbol{k}T]\, d\Omega} \tag{VII.10}$$

where $d\Omega = 2\pi \sin \gamma\, d\gamma$ is the element of solid angle. For a statistical element of length b_s, the separation of two neighboring charges is $b_s \sin (\gamma/2)$, and if the screened Coulomb potential is given by the Debye-Hückel limiting law, then

$$G_{j,j+1}^E(\gamma) = \frac{q^2 \exp [- \kappa b_s \sin (\gamma/2)]}{\mathscr{D} b_s \sin (\gamma/2)} \tag{VII.11}$$

While Katchalsky and Lifson assumed that the macrosocpic dielectric constant of water ($\mathscr{D} \approx 80$) may be used to obtain G_{el}^E, Rice and Harris pointed out that both the high concentration of nonpolar organic material and the high concentration of ionic charges in the vicinity of the polyion chain would tend to reduce the effective $\mathscr{D}$, and they suggested $\mathscr{D} = 5.5$ as an appropriate value.

At the same time that polyelectrolyte theories were being formulated in which the chainlike character of the polyion was taken explicitly into account, the problem of polyion expansion was also investigated by means of models in which the polyion was represented by a spherically symmetrical charge cloud. If the polymer coil is immersed in a large volume of a uni-univalent salt of molarity m_s, the local concentration of the mobile ions will be related to the local electrostatic potential ψ by the Boltzmann distribution law, and the local net charge density ρ becomes

$$\rho = \frac{\mathfrak{F}}{1000}[m_{fl} + m_s \exp(-e\psi/kT) - m_s \exp(e\psi/kT)] \tag{VII.12}$$

where $\mathfrak{F}$ is the Faraday constant and m_{fl} the molarity corresponding to the local concentration of the fixed charges on the polyion. A second relation between ρ and ψ is provided by the well-known Poisson equation:

$$\nabla^2\psi = -\frac{4\pi\rho}{\mathscr{D}} \tag{VII.13}$$

and a combination of (VII.12) with (VII.13) yields

$$\nabla^2\psi = -\left(\frac{4\pi\mathfrak{F}}{1000\mathscr{D}}\right)\left[-m_{fl} + 2m_s \sinh\left(\frac{e\psi}{kT}\right)\right] \tag{VII.14}$$

The Poisson-Boltzmann equation was first applied to flexible chain polyelectrolytes by Hermans and Overbeek (1948), who considered the case when $e\psi/kT \ll 1$, so that $\sinh(e\psi/kT) \approx e\psi/kT$. This corresponds to situations in which either the charge density on the polyion is relatively low or the salt concentration is relatively high. Calculations were carried out for a uniform charge distribution within a sphere with a molar volume V_e, which is hydrodynamically equivalent to the polyion, but the expansion of the coil is actually insensitive to the charge distribution. Hermans and Overbeek assumed that the polyion domain becomes increasingly asymmetric as it expands, but the application of their method leads to a much simpler result if we assume that the polyion domain retains a spherical shape. For coil dimensions very large compared to the thickness of the ion atmosphere (the usual case), the expression for the expansion coefficient becomes

$$\alpha_e^5 - \alpha_e^3 = \frac{3}{16}\left(\frac{\nu_p^2}{NV_{e0}m_s}\right) \tag{VII.15}$$

where ν_p is the number of charges carried by the polyion, and V_{e0} corresponds to the value of V_e in the uncharged state. We may note the analogy with the result for uncharged chains:

$$\alpha_e^5 - \alpha_e^3 = \frac{(\frac{1}{2} - \chi)V_2^2}{V_1 V_{e0}} \tag{III.29}$$

If we consider polyions with a fixed charge density, $\alpha_e^5 - \alpha_e^3$ will be predicted to be proportional to the square root of the chain length, just as in the case of uncharged chains.

The restriction of the Hermans-Overbeek theory to cases in which the electrostatic potential remains relatively small precludes its application to the systems of greatest interest. As Kimball et al. (1952) pointed out, it is more realistic to assume that the domain occupied by the polyion contains a sufficient excess of counterions over byions to render its net charge negligible. This assumption is the microscopic analog of the Donnan treatment of the equilibrium distribution of ions across a semipermeable membrane restricting the movement of polyions in macroscopic systems.

Flory (1953e) has pointed out that, whenever the Donnan assumption is applicable, the swelling of the polyion may be predicted without introducing the concept of electrostatic potential. This is physically obvious if we consider the swelling of a macroscopic gel, for which the Donnan condition must apply much more precisely. The driving force of the swelling process may then be assigned with equal justification to osmotic forces (due to an excess of mobile ions in the interior of the gel) or to electrostatic repulsion of ionic charges attached to the gel structure. Flory's theory of polyion expansion also takes account of the excess free energy resulting from the mixing of solvent and polymer segments. This treatment leads to the expression

$$\alpha_e^3[\alpha_e^2 - (\alpha_e^0)^2] = 1000\nu_p\left\{\frac{\nu_p}{1.16N\langle h_0^2\rangle^{3/2}\sum m_i\nu_i^2} + (\nu_b - \nu_c)\left[\frac{\nu_p}{0.81N\langle h_0^2\rangle^{3/2}\sum m_i\nu_i^2}\right]^2 + \cdots\right\} \quad \text{(VII.16)}$$

where α_e^0 is the expansion factor which would be observed with a solvated but uncharged chain, m_i and ν_i are molarities and valences of all mobile ions, and ν_b and ν_c are the valences of byions and counterions, respectively. The form of (VII.16) leads to the conclusion that chains which have already been expanded beyond the unperturbed dimensions by the solvating action of the medium will be less easily expanded further by the mutual repulsion of the fixed ionic charges.

Flory's theory is unrealistic in that the leading term in (VII.16) does not distinguish between the effects of the valence of counterions and that of byions on the polyion expansion. It is generally observed, however, that the properties of polyions are extremely sensitive to the valency of the counterion, ν_c, while a change in the charge of byions produces no measurable effect. This factor is accounted for in a theory formulated by Fixman (1964), which leads to

$$\alpha^3 = 1 + C_1 Z^{1/2} + \frac{C_2(Q/e\nu_c)}{\langle h_0^2\rangle^{1/2}\kappa} \quad \text{(VII.17)}$$

where C_1 and C_2 are adjustable parameters and the $C_1 Z^{1/2}$ term represents expansion due to causes other than the mutual repulsion of ionic charges carried by the chain molecule.

The effect of the attachment fo ionic charges to a polymer chain, which leads to a mutual repulsion of chain segments, may be represented as being due to an increase in the excluded volume of the chain segments. We shall see later that the charge of the polyion is almost compensated for by counterions within a short distance of the polyion chain, and we may then take as the domain excluded to other chain segments the wormlike volume represented schematically in Fig. VII.2, which is characterized by a high value of

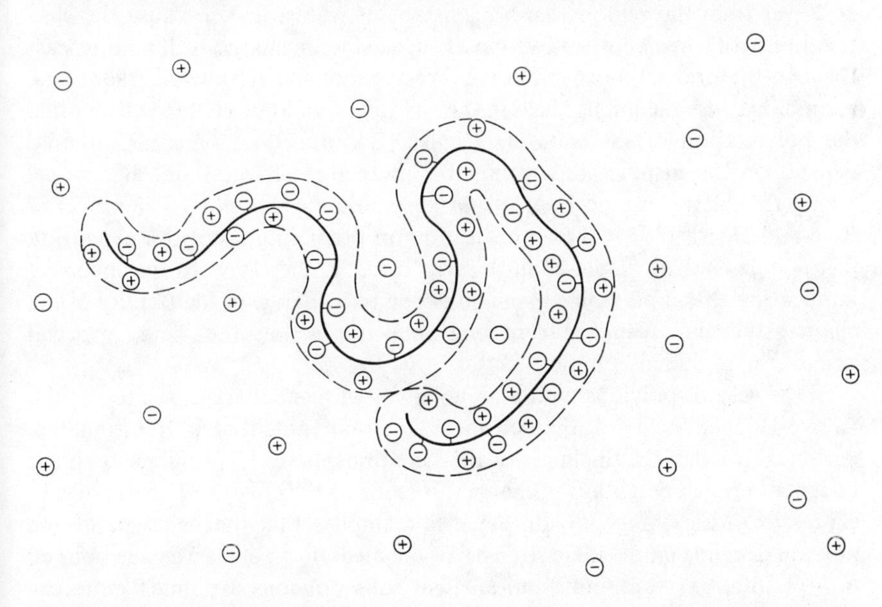

Fig. VII.2. Model of polyelectrolyte solution according to Strauss and Ander (1958).

the electrostatic potential. This volume will tend to shrink if the simple electrolyte concentration in the system is increased. If we can compute the distribution of electrostatic potential as a function of charge density on the polyion, simple electrolyte concentration, and nonelectrostatic forces between the polyions, the dependence of the excluded volume effect on these variables can be predicted. This method, which was used by Alexandrowicz (1967), seems to represent the most realistic approach to a theory of polyion expansion.

All theories of polyion expansion are based on idealized models and necessarily omit some features of the systems they are designed to represent. The difficulty concerns both the correct evaluation of the electrostatic interactions and the assumption that the local conformational distribution, governing the flexibility of the chain, is unaffected by the ionization of groups attached to the chain backbone.

A serious limitation of all theories in which the polyion is represented by a spherically symmetrical cloud of charge densities is their inability to account for the formation of ion pairs between the fixed charges on the polyion and the counterions. This can be understood only when one considers the close spacing of the charges along the macromolecular chain. However, even in theories taking account of the chain character of the polyion, a difficulty arises in the choice of the effective dielectric constant $\mathscr{D}_E$, which governs the interaction between the charges of the polyion. The value of $\mathscr{D}_E$ should be lower than the bulk dielectric constant of water, first, because the electrical lines of force between two closely spaced ionic charges will tend to pass through the organic macromolecule (Westheimer and Kirkwood, 1938) and, second, because the intense field in the vicinity of an ionic charge will saturate the polarization of the water molecules (Booth, 1951). Thus, $\mathscr{D}_E$ should depend on the distance separating the interacting charges, and any model taking this factor into account would become quite unwieldy. The model of Rice and Harris (1954) is also misleading in predicting a polyion expansion factor independent of the chain length. It leads, therefore, to the incorrect implication that a polyion expansion is due to a change in the rigidity of the chain, rather than resulting from long-range interactions (i.e., being a special case of the excluded volume effect).

In the case of polyions with a high density of fixed charges, the use of the Debye-Hückel form for the screened Coulomb potential is questionable, particularly when the thickness of the ion atmosphere, $1/\kappa$, is larger than the distance between neighboring charges attached to the polyion. The use of κ in expressions describing polyion expansion implies that the behavior of the polyion depends on the ionic strength of the medium—yet, as we have pointed out, in solutions containing uni-divalent salts polyions are much more expanded if the byion, rather than the counterion, is the doubly charged species (Flory and Osterheld, 1954). Even in Fixman's theory the effect of a multiply charged counterion is not given its proper weight.

Electrostatic factors affecting polyion expansion may involve both short-range and long-range interactions. In principle, the ionization of groups attached to a polymer backbone may affect the local conformational distribution because of a superposition of Coulombic interactions on the energies characterizing the various accessible conformations. This point is illustrated in Fig. VII.3 on a sequence in syndiotactic poly(acrylic acid), where a shift from the tt to the gg conformation produces an increase in the distance between the carboxylate charges. The effect of ionization of this polymer on its *unperturbed* dimensions was calculated by Lifson (1958) and by Allegra et al. (1972). Lifson assumed that the forces between the ionic

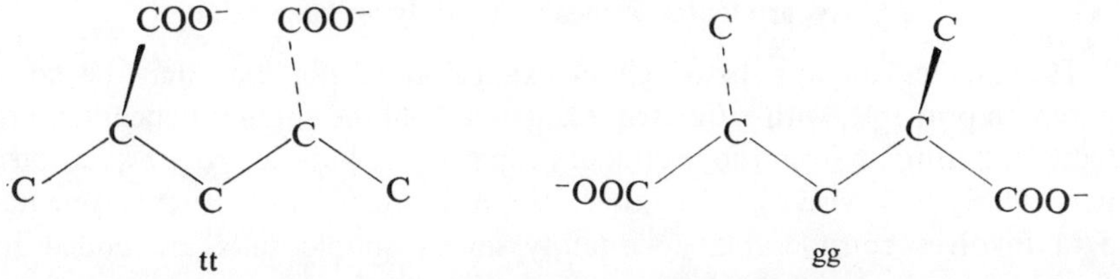

Fig. VII.3. Spacing of nearest-neighbor carboxylates in syndiotactic poly(acrylic acid), with the chain backbone in the *trans-trans* and the *gauche-gauche* conformations.

charges may be treated by the Debye-Hückel screened potential and concluded that the unperturbed dimensions should fall off with increasing ionization. Allegra et al., on the other hand, used a model in which hydrated counterions approached as closely as possible to the polyion, and they predicted that ionization should expand the unperturbed dimensions. The effect is thus found to be extremely sensitive to the counterion distribution, and it is doubtful whether a reliable theoretical analysis is feasible.

We have seen that the theories of Flory (1953e) and Fixman (1967) treat the excluded volume effect of polyelectrolytes as being produced additively by electrostatic and nonelectrostatic contributions. A similar idea is also incorporated in the treatment of Alexandrowicz (1967). All these treatments assume that the electrostatic contribution to the excluded volume of the chain segments should vanish only in the limit of infinite salt concentration. Thus, as Alexandrowicz points out, the fact that the Θ condition is reached at finite salt concentration would imply that the nonelectrostatic contribution to the excluded volume, assumed to be independent of the added electrolyte, is negative. These ideas may be criticized on two grounds.

(a) Since counterions are held close to the polyion and since ion pairs would be expected to attract one another, the electrostatic contribution to the excluded volume may well change from a positive to a negative quantity at finite electrolyte concentration. This concept is strongly supported by the behavior of polycarboxylic acids in media of lower dielectric constant, which would favor ion pairing. For instance, when poly(acrylic acid) is dissolved in methanol (a very good solvent), it will precipitate rapidly on being titrated with alkali.

(b) Since electrolyte addition will typically lead to the salting-out of organic solutes, the nonelectrostatic contribution to the excluded volume will change rapidly as salts are added to an aqueous polyelectrolyte solution, and its magnitude in Θ-solution cannot be taken as representing that of the polyelectrolyte in pure water after elimination of the effects of Coulombic interactions.

2. Experimental Evidence for Polyion Expansion

The theoretically predicted chain expansion of polyions may be compared, in principle, with estimates calculated from the angular dependence of light scattering or from the frictional properties of polyelectrolyte solutions, particularly their viscosity. In either case, interpretation of the experimental data involves considerable uncertainty unless simple salts are added to shield the charges of the polyions.

Theoretical aspects of light scattering from polyelectrolyte solutions and experimental results obtained in such studies have been admirably reviewed by Nagasawa and Takahashi (1972). In the absence of added salts, the theoretical interpretation of data is uncertain, since it cannot be assumed that the volume elements involved in concentration fluctuations can be treated as electroneutral. In addition, the very large mutual repulsion of the macroions tends to produce an ordered distribution of these highly charged species, even at relatively high dilution. This effect was observed by Doty and Steiner (1952) on solutions of serum albumin hydrochloride, which exhibited a strong dissymmetry of light scattering, although the dimensions of this globular protein are only of the order of 40 Å. In this case the phenomenon reflects the interference of light scattered by different macromolecules, and conclusions about the extension of the individual particle could be obtained only at dilutions which are too high for significant measurement. In the special case of high molecular weight double-helical DNA, interparticle forces were observed to be strong enough to develop an ordered structure gradually even in the presence of considerable added electrolyte (i.e., 0.02M sodium phosphate and 0.01M sodium chloride) if the solution was allowed to stand undisturbed. This structure leads to diffraction phenomena, with pronounced maxima in the scattered light intensity at angles characterizing the spacing of solute particles (Gruber et al., 1972).

However, in other cases complications arising from interparticle interference are effectively eliminated when polyelectrolyte solutions contain simple salts. In that case the theory outlined by Casassa and Eisenberg (1960) and Eisenberg (1962) is applicable, and the standard analysis of the data has to be modified only insofar as the refractive index increment $(\partial n/\partial c_2)_T$ in (V.27) is referred to the solution in dialysis equilibrium with the polymer solution, rather than the pure solvent. Typical plots of $(c_2/\Delta\mathscr{R}_\theta)_{c_2\to 0}$ against $\sin^2(\theta/2)$, from which $\langle s^2 \rangle$ may be obtained by applying (V.42), are reproduced in Fig. VII.4 from the work of Takahashi et al. (1967); they illustrate the increasing polyion expansion with decreasing concentration of added salt. It should be stressed, however, that for high charge densities on the polyion and low concentrations of simple electrolyte, deviations from ideal solution behavior become very large and that an extrapolation of the light

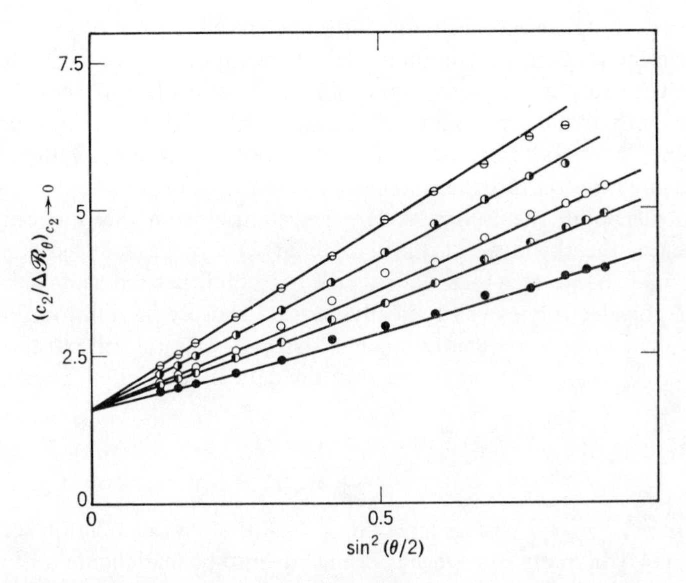

Fig. VII.4. Angular dependence of light scattering of sodium polystyrenesulfonate ($\bar{M}_w = 1.55 \times 10^6$) in solutions containing 0.005M (⊖), 0.01M (◑), 0.05M (○), 0.1M (◐), and 0.5M (●) NaCl.

scattering data to zero solution concentration may then be very difficult and uncertain.

We saw in Chapter VI that intrinsic viscosity is a most convenient experimental tool for the characterization of the dimensions of flexible polymer chains. However, here again the high charge of the polyion introduces some complicating factors. Even in the case of a suspension of rigid spheres, the intrinsic viscosity predicted by Einstein's theory should be modified if the spheres carry an electrostatic charge. This is the result of a lag of the counterion atmosphere behind a charged particle moving through a fluid, so that electrical work has to be added to the work dissipated by fluid friction if a steady-state motion of the particle is to be maintained. The magnitude of this "electroviscous effect" has been calculated by Booth (1950), and experimental results obtained with a sulfonated polystyrene latex (Stone-Masui and Watillon, 1968) are in agreement with his prediction. The effect is small except at extremely low ionic strength, and it may be safely assumed that for flexible chain polyions the relaxation effect is negligible compared to effects produced by the expansion of the macromolecular chain.

If the viscosity of polyelectrolyte solutions containing no added salt is to be investigated, the data cannot be interpreted in terms of polyion expansion, since consequences of the mutual interaction of the polyions cannot be properly taken into account. With uncharged chain molecules, effects of particle interaction are eliminated by an extrapolation of η_{sp}/c_2 to zero solute concentration, but this extrapolation is possible only because the shape of the individual polymer coil depends only slightly on solution concentration. The situation is very different in polyelectrolyte solutions: here an increasing dilution provides increasing volume for the counterions, which will then distribute themselves at larger distances from the polyion. As a consequence, the shielding of the fixed charges will be reduced, and the polyion will expand. It is then not surprising to find that reduced viscosities, η_{sp}/c_2, of polyelectrolytes will typically increase sharply on dilution. Plots of η_{sp}/c_2 against c_2 are frequently strongly curved, making an extrapolation impossible. Fuoss (1948) suggested that the data should follow a relation of the type

$$\frac{\eta_{sp}}{c_2} = \frac{A}{1 + B\sqrt{c_2}} \tag{VII.18}$$

which predicts $(\eta_{sp}/c_2)^{-1}$ to be linear in $\sqrt{c_2}$ Although this relation seems to be satisfactory in many cases, it has been shown to be inadequate if the polyion is very long (Alexander and Hitch, 1952). Another complication arises from the extremely large dependence of the viscosity of polyelectrolyte solutions on the rate of shear (Eisenberg, 1957). A careful study carried out by Eisenberg and Pouyet (1954) at very low rates of shear led to the results shown in Fig. VII.5. We see that in salt-free solution a plot of the reduced viscosity against the concentration of the polyelectrolyte passess through a

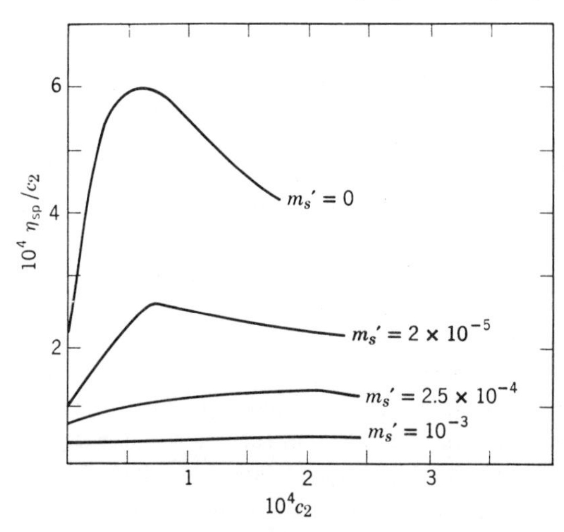

Fig. VII.5. Reduced viscosity plots of poly(*N*-butyl-4-vinylpyridinium) bromide with various concentrations of added sodium chloride (m_s').

maximum. This is so because the expansion of the polyion reaches an upper limit and the effects observed on further dilution merely reflect the decreasing interference between the expanded polyions. As salt is added to the system, the value of the intercept decreases, indicating the contraction of the polyions. At the same time, there is a marked decrease in the initial positive slope of the reduced viscosity plot, reflecting a decreasing interaction between the charged macromolecules.

In principle, it should be possible to dilute a salt-free polyelectrolyte solution with a salt solution of a concentration such as to keep the ionic atmosphere surrounding the polyions unchanged during the dilution process. Akkerman et al. (1952) suggested that this result should be achieved if the diluent contains uni-univalent electrolyte with a normality half as great as that of the salt-free polyelectrolyte, so that the concentration of the sum of mobile ions remains unchanged. This procedure, which has been called "isoionic dilution," yields linear plots of reduced viscosity against polyelectrolyte concentration quite similar to plots obtained with uncharged polymers. Intrinsic viscosities obtained in this manner are generally interpreted just as those of uncharged polymer chains, implying that, in spite of the considerable expansion, the polymer coil remains hydrodynamically impermeable and is not much more asymmetric than it is in the absence of

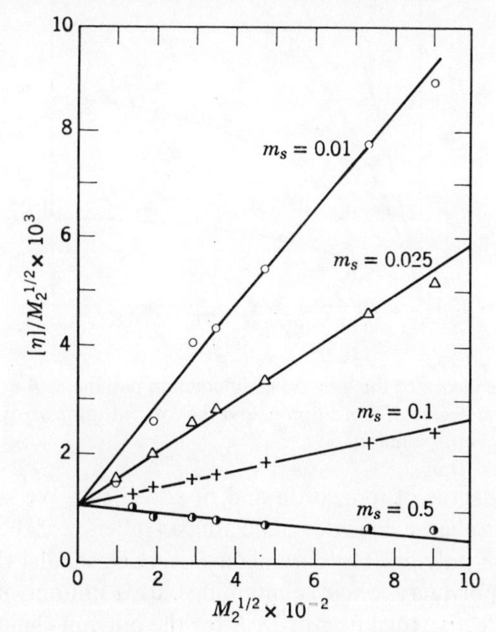

Fig. VII.6. Stockmayer-Fixman plots of poly(acrylic acid), 10% neutralized with NaOH in agueous solutions of NaBr.

Coulombic interactions. Nevertheless, for reasons which are not well understood, the Flory constant Φ' (VI.60) appears to depend on the polyion charge and may be as small as half of its usual value for uncharged polymer chains (Takahashi et al., 1967).

Noda et al. (1970) have carried out an extensive viscosimetric study of sodium polyacrylate, varying the degree of neutralization, the polymer chain length, and the concentration of added salt. Results plotted according to the Stockmayer-Fixman treatment (VI.64) are shown in Fig. VII.6. Plots of $[\eta]/M_2^{1/2}$ against $M_2^{1/2}$ are linear with slopes decreasing as the salt concentration is raised. Figure VII.7 shows the excluded volume parameter B of (VI.64)

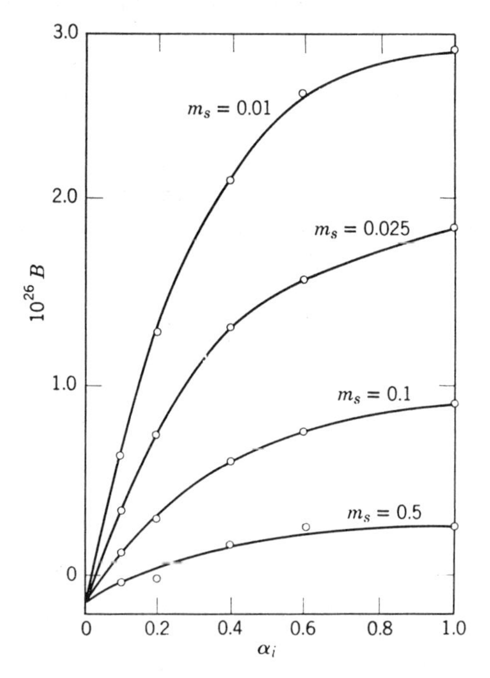

Fig. VII.7. Dependence of the long-range interaction parameter B of the Stockmayer-Fixman theory on the degree of ionization α_i and the concentration m_s of added salt. Poly(sodium acrylate) in NaBr solution.

as a function of degree of ionization and of added salt. We see that B tends to level off at high charge densities, reflecting a tendency of the counterion to associate with the polyion. The parameter B is much smaller than predicted; for fully ionized poly(acrylic acid) containing $0.01 M$ uni-univalent salt it corresponds to an effective thickness of 18 Å for the polyion chain. As predicted by Fixman's equation (VII.17), the dependence of $[\eta]$ on m_s is of the form $[\eta] = X + Y/m_s^{1/2}$ (Pals and Hermans, 1952; Noda et al., 1970) and for different types of polyelectrolytes Y is inversely proportional to the unperturbed chain dimensions (Smidsrød and Haug, 1971).

Whereas poly(acrylic acid) (PAA) expands smoothly with increasing density of ionic charges along the polymer chain, poly(methacrylic acid) (PMA) behaves in a strikingly different manner. During the initial stages of its ionization, the chain dimensions undergo little change; then, within a relatively narrow critical range, an increase in the charge density leads to an abrupt transition to a highly expanded state (Katchalsky and Eisenberg, 1951). This phenomenon has been the subject of a great deal of study. The nature of the transition has been clarified by Koenig et al. (1969) and Lando et al. (1973), whose Raman spectra show that the number of conformations avialable to the chain backbone increases during the transition. Thus, the interaction of charges attached to the polymer changes, in this case, the flexibility of the chain. The contrasting behavior of PAA and PMA illustrates the danger of applying idealized models to the behavior of real polymers, which may be profoundly altered by small changes in their chemical structures.

Long-range attractive interactions of hydrophobic groups attached to polyions may provide powerful resistence to chain expansion. As an example of such an effect, we may cite the difference in the behavior of alternating copolymers of maleic acid, where chain expansion is much more difficult when the comonomer is styrene rather than methyl vinyl ether (Ferry et al., 1951). An extreme case is represented by the "polysoaps" (Strauss et al, 1956; Medalia et al., 1959), in which the polyions carry long aliphatic hydrocarbon side chains. The cohesion of these side chains is analogous to the formation of soap micelles and is so strong that it may stabilize a compact globular form even at high charge densities.

B. THE DISTRIBUTION OF COUNTERIONS

We saw in the preceding section that the complexities associated with the behavior of flexible chain polyelectrolytes are caused largely by the interdependence of polyion expansion and the distribution of the mobile ions in the vicinity of the charged chain. The treatment of all the characteristics of such a system is extremely difficult and a number of investigators, therefore, have looked merely for a distribution of mobile ions consistent with some assumed distribution of fixed charges. Although this approach has obvious limitations, it provides qualitative insight into some important features of polyelectrolyte solutions.

The first study of this type was made by Alfrey et al. (1951) and Fuoss et al. (1951), who treated the polyion as a rigid rod of infinite length and with a uniform surface charge in a system containing only polyions and their univalent counterions, but no simple salt. The counterions were contained in a cylindrical space surrounding the reference polyion; at the boundary of this space the electrostatic forces of the surrounding polyelectrolyte were assumed to balance forces due to particles in the space reserved for the reference polyion and its counterions. The distribution of the electrostatic potential ψ is then given by the Poisson-Boltzmann equation, which is, for cylindrical symmetry,

$$\frac{1}{r}\frac{d}{dr}\left(r\frac{d\psi}{dr}\right) = \frac{4\pi \boldsymbol{N} m_f \boldsymbol{e}}{1000\mathscr{D}} \exp\left(\frac{\boldsymbol{e}\psi}{\boldsymbol{k}T}\right) \tag{VII.19}$$

where m_f is the molarity of the fixed charges averaged over the volume of the system. Whereas the Poisson-Boltzmann equation can be solved, for the spherically symmetrical case, only with the Debye-Hückel assumption that $\boldsymbol{e}\psi/\boldsymbol{k}T \ll 1$, for the cylindrically symmetrical charge distribution represented by (VII.19), an analytical solution may be obtained for the general case. Defining the potential so that $\psi = 0$ in regions where the local counterion concentration equals its volume-average value, we have

$$\psi(r) = \frac{\boldsymbol{k}T}{\boldsymbol{e}} \ln\left\{\frac{1000\delta^2\mathscr{D}\boldsymbol{k}T}{2\pi \boldsymbol{N} m_f \boldsymbol{e}^2 r^2 \cos^2[\delta(\ln r + \beta)]}\right\} \tag{VII.20}$$

where the integration constant δ and β depend on the thickness of the polyion and its charge density, as well as on the concentration of the system.

The distributions of electrostatic potential and of the counterions obtained from this result are illustrated in Figs. VII 8. and VII.9 for rodlike polyions with a radius of 6 Å and a charge density corresponding to half-ionized poly-

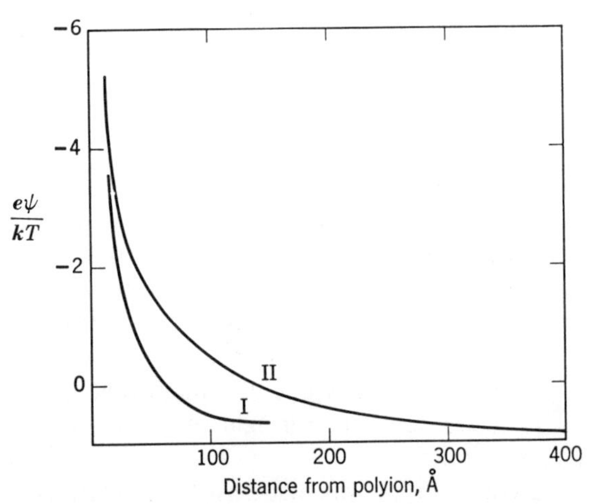

Fig. VII.8. Distribution of electrostatic potential around rodlike polyion with a radius of 6 A and a uniform surface charge of 0.2 electronic charges per A.(I) $m_f = 0.01$. (II) $m_f = 0.00125$.

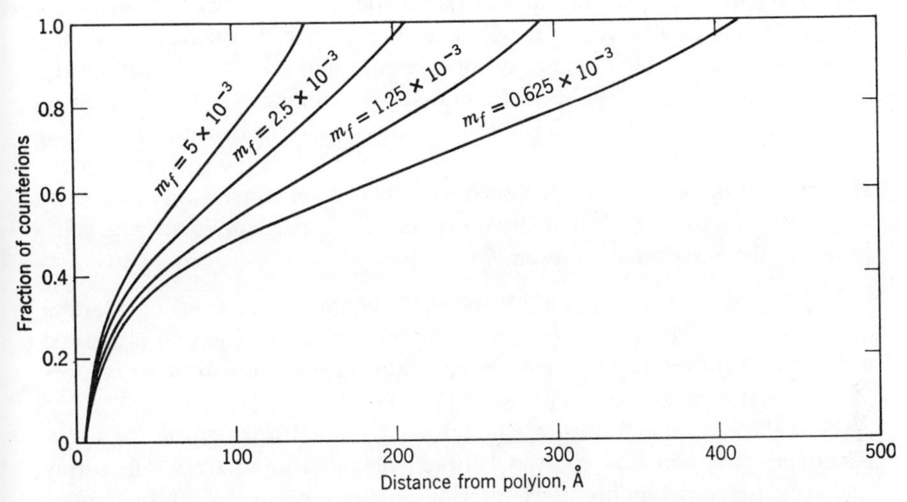

Fig. VII.9. Integral counterion distribution around rodlike polyions. Polyion model as in Fig. VII.8.

(acrylic acid). The most important conclusions are the following: (a) with polyions of high charge density the electrostatic potentials attain very high values over considerable regions of the system, so that the Debye-Hückel assumption that $e\psi/kT \ll 1$ is clearly inapplicable; (b) dilution of the polyelectrolyte solution has a surprisingly small effect on the concentration of counterions in the neighborhood of the polyion. In fact, even if the solution were diluted with an infinite volume of water, the reservoir of hydrogen and hydroxyl ions due to the self-ionization of the solvent medium would lead to appreciable counterion concentrations in regions close to the polyion.

The Poisson-Boltzmann equation for cylindrical symmetry, as given in (VII.19), can be solved analytically only for a system containing rodlike polyions and their counterions, but no added simple electrolyte. However, the distributions of the electrostatic potential for the general case of systems containing such polyelectrolytes with added salts have been obtained by an approximation technique (Alexandrowicz and Katchalsky, 1963) and by computational methods (Kotin and Nagasawa, 1962; Gross and Strauss, 1966). The charged rod model represents, of course, an extreme situation which exaggerates the expansion of the flexible chains, even if they carry a high charge density. However, the model is most useful since it corresponds to the highest possible dispersal of the fixed charges; for any less extended polyion the electrostatic interaction with the counterions will necessarily be even more pronounced.

A reconsideration of the cylindrical model for a polyelectrolyte led Imai and Onishi (1959) to the conclusion that a rodlike polyion cannot sustain a charge density exceeding a critical value. If a higher charge density is produced by closely spaced ionized groups, counterions will "condense" on the polyion. This concept has been elaborated by Manning (1969). With univalent counterions, his theory leads to a maximum charge density (in esu/cm) of $\mathscr{D}\boldsymbol{k}T/\boldsymbol{e}$, corresponding to one ionic charge per 7 Å of the rodlike polyion. However, although the conclusions of this theory are convincing in terms of the model employed, its quantiative application to real polyelectrolyte solutions may be questioned on several grounds.

(a) The "contour length of a fully extended chain" is a somewhat uncertain quantity. It will depend, in practice, on the maximum extension accessible without a prohibitive energy requirement; for instance, the all-*trans* conformation cannot be considered for isotactic poly (acrylic acid).

(b) If the ionizable groups are attached at some distance from the chain backbone, the increased spacing between neighboring charges will surely lead to a reduced binding of counterions for any given density of ionized groups per unit contour length of the chain backbone.

(c) Even in the simplest case, that of alkali salts of polymeric acids, counterion binding depends, to some extend, on the size of the cation and the nature of the ionized groups on the polyion.

(d) Manning's use of the macroscopic dielectric constant of water as governing Coulombic forces in the immediate neighborhood of polyions is unrealistic, even though it allegedly leads to calculated results in good agreement with experimental data.

Let us first consider the case of polyanions with alkali counterions. "Counterion condensation" involves then, from a chemical point of view, "site binding" of the cations to the anionic groups attached to the polymer. This may result either from ion-pair formation involving the hydrated cation, or from complex formation involving the displacement of water molecules in the hydration shell of the cation by the anionic group. Although most experimental methods do not allow a clear disinction between long-range Coulombic interaction of polyions with their counterions and site binding (Lyons and Kotin, 1965; Mandel, 1967), a comparison of the behavior of different alkali counterions will indicate the nature of the site binding, since long-range interactions will be the same for all counterions with the same charge. Such comparisons of ionic activity coefficients, polyion expansion, titration behavior, electrical conductance, and so forth, lead to the conclusion that site binding to polymers carrying strongly acidic goups [i.e., polysulfonic acids] follows the order $K^+ > Na^+ > Li^+$, so that binding is favored by a small size of the hydrated cation (Eisenberg and Mohan, 1959; Chu and Marinsky, 1967). This suggests that ion-pair formation is the dominant cause of site binding in this case†. On the other hand, with alkali salts of polycarboxylic acids site binding follows the opposite sequence: $K^+ < Na^+ < Li^+$ (Gregor and Frederick, 1957; Quadrifoglio et al., 1973; Rinaudo and Pierre, 1969; Rinaudo and Miles, 1969; Zana et al. 1971), and this is also the sequence observed with polyphosphates (Strauss and Ross, 1959). Thus, counterion binding must involve, in these cases, complex formation which is favored by a small *crystallographic radius* of the cation. Complex ion formation may be characterized quantitatively by phenomena which are directly related to the release of water coordinated to the cations. Since the hydration shell is compressed by electrostriction, this release leads to an increaes of volume (Strauss and Leung, 1965) and a characteristic decrease in refractive index (Ikegami, 1964). Independent measurements of transport numbers of polyions and their counterions yields the total number of bound counterions whatever the mechanism of site binding (Nagasawa et al., 1972).

Much more stable complexes are formed by polyanions with divalent cations. In this case, several of the functional groups attached to the polymeric chain may associate simultaneously with a given cation to form a chelate complex. Since the associating cation competes with hydrogen ions for the carboxylate groups, the presence of these cations will produce a shift in the titration curve of the polymeric acid. We shall see in Section VII.C.4 that the acidity of the carboxyls in a polycarboxylic acid decreases with increasing degree of ionization, thus complicating the interpretation of the pH shift produced by chelate formation. However, if we assume that the acidity of polyions carrying complexed cations is identical to that of the partially ionized uncomplexed polymeric acid with the same net charge density, the number of carboxylates engaged in chelate formation may be calculated from the titration data. This technique was used by Morawetz et al. (1954) to study the binding of alkaline earth ions by maleic acid copolymers, and Mandel and Leyte (1964) applied it to investigate the chelation of Cu^{2+}, Cd^{2+}, Zn^{2+}, Ni^{2+}, Co^{2+}, and Mg^{2+} by poly (methacrylic acid). Typical examples of their results are shown in Fig. VII.10. In all these cases, two carboxylates of the polyanion are involved in chelate formation. The nature of the complex depends only on the degree of ionization of the polymeric acid, and not on its concentration, as may be understood from the following consideration. The coordination of the first carboxylate to the cation depends on the stoichiometric concentration of carboxylate groups; however, once the cation is bound to a polyanion. its further interaction with a ligand group involves, with an overwhelming probability, a group carried by the same chain, as

†Cryoscopic measurements of the osmotic coefficients of alkali polystyrenesulfonates (Kozak et al., 1971) indicate that cation binding follows the order of decreasing size of the hydrated cation at concentrations above 0.056 *N*. At higher dilution Na^+ behaves anomalously, being bound even less than Li^+ and H^+.

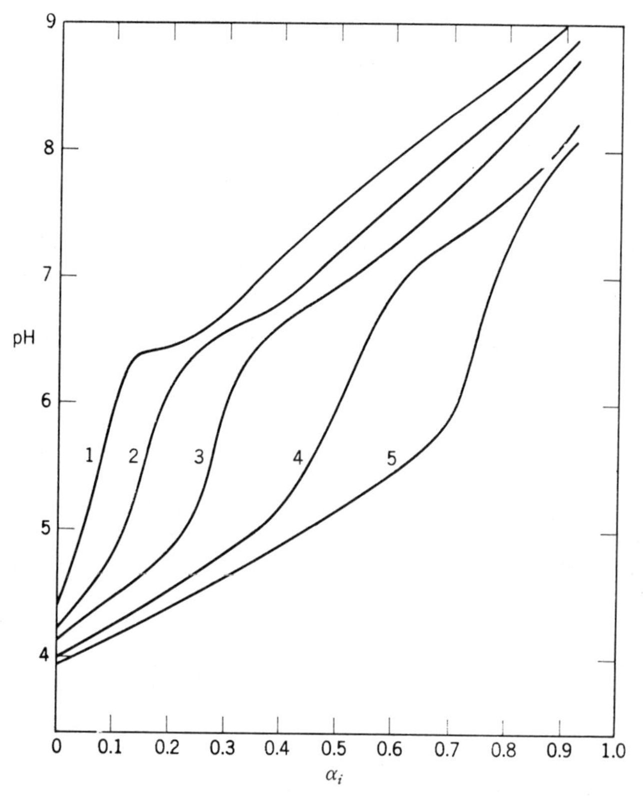

Fig. VII.10. Titration of 0.00173N poly(methacrylic acid) with NaOH in the presence of the following molarities of $Cu(NO_3)_2$: (1) 0; (2) 8.28×10^{-5}; (3) 2.07×10^{-4}; (4) 4.14×10^{-4}; (5) 6.21×10^{-4}.

long as the polymer solution is reasonably dilute. The consequences of this high local ligand concentration were studied by Morawetz and Sammak (1957) on poly(ε-methacrylyl-L-lysine):

$$
\begin{array}{l}
\quad\ CH_3 \\
\quad\ \ | \\
(\!-\!C\!-\!CH_2\!-\!)_n \\
\quad\ \ | \\
\quad\ CO \qquad\qquad COO^- \\
\quad\ \ | \qquad\qquad\qquad | \\
\quad\ NH(CH_2)_4\!-\!CH \\
\qquad\qquad\qquad\quad\ \ | \\
\qquad\qquad\qquad\quad NH_3^+
\end{array}
$$

Cupric ions may chelate with the α-aminocarboxylic acid groups at the end of the side chains, producing complexes with distinct spectra depending on whether one or two of these groups are coordinated to the cation. The chelate containing two amino acid groups is heavily favored under conditions where it would not form to any appreciable extent in a solution of a monomeric analog.

The high local concentration of chelated metal ions within the swollen polymer coil may also lead to some interesting effects. Leyte et al. (1968) studied the binding of Cu^{2+} to poly(methacrylic acid) by UV and IR spectroscopy and by magnetic moment measurements. Their data are consistent with the formation of a diamagnetic species in which two cupric dicarboxylate chelates are joined by a Cu-Cu covalent bond as long as the expansion of the polyanion is relatively low.

Site binding by ion-pair formation is demonstrable by spectroscopy only in some special cases. The absorbance of $Co(NH_3)_6^{3+}$ in the region of its charge transfer band (235.8 nm) is intensified when it forms an ion pair, and this has been used to study the binding of this ion to polyacrylate (Eldridge and Treloar, 1970). It should be emphasized that this binding does not involve displacement of the coordinated ammonia molecules by carboxylate groups.

Specific binding of anions to a cationic polymer is observed with partially ionized polyvinylpyridine, where the polymer acquires a negative charge in the presence of high bromide concentrations (Strauss et al., 1954). This phenomenon is due to the formation of a charge transfer complex of the bromide with the pyridine nucleus, which may be observed by a characteristic shift in the UV spectrum (Slough, 1959).

C. THERMODYNAMIC PROPERTIES

1. Ionic Activity Coefficients

The powerful electrostatic forces between polyions and their counterions are clearly revealed in the unusually low ionic activity coefficients characterizing polyelectrolyte solutions. However, none of the methods by which these activity coefficients are determined can provide a quantitative distinction between counterion inactivation by long-range Coulombic interactions and by site binding.

The activity of a solute species is defined in terms of its partial molar free energy. If the solute is an electrolyte, it is impossible to vary the concentration of only one of its ions and a single ion activity cannot be defined unambiguously in terms of an experimentally measurable quantity. Nevertheless, extensive data on the apparent byion and counterion activities of polyelectrolytes have been collected using concentration cells without transference (Nagasawa et al., 1959a), but the results should be accepted with some reserve, since the measured emf contains an unknown (and possibly important) contribution from a liquid junction potential. The results are qualitatively interesting in that they demonstrate, as would be expected, that polyions have a large effect in reducing the activities of counterions, but are without major influence on the activities of byions.

On the other hand, the activity of an electrolyte related to the partial molar free energy of the electroneutral species consisting of cations and anions is a perfectly well-defined quantity. However, with salt-free polyelectrolytes a conceptual difficulty arises in its interpretation, since we cannot take as a frame of reference a highly dilute solution in which all interactions of charged species are eliminated, so that ideal solution behavior is attained. As shown in Fig. VIII.9, a considerable fraction of the counterions remains closely associated with the polyion even at high dilution, and we also have to take account of the counterion condensation on polyions with very high charge densities. Therefore, the ionic activity coefficients lie far below unity even at the highest dilution which allows meaningful measurements.

Theoretical and experimental aspects of ionic activities in polyelectrolyte solution have been reviewed by Katchalsky et al. (1966), Katchalsky (1971), and Ise (1971). The experimental methods include (a) the measurement of emf in concentration cells with transference, (b) the determination of dialysis equilibria, and (c) the measurement of colligative properties.

The first method was used, for instance, by Ise and Okubo (1965) to study sodium polyacrylate (NaPAA) in the absence of added salt. The concentration cell was of the type

Na^+-selective glass electrode	NaPAA $(a_\pm)_I$	NaPAA $(a_\pm)_{II}$	Na^+-selective glass electrode

If the transference number of the polyion is t_p, the passage of 1 faraday ($\mathfrak{F}$) of electricity will lead to the transfer of (t_p/ν_p) moles of polyions and their t_p moles of counterions from the solution with mean ionic activity $(a_\pm)_{II}$ to one with mean ionic activity $(a_\pm)_I$. The emf of the cell will then be

$$\mathscr{E} = \left(1 + \frac{1}{\nu_p}\right)\frac{RT}{\mathfrak{F}}\int_{(a\pm)_{II}}^{(a\pm)_I} t_p d\ln a_\pm \qquad \text{(VII.20)}$$

The interpretation of dialysis equilibria is based on the following considerations:

When a solution containing a polyelectrolyte and a simple salt is separated from a compartment filled with the pure solvent by a membrane which will allow the small ions to diffuse but will not permit passage of the polyions, the distribution of the salt between the two portions of the system must satisfy two conditions (Donnan, 1934):

(a) The solutions on either side of the membrane must be electroneutral. Thus, if m_f is the molarity of the fixed charges in the solution containing the polyelectrolyte and m_s is the molarity of an added uni-univalent salt, the molarity of byions will be $m_b = m_s$ and that of counterions will be $m_c = m_f + m_s$. In the dialyzate both ions will be at the same molarity, $m'_b = m'_c = m'_s$.

(b) At equilibrium, the diffusible electrolyte must have the same thermodynamic activity on both sides of the dialysis membrane,

$$m_b m_c \gamma_\pm^2 = m'_b m'_c (\gamma'_\pm)^2$$
$$(m_f + m_s) m_s \gamma_\pm^2 = (m'_s)^2 (\gamma'_\pm)^2 \tag{VII.21}$$

where $\gamma_\pm$ and $\gamma'_\pm$ are the mean ionic activity coefficients in the presence and in the absence of polyelectrolyte, respectively.

Let us first consider the result which would be obtained if the activity coefficients could be neglected. In that case (VII.21) gives for a relation between m_s and m'_s

$$m_s = -\frac{m_f}{2} + \sqrt{(m_f/2)^2 + (m'_s)^2} \tag{VII.22}$$

If $m'_s/m_f \ll 1$, we obtain $m_s = m'_s(m'_s/m_f)$, that is, the added salt is largely excluded from the polyelectrolyte solution. In the other extreme case, when $m'_s/m_f \gg 1$, an expansion of (VII.22) yields

$$m_s = m'_s - \frac{m_f}{2} + \frac{m_f^2}{8m'_s} - \cdots \tag{VII.23}$$

so that the difference in the total molarity of mobile ions, Δm_i, on the two sides of the membrane is

$$\Delta m_i \equiv m_c + m_b - m'_c - m'_b = \frac{m_f^2}{4m'_s} - \cdots \tag{VII.24}$$

(VII.24) shows that Δm_i is minimized by increasing the concentration of added salt.

As pointed out above, the large forces between polyions and counterions will inevitably depress the ionic activity coefficients, so that $\gamma_\pm < \gamma'_\pm$. If the ionic concentrations are determined analytically, (VII.21) may be used to calculate $\gamma_\pm$, since $\gamma'_\pm$, corresponding to a simple salt solution, can be determined by conventional methods or obtained from tabulated data. Outstanding studies using this method are those of Strauss and Ross (1959),

Nagasawa et al. (1959b), and Alexandrowicz (1960, 1962).† Although dialysis equilibrium cannot be used to determine the activity of a salt-free polyelectrolyte, the dilution of the added salt, whose activity is being measured, is limited only by the sensitivity of our analytical techniques and our confidence that none of the polyions can pass through the membrane. If the salt concentration is very small compared to that of the polyelectrolyte, $\gamma_\pm$ may be related to the distribution of the electrostatic potential ψ in the salt-free polyelectrolyte (Marcus, 1955). For instance, with uni-univalent electrolyte, the mean ionic activity coefficient $\gamma_\pm$ is given by

$$\gamma_\pm^2 = \frac{V^2}{\int \exp(-e\psi/\boldsymbol{k}T)\, dV \int \exp(e\psi/\boldsymbol{k}T)\, dV} \tag{VII.25}$$

which allows a comparison of experimentally determined activity coefficients with predictions based on theoretical models. Here it is reasonable to treat even flexible polyions as being *locally* rodlike, so that ψ is cylindrically symmetrical.

In using colligative properties for the study of activities in polyelectrolyte solutions, freezing point depression (Mock, 1960 Kozak et al., 1971) and isopiestic measurements (Ise and Okubo, 1967, 1968) are particularly useful for characterizing solutions containing no simple electrolyte, while osmometry has been used for the study of ternary systems containing both a polyelectrolyte and a simple electrolyte solute (Pals and Hermans, 1952; Alexandrowicz, 1962). If the solution were ideal, the freezing point depression and vapor pressure depression would depend only on the number of particles in solution that is, $\nu_p + 1$ particles for each polyion and its counterions. Because of the interactions of the charged particles, the magnitude of these colligative properties is reduced by a factor ϕ, the "osmotic coefficient" below its ideal value. The mean ionic activity coefficient at a molarity m is then related (Harned and Owen, 1958) to the osmotic coefficient by

$$\ln \gamma_\pm = (\phi - 1) + \int_0^m (\phi - 1)\, d \ln m \tag{VII.26}$$

In osmotic pressure measurements involving solutions which contain both a polyelectrolyte and a simple salt, the ideal solution law (IV.13) has to be modified to take account of the unequal distribution of the mobile ions as given in (VII.24), so that

†In the case of dialysis equilibrium measurements with globular proteins many authors have interpreted a deviation of $\gamma_\pm$ from $\gamma_\pm'$ in terms of ion binding by the macromolecule. As long as it is clearly understood that "binding" is defined operationally in this manner, there is no objection to this practice. Since the charge density on the proteins is usually not too high, site binding probably accounts for most of the deviation of $\gamma_\pm/\gamma_\pm'$ from unity.

$$\Pi_{\text{ideal}} = \boldsymbol{RT}\left(\frac{c_2}{M_2} + \frac{m_f^2}{4000m_s'} + \cdots\right) \tag{VII.27}$$

while the real osmotic pressure, reflecting the osmotic coefficient, would become

$$\Pi = \boldsymbol{RT}\left(\frac{c_2}{M_2} + \frac{m_f^2\phi^2}{4000m_s'} + \cdots\right) \tag{VII.28}$$

if effects due to the mutual interaction of the polyions could be neglected. Comparing (VII.28) with the definition of the second virial coefficient in (IV.14) yields then, under these conditions,

$$\mathrm{A}_2 = \frac{1000(\nu_p\phi/M_2)^2}{4m_s'} \tag{VII.29}$$

One of the most interesting results for polyelectrolytes in salt-free solution concerns the behavior of the osmotic coefficient as the charge density of the polyion is being increased (Alexandrowicz and Katchalsky, 1963): the product of the osmotic coefficient and the charge density along the polyion becomes constant for high charge densities, so that the polyelectrolyte behaves as an "osmotic buffer." This is consistent with the concept of "counterion condensation" of Imai and Onishi (1959) and Manning (1969), which we have previously discussed, but we have no way of proving that site binding is solely responsible for the observed behavior.‡

The electrochemical properties of polyelectrolyte solutions are generally insensitive to the chain length of the polyion once it carries more than about 20 charges. This important fact demonstrates the weakness of all theoretical models in which the fixed charges carried by the polyion chain are smeared out to a spherically symmetrical charge cloud. Since the volume occupied by the polyion coil increases more rapidly than the chain length, the smeared-out fixed charge density will decline. A much more realistic model represents the polyion domain as a sausage-shaped region enclosing the coiled macromolecule, as shown schematically in Fig. VII.2. If we think then of the polyion region as separated by a membrane from the rest of the solution, the byions will tend to be excluded from this region because of the high local concentration of fixed charges. The fraction of the volume from which the salt is excluded in the presence of polyelectrolyte will then reduce its average concentration and thus determine the dialysis equilibrium with a salt solution containing no polyions.

Theoretical considerations of the behavior of dilute systems have led to

‡In equilibrium ultracentrifugation, $d\ln c_2/dr^2$ is inversely proportional to $d\Pi/dc_2$; thus, $d\ln c_2/dr^2$ in salt-free polyelectrolyte solutions is too small to be detected. Eisenberg (1962) has treated ultracentrifuge equilibria of polyelectrolyte solutions with added salts.

the conclusion that the addition of simple salts has relatively little effect on the counterion atmosphere in the immediate vicinity of a highly charged polymeric chain. Thus, it seems reasonable that the activities of ions in a system containing both polyelectrolytes and simple salts should be the sums of activities of the ionic species to be expected in a salt-free polyelectrolyte and a polyelectrolyte-free salt solution, respectively. This simple superposition principle was verified experimentally by pH measurements on solutions containing polysulfonic acids and HCl (Mock and Marshall, 1954) and by dialysis equilibrium studies (Nagasawa et al., 1959a; Alexandrowicz, 1960, 1962). Table VII.2 lists the values given by Alexandrowicz

TABLE VII.2
Activity Coefficients of Sodium Counterions in Partially Ionized Poly(acrylic acid)

Polymer Concentration	Activity Coefficient				
(g/l)	$\alpha_i = 0.05$	$\alpha_i = 0.1$	$\alpha_i = 0.3$	$\alpha_i = 0.5$	$\alpha_i = 0.8$
36	0.68	0.56	0.41	0.37	0.26
7.2	0.79	0.54	0.41	0.28	0.17
1.44	0.73	0.58	0.35	0.21	0.12

(1959) for the apparent activity coefficients of sodium ions in poly(sodium acrylate) as a function of the polymer concentration and its degree of ionization. However, experimental data obtained by Strauss and Ander (1958) with polyphosphate solutions show an increasing contribution of the polyelectrolyte to the ionic activity at higher salt concentrations. Boyd (1974) has carried out a detailed study of osmotic coefficients in solutions containing sodium polystyrenesulfonate and NaCl and he has found that the ionic activities are generally larger than predicted from simple additivity. The discrepancy increases with the solute concentration and is at a maximum if polyelectrolyte and simple electrolyte are present at equivalent concentrations.

Typical data on osmotic pressures of a polyelectrolyte in solutions containing varying concentrations of a simple salt are those of Pals and Hermans (1952) for sodium pectinate, plotted in Fig. VII.11. These data agree with the linear dependence of A_2 on $1/m_s'$ predicted by (VII.29) and yield an osmotic coefficient of about 0.16. Takahashi et al. (1967) carried out an extensive light scattering study of the second virial coefficient in polystyrenesulfonate solutions and found A_2 to be only slightly molecular weight dependent over a fourfold range of M_2. On the other hand, the A_2 values obtained at high concentrations of added salt would correspond, according to (VII.29) to $\phi > 1$, so that mutual interactions between the polyions obviously made a significant contribution to the second virial coefficient under these conditions. An osmometric study of this system (Takahashi et al., 1970) shows that, at high values of m_s', A_2 is linear in $1/\sqrt{m_s'}$ rather than $1/m_s'$.

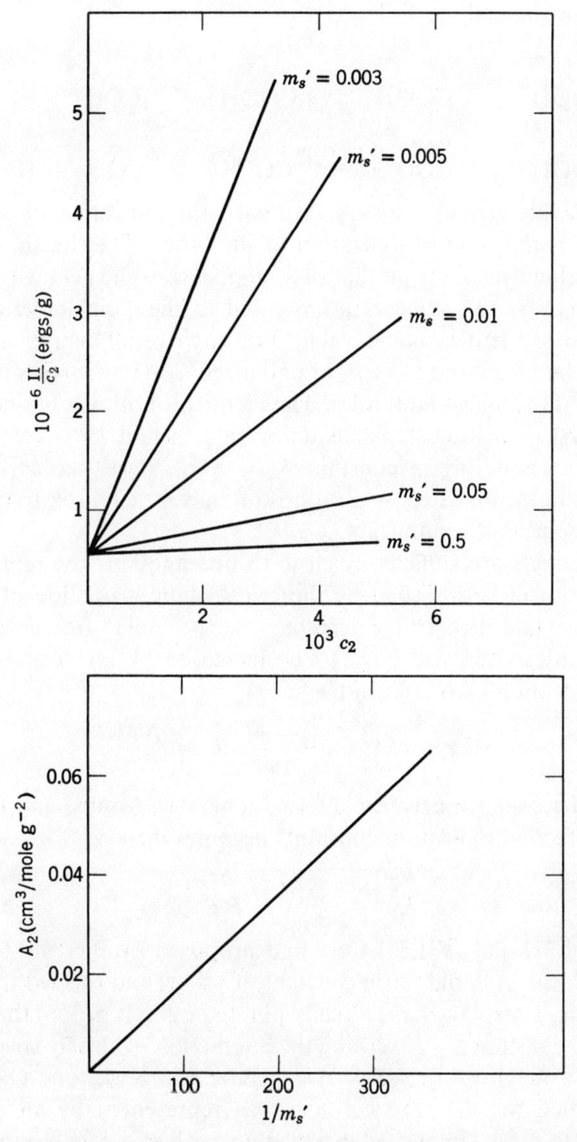

Fig. VII.11. Osmotic behavior of sodium pectinate at various concentrations of added sodium chloride (m_s').

2. Ionization Equilibria

Before considering the ionization equilibria of polymeric species with a large number of ionizable groups, let us deal briefly with the titration of materials such as dicarboxylic acids. Their ionization equilibria may be represented schematically as follows:

$$\overbrace{\text{COOH} \qquad \text{COOH}} \overset{K_1}{\rightleftharpoons} \overbrace{\text{COOH} \qquad \text{COO}^-} + \text{H}^+ \quad \text{(VII.30a)}$$

$$\overbrace{\text{COOH} \qquad \text{COO}^-} \overset{K_2}{\rightleftharpoons} \overbrace{\text{COO}^- \qquad \text{COO}^-} + \text{H}^+ \quad \text{(VII.30b)}$$

If the ionizable groups are very far apart, the ionization of each will be independent of the state of ionization of the other. The titration curve will then be indistinguishable from that of a monocarboxylic acid with an ionization constant K° and a concentration equal to the stoichiometric carboxyl concentration of the dicarboxylic acid. For any overall degree of ionization α_i, there will be a fraction $(1 - \alpha_i)^2$ of unionized, $2\alpha_i(1 - \alpha_i)$ of singly ionized and α_i^2 of doubly ionized molecules. The identical result is obtained if we assign to the first ionization step an equilibrium constant $K_1 = 2K^\circ$ and to the second step an equilibrium constant $K_2 = K^\circ/2$, since two carboxyls may lose a proton in the first step, while a proton may attach itself to two alternative sites in the second equilibrium.

If the carboxyls are sufficiently close to one another, the removal of the second proton will be impeded by the electrostatic attraction of the neighboring carboxylate group. To overcome it, the molar free energy for the second ionization step will have to be increased by an "electrostatic free energy of ionization" ΔG_{el}^i, given by

$$\Delta G_{\text{el}}^i = N \int_{r=r^e}^{r=\infty} - \frac{e^2}{\mathscr{D} r^2}\, dr = \frac{Ne^2}{\mathscr{D} r_e} \qquad \text{(VII.31)}$$

where r_e is the distance between the two ionizable groups, and the ratio of the first and second ionization constants becomes then

$$\frac{K_2}{K_1} = \tfrac{1}{4} \exp\left(\frac{-\Delta G_{\text{eI}}^i}{RT}\right) \qquad \text{(VII.32)}$$

Relations (VI.31) and (VII.32) were first proposed by Bjerrum (1923), who used $\mathscr{D} = 80$, the bulk dielectric constant of water, and this led to unreasonably low values for r_e. We have already pointed out (cf. p. 356) that the effective dielectric constant $\mathscr{D}_E$ governing the interaction of closely spaced charges should have a much lower value. Westheimer and Kirkwood (1948) used a model in which the dicarboxylic acid was represented by an ellipsoid of revolution with $\mathscr{D} = 2$ immersed in a medium with $\mathscr{D} = 80$. The ionic charges were located at the foci of the ellipsoid, and the interfocal distance r_e was calculated which would yield a ΔG^i_{el} value consistent with titration data as interpreted by (VII.32). This procedure yielded for a number of dicarboxylic acids r_e values intermediate between that corresponding to the rms charge separation assuming free rotation and that corresponding to the molecular conformation in which the charges are most distant from each other (Westheimer and Shookhoff, 1939). With modern methods of conformational analysis a more detailed interpretation of K_2/K_1 should be possible (Morawetz, 1973). The sensitivity of K_2/K_1 to the conformational distribution has been strikingly illustrated by Eberson (1959) in studies of α,β-dialkylsuccinic acids. When the substituents R are very bulky, so that they have to lie *trans* to each other:

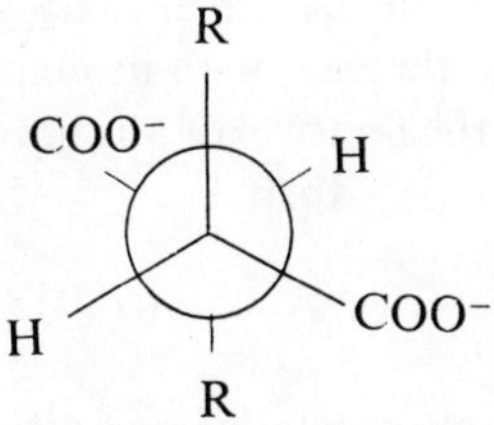

R, COO⁻, H, H, COO⁻, R

Meso form

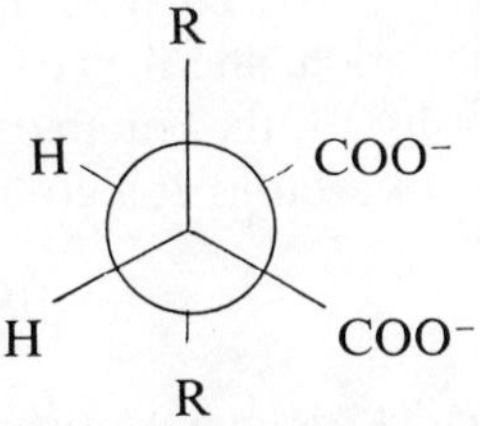

R, H, COO⁻, H, COO⁻, R

Racemic form

the charges of the doubly ionized *meso* form are much further from each other than those of the racemic form. As a result, for R = $-CH(CH_3)_2$ Eberson found K_2/K_1 to be smaller by a factor of 4×10^5 in the racemic acid.

The enthalpic and entropic contributions to ΔG^i_{el} for a number of dicarboxylic acids were determined by Christensen et al. (1967). If we can neglect effects due to a change in conformational distribution accompanying the ionization, $\Delta G^i_{el} = C/\mathscr{D}_{E'}$, where C is a constant. We have then

$$\Delta S^i_{el} = -\left(\frac{\partial\,\Delta G^i_{el}}{\partial T}\right)_P = \frac{C}{\mathscr{D}_E^2}\left(\frac{\partial \mathscr{D}_E}{\partial T}\right)_P = \Delta G^i_{el}\left(\frac{\partial \ln \mathscr{D}_E}{\partial T}\right)_P \quad \text{(VII.33)}$$

$$\Delta H^i_{el} = \Delta G^i_{el} - T\,\Delta S^i_{el} = \Delta G^i_{el}\left[1 + \left(\frac{\partial \ln \mathscr{D}_E}{\partial \ln T}\right)_P\right] \quad \text{(VII.34)}$$

For water at 25°C, $\partial \ln \mathscr{D}/\partial \ln T = -1.37$, so that $\Delta H^i_{el}/\Delta G^i_{el} = -0.37$ would be expected if the medium separating the ionic charges had the properties of water in bulk. The data of Christensen et al. lead mostly to more negative values, for example, -0.63 for fumaric, -0.96 for succinic, -0.70 for glutaric, and -0.63 for adipic acid. A major difference is observed with the doubly substituted β,β,-dimethylglutaric acid, for which $\Delta H^i_{el}/\Delta G^i_{el} = +0.2$. Analogous studies of the ionization of diamines (Schwarzenbach, 1970) yielded $\Delta H^i_{el}/\Delta G^i_{el}$ values of $+0.55$ and -0.13 for ethylene diamine and 1,3-diaminopropane, suggesting that the dielectric medium separating two cationic charges has properties different from those of the medium between two ionized carboxyls of unbranched dibasic acids.

With a polymer carrying a large number of ionizable groups, it is obviously impracticable to specify the successive ionization constants. Instead, we define the apparent ionization constant K_{app} of an average ionizable group carried by the polyion in the usual manner by

$$\frac{(\mathrm{H}^+)\alpha_i}{1-\alpha_i} = K_{app} \tag{VII.35}$$

where K_{app} will, of course, vary with the degree of ionization since the charged polymer will interact with the hydrogen ions. With polymeric acids the polyanion will attract the hydrogen ions and $dK_{app}/d\alpha_i < 0$; with polymeric bases, on the other hand, the hydrogen ions will be repelled by the polycation, and the acid strength of the polymer will increase with its charge density.† If the required electrostatic free energy for the removal of an equivalent of protons at a given degree of ionizations is $\Delta G^i_{el}(\alpha_i)$, then

$$\mathrm{p}K_{app} = \mathrm{p}K^\circ + \frac{0.43\Delta G^i_{el}(\alpha_i)}{RT} \tag{VII.36}$$

where K° is characteristic of the ionizing group when electrostatic interactions with other ionizing groups are absent. The value of $\Delta G^i_{el}(\alpha_i)$ is related to the excess electrostatic free energy, which we used previously in our discussion of the expansion of flexible polyions, by

$$\Delta G^i_{el}(\alpha_i) = \frac{N\,\partial G^E_{el}}{\partial \nu_p} \tag{VII.37}$$

It is customary to combine (VII.36) and (VII.37) and write the result in logarithmic form,:

$$\mathrm{pH} = \mathrm{p}K^\circ + \log\left(\frac{\alpha_i}{1-\alpha_i}\right) + \frac{0.43\,\Delta G^i_{el}}{RT} \tag{VII.38}$$

For spherical particles with a relatively low density of surface charge, ΔG^i_{el} may be estimated using for a model a sphere with radius R_e and a uniform surface charge. The application of the Debye-Hückel treatment to this model (Scatchard, 1949) gives

$$\Delta G^i_{el} = \frac{Ne\nu_p}{\mathscr{D}}\left(\frac{1}{R_e} - \frac{\kappa}{1+\kappa R_e}\right) \tag{VII.39}$$

This treatment is employed for globular proteins, and it is customary to use

† During the titration of poly(2-vinylpyridine) or poly(4-vinylpyridine) in salt-free 45% aqueous ethanol, $\mathrm{p}K_{app}$ first decreases, as expected, with an increasing degree of protonation, but the trend is reversed if this degree exceeds 0.3 (Kirsh *et al.*, 1973). The anomalous behavior of these systems is also reflected in the solution viscosity, which passes through a maximum at degrees of ionization of 0.2 and 0.1, respectively.

the bulk dielectric constant of water for $\mathscr{D}$. As for the specification of ν_p, the number of unit charges carried by the particle, this requires not only a knowledge of the number of cationic and anionic groups, but also information about the number of ions of simple electrolytes which may have associated with the protein molecule. The relation in (VII.38) applies then to every type of ionizable group, where K° is its intrinsic ionization constant and α_i the fraction of that type of group which is present in the form of its conjugate base. For a typical case of a particle with an equivalent radius $R_e = 2 \times 10^{-7}$ cm, we obtain for ΔG^i_{el} at an ionic strength of 0.1 a value of 40 cal/mole per electronic charge carried by the macromolecule. However, at high charge densities, when ΔG^i_{el} is no longer small compared to $\boldsymbol{RT}$, the Debye-Hückel treatment which leads to (VII.39) is not justified and ΔG^i_{el} can be obtained only by numerical methods with the aid of an electronic computer (Nagasawa and Holtzer, 1964a).

For quantitative interpretation of the titration data obtained with globular proteins, appropriate K° values corresponding to the various titratable groups must first be chosen. According to Tanford and Hauenstein (1956), data obtained with low molecular weight analogs would lead one to expect a $\text{p}K^\circ = 3.75$ for the α-COOH group at the polypeptide chain end, $\text{p}K^\circ = 4.6$ for side-chain COOH, $\text{p}K^\circ = 7.0$ for imidazole, $\text{p}K^\circ = 7.8$ for the α-amino group, $\text{p}K^\circ = 9.6$ for the phenolic groups of tyrosine, and $\text{p}K^\circ = 10.2$ for side-chain amino groups. Guanidyl groups of arginine residues would be expected to be titrated above pH 12, where irreversible denaturation usually occurs, so that a quantitative interpretation of the data is no longer possible. It is often found that the $\text{p}K^\circ$ calculated from protein titration data deviates significantly from these expected values. There may be several reasons for such a discrepancy. The smearing-out of the protein charge over the surface of the globular molecule obscures possible effects of spatial proximity of a given ionizable group to another positively or negatively charged function. Also, the effective local dielectric constant in the neighborhood of an ionizable group may differ significantly from the bulk value for water. Finally, conformational transitions (reversible or irreversible) of the protein molecule with a change of its net charge would be expected to influence ionization equilibria. In favorable cases, the ionization of specific residues can be followed. This has been accomplished by NMR spectroscopy for the imidazole groups of the four histidine residues of ribonculease (King and Bradbury, 1961).

With synthetic polyelectrolytes the situation may at first appear simpler, since the polyion usually carries only one type of ionizable groups and we do not have to take account of the consequences of a specific macromolecular conformation. However, these advantages are far outweighed by the complexities introduced by increasing chain expansion and ion-pair formation as the charge density of the polyion is increased. Mandel (1970) has studied carefully the titration behavior of poly(acrylic acid), and his results, expressed as the dependence of pK_{app} on α_i, are shown in Fig. VII.12. The curves are closely fitted by $pK_{app} = pK^\circ + b_1\alpha_i + b_2\alpha_i^2$, and the dependence of the three parameters on the concentration of added $NaNO_3$ is shown in

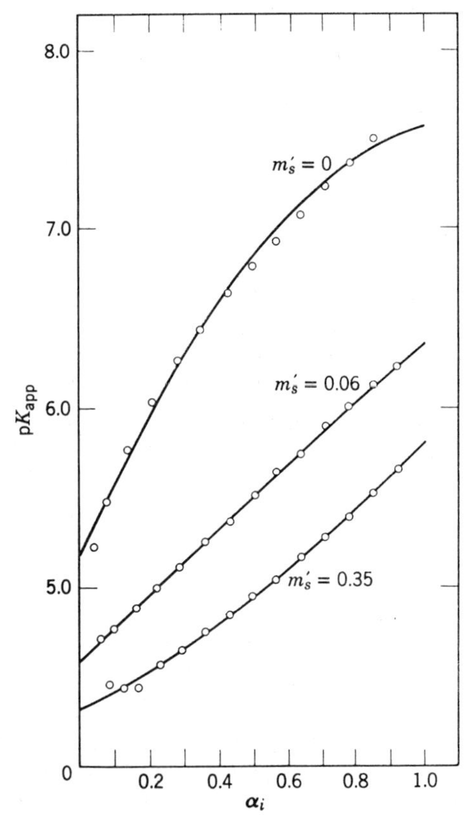

Fig. VII.12. Dependence of pK_{app} of poly(acrylic acid) (M.W. = 7.9×10^5; conc. $4.7 \times 10^{-3}N$) on the degree of ionization α_i.

Fig. VII.13. The change of pK° with the concentration of added salt is similar to that observed with monocarboxylic acids. The gradual reduction of b_1, the initial value of $dpK_{app}/d\alpha_i$, with increasing salt concentration, reflects the shielding of the polyion charge by the counterions. An interesting feature of Mandel's data is the gradual change of b_2 from a negative to a positive value as simple electrolyte is added to the solution. Mandel also claims small but significant differences in the titration of poly(acrylic acid) samples with molecular weights of 1.2×10^5 and 7.9×10^5, a result which seems surprising, since $\Delta G_{el}^i(\alpha_i)$ would be expected to be determined by interactions between ionized groups which are close to each other compared

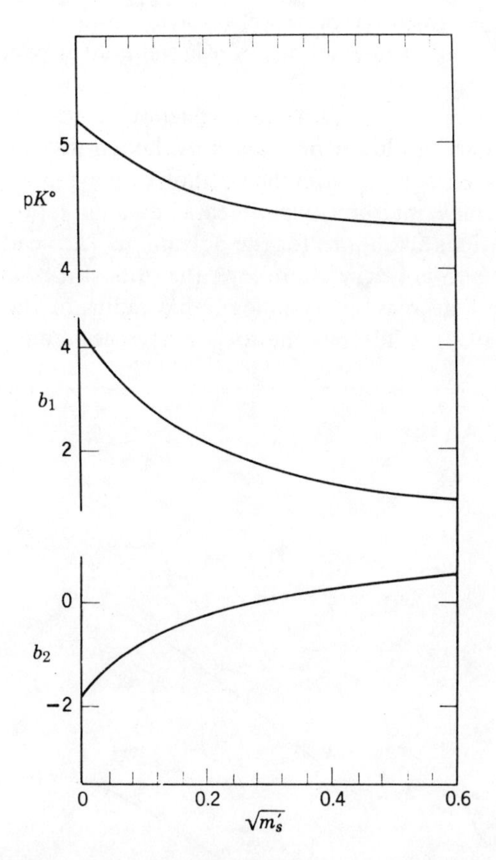

Fig. VII.13. Dependence of the parameters $pK°$, b_1, and b_2 in the titration of poly-(acrylic acid) on the molarity of added $NaNO_3$.

to the overall dimension of the polyion. Finally, it is instructive to compare the range of ΔG_{el}^i observed in poly(acrylic acid) with the electrostatic free energy of ionization of a dicarboxylic acid analog. Glutaric acid is characterized by $\Delta G_{el}^i = 0.66$ kcal/mole. As we approach full ionization of poly-(acrylic acid), a hydrogen ion removed from a carboxyl interacts with two neighboring carboxylates, and the contribution to ΔG_{el}^i from interactions with the two nearest neighbors may then be estimated as 1.3 kcal/mole. Mandel's data for the fully ionized polymeric acid yield ΔG_{el}^i values of 3.5 and 2.1 kcal/mole in the absence of salt and in 0.33M $NaNO_3$, respectively. This suggests that only a minor contribution to ΔG_{el}^i is due to nearest-neighbor interactions in the absence of salt and that, even in the presence of considerable electrolyte, interactions with more distant charges remain significant. Also, model compounds with only two ionizing groups show only a weak dependence of ΔpK on electrolyte concentration (Fabrizzi et al.,

1973), so that the sensitivity of polyelectrolyte titration curves to the ionic atmosphere must reflect interactions of a large number of the fixed charges with the counterions.

Any counterion binding will lead to a reduction of ΔG^i_{el} and thus to a shift of the titration curve to lower pH values. Although we cannot evaluate the extent of counterion binding from the titration of polymeric acids with alkali hydroxides, we may interpret experimental data in terms of the *relative* affinity of the various alkali ions for the polyanion. This leads to the conclusion that binding to polyacrylate follows the order $K^+ < Na^+ < Li^+$, that is, it increases with a decreasing crystallographic radius of the cation (Gregor and Frederick, 1957), while binding to poly(styrenesulfonic acid) increases

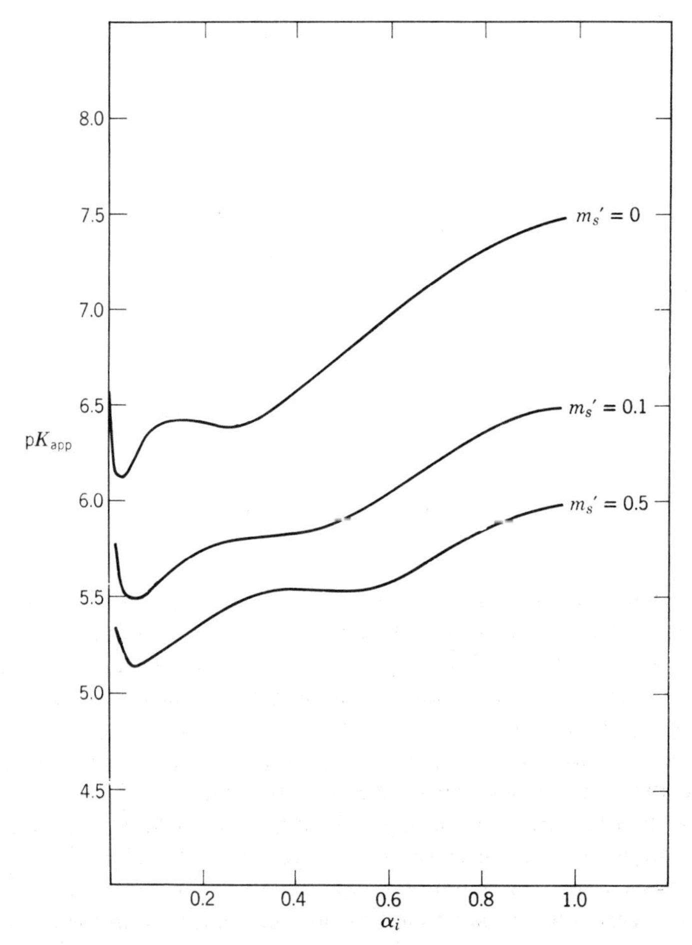

Fig. VII.14. The dependence of the apparent ionization constant of poly(methacrylic acid) on the degree of ionization and the concentration of added KNO_3.

in the reverse order, being favored by a small radius of the hydrated cation (Lapanje, 1962).†

With poly(methacrylic acid) the dependence of pK_{app} on the degree of ionization presents a much more complex pattern. Data of Kotliar and Morawetz (1955), plotted in Fig. VII.14, show that two regions characterized by a pK_{app} rising with an increasing charge density of the polyion are connected by a range within which pK_{app} is independent of α_i. This phenomenon is related to the conformational transition in poly(methacrylic acid), which we discussed previously (cf. p. 363). In this context, a comparison of the heat effects accompanying the ionizations of poly(acrylic acid) and of poly(methacrylic acid) is particularly valuable. Data of Crescenzi et al. (1972), plotted in Fig. VII, 15, show that ΔH_{diss} characterizing the ionization of a carboxyl group in poly(acrylic acid) tends to become more negative with increasing α_i, as would be expected from the behavior of dicarboxylic acids [cf. (VII.34)]. With poly(methacrylic acid), a sharp endotherm is superimposed on the decreasing ΔH_{diss} in the range of the conformational transition. Since ΔG_{diss} remains constant in this range, ΔS_{diss} must be characterized by

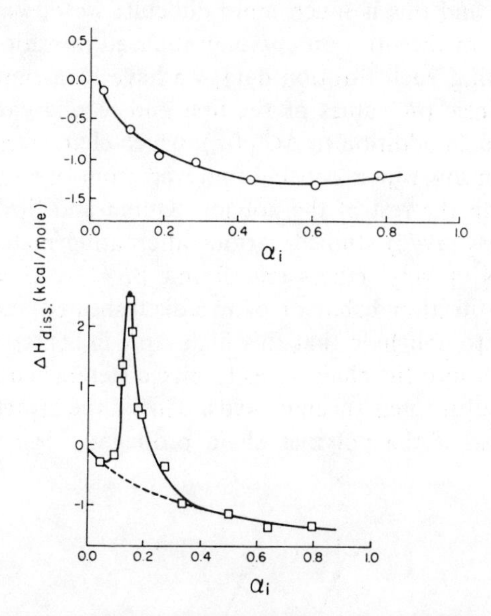

Fig. VII.15. Dependence of the enthalpy of dissociation of polycarboxylic acids on the degree of ionization α_i in the absence of added salts. (○) 0.065N poly(acrylic acid), (□) poly(methacrylic acid).

†A potentiometric and calorimetric study of the titration of carboxymethylcellulose with different bases (Rinaudo and Milas, 1973) allowed a determination of the thermodynamic functions characterizing an exchange of counterions. For instance, replacing K^+ by Li^+ led to $\triangle H = 845 \pm 15$ cal/mole, $\triangle S^\circ = +3$ cal/mole-deg.

a large peak. This is consistent with the spectroscopic observations of Koenig et al. (1969) that the transition of the polyion leads to a broader distribution of accessible conformations.

With both poly(acrylic acid) and poly(methacrylic acid), titration curves have been found to depend to some extent on the stereoregularity of the polymer (Barone et al., 1965; Nagasawa et al., 1965; Kawaguchi and Nagasawa, 1969). For both polymeric acids, the isotactic species is characterized by higher pK_{app} values.

So far, we have tacitly assumed that the ionizable groups are evenly spaced along the backbone of the polymer molecule. As stated earlier, it is also possible to prepare copolymers of maleic anhydride that after hydrolysis yield chain molecules, in which pairs of closely spaced carboxyl groups are separated from one another by the comonomer unit. Typical examples are the maleic acid copolymers with styrene or methyl vinyl ether shown in Table VII.1. When such polymeric acids are titrated, one carboxyl of each pair ionizes relatively easily. However, if the neutralization is carried beyond 50%, hydrogen ions have to be removed from a carboxyl lying very close to an anionic group, and this is much more difficult; we observe, therefore, a pronounced break in the titration curve at half-neutralization (Ferry et al., 1951). In interpreting such titration data, we have to assign values to pK_1° and pK_2°, the intrinsic pK values of the first and second carboxyls in each maleic acid residue, in addition to $\Delta G_{el}^i(\alpha_i)$, which characterizes the interaction of a hydrogen ion which is being removed from one of these residues with the charges on the rest of the polyion. Dubin and Strauss (1970) and Schultz and Strauss (1972) studied various alternating maleic acid copolymers with a series of vinyl ethers and found $pK_2^\circ - pK_1^\circ \approx 3$. From our discussion of the titration behavior of α,β-disubstituted succinic acids (p. 377) it seems safe to conclude that this high ΔpK indicates that the maleic acid residue is built into the chain so as to give a stereochemistry analogous to the racemic disubstituted succinic acids. This is the structure which will result on hydrolysis if the polymer chain propagates *trans* to the maleic anhydride ring;

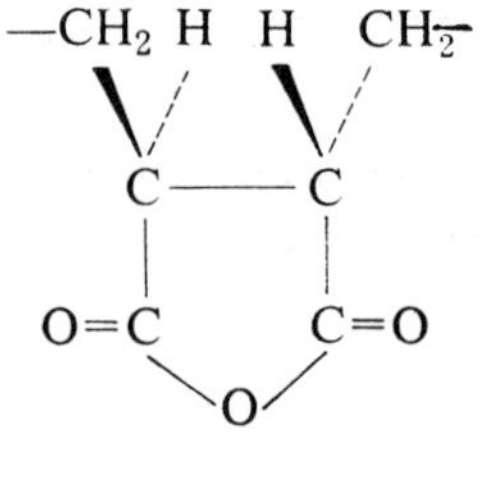

```
—CH2 H   H   CH2—
    \ /     \ /
     C ——— C
     |       |
   O=C       C=O
      \     /
         O
```

a mode of propagation which should be favored on steric grounds and which is also consistent with the spectroscopic behavior of a maleic anhydride copolymer (Bacskai et al., 1972). Dubin and Strauss have also shown that in the case of maleic acid copolymers with long hydrophobic side chains the titration curve reflects a transition from a compact to an expanded structure, similar to that observed with poly(methacrylic acid).

If should be pointed out, however, that a sharp rise of pK_{app} at half-ionization may also occur with a polymer chain carrying *evenly spaced* ionizable substituents. Such behavior would be expected if the ionizable groups are spaced so close to each other that the interaction energy of two ionized nearest-neighbor groups is large compared to $\boldsymbol{k}T$ (Marcus, 1954). In that case, alternate groups will tend to ionize first. Only when $\alpha_i > \frac{1}{2}$ will the intermediate groups be forced to ionize, and this will require a sharp increase in pK_{app}. Barone and Rizzo (1973) have observed behavior of this kind with poly(maleic acid), in which every carbon of the chain backbone carries a carboxyl substituent.

In polypeptides carrying ionizable side chains, an increasing density of ionic charges destabilizes the helical form and leads to an abrupt transition to the random coil. A theoretical analysis of the titration curves to be expected in such a case was carried out by Zimm and Rice (1960). Figure VII. 14*a* shows experimental data obtained by Nagasawa and Holtzer (1964b) for poly(α-L-glutamic acid) with various concentrations of added salt. Region I, in which pK_{app} first decreases with increasing α_i, is believed to reflect the molecular association of the polymer at low charge densities, which may be demonstrated by a variety of techniques (Jennings et al., 1968). In region II, the helical form is stable, region III represents the helix-coil transition, and in region IV the polyion exists as a random coil. The analysis proposed by Nagasawa and Holtzer assumes that the transition region may be treated as representing a mixture of chains which are either fully helical or in which the helix has been completely obliterated. This is admittedly inexact, but the error introduced by this assumption into the interpretation of the data is negligible. Nagasawa and Holtzer extend the curves in regions II and IV to estimate the appearance of the titration curves *h* and *c*, which would be obtained if the polymer could be kept fully helical or as a random coil over the entire range of α_i. This device allows them to estimate the fraction f_h of the polypeptide residues in the helical form at any given pH, since the experimentally observed $\bar{\alpha}_i$ must be the weighted average of α_i values corresponding to *h* and *c*, respectively. In addition, it can be shown that the standard free-energy change ΔG°_{h-c} corresponding to a helix-coil transition per amino acid residue *with both forms in the uncharged state* can be estimated from

$$\Delta G^{\circ}_{h\to c} = -\boldsymbol{RT} \ln\left(\frac{1 - f_h}{f_h}\right)_{\bar{\alpha}_i} - 2.3\boldsymbol{RT}\int_{\bar{\alpha}_i=0}^{\bar{\alpha}_i} [(\alpha_i)_h - (\alpha_i)_c]\, d\text{pH}$$

$$f_h = \frac{(\bar{\alpha}_i) - (\alpha_i)_c}{(\alpha_i)_h - (\alpha_i)_c} \tag{VII.40}$$

where $(\alpha_i)_h$, $(\alpha_i)_c$, and $\bar{\alpha}_i$ have the significance illustrated in Fig. VII.16*b*.

In poly(α-L-glutamic acid) and other synthetic polypeptides which undergo

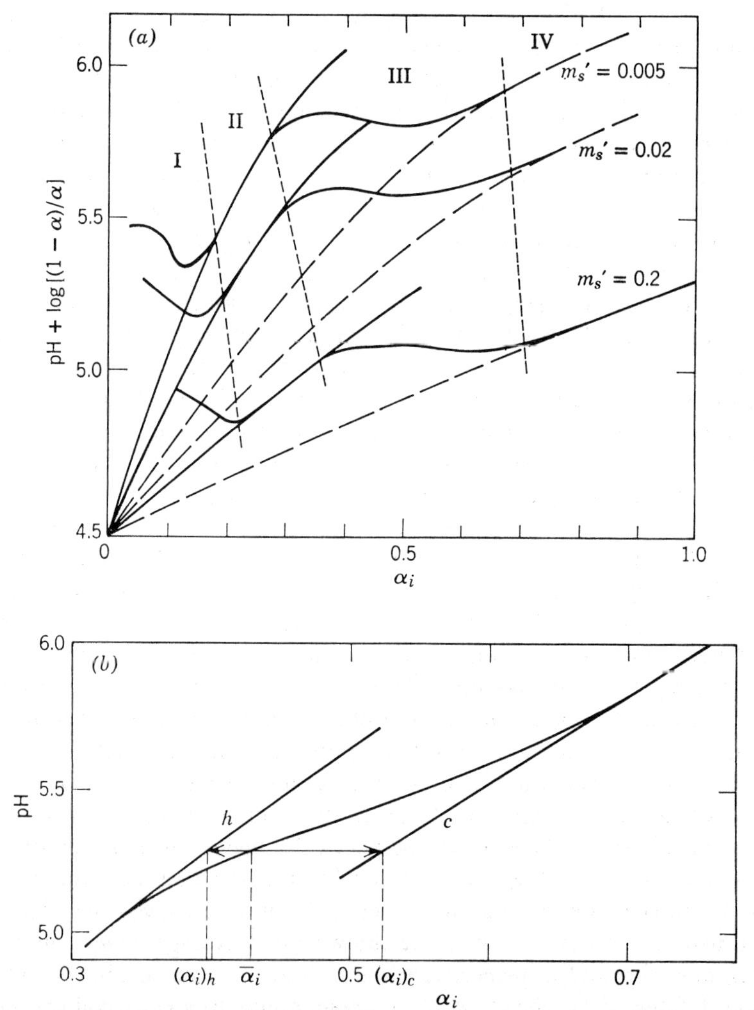

Fig. VII.16. Interpretation of titration data on poly(α-L-glutamic acid). (A) Regions reflecting I, polymer association; II, titration of the helical form; III, the helix-coil transition; and IV, titration of the random coil form at various concentrations, m_s', of added NaCl. (B) Graphical method for estimating f_h.

helix-coil transitions during titration, this change is fully reversible. This is not the case, however, with DNA, where the double-helical form is denatured irreversibly at low or high pH. As a result, the titration curve of native DNA is also irreversible (Cox and Peacocke, 1957).

D. TRANSPORT PROPERTIES

1. Electrophoresis

If an ionized macromolecule carrying an electrical charge Q is suspended in a viscous medium with an electrical potential gradient X, the particle will accelerate until the electrical force is equal to the frictional resistance. Since the system as a whole is electroneutral, the charge of the macromolecule must be counterbalanced by a similar charge of opposite sign carried by the counterions distributed in the surrounding medium. If the macromolecule has the form of a spherical particle with an effective radius R_e and the charge density on its surface is not too high, the theory of Debye and Hückel (1923) may be used to describe the distribution of counterions, which is then best characterized by the "mean thickness of the ion atmosphere" $1/\kappa$, defined by (VII.4). For macromolecular solutions containing little added salt, the thickness of the counterion atmosphere may be much larger than the radius of the colloidal particle, and in that case the electrical driving force will be given simply by

$$\mathrm{F} = \tfrac{1}{300}\, QX \qquad (\kappa R_e \ll 1) \qquad \text{(VII.41)}$$

where the numerical factor is required to obtain F in dynes when Q is in electrostatic units and X in volts per centimeter. Equating this driving force to the viscous drag as given by Stokes's law [(VI.2)], we obtain for the steady-state velocity

$$u = \frac{\tfrac{1}{300} QX}{\mathrm{f}} = \frac{\tfrac{1}{300} QX}{6\pi\eta_0 R_t} \qquad (\kappa R_e \ll 1) \qquad \text{(VII.42)}$$

where R_t is the effective radius determining viscous resistance to translation.† This result was first obtained by Hückel (1924). On the other hand, Smoluchowski (1921) considered the case of charged colloidal particles in a medium containing a high concentration of small ions, so that the thickness of the double layer is much smaller than the dimensions of the charged sphere. In that case the double layer may be treated as a parallel plate condenser with an area A and a plate separation δ. The viscous resistance to the slippage of the two plates past one another is $\eta_0 Au/\delta$; and

†In practice we may disregard the distinction between R_e and R_t.

equating this to the electrostatic driving force as given in (VII.41), we obtain for the steady-state velocity

$$u = \frac{\frac{1}{300} Q X \delta}{\eta_0 A} \qquad (\kappa R_e \gg 1) \tag{VII.43}$$

The Hückel and Smoluchowski results may be compared in terms of the surface potential ψ_s, which is given by $\psi_s = Q/\mathscr{D}R_e$ for an infinitely thick double layer and by $\psi_s = 4\pi Q\delta/\mathscr{D}A$ for the parallel plate condenser. We obtain then

$$u = \frac{\frac{1}{300} X \mathscr{D} \psi_s}{6\pi\eta_0} \qquad (\kappa R_e \ll 1) \tag{VII.44a}$$

$$u = \frac{\frac{1}{300} X \mathscr{D} \psi_s}{4\pi\eta_0} \qquad (\kappa R_e \gg 1) \tag{VII.44b}$$

If we want to express the electrophoretic velocity u in terms of the charge of the spherical particle, we may use the results of the Debye-Hückel theory in the form

$$Q = \mathscr{D} \psi R_e \frac{1 + \kappa(R_e + r_i)}{1 + \kappa r_i} \tag{VII.45}$$

where r_i is the radius of the small ions in the solution. The velocity of the spherical macromolecules becomes then, assuming $R_e = R_t$,

$$u = \frac{1}{300} \frac{QX(1 + \kappa r_i) f(\kappa R_e)}{6\pi\eta_0 R_e[1 + \kappa(R_e + r_i)]} \tag{VII.46}$$

where $f(\kappa R_e) \to 1$ as $\kappa R_e \to 0$ and $f(\kappa R_e) \to \frac{3}{2}$ as $\kappa R_e \to \infty$. For intermediate values of κR_e (i.e., for the case where the thickness of the counterion atmosphere is comparable to the radius of the charged macromolecule) we have $1 < f(\kappa R_e) < \frac{3}{2}$, where the values of $f(\kappa R_e)$ have been computed by Henry (1931).

Two complicating features tend to reduce the velocity u below the value predicted by (VII.46). The first effect, referred to as the electrophoretic retardation, is due to the hydration of the counterions, so that they carry a considerable number of water molecules in a direction opposite to that in which the charged macromolecule is moving. The macromolecule is then transported against the flow of the streaming solvent, and its velocity should be correspondingly reduced. The second effect is caused by the finite time required to re-establish the counterion atmosphere when the charged particle is changing its location. As a result of this "relaxation effect" the center of the counterion atmosphere tends to lag behind the center of the moving particle, resulting in a backward pull. Experimental data on serum albumin

(Möller et al., 1961) indicate that most of the reduction in electrophoretic velocity is accounted for by electrophoretic retardation, while the relaxation effect plays a minor role. Figures VII.17 and VII.18, taken from the results of these investigators, show how the electrophoretic mobility, $\bar{u} = u/X$, of salt-free solutions of serum albumin titrated with base changes with increasing charge at constant protein concentration and with increasing protein concentration at constant charge. As would be expected for the various reasons listed above, a plot of the mobility of the protein against its charge

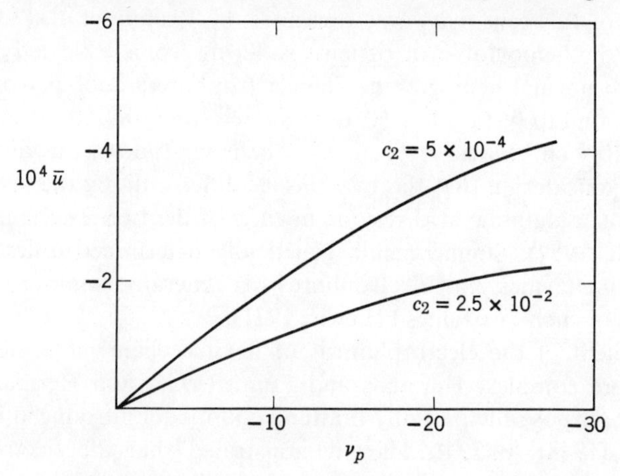

Fig. VII.17. Dependence of the electrophoretic mobility of bovine serum albumin on the net number of charges ν_p.

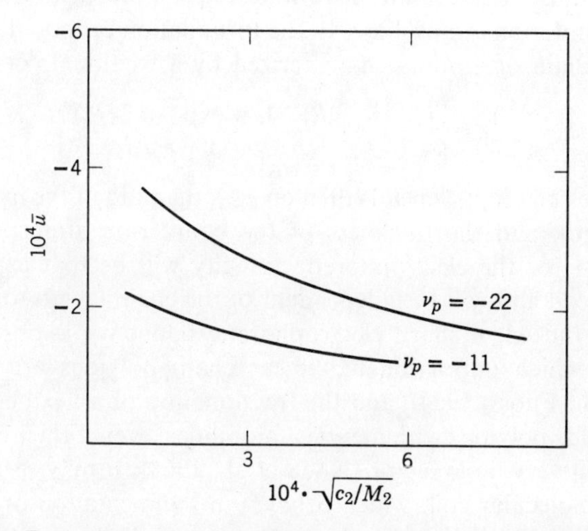

Fig. VII.18. Dependence of the electrophoretic mobility of bovine serum albumin on the concentration of the protein.

has a downward curvature and the mobility decreases also with increasing protein concentration. In the presence of a fairly high concentration of added salt, the mobility of the protein molecule seems to be proportional to its charge, as found by Longsworth (1941) for ovalbumin in solutions with an ionic strength of 0.1. The mobilities which he found under these conditions were about 60% of the values predicted by (VII.46).

Electrophoresis is unique among solution properties in its sensitivity to small differences between related protein species. The most dramatic demonstration of this sensitivity was presented by Pauling et al. (1949), who showed that the hemoglobin of patients suffering from sickle cell anemia is different from normal hemoglobin. The electrophoretic mobility of the two species was found to be (at pH 6.52 and an ionic strength of 0.1) 2.63×10^{-5} and 2.23×10^{-5} cm^2/V sec, respectively. The magnitude of this difference is remarkable, considering that the two species differ only by the substitution of a valine for a glutamic acid residue in each of the two β subunits (see p. 421) (Ingram, 1959). Similar small, genetically determined differences are observed with enzymes, and electrophoresis is generally employed to reveal the existence of such "isozymes" (Lewis, 1971).

The treatment of the electrophoresis of flexible chain molecules is considerably more complex. Hermans and Fujita (1955) and Hermans (1955) used for this purpose the partially draining coil model introduced by Debye and Bueche (1948) (p. 294). They demonstrated that the electrophoretic velocity depends not only on hydrodynamic factors (i.e., on the ratio of the radius of the hydrodynamically equivalent sphere R_t to the hydrodynamic shielding length L_f) but also on the ratio of the coil dimensions to the thickness of the counterion atmosphere. In the formulation proposed by Hermans (1955) for a chain of Z units characterized by a frictional coefficient f_0,

$$\bar{u} = \frac{1}{300} \frac{Q}{Zf_0} \left[1 + \frac{R_t^2}{3L_f^2} \frac{1 + \exp(-2\kappa R_t)}{3 + \kappa^2 R_t^2} \right] \qquad \text{(VII.47)}$$

For high values of κR_t, $\bar{u}$ depends then on κL_f, the ratio of the hydrodynamic shielding length, and the thickness of the counterion atmosphere. In the limit of $\kappa L_f \to \infty$, the electrophoretic velocity will be that expected for a free-draining coil and will be independent of the chain length for any charge density Q/Z. Indeed, a sharp electrophoretic boundary is observed in experiments in which unfractionated linear chain polyions are investigated (Fitzgerald and Fuoss, 1954), and the fractionation of a flexible chain polyelectrolyte yields polyions with identical mobilities, even if their chain lengths are varied by a very large factor (Noda et al., 1964). It may seem surprising that a macromolecular coil, which behaves in sedimentation or diffusion as impermeable, should behave in electrophoresis as if it were free draining. However, Hermans and Fujita point out that in the limit of high ionic strength the net charge density vanishes at all points, so that the net force acting on any volume element is zero. Under these circumstances there is no reason to expect interaction effects which would render the coil impenetrable. Experimental results are in agreement with the conclusion of the Hermans-Fujita theory that polyions behave in electrophoresis as free-draining coils in the presence of moderate concentrations of added salt (Nagasawa et al., 1958; Noda et al., 1964), and this behavior appears to extend to even lower electrolyte concentrations than the theory would predict (Nagasawa et al., 1972). At high salt concentration, the mobility of the polymer is similar to that expected for a chain segment with a single ionic charge, as predicted by (VII.47).

Strauss et al., (1954) and Varoqui and Strauss (1968) found that the attachment of long hydrophobic side chains has no effect on the electrophoretic mobility of polycations or polyanions, although the cohesion of these side chains in an aqueous medium converts a highly expanded polyion to a compact "polysoap" particle. This phenomenon is not well understood. In the case of the compact particles, the polyion charges in the interior would be expected to be compensated for by counterions, so that the electrophoretic mobility would be determined by the surface potential ψ_s [cf. (VII.44)], which depends on the charge density at the particle surface and the thickness of the ionic atmosphere. Thus, the compact polysoap particle has $\bar{u}$ independent of particle size, just as is the case with expanded polyion chains.

The electrophoretic mobility of tetramethylammonium polyphosphate was found to remain unchanged on addition of $1M$ tetramethylammonium bromide but was cut in half in the presence of $0.8M$ NaBr (Strauss et al., 1957). This difference was interpreted as reflecting the binding of sodium counterions. On the assumption that the tetramethylammonium counterions were unassociated with the polyion, the electrophoretic mobility of the polyphosphate was found to be proportional to its effective charge, so that electrophoretic data were used to estimate the degree of association with various counterions (Strauss and Bluestone, 1959; Strauss and Ross, 1959). On the other hand, the electrophoresis of polystyrenesulfonate yielded identical data, whether the counterion was $(CH_3)_4N^+$, Li^+, Na^+, K^+, or Cs^+, suggesting that none is associated with the polyion (Nagasawa et al., 1972).

Theoretical and experimental aspects of electrophoresis are discussed in an excellent monograph edited by Bier (1959).

2. Conductance and Dielectric Increment

The conductance of solutions containing polyelectrolytes as well as added simple electrolytes is the sum of contributions made by the small ions and the polyion. Huizenga et al. (1950) suggested that measurements of the specific conductivity l_{sp} and the electrophoretic mobility of the polyion $\bar{u}_p$ may be interpreted to yield the fraction f of polyion charges dissociated from their counterions, provided it can be assumed that the mobilities of byions $\bar{u}_b$ and of counterions $\bar{u}_c$ are the same as in solutions of simple electrolytes. If m_f and m_s are the molar concentrations of fixed charges and of a uni-univalent salt, then

$$l_{sp} = \frac{\mathfrak{F}}{1000}[m_f \mathrm{f}\bar{u}_p + m_s\bar{u}_b + (m_s + m_f\mathrm{f})\bar{u}_c] \quad \text{(VII.48)}$$

On the basis of this analysis, Huizenga et al. were the first to conclude that a large fraction of counterions must be bound to the polyion at high charge densities.

Although this conclusion is undoubtedly qualitatively correct, the assumptions on which the quantitative interpretation of experimental data is based are highly questionable. We have seen that the net charge of the polyion is almost fully compensated for by counterions lying very close to the polymer chain, and it would then be expected that electrostatic interactions should lead to a substantial reduction in the mobility of "free" counterions. This expectation has been verified in experiments in which the transference numbers of polyions, byions, and counterions were all determined in a Hittorf apparatus (Nagasawa et al., 1972). For instance, whereas Na^+ in 0.05N NaCl has a mobility of 4.5×10^{-4} cm^2/V sec, in solutions containing 0.043N poly(acrylic acid) and 0.049N NaCl this mobility is reduced to 4.4, 3.1, and 2.05×10^{-4} cm^2/V sec as the degree of ionization of the polyion, α_i, is increased to 0.2, 0.6, and 1.0. On the other hand, the mobility of byions is, as expected, quite insensitive to the presence of a polyelectrolyte. Use of the procedure of Nagasawa et al. leads to the estimate that 20% of the polyion charges are associated with counterions at $\alpha_i = 0.2$, while 70% are associated with counterions at $\alpha_i = 1$. The extent of this association is almost independent of the amount of added salt. It appears that the counterion mobility is the weighted average of the mobility of the free counterions of the polyelectrolyte and the mobility characterizing these ions in the added salt. The conductance of solutions containing polyelectrolytes and simple electrolytes follows then an additivity rule (Mock et al., 1954) analogous to that which is a good approximation to the behavior of ionic activity coefficients.

Although the conductance of long-chain polyelectrolytes is independent of the chain length of the polyion, a chain length dependence should be observed at relatively low degrees of polymerization. The nature of this dependence was studied by Schindewolf (1954b) on sodium polyphosphate. His results (Fig. VII.19) show that the equivalent conductance of the polyelectrolyte first increases with the degree of polymerization and later decreases to approach an asymptotic value. This behavior may be interpreted as reflecting the superposition of two effects. First, The ratio of the charge

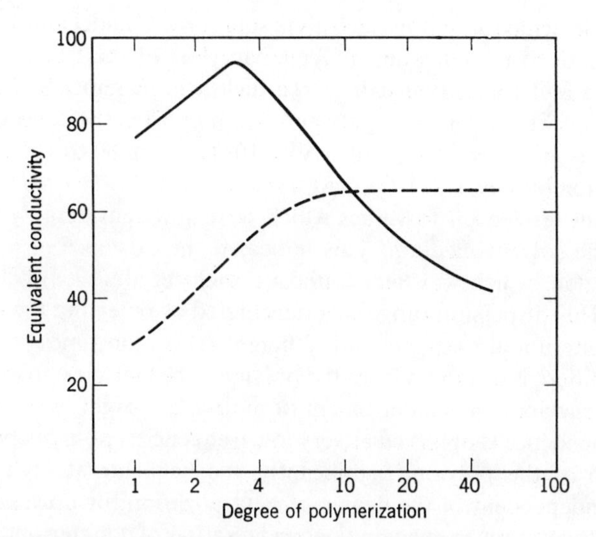

Fig. VII.19. Equivalent conductance of sodium polyphosphate (——) and of the polyphosphate ion(– –) as a function of the degree of polymerization.

of the polyion to its frictional coefficient tends to increase with increasing chain length. Second, counterion binding increases as the polyion chain increases in length. Using measurements of the transference number of the polyphosphate and assuming that the mobility of the free sodium ions is the same as in simple electrolytes, Schindewolf estimated equivalent conductances of the polyphosphate species as shown by the dashed line in Fig. VII.19. Although we now know that the free sodium ions are less mobile than assumed in this treatment, the indicated approach to a limiting mobility of the polyion is probably fairly close to its actual behavior.

When polyions exist in a highly extended form, their mobility becomes significantly larger in the direction of maximum chain extension than in other directions. Thus, if the polyions are induced to assume a preferred spatial orientation, the electrical conductivity of the solution becomes dependent on the direction in which it is being measured. This fascinating effect has been observed in solutions of polyelectrolytes oriented by high shear gradients (Jacobson, 1953; Schindewolf, 1954a). The same effect may be produced when a flexible chain polyelectrolyte solution is placed in a strong electrical *alternating* field (Eigen and Schwarz, 1957; Schwarz, 1959). In this case the orientation of the chains is due to the polarizability of the counterion cloud, which is at a maximum when the chain-end displacement lies parallel to the electrical field. This experiment provides, therefore, a striking demonstration of the large anisotropy of the polyion.

The high polarizability of the counterion atmosphere is also apparent from the dielectric behavior of polyelectrolyte solutions. Mandel and Jenard (1963) studied solutions of potassium polymethacrylate in the frequency range of 10^4–10^6 Hz and found unusually large dielectric increments. For instance, for a polymer with a degree of polymerization of 5200, the dielectric constant of water was increased at a frequency of 10^4 Hz from 80 to 155 at a polymer concentration of only $5 \times 10^{-3}N$. At a frequency of about 1 MHz, the dielectric constant leveled off to values which were still considerably higher than those of the solvent medium. This indicated the existence of a second dispersion region, which was later found by Sachs et al. (1969) between 1 and 10 MHz. This dispersion range was interpreted as reflecting the polarization of the counterion atmosphere with different relaxation times of counterions placed at different distances from the polyion. The dielectric increment in the high-frequency region is independent of molecular weight, while a molecular weight dependence is observed at very low frequencies, possibly because of an orientation of the polyion. Significantly, the behavior at high frequencies becomes independent of the degree of neutralization for polyacrylate above $\alpha_i = 0.3$, illustrating once again the condensation of counterions on polyions above a critical charge density.

3. Diffusion and Sedimentation

The attachment of ionic charges to a macromolecular particle results in a reduction of its sedimentation coefficient. This reduction may be minimized by the addition of simple salts to the polyelectrolyte solution. The effect is observed even with globular molecules, which would not be expected to change their shape when their degree of ionization is altered (Tiselius, 1932). Of course, the dependence of the sedimentation coefficient on the charge of the particle becomes more pronounced with flexible chain polyelectrolytes (Rosen et al., 1951; Howard and Jordan, 1954), in which the expansion of the polyion results in an increase of the frictional coefficient.

The theory of these phenomena was first formulated by Svedberg and Pedersen (cf. Pedersen, 1958), who considered the sedimentation rate of the polyion as the sum of two virtual motions, the first due to the gravitational field without consideration of electrostatic interactions and the second an electrophoretic flow caused by the electrical field X which is produced by the unequal sedimentation rate of the polyions and the small ions. The aggregate velocity u_i of each species i is then

$$u_i = \mathbf{s}_i^* \, \omega^2 r - \frac{\bar{u}_i X}{300} \tag{VII.49}$$

where $\mathbf{s}_i^*$ is the sedimentation constant for a hypothetical condition in which the effects of electrical charge interactions are eliminated, and u_i is the electrophoretic mobility of species i. Since the total current flow $\sum u_i m_i e \nu_i$ must vanish, we have

$$\omega^2 r \sum m_i e \nu_i \mathbf{s}_i^* - \frac{X}{300} \sum m_i e \nu_i \bar{u}_i = 0$$

$$\frac{X}{300} = \frac{\omega^2 r \sum m_i \nu_i \mathbf{s}_i^*}{\sum m_i \nu_i \bar{u}_i}$$

$$\mathbf{s}_i = \mathbf{s}_i^0 - \frac{\bar{u}_i \sum m_i \nu_i \mathbf{s}_i^*}{\sum m_i \nu_i \bar{u}_i} \tag{VII.50}$$

where $\mathbf{s}_i^0$ is the sedimentation constant of species i as its concentration approaches zero. As a result, we should have for the polyion $\mathbf{s}^0 = \mathbf{s}^*$ in the limit of zero polyelectrolyte concentration in a solution of uni-univalent salt whose cation and anion have the same frictional coefficient. In practice, Pedersen noted that $\mathbf{s}^0$ for proteins bearing a net charge is always significantly smaller than it is at the isoelectric point. He ascribed this discrepancy to a swelling of the globular proteins, but such swelling would have to be improbably large to explain his data.

According to Nagasawa and Eguchi (1967), the fallacy in Pedersen's treatment is a result of the implied assumption that the small ions are uniformly distributed in the polyelectrolyte solution. In fact, the counterions must, of course, remain in the immediate vicinity of the polyion, and the sedimenting polyion will necessarily have to pull its counterions with it. If the polyion is a hydrodynamically impermeable coil, all ions located within the coil will be carried with it without any additional dissipation of energy by viscous friction. On the other hand, counterions which are outside the coil (or outside a globular protein particle carrying a net electrical charge) will add a relatively large term to the frictional coefficient of the polyelectrolyte and will thus lead to an appreciable decrease of the sedimentation rate. We may note that this effect should decrease with increasing salt concentration, since the polymer coil with the ions located in its interior becomes more nearly a Donnan system with vanishing net charge. Nagasawa and Eguchi found that R_η/R_t as derived from the sedimentation coefficient and the intrinsic viscosity [cf. (VI.68)] has unusually low values and decreases with increasing dilution of added salt or increasing molecular weight of the polyelectrolyte. This is consistent with the assumed effect of the counterion atmosphere, which would lead to an apparent increase in R_t (as reflected in the sedimentation rate) while having no significant effect on R_η.

The conversion of uncharged macromolecules to polyelectrolytes increases the number of osmotically active particles and the driving force for diffusion (Kedem and Katchalsky, 1955). The forces between the polyions and the counterions produce in sedimentation and in diffusion effects which work in opposite directions. In sedimentation, the driving force on the light counterions is small and they retard the polyion, whereas in diffusion the driving force is independent of particle size, so that the counterions increase the driving force by a larger factor than the frictional resistance to translation. One interesting effect deserves special attention. Since the expansion of the polyions in salt-free solutions of flexible chain polyelectrolytes increases sharply with the dilution of the system, the frictional coefficient of these polyions is correspondingly increased. This effect and the increasing electrophoretic effect, as discussed above, cooperate in producing a diffusion coefficient which drops sharply as the polyelectrolyte concentration is reduced, instead of increasing slowly as with uncharged polymers. A polyelectrolyte solution diffusiong into the pure solvent tends, therefore, to produce a front with an extremely high concentration gradient, frequently giving the appearance of the swelling of a crosslinked gel. A detailed investigation of this phenomenon has been reported by Nagasawa and Fujita (1964).

Chapter VIII

MOLECULAR ASSOCIATION

Dissolved macromolecules frequently form molecular association complexes either with species of low molecular weight or with other macromolecules. We may cite as typical examples of the first class the binding of counterions to polyions (disscussed in Chapter VII), the binding of iodine to amylose, and the association of enzymes with substrates, inhibitors, coenzymes, and activating ions. Molecular association of macromolecules with one another encompasses an even broader range of phenomena, such as the aggregation of poly (vinyl chloride) molecules in certain solvent media, nonspecific associations of cationic and anionic polymers, the formation of hemoglobin and various enzymes from separate protein subunits, the interactions of antigens with antibodies, and the spontaneous formation of the tobacco masaic virus particle from its nucleic acid and protein. The formation of multistrand helices from polypeptide or polynucleotide chains (cf. Sections III. C.2 and III. C.3) also falls into this category.

The study of such phenomena may be motivated by a number of different objectives. It is clearly necessary to understand molecular aggregation of synthetic or natural macromolecules if the concept of a molecular weight is to have an unambiguous significance, and such aggregation is frequently related to the ability of a polymer-solvent system to form a thermally reversible gel. However, the most intriguing problems in this field concern the high degree of specificity characteristic of the molecular association behavior of certain biologically important macromolecules. For instance, an antibody will associate much more powerfully with one particular macromolecule (its "homologous antigen") than with a number of very similar substances, and enzymes may be highly specific in their association with substrates or inhibitors. This specificity depends on a complementarity of the "molecular surfaces," particularly closely defined in proteins which retain in solution a highly specific tertiary structure (see pp. 170–181). Although the discovery of the retention of helical conformation in solutions of synthetic polypeptides and polynucleotides demonstrates the possibility of producing synthetic polymers which are characterized in solution by a high degree of conformational rigidity, it is unknown how a synthetic polymer could be induced to fold in a unique way into a globular particle, and such folding may be essential to a high specificity in molecular association, such as is exhibited by antibodies and enzymes.

Many of the methods discussed in preceding chapters are applicable to a study of association. When a small molecular species associates with a macromolecule, the equilibrium may frequently be studied by the distribution of the small species across a semipermeable membrane (dialysis equilibrium). Equivalent information may be obtained from an ultracentrifuge equilibrium measurement, if the concentration of the small species is spectroscopically determinable. The concentration of unassociated small molecules will be essentially constant within the ultracentrifuge cell, but the *total* concentration of these species will vary because of the variation in the concentration of the macromolecules with which they associate (Steinberg and Schachman, 1966). Alternatively, association equilibria can be derived from the concentrations of small molecules on the two sides of the boundary in an ultracentrifuge sedimentation experiment (Lloyd et al., 1968). In some cases, association is accompanied by a spectral shift or a change in fluorescence, which may be used to estimate the association equilibrium.

Similarly a variety of methods may be employed in studying the association of macromolecules with one another. Work in this area up to 1963 has been admirably summarized by Nichol et al. (1964). If deviations from solution ideality can be neglected, the osmotic pressure, light scattering, and ultracentrifuge equilibrium over a range of solution concentrations yield the concentration dependence of the average molecular weight, which may be interpreted in terms of association equilibrium constants. However, in many cases effects due to solution nonideality are not negligible compared to effects resulting from molecular association, and they must be taken into account. This can be done for solutions of associating globular particles, where the excess partial free energies per unit weight of all solute species may be assumed to be identical, depending only on the total solution concentration (Adams and Fujita, 1963; Adams, 1965). The utilization of frictional properties of macromolecules is often useful, although the quantitative interpretation in terms of association equilibria is uncertain. For instance, the translational frictional coefficient or the intrinsic viscosity of a dimer formed from two elongated particles will be much larger if the particles aggregate end to end than if they are joined to each other side by side. Ultracentrifuge sedimentation will, in general, be accelerated by molecular association, since the frictional coefficient will increase less rapidly than the weight of the particle.‡ On the other hand, association of identical particles

‡The association of two polynucleotide strands to the DNA double helix is an exception to this rule. Here the increase in the frictional coefficient outweighs the effect of the increase in particle mass and the sedimentation rate is reduced (F. W. Studier, *J. Mol. Biol.*, **11**, 373 (1965)).

may or may not change their electrophoretic mobility. However, electrophoresis is most useful in studying the association of dissimilar charged macromolecules.

In interpreting experimental data, it is essential to take into account the relation between the rate of the molecular dissociation or association processes and the time scale of the experimental method. For instance, if the association of macromolecules and the dissociation of their complex occur many times during an ultracentrifuge sedimentation run, a single sedimentation boundary will be observed and the velocity with which this boundary moves will reflect a weighted average of the velocities characteristic of the isolated particles and their complex. Conversely, if the association and dissociation processes are slow, separate boundaries will be observed for the unassociated macromolecules and their aggregates. The unique advantage of optical methods (light scattering, spectroscopy), is that they make it possible to study the instantaneous state of the system and thus enable the investigator to follow the time dependence of a molecular association process up to extremely high velocities.

A. ASSOCIATION OF MACROMOLECULES WITH LOW MOLECULAR WEIGHT SPECIES

1. Association of Small Molecules and Ions with Flexible Chain Molecules

The association equilibria of chain molecules with species of low molecular weight may be of two types. In the first, the functional groups attached to the molecular backbone interact with various reagents in much the same manner as would be expected of low molecular weight analogs of the repeating unit of the macromolecule. A characteristic case of this type is the formation of highly colored charge transfer complexes of poly (α-vinylnaphthalene) with strong electron acceptors (Slough, 1962). For instance, a solution of this polymer will develop, on addition of tetracyanoethylene, an absorption spectrum with maxima at 5900 and 4500 Å, quite similar to spectra observed in solutions containing tetracyanoethylene with naphthalene. Of much more interest, however, are macromolecular complexes of the second type, in which cooperative phenomena enhance the stability of the complex. Thus, Litt and Summers (1973) found that the polymeric donor with the following structure:

$$\begin{array}{l}(\text{—N—CH}_2\text{CH}_2\text{—})_n\\ \quad | \\ \quad \text{CO} \\ \quad | \\ \quad \text{CH}_2\text{CH}_2\text{CH}_2\text{—O—C}_6\text{H}_4\text{—SCH}_3\end{array}$$

formed charge transfer complexes which were characterized by formation constants up to 50 times higher than those observed with monomeric donor model compounds. This effect was ascribed to a favorable spacing of the donor groups in the polymer, allowing intercalation of the acceptor between two donor groups.

The best-known example of this kind of molecular aggregate is the formation of the blue starch-iodine complex. It was reported only 3 years after the discovery of iodine (Colin and Gaultier de Claubry, 1814), and in reading the account of the experiments describing that "*la couleur est d'un bleu superbe si ces substances sont en proportions convenable*" one may relive the excitement caused by the first observation of this beautiful phenomenon. The first quantitative study seems to have been that of Küster (1894), who showed that the iodine content of the complex increases with the concentration of free iodine in solution. In modern times, it has been demonstrated that starch consists generally of two fractions, the linear chain amylose and the highly branched amylopectin, and that only amylose forms the blue iodine complex, the complex of amylopectin being reddish.

The formation of the amylose-iodine complex involves a helical conformation of the amylose chain, which leaves in the center of the helix a channel-like cavity with dimensions appropriate for the accommodation of the iodine. (See Fig. VIII.1). This model was first proposed by Hanes (1937) and by

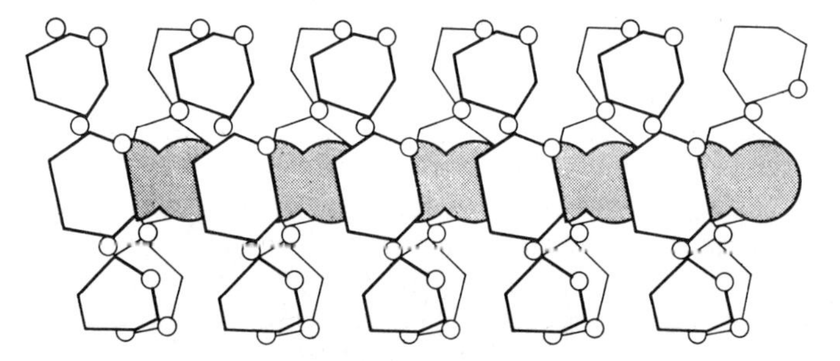

Fig. VIII.1. Model of an amylose helix with iodine molecules stacked up in the helical cavity.

Freudenberg et al. (1939) before the discovery of the urea clathrate compounds, whose structure reflects the same general principle. It was substantiated by the finding that the starch-iodine complex forms hexagonal crystals (Rundle and French, 1943), which would be difficult to account for with any other geometry of the amylose chains, and was strengthened further by the discovery of flow dichroism (Rundle and Baldwin, 1943) proving that the iodine chains lie parallel to the long dimension of the amylose chain.†

†This appears to have been the first demonstration that chain molecules can exist in solution in a helical conformation.

The formation of the iodine complex has been found to leave the viscosity of amylose solutions unchanged; it apears that, in both the presence and the absence of iodine, the amylose chain contains helical and randomly coiled sections (Hollo and Szejtli, 1958). With carboxymethylamylose, the iodine complex can form only at low pH, since the helical sections of the chain are disrupted at high pH by the mutual repulsion of the ionized carboxyls (Rao and Foster, 1965).

Since synthetic amylose can be prepared with an extremely narrow distribution of chain lengths (cf. p. 18), studies of its iodine complexation equilibria and of the optical properties of these complexes (Pfannemüller et al., 1969, 1971) are particularly valuable. Potentiometric titrations show that the affinity of amylose for iodine increases with the polymer chain length, leveling off for degrees of polymerization above 100. The inverse of the wavelength at which the absorption of the complex has its maximum is linear in the inverse of the polymer chain length. After addition of iodine to an amylose solution, the optical density changes with time, suggesting that the rearrangement of helical and randomly coiled sections of the amylose chain to an equilibrium conformation is a slow process. Since the iodine chains enclosed in the amylose helix are in an asymmetric environment, their absorption exhibits circular dichroism—and this phenomenon is an important piece of evidence in the interpretation of the structure of the amylose-iodine complex. The intensity of the CD band passes through a maximum at degrees of polymerization of about 50, and this anomalous behavior may be related to the anomalous molecular weight dependence of amylose solubility (cf. p. 61).

It has been reported that the characteristic color of the amylose-iodine complex does not appear if iodide is rigorously excluded from the system; the absorption intensity first increases with increasing iodide concentration, but at high concentrations of iodide the absorption is again reduced, with a shift of the absorption maximum to shorter wavelengths. There is also a tendency for the absorption to be intensified and the absorption maximum to be shifted to longer wavelengths by an increase in the ionic strength of the solution (Kuge and Ono, 1960). These findings suggest that the complex contains chains with a preferred ratio of I and I^-; this ratio is 3 : 2 according to the potentiometric studies of Gilbert and Marriott (1948) and 1 : 1 according to the spectroscopic data obtained by Kontos (1959). At any rate, the amylose acquires an increasingly negative charge during complexation with iodine (Beckmann, 1964). Such a process would, of course, be favored by an increase in the ionic strength, which reduces the electrostatic free energy of the complex formation. A theory for amylose-iodine complexation must account mainly for two phenomena: the increasing stability of the complex with an increasing chain length of the polymer, and the observed absorption spectrum. Stein and Rundle (1948) believed that the forces responsible for the formation of the complex originate in the large dipole of the amylose helix; this induces dipoles in the iodine molecules, and the effect of dipole interactions increases cooperatively with the number of interacting iodine molecules. The alignment of the iodine within the helix leads to electron delocalization, which shifts the absorption maximum to longer wavelengths as the iodine chain is extended (Ono et al., 1953). The iodine chain may be described as a "one-dimensional metal," and it is significant that the solid amylose—iodine complex gives an electron spin resonance spectrum similar to that observed in metallic lithium or sodium (Bersohn and Isenberg, 1961). Cyclohexaamylose, a cyclic analog of amylose, forms an iodine inclusion compound with a spectrum similar to that of amylose—iodine, and single crystals of this compound may be obtained for crystallographic analysis. This shows (Dietrich and Cramer, 1954) that iodine *atoms* are stacked up within the cylindrical cavity at a uniform distance of 3.06 Å, making the description of the arrangement as a unidimensional metal even more appropriate.

One would expect the amylose helix to be able to accommodate a variety of linear molecules with polarizable groups which can interact with the anhydroglucose residues. In agreement with this concept, it has been shown by potentiometric titration that fatty acids may displace iodine from its amylose complex (Mikus et al., 1946).

Poly(vinyl alcohol) forms with iodine (or possibly with triiodide) a highly colored complex similar to that observed with amylose (Staudinger et al., 1927). This is somewhat surprising, since poly(vinyl alcohol) would not be expected to exist in helical conformations which would enclose a cavity of sufficient dimensions to contain the iodine. The extinction coefficient of poly(vinyl alcohol)—iodine increases with the chain length of the polymer (suggesting again that a cooperative phenomenon is responsible for complexation) but, in contrast to the behavior of the amylose complex, the absorption maximum at 620 nm is independent of molecular weight. Imai and Matsumoto (1961) found that the intensity of the color produced by the complex was enhanced when the polymer was prepared under conditions which would favor a syndiotactic configuration, while Nozakura et al., (1973) reported that iodine complexation was dramatically increased with a poly-(vinyl alcohol) sample estimated to have 65% heterotactic and 29% syndiotactic triads. A thorough study of the poly(vinyl alcohol)—iodine system by Zwick (1965) showed that the location of the absorption maximum is shifted to longer wavelengths by reducing the iodine / polymer ratio, and may be displaced up to 700 nm by the addition of boric acid. However, the spectrum of the complex depends on the sequence in which the reagents are added to each other, so that the properties of the system may frequently not reflect equilibrium conditions. The data are interpreted by a model in which helically wound sections of the polymer, enclosing chains of iodine atoms, are separated by randomly coiled portions of the chain, with the helical sections tending to associate with one another. Apparently, the resulting structures rearrange exceedingly slowly when the conditions are changed. At full saturation, the complex contains one iodine atom for 12 monomer residues of the chain molecule. It has also been found that the ability of poly(vinyl alcohol) to bind species of low molecular weight is not limited to iodine. A variety of small molecules, such as polyhydric phenols, naphthylsalicylamide, and the diamide of salicylic acid and benzidine cause gelation of poly(vinyl alcohol) solutions (McDowell and Kenyon, 1941; Lowe, 1943). Soluble complexes must be formed before the system sets to a gel, but no studies of the association equilibria have been reported.

Another uncharged polymer, poly(*N*-vinylpyrrolidone):

```
(—CH2—CH—)n
       |
      ,N\
     /   \
   CH2   CO
    |     |
   CH2—CH2
```

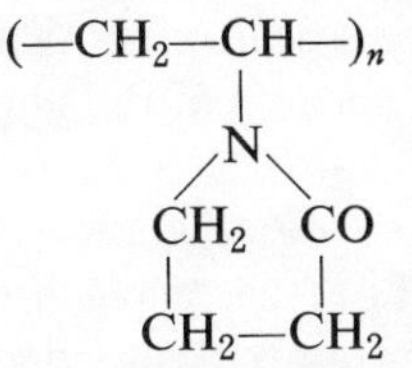

has been observed to bind a wide variety of small molecules and ions (Scholtan, 1953). If the polymer is added to a solution of iodine in potassium iodide, the spectral absorption is strongly intensified but the absorption maximum shifts only slightly from 350 to 360 nm, that is, the complex is quite different from the blue species which iodine forms with amylose or poly(vinyl alcohol). A cooperative phenomenon may not be required to account for the complex formation, since iodine and ethylpyrrolidone also seem to form an association complex, as indicated by a shift in the frequency of the carbonyl stretching vibration (Néel and Sébille, 1961). Dialysis equilibria show that both triiodide and iodine are bound to the polymer with the I_3^-/I_2 ratio in the complex rising with an increasing concentration of free iodide in the solution (Barkin et al., 1955). In spite of the charge acquired by the poly(*N*-vinyl pyrrolidone) in this process, the solution viscosity drops sharply, presumably because of poor solvation of the complex.

Poly(*N*-vinylpyrrolidone) also forms complexes with various dyes (Scholtan, 1953; Saito, 1957; Frank et al., 1957). A very detailed study of the association of a variety of aromatic molecules to poly(*N*-vinylpyrrolidone) was published by Molyneux and Frank (1961), who found that the association process is typically endothermic, suggesting that hydrophobic bonding (cf. Section II.A.6) is the main driving force. In agreement with other investigators, they found that only neutral and anionic materials are bound, while species such as the anilinium ion do not interact with the polymer. In this respect poly(*N*-vinylpyrrolidone) resembles poly(vinyl alcohol), but the cause for the preference of these uncharged macromolecules for anions over cations is completely unknown. The association equilibria approximate the form of a Langmuir adsorption isotherm with a polymer segment of 10 monomer units acting as an adsorption site for all of the small molecules investigated, in spite of their rather pronounced difference of size.

Anionic detergents are bound by poly(vinyl alcohol) (Saito. 1953, 1954, 1957; Isemura and Imanishi, 1958) and poly(*N*-vinylpyrrolidone) (Saito, 1957; Barkin, 1957), and some water-insoluble polymers, such as poly(vinyl acetate) and poly(vinyl formal) are solubilized by these detergents (Saito, 1953, 1954; Isemura and Imanishi, 1958). The solutions are optically clear and have a relatively high specific viscosity, so that they obviously cannot be considered colloidal dispersions of relatively large polymer particles. The behavior of such systems suggests that the individual polymer chains adsorb a large number of the detergent anions, so that they behave like polyanions with chains highly expanded because of the mutual electrostatic repulsion of the adsorbed species.

The binding of cationic dyes to polymeric acids is frequently accompanied by a characteristic spectral shift (the "metachromatic effect"), similar to the shift which occurs when dye molecules aggregate with one another in the absence of polymer. We may, therefore, think of the polymer as an agent which facilitates the aggregation of the dye molecules. The fraction of the dye molecules occupying neighboring sites on the polymer (and hence characterized by the metachromatic shift) depends not only on the dye / polymer ratio, but also on the nature of the polymer (Bradley and Wolf, 1959). The experimental data may be accounted for quantitatively by assigning to every dye-polymer system a "stacking tendency" which expresses the relative preference of a dye to occupy a site next to another bound dye molecule rather than an isolated site on the polymer chain. In the case of systems containing a polymer and two dyes, the spectra may not be accountable by the metachromatic shifts of the individual dyes, and the "complex metachromasis" reflects then the tendency of dissimilar dye molecules to occupy neighboring positions on the polymer chain (Pal and Schubert, 1963).

In the binding of low molecular weight species to DNA and to synthetic polynucleotides, a variety of interesting phenomena are observed. Cations may be complexed either by the phosphate residues or by the purine and pyrimidine bases. The binding of Mg^{2+} occurs exclusively by interaction with the phosphate, and it leads to a sharp increase in the stability of the double-helical structure of DNA, so that T_m is raised by up to 45°C (Dove and Davidson, 1962a). Binding of Hg^{2+} involves first the phosphate groups; at higher Hg^{2+} concentrations, the cation is complexed by two of the heterocyclic bases, with the A-T pairs forming more stable complexes than the G-C (Nandi et al., 1965). Coordination of Ag^+ seems to occur only to the purine and pyrimidine bases, and in this case complexation with G-C pairs is preferred (Jensen and Davidson, 1966). Binding of either Hg^{2+} or Ag^{+} leads to a sharp increase in the buoyant density of the nucleic acid, and the difference in the affinity to A-T and G-C pairs can then be utilized for a separation of DNA species differing in base composition. The binding of aminoacridines to nucleic acids (Blake and Peacocke, 1968; Armstrong et al., 1970) is of special interest, since it correlates with the biological activity of aminoacridines as antibacterial and mutagenic agents. A variety of experimental evidence shows that at low concentrations these dyes are bound by intercalation between the base pairs of the DNA, with a partial untwisting and an extension of the double helix. Dye molecules cannot be accommodated in adjacent "slots," and additional aminoacridines are bound more weakly outside the double-helical structure. Finally, there is the fascinating phenomenon of the association of purines and purine nucleosides with the complementary polynucleotides to form double- or triple-helical structures analogous to those formed from two or three polynucleotide chains (Howard et al., 1966; Huang and Ts'o, 1966). Figure VIII.2, representing data of Davies and Davidson (1971), shows the extent of complex formation between deoxyadenosine and two chains of polyuridylic acid or of deoxyguanosine and two chains of polycytidylic acid as a function of the logarithm of the

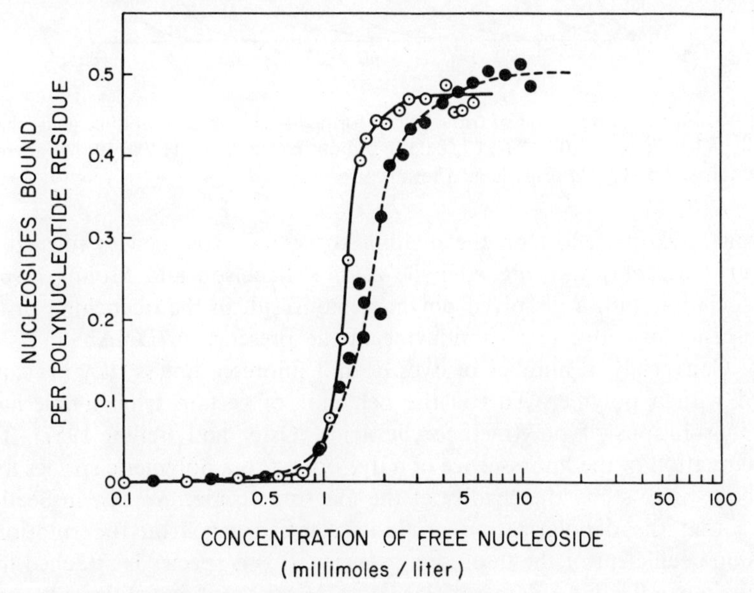

Fig. VIII.2. Binding isotherms for deoxyadenosine (-●-) to two poly-U chains and deoxyguanosine (-○-) to two poly-C chains (temp. 5°C, pH 7.1, $0.15M$ Na^{+}, conc. of polynucleotide residues $0.003M$).

concentration of the nucleoside. The cooperativity of the complexation is characterized by the slope of the plot at half-saturation; this is seen to be steeper for the formation of the GC_2 triple helix, in conformity with the general observation that G-C pairs lead to more stable helices. As would be expected, the interactions of polynucleotides with complementary oligomers lead to helix formation at much higher dilutions, and the stability of the helical structure increases with the length of the oligomer. Figure VIII.3

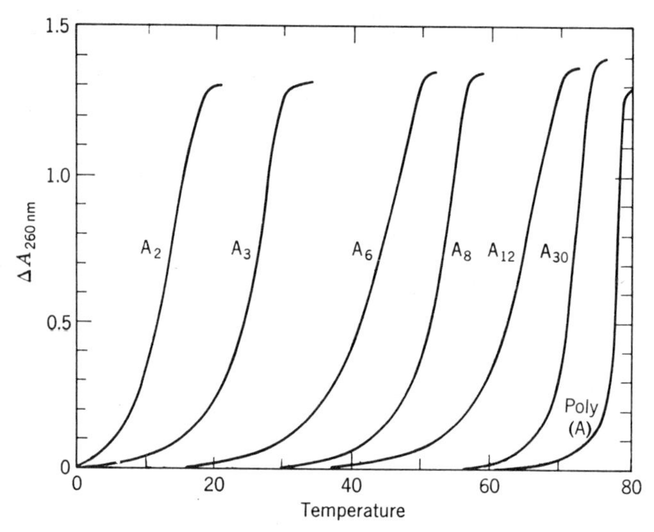

Fig. VIII.3. Melting curves of triple helices formed from two poly(uridylic acid) chains and oligoadenylic acids (0.1*M* NaCl, 0.05*M* sodium cacodylate, pH 7.0, $10^{-4}M$ adenylic acid residues, $2 \times 10^{-4}M$ uridylic acid residues).

demonstrates this effect on the melting curves of triple helices formed by poly(uridylic acid) and oligoadenylic acids (Michelson and Monny, 1967).

Association with a dissolved polymer may result in the quenching of the fluorescence of a dye (e.g., acriflavine in the presence of DNA; cf. Oster, 1957). Conversely, a number of dyes do not fluoresce unless they are complexed with a polymer. This is the behavior of certain triphenylmethane dyes in solutions of poly(methacrylic acid) (Oster and Bellin, 1957). The depolarization of the fluorescence of a dye bound to a polymeric species may also be used to study the rigidity of the macromolecule. We saw in Section VI.B.5 that the depolarization of fluorescence depends on the rotational diffusion coefficient of the fluorescing species. If this species is attached to a side chain of a flexible polymer molecule, rotational motions of the side chain are possible without significant motion of the backbone. However, if the polymer forms a rigid structure, rotation of the fluorescing species may involve rotation of the whole macromolecule; under these conditions the motion is greatly slowed down, and the depolarization of fluorescence correspondingly reduced. This principle was used by Ermolenko and Katibnikov (1962) to demonstrate the stiffening of poly(methacrylic acid) chains with an increasing degree of ionization.

When a small molecule which has no center of asymmetry is bound to an asymmetric macromolecule, optical activity may be induced in the chromophore of the complexed species by the asymmetric environment. These "extrinsic optical activities" are generally enhanced when the chromophore is bound to a polymer chain in a helical conformation. A number of studies of such induced optical activities involving complexes of acridine derivatives with polypeptides, polysaccharides, or nucleic acids have been reported (Blout, 1964; Blake and Peacocke, 1966; Eyring et al., 1968). The phenomenon is complicated by differences in the optical properties of "stacked" and isolated dye molecules; in addition, the asymmetric environment may involve the relative positioning of the chromophores adsorbed onto a helical structure. For instance, Blake and Peacocke found that nearest-neighbor interaction accounts for most of the extrinsic optical activity of proflavine bound to native DNA, although a small value of circular dichroism is observed, according to Li and Crothers (1969), even when the intercalated proflavine molecules are distant from one another.

Just as an asymmetric environment of a molecule in the ground state induces circular dichroism, such an environment of the excited molecule leads to circular polarization of fluorescence (CPF). This phenomenon is simpler to interpret than CD, since it involves a single electronic transition of the chromophore. Schlessinger and Steinberg (1972) studied dyes bound to chymotrypsin both by CD and by CPF and found that the nature of the environment changes when the chromophore is excited.

2. Binding of Small Molecules and Ions to Globular Proteins

Proteins contain a variety of functional groups which are known to form complexes with multivalent cations. Hence it is not surprising that these cations form protein association complexes and that in many cases the stability of such complexes is in reasonable agreement with predictions made on the basis of complexation data with low molecular weight analogs, taking proper account of the electrostatic free energy characterizing the association with the charged macromolecule. The nature of the groups participating in the complex formation may be deduced from the pH dependence of the association constant and from spectroscopic data. Typical examples of cation binding to proteins falling into this category are the binding of cupric ion (Klotz and Fiess, 1951) and of zinc ion (Gurd and Goodman, 1952) to serum albumin. In the case of enzymes, inhibition by traces of heavy metals is encountered if the thiol group of a cysteine residue forms part of the catalytically active site, and this inhibition may then be taken as a measure of complex formation by this thiol group. For instance, with the enzyme urease half of the activity is lost in solutions containing 10^{-10} *M* free silver ion, and this result is reasonable in view of the known affinity of low molecular weight mercaptans for Ag^{+} (Ambrose et al., 1951).

In many cases, however, the binding of ions to globular proteins involves a cooperative effect of several ligand groups. This may lead to unusual specificities, determined by the spatial disposition of the ligands in the rigid protein structure. For instance, the enzyme enolase is activated by Zn^{2+}, which generally forms stable complexes, as well as by Mg^{2+} and Mn^{2+}, which are weak complex formers, but by no other ions (Malmström, 1955). The association constants of ion complexes with globular proteins may also have strikingly high values, reflecting the favorable placement of the ligand groups for the coordination of the cation. In a number of cases the geometry of the complex has been clarified by crystallography. For instance, the binding of the activating Zn^{2+} to carbonic anhydrase involves coordination with imidazole groups of the histidine residues in positions 93, 95, and 117 along the polypetide chain (Liljas et al., 1972), and the cooperation of the three groups leads to an association constant of 10^{12} at pH 7.5 (Lindskog and Malmström, 1962). Another example is the binding of the activating Zn^{2+} to carboxypeptidase A, involving histidines in positions 69 and 196 and glutamic acid in position 72 (Lipscomb, 1970). It should be emphasized that the residues cooperating in the cation binding are far from one another along the polypeptide chain backbone and that the efficiency of the chelation is a result of the specific chain folding in the native protein structure. On the other hand, the steric disposition of the ligand groups in this native structure may lead to a pronounced distortion of the normal geometry of the ion complex. Vallee and Williams (1968) have pointed out that such distortions may serve an important biochemical function. For instance, if the coordination of a copper ion is intermediate between the square planar characteristic of Cu (II) and the tetrahedral favored by Cu(I), a change in the oxidation state should be facilitated, and this may be important for the mechanism of certain enzymes catalyzing oxidation-reduction processes. In the case of rubredoxin the protein structure is known to a high precision, and the coordination of Fe(III) to four sulfur atoms has been found to be strikingly distorted from the usual tetrahedral symmetry, with one Fe—S bond 0.4 Å shorter than the other three and the S—Fe—S angles ranging from 102° to 117° (Watenpaugh et al., 1972). As would be expected, removal of an activating ion leads to a reduction in the stability of the native structure of enzymes (Rosenberg and Lumry, 1964); in some cases, the chelation of a cation appears to have no other function than the stabilization of the native conformation (Feder et al., 1971; Edelman et al., 1972).

Most surprisingly, the activity of enzymes depends in a number of cases on the presence of *monovalent* ions. Specific activation by either K^+ or Na^+ has been reported (Suelter, 1970). Dependence on monovalent anions is rarer, but amylase was reported to require chloride, with as little as $5 \times 10^{-4}N$ Cl^- sufficient for half-activation (Myrbäck, 1926). These phenomena seem to imply that the protein conformation may be modified by complexation with these ions. For reasons unknown, serum albumin binds many anionic species; the equilibria of chloride and thiocyanate complexation have been studied in detail (Scatchard et al., 1949).

The association of enzymes with their substrates is frequently highly specific. The origin of this specificity has been described in a classical metaphor as analogous to the fit of a key into a lock (Fischer, 1894), suggesting a close complementariness of two rigid molecular surfaces. Crystallographic analysis of a number of enzymes (Table III.1) showed that the structures of these globular proteins have a groove containing the active site, and the dimensions of this groove are such as to accommodate the substrate molecule. However, the formation of the enzyme-substrate complex is accompanied, at least in some cases, by a readjustment of the molecular conformation of the enzyme. Such an "induced fit" was first inferred from the behavior of β-amylase, which catalyzes the hydrolysis of the glucosidic bond two glucose units from the end of the amylose chain. The enzymic catalysis must, therefore, involve a mechanism for recognizing the distance of the reactive bond from the amylose chain end. Yet cyclohexaamylose, which has no chain end, binds to the active site of the enzyme as a competitive inhibitor, implying that the conformation of the active site must be different in the enzyme-substrate and the enzyme-inhibitor complex (Thoma and Koshland, 1960).

Citri (1973) has discussed critically the various observations bearing on the conformational adaptability of enzymes. In some cases, it is possible to analyze crystallographically enzyme complexes with inhibitors or poor substrates and to compare the protein structures in these complexes with the structures of the free enzymes. Results obtained in this manner range from very extensive conformational transitions in lactate dehydrogenase (Adams et al., 1973) through more limited transitions in carboxypeptidase A, lysozyme, and nuclease (Lipscomb, 1970; Blake et al., 1967b; Arnone et al., 1969) to no detectable conformational change in chymotrypsin (Steitz et al., 1969). However, there are several reasons why results obtained in this manner may not necessarily reflect the behavior of enzymes in solution. For instance, when a substrate is allowed to diffuse into an enzyme crystal to form the complex, a conformational transition which would take place in solution may be prevented from occurring because of the rigidity of the crystal. A variety of spectroscopic techniques have been employed to demonstrate conformational transitions of enzymes in solution when the enzyme-substrate or enzyme-inhibitor complex is formed. Such data must be interpreted with great care, since the change in the environment of the active site of the enzyme when it is shielded from the aqueous medium by an adsorbed small molecule may lead to spectral changes even in the absence of a conformational transition. Two outstanding examples of clear proof of a conformational change may be cited. McClure and Edelman (1967) used a dye which fluoresces very slightly in water solution, but has its fluorescence enhanced 200-fold when it is adsorbed to hydrophobic regions of proteins. Chymotrypsin has the same affinity for the dye whether or not a substrate for the enzyme is present, so that the dye is not bound close to the active site. Yet the fluorescence efficiency and the emission spectrum are altered when the enzyme-substrate complex is formed, so that a conformational change must have occurred near the dye-binding site. An even more revealing study (Markley et al., 1970; Markley and Jardetzky, 1970) used NMR spectroscopy to characterize changes accompanying substrate binding by nuclease. One of the histidine hydrogens of the enzyme is characterized by two NMR peaks, suggesting that the free enzyme exists in two conformations in equilibrium with each other. Binding of an inhibitor changed the relative intensity of these peaks, reflecting a shift in the equilibrium of the two conformations. Such a shift may be the mechanism by which "induced fit" phenomena operate; according to Markley and Jardetzky, the NMR method may even be able to distinguish between a larger number of conformational states.

Very extensive work has been carried out in attempts to define the geometry of the active site of enzymes and to gain information about the spatial disposition of its various functional groups. In some cases, this has been done without the benefit of crystallographic information on the conformation of the protein, while other studies have aimed at a comparison of the structure existing in solution with that characterizing enzyme crystals. Very revealing data are obtained from enzyme-substrate (or enzyme-inhibitor) association constants for a series of judiciously chosen substances. Table VIII.1 reproduces some of the data on lysozyme (Chipman and Sharon, 1969), which has been the subject of a particularly extensive study. They show clearly that the adsorption site on the enzyme is quite extensive; we may note, for instance, that the association constant of GNAc—MNAc—GNAc is reduced by two orders of magnitude if another MNAc unit is added to the chain or if the acetamido group is eliminated from the terminal glucose unit. Another striking result was obtained with the enzyme papain (Schechter and Berger, 1968). If an oligopeptide contained a phenylalanine residue, the *next-but-one* peptide bond in the direction toward the carboxy terminal was split, showing that a binding site for the aromatic ring is located a considerable distance from the groups involved in the catalytic action. In studies of this type it is particularly advantageous to use conformationally rigid substrates or inhibitors (Erlanger, 1967; Cohen and Schultz, 1968).†

TABLE VIII.1.
Lysozyme Association Constants with Saccharides Containing *N*-Acetyl-D-glucosamine (GNAc), *N*-Acetylmuramic acid (MNAc), and D-Glucose (G) Residues

Saccharide	*Association Constant ?M^{-1}?*
GNAc	20
GNAc—GNAc	5,000
GNAc—GNAc—GNAc	100,000
GNAc—GNAc—GNAc—GANc	100,000
GNAc—MNAc	20
MNAc—GNAc	10,000
GNAc—MNAc—GNAc	300,000
GNAc—MNAc—GNAc—MNAc	2,000
GNAc—MNAc—G	1,800
GNAc—MNAc—G—G	170
GNAc—GNAc—GNAc—G—G	30,000

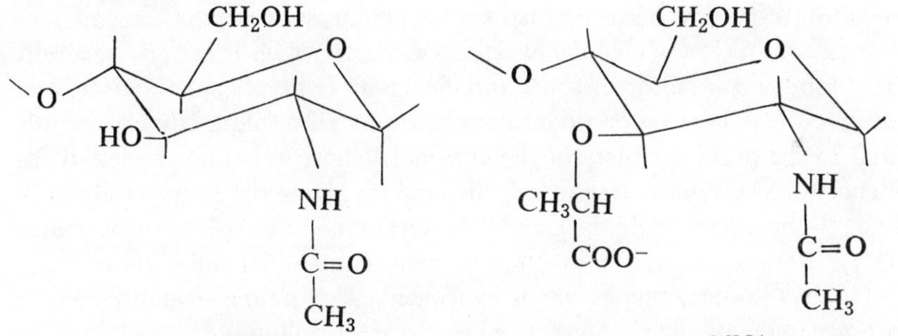

CH_2OH O O HO NH C=O CH_3

GNAc

CH_2OH O O O CH_3CH COO^- NH C=O CH_3

MNAc

Various spectroscopic techniques are also being employed to explore the environment in which small molecules find themselves when they are adsorbed to the active site of enzymes. For instance, when trifluoroacetyl-D-glucosamine oligomers are bound to lysozyme, the ^{19}F peaks in the NMR spectrum are shifted because of their proximity to the ionized aspartic acid and glutamic acid residues of the enzyme (Millet and Raftery, 1972). A very ingenious method was used by Latt et al., (1972) to define characteristic

†Fruton (1974) has shown that the activity of pepsin towards a series of synthetic substrates may be increased by several orders of magnitude by substituents located a considerable distance from the bond susceptible to enzymatic attack. He showed also that this effect is not due to tighter substrate binding but rather to a pronounced influence on the unimolecular rate constant characterizing the conversion of the enzyme-substrate complex. These results suggest that portions of the substrate far from the susceptible bond serve to effect a conformational transition of the enzyme which may greatly enhance its reactivity.

distances in carboxypeptidase-A complexes with its substrates. When a fluorescent substrate which had an absorption band overlapping the emission spectrum of tryptophan residues was used, the distance from a tryptophan located in the active site to the chromophore of the substrate could be calculated from the nonradiative energy transfer (cf. p. 241). Moreover, since the enzyme can be activated both by Zn^{2+} and by Co^{2+}, the reduction in the intensity of the substrate fluorescence due to energy transfer to the cobalt defined also the distance of this chromophore to the activating ion. Yet another powerful method utilizes Mn(II) or Gd(III) attached to a specific location in the active site; this produces a broadening of NMR peaks of adsorbed molecules, from which the distances of the various nuclei to the paramagnetic ion can be calculated (Butchard et al., 1972). With a method of recording the *difference* between the NMR spectra of two samples of a protein, one of which contains the paramagnetic ion, the complex spectrum of the protein is simplified to the spectrum of the vicinity of the paramagnetic probe (Campbell et al., 1973). This may develop into the most powerful method for mapping the geometry of enzymatic active sites in solution.

In a number of enzymatic reactions the catalytic process involves an association complex of three distinct molecular species: the enzyme, the substrate, and a low molecular weight coenzyme which functions generally as the donor or acceptor of some specific group (e.g., phosphate, acetyl) or as an intermediate in redox processes. Coenzymes are frequently very tightly bound to the protein moiety of the enzyme. Velick (1954) has reviewed the evidence for two typical systems of this kind involving the enzymes glyceraldehyde-3-phosphate dehydrogenase (GDH) and alcohol dehydrogenase (ADH). Both these enzymes employ nicotinamide adenine dinucleotide (NAD^+) as the coenzyme acting as hydrogen acceptor; the structures of this substance in the oxidized and reduced forms are as follows:

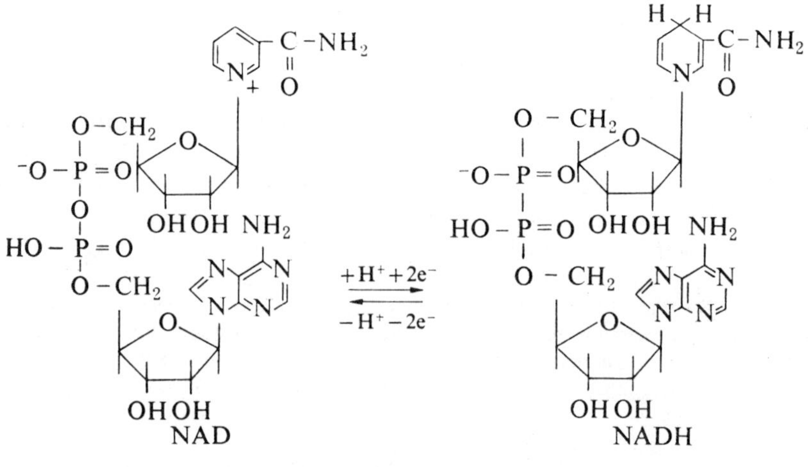

$$\text{NAD} \underset{-H^+ - 2e^-}{\overset{+H^+ + 2e^-}{\rightleftharpoons}} \text{NADH}$$

NAD NADH

The dissociation contant of the GDH complex with NAD^+ has been obtained from ultracentrifuge sedimentation data (since the free coenzyme does not sediment appreciably); it lies in the neighborhood of 10^{-5} (Velick, 1953). With alcohol dehydrogenase (Theorell and Chance, 1951) there is a remarkable difference in the affinity of the enzyme for the reduced and the oxidized form of the coenzyme. The equilibrium may be studied spectroscopically; the data yield a dissociation constant of 10^{-7} for the ADH-NADH complex, while the dissociation constant for ADH-NAD^+ is found to be 100 times higher. This large difference in the affinity of the enzyme for two such similar species illustrates rather dramatically the sensitivity of specific association phenomena to minor structural changes of the interacting molecules.

Some enzymes require for their catalytic activity the simultaneous presence of both a coenzyme and an activating ion. In that case, a very interesting method is available for the study of the association complex, provided that the activating ion is a paramagnetic species such as Mn^{2+}. This ion produces a very large acceleration of the nuclear magnetic relaxation rate of the water protons, but the efficiency of Mn^{2+}, in producing this effect, is quite sensitive to its state of complexation. It was found that the effect of Mn^{2+} is enhanced much more in ternary complexes with enzyme and coenzyme than in binary complexes with either of them alone (Cohn and Leigh, 1962). Therefore, the technique may be used to decide whether the paramagnetic ion in the ternary complex is coordinated simultaneously with both the enzyme and the coenzyme (Cohn, 1963). The use of paramagnetic probes for studies of this type has been reviewed by Mildwan and Cohn (1970).

The specific association of globular proteins with small molecules may serve a number of functions other than enzymatic catalysis. For instance, many low molecular weight species, such as sulfate, sugars, and amino acids, require protein carriers for their transport across the membranes of living cells, and a number of these proteins have been isolated and characterized (Pardee, 1968). Proteins which change their conformation on association with small molecules are also involved in the mechanism for the detection of taste and odor (Dastoli et al., 1968; Ash, 1968). Proteins binding specific sugars and polysaccharides have been isolated from many sources and studied in great detail (Sharon and Lis, 1972). Yet another phenomenon of this type is involved in the biosynthesis of certain enzymes which is induced by the binding of cyclic adenosine 3′,5′-phosphate (CAP) to a specific protein receptor on the cell membrane. This receptor has been isolated, and its association equilibrium with CAP has been determined (Anderson et al., 1971).

We shall return to interactions of proteins with low molecular weight species in later sections of this chapter in discussing the interaction of hemoglobin with its ligands, the allosteric control of enzyme activity, the association of antibodies with haptens, and the role of gene repressors.

B. MOLECULAR AGGREGATION OF MACROMOLECULES

1. Aggregation of Flexible Chain Molecules

For each type of molecular association of small molecules, an analogous complexation must exist with polymers carrying the appropriate interacting groups. In fact, the requirements for the association of chain molecules carrying a large number of interacting groups are much less stringent, since the complexes formed by their monofunctional analogs may be too unstable to be experimentally observable.

Typical examples of associating chain molecules are polystyrenes carrying a small number of hydroxyl or carboxyl groups (Trementozzi et al., 1952; Chang and Morawetz, 1956). The apparent molecular weights of such materials are found to depend on the solvent, and the high values observed in media such as toluene and carbon tetrachloride are due to hydrogen bond formation between the macromolecules. As expected, the macromolecular association may be eliminated by the addition of a low molecular weight hydrogen bond acceptor. This effect is illustrated in Fig. VIII,4, where the

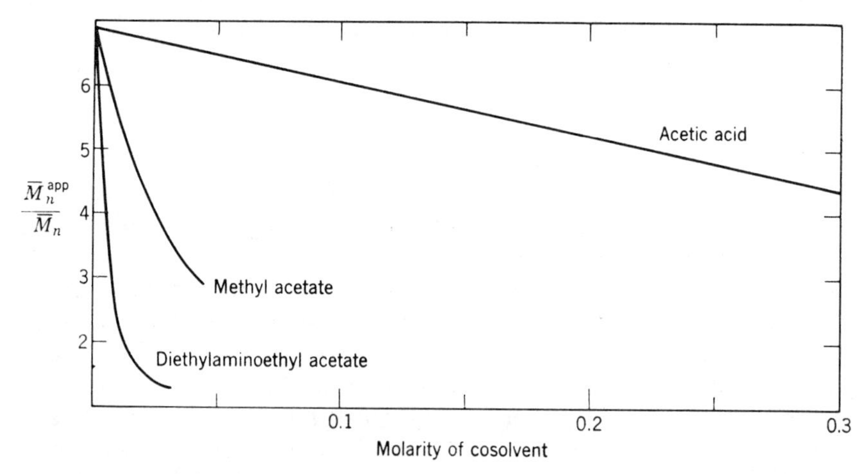

Fig. VIII.4. Effect of cosolvent on the apparent molecular weight of a methyl methacrylate copolymer with 4.9 mole % methacrylic acid in benzene solution at 30°C, ($\bar{M}_n$ = 34,500).

ratio of the apparent to the true molecular weight of a methyl methacrylate—methacrylic acid copolymer in benzene solution is given as a function of the concentration of various cosolvents (Morawetz and Gobran, 1955). The relative efficienty of a cosolvent in preventing molecular aggregation of the polymer is a measure of its affinity for the interacting carboxyl groups.

Molecular association is also observed in aqueous solutions of polymers carrying hydrophobic groups. This phenomenon was first described for incompletely hydrolyzed poly(vinyl acetate), which has an apparent light-scattering molecular weight increasing with temperature (Nord et al., 1951). Similar behavior characterizes aqueous solutions of methylcellulose (Neely, 1963)—decreasing solvation with increasing temperature is, in these cases, the expected behavior of phenomena due to hydrophobic bonding. In aqueous solutions of poly(α-L-glutamic acid) at alow pH, aggregation is favored by a lowering of the temperature (Jennings et al., 1968), so that the process must be exothermic; this suggests that molecular association results, in this case, from hydrogen bonding between the carboxyl groups.

In a number of cases, block copolymers in media in which only one of the blocks is well solvated form micelle-like aggregates. Such phenomena are exhibited, for instance, by chains with blocks of styrene and ethylene oxide (Franta, 1966) or styrene and methyl methacrylate residues (Kotaka et al., 1972).

A different type of aggregate is formed by poly(vinyl chloride) (Kratochvil et al., 1967, 1968). Polymer association takes place, in this case, even in thermodynamically strong solvents, and it is doubtful whether it can be entirely eliminated below temperatures at which the polymer degrades chemically. Even small amounts of aggregates produce characteristic distortions of Zimm light scattering plots, and larger amounts, in poorer solvent media, can be demonstrated by ultracentrifuge sedimentation. Yet these large aggregates are characterized by surprisingly low intrinsic viscosities, suggesting a very compact structure. The extent of aggregation increases when the temperature at which the polymer is prepared is lowered, and this indicates that incipient crystallization of syndiotactic sequences in the PVC chain is the cause of the polymer association (Crugnola and Danusso, 1968). Another interesting system in which macromolecular association is of this type is formed by mixtures of isotactic and syndiotactic poly(methyl methacrylate) in good solvent media. We noted previously (see p. 95) that the mixing of solutions of the two stereoregular polymers may lead to gel formation. In more dilute systems the growth of molecular aggregates can be studied by following the increase in the intensity of scattered light, and this technique may be used to determine the stoichiometry of the complex. Liquori et al., (1965) obtained the same results by this method and by using as an index of complex formation the intensification of the optical density at 212.5 nm. They prepared a fiber from the isotactic—syndiotactic complex and proposed its crystal structure. Liu and Liu (1968) found that the complex forms only in poor solvent media and that it leads to a pronounced broadening of the NMR peaks of the polymer. This indicates that the association must involve long sequences of the interacting chains, leading to structures in which conformational transitions are strongly impeded. This case is of special interest since the two interacting polymers are chemically very similar. Molecular association is here a consequence of the steric complementariness of the two interacting species, utilizing a principle similar to that which accounts for such striking specific interactions as those of an enzyme with its substrate or an antigen with its antibody.

While the mixing of solutions containing polymeric acids and polymeric bases leads to mutual precipitation even in extremely dilute systems (Fuoss and Sadek, 1949), the mixing of copolymers carrying relatively low densities of acidic and basic groups produces soluble complexes (Morawetz and Gobran, 1954). Mixtures of poly(ethylene oxide) and poly(acrylic acid) in aqueous solution lead to association due to the interaction of the acidic carboxyls with the basic ether oxygens. An analogous complex of a monofunctional acid with a simple ether would not be observable in an aqueous medium, where water competes as both a hydrogen bond donor and acceptor. Here the high stability of the polymer aggregate is the result of the cooperative effect of a large number of interacting groups. At low pH the complex is very insoluble, but as the pH is raised the extent of the molecular association decreases and the complex dissolves (Bailey et al., 1964).

Some interesting generalizations may be made about these various systems exhibiting molecular aggregation of flexible polymer chains.

(a) If the polymer carries only one kind of interacting groups which tend to dimerize (e.g., carboxyl), it may happen that in dilute solution very little macromolecular aggregation is observed, even though spectroscopic evidence shows that most of the interacting groups are associated. This means that most of the dimerization involves interacting groups attached to the same polymer chain back bone, as would be expected since the local density of the interacting groups is much higher within the domain of the isolated polymer coil than in the system as a whole. Typical results, comparing the extent of carboxyl dimerization in solutions of styrene-methacrylic acid copolymers and solutions of pivalic acid, their monocarboxylic acid analog (Chang and Morawetz, 1956), are shown in Fig. VIII.5. We can see that carboxyl dimerization is independent of the concentration of the copolymers, merely reflecting conditions within the isolated macromolecule.

(b) If we compare the behavior of different molecular weight fractions with a constant density of interacting groups, the extent of aggregation increases with the polymer chain length, since a larger number of associating groups are available to stabilize the complex. As a result, molecular association leads to an increase in the apparent breadth of the molecular weight distri-

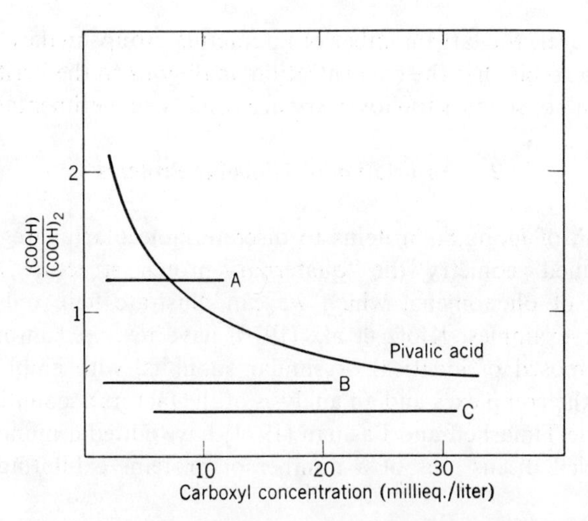

Fig. VIII.5. Carboxyl dimerization in styrene-methacrylic acid copolymers in 1,1',2,2'-tetrachloroethane. (A) 0.79 mole % methacrylic acid. (B) 6.5 mole % methacrylic acid. (C) 11.7 mole % methacrylic acid.

bution. In particular, the ratio of the apparent weight-average and number-average molecular weights may attain anomalously high values.

(c) The dissociation of polymer aggregates is usually much more rapid than the association of macromolecules, but both processes may require a long time for an approach to equilibrium conditions (Doty et al., 1947; Strauss et al., 1956; Neely, 1963). A striking demonstration of this effect was provided in studies of poly(vinyl chloride) (Hengstenberg and Schuch, 1964). When the polymer was dissolved in cyclohexanone (a good solvent) and diluted with the poorly solvating butanone, sedimentation diagrams gave no evidence of molecular aggregation. If, however, the same system was prepared by diluting a butanone solution of the polymer with cyclohexanone, molecular aggregation was as pronounced as if butanone had been the sole solvent.

(d) Although light scattering studies of poly(γ-benzyl-L-glutamate) reflect the dissociation of polymer aggregates so that extrapolation to zero concentration yields the correct $\bar{M}_w$ for the molecularly dispersed species (Elias and Gerber, 1968; Gerber and Elias, 1968), the size of other molecular aggregates is frequently constant in the concentration range in which measurements are feasible (Morawetz and Gobran, 1954; Hengstenberg and Schuch, 1964). Data obtained in a single solvent and at a single temperature may, therefore, give no warning that the polymer molecules are associated. The constancy of the size of the molecular aggregate over a concentration range is reminiscent of the formation of micelles from low molecular weight species. However, the large number of interacting groups makes the polymer aggregates so stable that the concentration analogous to the "critical micelle concentration" of soaps is too low to be accessible to experimental study.

2. Association of Globular Proteins

Association of globular proteins to discrete molecular aggregates with a precisely defined geometry (the "quaternary protein structure") leads to a wide variety of phenomena, which we can illustrate here only by a few characteristic examples. Klotz et al., (1970) have reviewed information on proteins composed of identical or similar subunits, with emphasis on the geometry of the complexes and an analysis of the factors accounting for their stability, while Timasheff and Fasman (1971) have edited a monograph containing detailed discussions of a number of proteins exhibiting molecular association.

The behavior of insulin is characteristic of one type of protein association. With its molecular structure completely clarified, its molecular weight is known as 5733, but under most experimental conditions its apparent molecular weight is twice or even six times as high as this value (Oncley et al., 1952; Fredericq, 1956; Marcker, 1960). The dissociation constant of the dimer at 25 °C, pH 2, and an ionic strength of 0.1 has been estimated from the concentration dependence of the apparent molecular weight as $10^{-4}M$ (Jeffery and Coates, 1966). It is undoubtedly significant that the unit cell of insulin crystals contains three dimers (Adams et al., 1969), so that the observation of dimers and hexamers in solution probably involves the same interactions between the insulin molecules which determine their packing in the crystal lattice.

In the case of hemocyanin, the copper-containing oxygen carrier in the bloodstreams of molluscs and arthropods, the aggregation behavior is much more complex. For instance, the hemocyanin of *Helix pomatia* forms at neutral pH complex aggregates with a molecular weight of 9×10^6. These have been shown, by electron microscopy, to be cylindrical with a diameter of 350 Å and a height of 150–180 Å (cf. Fig.I.6). As the pH is varied, the heavy particles dissociate into halves, tenths, and twentieths, but no intermediate species are observed. The pH regions over which the various aggregates are stable have been defined by Konings et al. (1969) by sedimentation analysis (Fig. VIII.6). Even the smallest species revealed in this experiment can, however, be further dissociated by drastic conditions into 18 polypeptide chains.

In several important cases, a large number of globular protein molecules associate to linear chains of indefinite length in a process frequently referred to as a "polymerization." A particularly interesting example of this type is

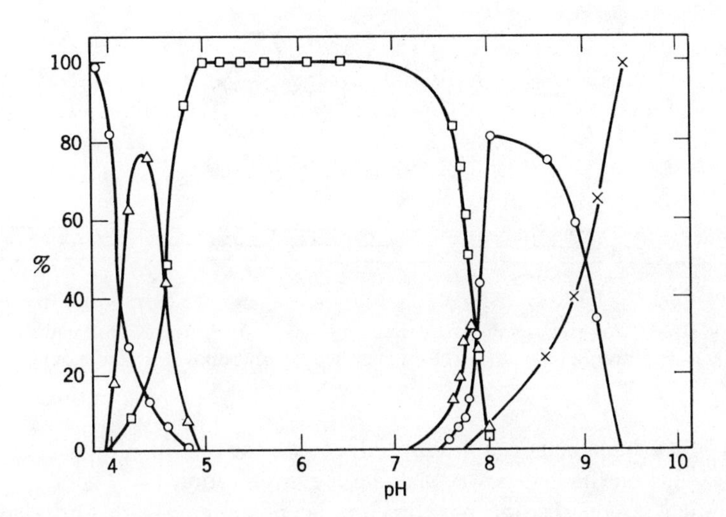

Fig. VIII.6. The pH dependence of the state of aggregation of hemocyanin from *Helix pomatia*.(□)Particles with M.W. = 9×10^6; (△)halves; (○)tenths; (×)twentieths. Concentration 2.8 mg/ml; ionic strength 0.1.

provided by actin, one of the components of the contractile system of muscle tissues. This protein consists of globular molecules ("G-actin") which have a high tendency to associate in the presence of certain salts in a very specific manner to form filamentous aggregates ("F-actin"). Each G-actin molecule is associated with one molecule of adenosine triphosphate, which is dephosphorylated during the polymerization process (Mommaerts, 1952), but the mechanism which relates F-actin formation with the breakdown of ATP is unknown. A detailed study of actin polymerization by flow birefringence (Oosawa et al., 1959) led to the conclusion that no F-actin forms below a critical actin concentration and that the concentration of G-actin is independent of the concentration of F-actin in the system. Thermodynamic and kinetic data led Oosawa and Kasai (1962) to the conclusion that the globular protein molecules associate to form a helical array in which each unit lies next to four nearest neighbors and that the cooperative nature of this process makes it the intermolecular analog of intramolecular helix-coil transitions. Electron microscopic observations of F-actin (Hanson and Lowy, 1963) show that it consists of two intertwined strings of G-actin molecules, as represented schematically in Fig. VIII.7a. This model is also consistent with the spectacular observation that solutions of F-actin catalyze the dephosphorylation of ATP when they are subjected to sonic vibrations (Asakura, 1961). The vibrations produce breaks in the double-helical aggregate of F-actin, and the "annealing" of these breaks is attended by the ATP breakdown characterizing actin polymerization. Since the repair of such a break in one

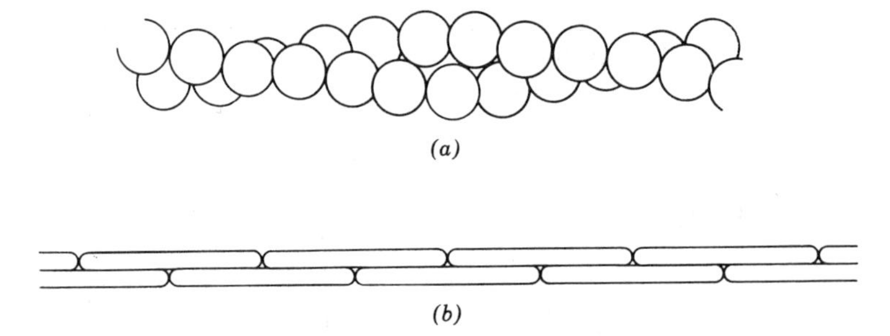

Fig. VIII.7. "Polymerization" of globular proteins. (*a*) Structure of F-actin as suggested by its appearance under the electron microscope. (*b*) Proposed structure of fibrinogen polymers.

strand of a double helix is a first-order process, the rate of ATP decomposition depends on the first power of F-actin concentration.

A similar polymerization phenomenon is encountered with fibrinogen, a globular protein responsible for the clotting of blood. The extensive literature dealing with this subject has been reviewed by Doolittle (1973). The native fibrinogen appears under the electron microscope as a linear array of three nodules held together by a relatively thin thread (Hall and Slayter, 1959). In this form the protein has no tendency to form molecular aggregates. However, when it is subjected to attack by the proteolytic enzyme thrombin, a small peptide is split off from fibrinogen, and the residual protein ("fibrin monomer") has then a high tendency to associate to large structures which eventually form a crosslinked gel. The crosslinking process may be inhibited by various reagents, and under these conditions it is possible to study the association of fibrin monomer to rodlike polymers. Light scattering studies suggest that the cross section of these rods is twice as great as that of fibrin monomer, and this result led to the proposal (Ferry et al., 1952; Casassa, 1956) that the linear "polymer" of fibrin grows by a staggered overlap of the elongated monomer units, as represented schematically in Fig. VIII.7b. However, electron microscopic studies yield a repeat period of 225 Å for fibrin and a length of 475 ± 25 Å for fibrinogen (Hall and Slayter, 1959), and these data suggest either a shrinkage of the monomer unit during polymerization or a different pattern of staggering.

Whereas the polymerization of myosin and fibrinogen to filamentous aggregates is obviously related to the function of these proteins, the significance of such associations may, in other cases, be quite obscure. Thus, the enzyme glutamate dehydrogenase, which consists of six subunits tightly bound to each other, will polymerize to rodlike aggregates with a length which depends on the enzyme concentration (Eisenberg, 1971). The polymerization may be described by a single equilibrium constant, $K = c_{i+1}/c_i c_1$, where the subscripts of the concentration terms indicate the number of enzyme molecules in the aggregate. It can be shown that the weight-average degree of polymerization $\bar{P}_w$ is then related to the total enzyme concentration, $c = \sum c_i$, by $(\bar{P}_w)^2 = 1 + 4Kc$. A number of other enzymes have also been observed to polymerize (Frieden, 1971), although the equilibria governing the size of the aggregate have not been defined in detail. Various reagents which presumably bind to the protein and effect a conformational transition may either stabilize or destabilize the polymeric form.

A very complex situation is encountered with casein, which contains a number of distinct components interacting with one another to form large micellar aggregates (Waugh, 1958, 1961). Partial proteolysis of one fraction ("κ-casein") by the enzyme rennin appears to lead to changes analogous to the conversion of fibrinogen to fibrin monomer, and the altered κ-casein may initiate, in the presence of Ca^{2+}, the polymerization of the protein leading eventually to the formation of an insoluble clot.

Yet another fascinating illustration of the effects which may result from the molecular association of globular proteins is provided in the structure of ferritin, utilized in higher organisms for the storage of iron. This substance has a molecular weight of 465,000, but more than 20% of this weight is due to ferric oxide and phosphate, which seems to be rather loosely associated with the protein. The iron may be removed, after reduction to the ferrous state, without destroying the protein structure, and this residual "appoferritin" dissociates in detergent solutions to subunits with a molecular weight of about 18,500. Crystallographic evidence suggests that these subunits are fitted together into the shape of a hollow sphere, which acts as a receptacle for the inorganic material (Crichton, 1973).

The buildup of protein molecules from separate polypeptide subunits may lead to cooperative effects in the binding of small molecules. This phenomenon, which is of crucial importance in biology, has been studied in particular detail on hemoglobin. Early studies of this protein, which started as far back as 1840, have been beautifully related by Edsall (1972): summaries of modern researches were published by Antonini and Brunori (1970, 1971). Hemoglobin consists of four subunits, two α and two β chains, which are similar but not identical. Each α chain has a special affinity for one of the β chains, so that the tetrameric $\alpha_2\beta_2$ protein may be relatively easily dissociated into $\alpha\beta$ halves. After blockage of a specific thiol group, the α and β chains may be separated chromatographically; the blocking group may then be removed, to allow comparative studies of the behavior of the subunits and the tetramer. Each subunit carries a heme group in which ferrous iron is coordinated with four nitrogen atoms of protoporphyrin. A fifth coordination bond is formed between the iron and a histidine residue of the protein, while a sixth coordination bond is formed with a water molecule. The water may be replaced by a variety of ligands; the reversible binding of oxygen is the physiological function of hemoglobin, while the much tighter binding of carbon monoxide is the cause of the high toxicity of that gas.

Hemoglobin has a number of most remarkable properties. The coordination of oxygen to the heme group in the intact protein leaves the iron in the Fe (II) state, while the iron of the free heme in aqueous solution is rapidly oxidized to Fe (III). The resistance of hemoglobin to oxidation, which is essential to the reversibility of its interaction with oxygen, may be ascribed to the location of the heme in a nonpolar environment (Wang, 1958), and it was demonstrated that a dye which fluoresces more intensely in hydrocarbon than in water solutions has its fluorescence dramatically enhanced when it replaces heme groups at their hemoglobin binding sites (Stryer, 1965). The oxygen-binding equilibrium of hemoglobin (Fig. VIII.8) is characterized

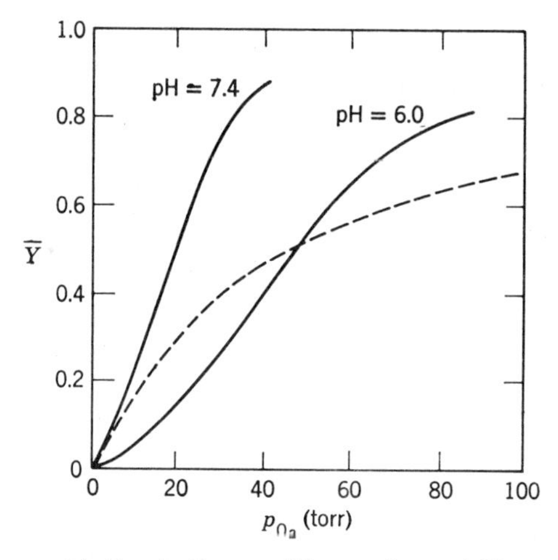

Fig. VIII.8. Oxygen-binding isotherms of human hemoglobin at 37°C (Adair, 1925). See text for the significance of the dashed curve.

by sigmoid isotherm curves, showing that oxygen binding by one subunit increases the affinity of oxygen for the other subunits. The affinity of the protein for oxygen is also increased when the pH is raised ("Bohr effect"). Both the cooperativity of oxygen binding and its pH dependence are absent in the monomeric subunits and in the tetramer of β chains (Bucci et al., 1965), so that they depend on interactions between α and β subunits. These two effects are also of great physiological advantage. The sigmoid shape of the binding isotherm causes the degree of oxygenation of hemoglobin to change abruptly within a modest range of oxygen pressures. Thus, a hypothetical oxygen carrier, which is half saturated at the same oxygen pressure as hemoglobin at pH 6.3, but in which the oxygen-binding sites are independent of each other (dashed line in Fig. VIII.8), will have its degree of oxygenation $\bar{Y}$ change only from 0.25 to 0.60 when the oxygen pressure is raised from 20 to 80 torr, whereas for hemoglobin $\bar{Y}$ rises from 0.12 to 0.80. The Bohr effect causes the high concentration of carbon dioxide in venous blood to favor the release of oxygen from hemoglobin.

Since the X-ray structures of both oxyhemoglobin and deoxyhemoglobin are known, a detailed interpretation of the cooperative ligand binding has become possible (Perutz, 1970). The heme groups in the tetrameric protein are too distant from each other for direct interaction. However, although the coordination of oxygen to a heme group leads to very small changes in the structure of the subunit, the subunit interactions and the quaternary protein structure are profoundly affected. Hemoglobin subunits have a much higher affinity for oxygen than the tetrameric protein, so that the cooperativity in the

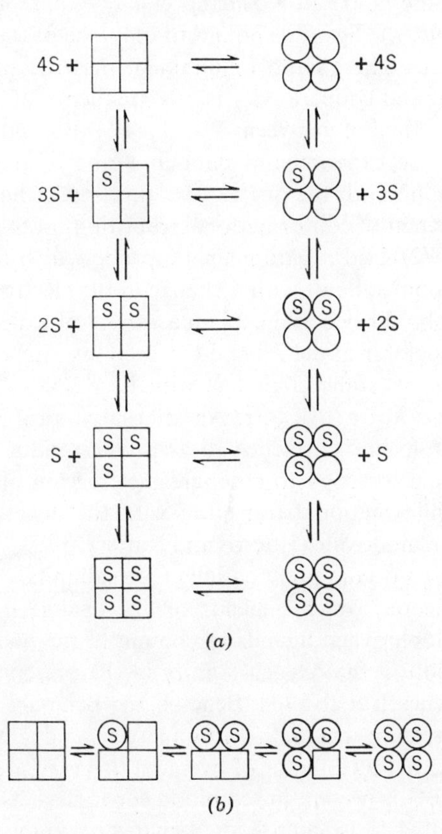

Fig. VIII.9. Models for the cooperative effect of ligand binding. Squares and circles symbolize the T and R conformational states, respectively; S symbolizes the ligand. (*a*) Model of Monod et al. (1965). (*b*) Model of Koshland et al.(1966).

oxygen binding may be thought of as a consequence of a low accessibility of the heme groups resulting from the quaternary structure of deoxyhemoglobin.

Cooperative phenomena, such as are shown in the binding of ligands by hemoglobin, are usually analyzed in terms of two alternative models, represented schematically in Fig. VIII.9. In the model of Monod et al. (1965) (Fig. VIII.9*a*) the subunits are assumed to exist in two alternative conformations, T and R, with different dissociation constants, K_T and K_R, for the association complex with a ligand. Cooperativity of ligand binding is postulated to be a consequence of a requirement that all subunits exist in the same conformational state. If we denote the equilibrium constant for the conformational transition of the unliganded protein, $R_0 \rightleftharpoons T_0$, by **L** and set $K_R/K_T = \mathbf{c}$, then cooperativity will increase with increasing **L** and decreasing **c** (Fig. VIII.10). In the model of Koshland et al (1966), represented schematically in Fig. VIII.9*b*, the ligand is bound to only one of the conformational states of the subunits. There is no requirement that the subunits be in the same conformation, and cooperativity is a consequence of differences in the free energies of interaction between T + T, T + R, and R + R nearest neighbors. To analyze experimental data in terms of these models, it is necessary to determine both the degree of saturation of hemoglobin by the ligand, $\bar{Y}$, and the extent of conformational transitions of the subunits. Ogata and McConnell (1972) used a paramagnetic probe which binds only to the unliganded hemoglobin subunit with a change in the electron spin resonance spectrum of the probe. They could explain a variety of experimental data on CO binding to hemoglobin at pH 7.3 and 13°C, taking into account the different ligand affinities of α and β chains, with $\mathbf{L} = 3000$, $K_T^\alpha = 24.2$, $K_R^\alpha = 0.36$, $K_T^\beta = 32.8$, and $K_R^\beta = 0.18$ torr. An alternative probing technique, in which a fluorinated group is attached to a specific residue of the β chains, utilizes ^{19}F NMR spectroscopy to compare the fraction of β chains which have undergone conformational transition with the degree of hemoglobin saturation by carbon monoxide (Huestis and Raftery, 1973). The results show that ligand association occurs preferentially to α subunits.

Another phenomenon, which depends on the change in the quaternary structure of hemoglobin when ligands are bound to the heme groups, is the pronounced reduction in the oxygen affinity in the presence of diphosphoglycerate (DPG) (Benesch et al. 1968; Benesch and Benesch, 1969). At $10^{-3}M$ DPG, the oxygen pressure required to attain any given degree of hemoglobin oxygenation is increased by a factor of five, and this is of great physiological importance since DPG is present in red blood corpuscles. Yet DPG does not modify oxygen binding to isolated α or β chains or to the β tetramer. One molecule of DPG associates with one molecule of deoxyhemoglobin, and the reduced oxygen affinity results from the free-energy requirement for the dis-

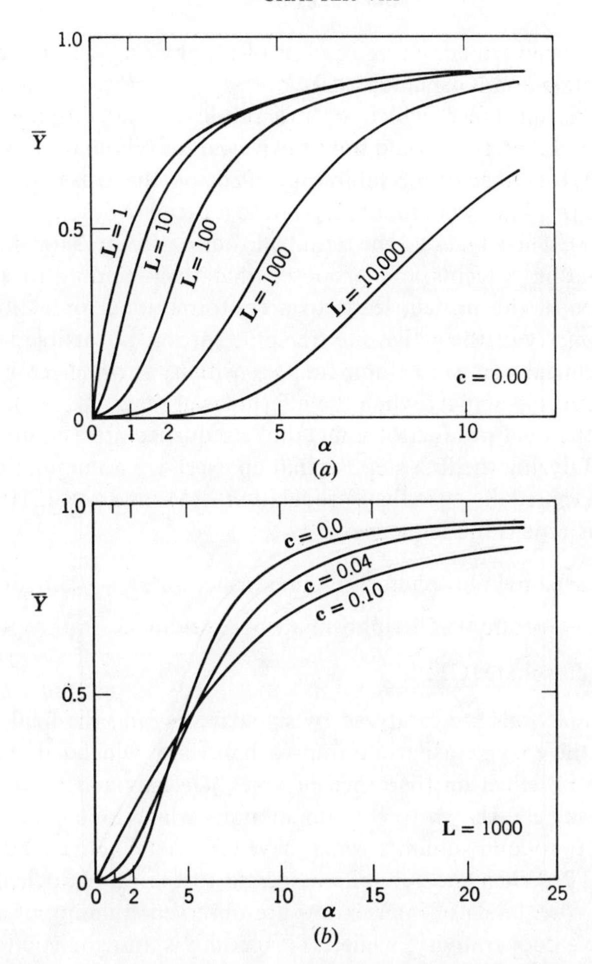

Fig. VIII.10. Dependence of the cooperativity of ligand binding on the parameters **L** and **c** according to the model of Monod et al. (1965) for a protein containing four subunits; $\alpha = (S)/K_R$.

sociation of this complex before a transition to the quaternary structure of oxyhemoglobin is possible.

With enzymes composed of several subunits with catalytic sites, substrate binding to one subunit may increase or decrease the substrate affinity to the other subunits (positive or negative cooperativity). It is commonly assumed that velocity **v** the enzymatically catalyzed reaction is proportional to the degree of saturation $\bar{Y}$ of the enzyme with substrate at a given free substrate concentration (S). If $\mathbf{v}_{max}$ is the reaction velocity attained at large (S) (i.e., at $\bar{Y} = 1$). a sigmoid shape of a plot of $\mathbf{v}/\mathbf{v}_{max}$ against (S), similar to the plots in Fig. VIII.8, is indicative of positive cooperativity. Negative cooperativity, revealed by a downward curvature of plots of $1/\mathbf{v}$ against $1/(S)$, has also been observed (Levitzki and Koshland, 1969).

In many cases, substances which are structurally unrelated to the substrate of an enzyme and therefore would not be expected to be bound to its catalytic site have a large enhancing or inhibitory effect on the enzymatic activity. Phenomena of this kind were first subjected to a detailed description by Monod et al. (1963), who suggested the term "allosteric effectors" for the activating or deactivating reagents and proposed that their binding to a distinct "allosteric" site of the protein leads to a conformational transition which affects the geometry of the active site. In some cases it is possible to modify an enzyme chemically so as to eliminate its sensitivity to an effector, without changing the enzyme activity when the effector is absent.

Sometimes the final product of a metabolic sequence of reactions inhibits the enzyme catalyzing the first step, so that an excessive accumulation of the product is prevented by "feedback regulation" (Monod et al., 1963). An example of this type is the sequence

$$\text{Aspartate} + \text{carbamyl phosphate} \xrightarrow{1} \text{ureidosuccinate} \xrightarrow{2} \text{dihydroorotate} \xrightarrow{3} \text{orotate} \xrightarrow{4} \text{orotidine-5}'\text{-triphosphate} \xrightarrow{5} \text{uridine-5}'\text{-triphosphate} \xrightarrow{6} \text{cytidine-5}'\text{-triphosphate (CTP)}$$

where the six reactions are catalyzed by six enzymes, and the final product, CTP, inhibits the enzyme aspartate transcarbamylase, which catalyzes reaction 1 but has no effect on the other enzymes (Gerhart and Pardee, 1962). The enzyme has been shown to contain subunits which bind CTP but have no catalytic activity and subunits which have enzymatic activity but are unaffected by CTP. When the subunits associate to the physiologically functional enzyme, two kinds of interactions are observed: binding of substrate exhibits positive cooperativity, while CTP binding is antagonistic to the association with substrate (Changeux et al., 1968). Crystallographic analysis of the enzyme structure shows that it contains six catalytic and six regulatory subunits, and it is hoped that at higher resolution the mechanism of the allosteric regulation can be clarified (Evans et al., 1973). With other enzymes, the same subunit contains both the catalytic site and the site binding activating or inhibitory effectors. In some cases, binding of the effectors changes the degree of subunit association which may enhance or depress the catalytic activity, but this is not necessarily the mechanism by which effectors fulfill their function (Stadtman, 1966).

Molecular association complexes may sometimes be formed by distinct enzymes catalyzing the same reaction. A typical example is lactate dehydrogenase, which occurs in two variants, one (M) predominant in skeletal muscle, and the other (H) most plentiful in heart tissue. The enzyme contains four subunits, and all possible subunit combinations, M_4, M_3H, M_2H_2, MH_3, and H_4, occur in the living tissues (Markert, 1969); their behavior is very similar and they are referred to as "isozymes." Of much greater importance is the frequent association of enzymes, catalyzing successive metabolic processes, to "multienzyme complexes." For tryptophan synthetase, which has an $\alpha_2\beta_2$ structure, the catalytic activity of the isolated subunits is relatively small (Crawford and Yanofsky, 1958; Crawford and Ito, 1964), but in other cases the isolated enzymes are fully functional. The field has been reviewed by Reed and Cox (1970). The multienzyme complexes may also contain regulatory units, and the components of the particles are arranged in a precise geometric order which has been studied by electron microscopy and x-ray diffraction. If a single particle can provide an "assembly line" on which successive reactions of a metabolite can be performed, the free-energy expenditure is saved which would otherwise be required for the concentration of the substrate in the vicinity of the enzyme at each step of the reaction sequence. An interesting example of the use to which evolutionary development has put this principle is the biosynthesis of cyclic polypeptide antibiotics such as gramicidin S (Lipmann, 1973).

In a number of cases, organisms synthesize proteins which are specific enzyme inhibitors. The best-known examples of this phenomenon are proteins from a variety of biological sources which inhibit trypsin by forming with it stoichiometric complexes. The equilibrium constant for the formation of one such complex has been estimated to exceed 10^{14} (Vincent and Lazdunski, 1972), and this astounding value can be understood as a reflection of the close steric fit of the two proteins, as revealed also by x-ray crystallographic analysis (Rühlman et al., 1973).† A related phenomenon is the association of oxyhemoglobin with the glycoprotein haptoglobin. Nagel et al. (1965) showed that this association eliminates the cooperativity of oxygen binding and decreases the oxygen dissociation pressure. Deoxyhemoglobin does not form a complex with haptoglobin, but deoxygenation of the oxyhemoglobin complex leaves the haptoglobin-hemoglobin complex intact (Chiancone et al., 1966). These results indicate that haptoglobin complexation requires the quaternary structure of liganded hemoglobin and that this structure is "frozen" once the complex is formed, so that transition to the structure of deoxyhemoglobin is inhibited even if the ligand is removed.

Another class of important molecular interactions based on the principle of stereochemical complementarity involves the physiological action of hormones. This action requires the hormones to associate with a specific "receptor protein" on the cell membrane. In the case of insulin the receptor

† Means *el al* (1974) have recently reviewed studies of the association of trypsin and other proteolytic enzymes with proteins functioning as their specific inhabitors.

has been isolated and has been found to associate specifically with insulin (Cuatrecasas, 1972).

A detailed discussion of the extremely complex interactions between a number of protein species which are responsible for muscular contraction is beyond the scope of this book.† Essentially, the phenomenon involves the regulation of the activity of the enzyme ATP-ase, which catalyzes the hydrolysis of adenosine triphosphate, the reaction which provides the energy for muscular work. The ATP-ase activity is located in the globular ends of myosin (cf. p. 155), and the enzyme is activated when myosin associates with F-actin (cf. Fig. VIII.7). If, however, actin forms a complex with the rodlike protein tropomyosin and a small globular protein, troponin, in the absence of Ca^{2+}, the F-actin appears to undergo a structural change which prevents it from activating myosin. Extremely small concentrations of Ca^{2+} are sufficient to bring about a conformational change in the regulatory protein troponin, so that the activation of the ATP-ase by myosin is restored (Ebashi and Endo, 1968; Hitchcock, 1973).

3. The Antigen-Antibody Reaction

When a globular protein from another species is injected into the bloodstream of mammals or birds, the organism produces a new form of protein (an "antibody") which interacts in a highly specific manner with the foreign substance. Materials which may evoke this response are called antigens; they include, in addition to proteins, certain polysaccharides as well as some synthetic polypeptides. Early work in this field is beautifully related in Landsteiner's (1946) classical monograph. The spectacular developments in later years are summarized in Vol. 32 (1967) of the *Cold Spring Harbor Symposia on Quantitative Biology*, in a review by Dorrington and Tanford (1970), and in Vol. 190 (1971) of the *Annals of the New York Academy of Sciences*.

When antigen solutions are mixed in certain proportions with solutions of their antibodies, the two substances form an insoluble complex; when one or the other component is in large excess, the formation of soluble complexes may be demonstrated by light scattering, ultracentrifuge sedimentation analysis, electrophoresis, and other techniques. The most striking aspect of this interaction is its very high degree of specificity. A spectacular example is the isolation of an antibody specific for hemoglobin from patients of sickle cell anemia, although this differs from normal human hemoglobin in only 2 of its 574 amino acid residues (Noble et al. 1972). Antibodies react most

†An excellent review was written by Huxley (1972), and an extensive account of the status of this field is contained in Vol. XXVII (1972) of the *Cold Spring Harbor Symposia on Quantitative Biology*.

strongly with the "homologous antigen" which induced their formation, but they frequently "cross-react" to some extent with closely related species. For instance, the antibody to chicken egg albumin will precipitate duck egg albumin, but after completion of this reaction the supernatant will still precipitate the homologous antigen. This suggests that some of the antibody molecules interact with portions of the egg albumin which are similar in the proteins originating from chickens and ducks, while other antibody molecules are specific to regions characteristic of the chicken protein alone. This heterogeneity of antibody preparations was for many years a serious obstacle to studies of their molecular structure.

The antigen-antibody reaction is eliminated if the interacting proteins are denatured, and the interaction must then depend on a complementarity in the "molecular surfaces" of the two native proteins. A powerful method for the study of the specificity of such interactions was developed by Landsteiner, who introduced antigens produced by coupling proteins with a variety of diazotized aniline derivatives. The specificity of antibodies produced by such conjugated antigens is largely determined by the substituent groups (haptens). Since the antibody will also bind haptens attached to small molecules, the association equilibrium may be determined in a dialysis experiment (Haurowitz and Breinl, 1933). This method establishes also that antibodies of the most important class, designated as IgG (immunoglobulin G), associate with two hapten groups.

The structure of IgG antibodies is known today in such detail as would have seemed beyond hope only a few years ago. The sequence of discoveries leading to such understanding has been described in the Nobel Prize addresses of Porter (1972) and Edelman (1972). When antibodies isolated from a rabbit were treated with the proteolytic enzyme papain, three fragments were obtained. One of them (now designated as Fc) had no affinity for the antigen but was found to crystallize; this had not been observed with intact antibody because of its molecular heterogeneity. The other two fragments (now designated as Fab) behaved like univalent antibodies, that is, they could not precipitate antigen but could block the antigen-antibody precipitation. An alternative method of antibody degradation, in which interchain disulfide bonds were reduced in a medium favoring the separation of polypeptide chains, yielded two identical heavy chains with a molecular weight of about 50,000 and two identical light chains with a molecular weight of about 20,000.

The next major advance exploited an unexpected accident of nature. It had long been known that patients suffering from a cancerous condition called multiple myeloma excrete into their urine a substance called Bence-Jones protein. Edelman and Gally (1962) showed that this molecularly homogeneous protein is similar to the light chain of normal IgG; thus, the homogeneity of Bence-Jones protein reflects the pathological proliferation of cells capable of synthesizing a single antibody species. Suddenly, large quantities of molecularly homogeneous IgG, characteristic of each myeloma patient, became available. This provided a Rosetta stone which made it possible to

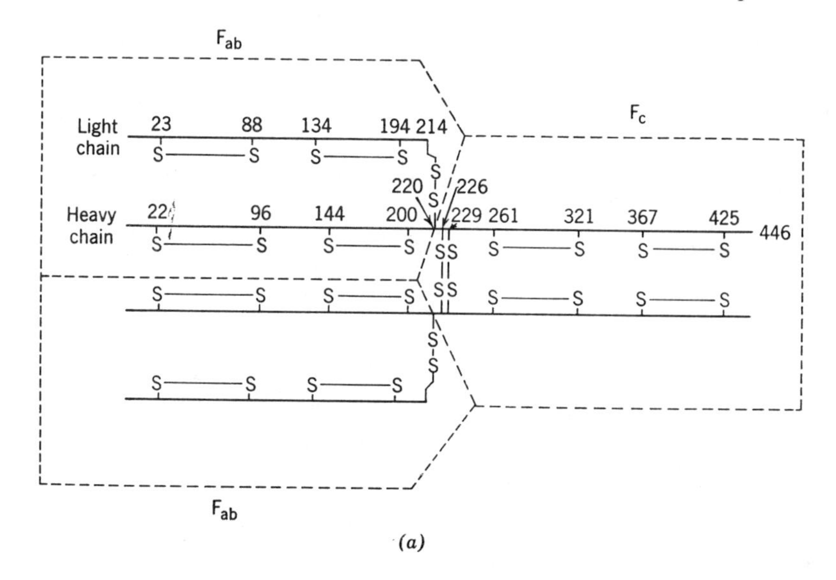

(a)

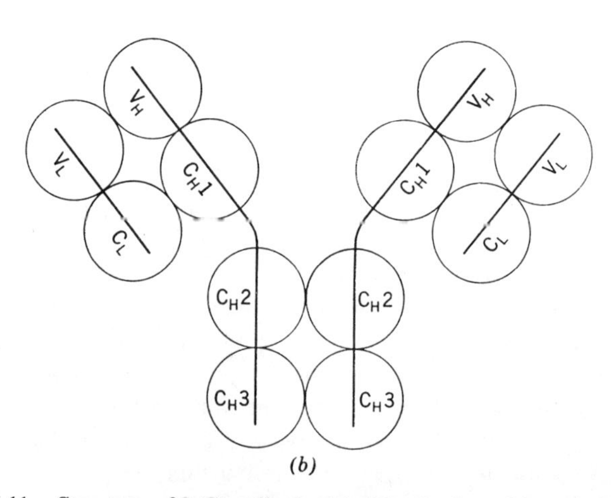

(b)

Fig. VIII.11. Structure of IgG antibody. (*a*). Distribution of intrachain and interchain disulfide bonds in IgG from a myeloma patient. The first 108 residues of the light chains and the first 114 residues of the heavy chains constitute the variable region responsible for the antibody specificity. Trypsin attack on residue 222 of the heavy chain (bonded in the "hinge" between residues 220 and 226) degrades the IgG into a Fc and two Fab fragments. (*b*). Schematic representation of the buildup of IgG from constant (C) and variable (V) globular regions of light (L) and heavy (H) chains.

derive details of antibody structure and, it was hoped, could lead to a clarification of the mechanism of its action.

Soon after the report of Edelman and Gally was published, two laboratories reported that the 212 amino acid residues of Bence-Jones protein consist of two regions of equal length, one of them constant, the other highly variable from patient to patient (Hilschmann and Craig, 1965; Titani et al., 1965). A variable region of similar length was later found also in the heavy chains of IgG, giving rise to the hypothesis that the variable regions of the light and heavy chains are responsible for the specificity of the antibody. The availability of homogeneous IgG samples from a myeloma patient resulted in the solution of the complete amino acid sequence and the location of all intrachain and interchain disulfide bonds (Gall and Edelman, 1970). The structure is shown schematically in Fig. VIII.11*a*. There is a great deal of evidence that the regions in both the light and heavy chains which contain an intrachain disulfide bond form compact globular structures (Fig. VIII. 11*b*), and such an arrangement is also indicated by crystallographic analysis (Padlan et al., 1973). The Fab fragment can be broken to separate the constant and variable regions; the variable regions of the light and heavy chains are strongly associated by noncovalent bonds to yield a particle which retains all of the specificity of the intact IgG (Hochman et al., 1973). The hinge by which the Fab fragments are attached to Fc appears to be quite flexible. This follows, for instance, from an experiment in which a fluorescent hapten was associated with its antibody and the decay of the polarization of fluorescence was followed after a light flash. This decay was biphasic with relaxation times of 33 and 168 nsec. The longer relaxation time was interpreted as reflecting rotational diffusion of the antibody as a whole, while the fast process was attributed to the flexibility of the antibody hinge (Yguerabide et al., 1970)

Antigen precipitation by IgG may be interpreted as the formation of a gel network through the interaction of divalent antibody with multivalent antigen. The growth of such a network may be inhibited by an excess of either reagent, as represented schematically in Fig. VIII.12. Unless the antigen consists of a number of identical subunits, its various antigenic determinants will all be different and a single discrete antibody species will then be able to bridge only two antigen molecules—it will not be able to produce antigen precipitation. However, a large excess of such a pure antibody is able to interfere with the precipitation by pooled antibody. Similarly, fragments of antigenic proteins have been shown to act as single antigenic determinants, incapable of precipitating antibody but able to prevent antibody precipitation by the intact protein antigen. Arnon and Sela (1969) have prepared the antibody to the oligopeptide containing amino acid residues 64–83 on the 120-residue chain of lysozyme. As shown schematically in Fig. III.32*a*, a disulfide bond bridges residues 65 and 72 and the conformationally rigid loop is

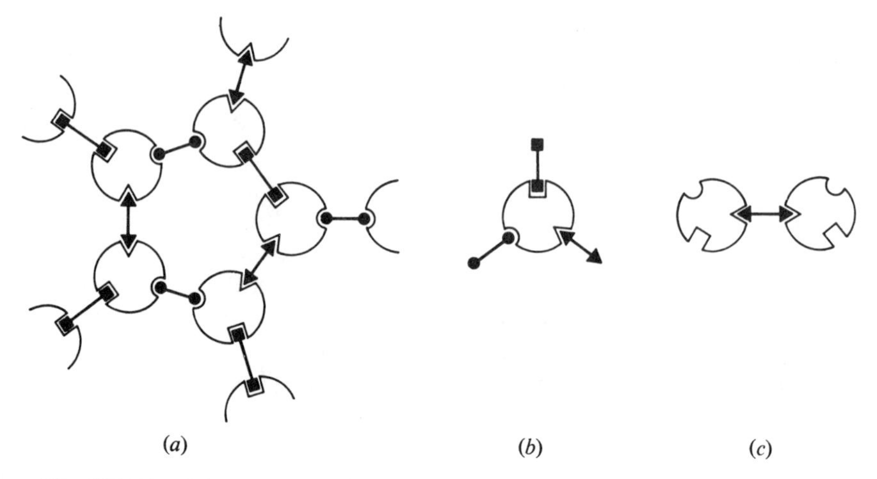

Fig. VIII.12. Schematic representation of an antigen-antibody reaction. (*a*) Association of antigen with three antigenic determinants with antibody in proportions such as to favor the formation of a large gel network. (*b*) Soluble particles formed with excess antibody. (*c*) Bridging of an excess of a multivalent antigen by an antibody directed against a single determinant.

apparently highly effective as an antigenic determinant. Even so, the antibody to the "loop peptide" does not appear to be entirely homogeneous, and its affinity is distinctly less for the peptide than for the protein antigen (Maron et al., 1971).

Evaluations of the association constant characterizing the interaction of antigen and antibody sites may be obtained in a number of ways. For instance, system containing a stoichiometric excess of antigen (AG) over antibody (AB) in terms of interacting sites, will contain mostly the species AG, AG–AB, and AG–AB–AG. After determination of the concentration of one of these species by electrophoresis or ultracentrifugation (Singer and Campbell, 1955), the concentrations of all the other species was calculated neglecting the antibody heterogeneity and assuming that the affinity of two interacting sites for one another is independent of the size of the aggregate. A particularly elegant technique for evaluating the association equilibrium utilizes the quenching of the tryptophan fluorescence of antibodies when they combine with haptens absorbing in the 300–400 nm range (Velick et al., 1960). Since fluorescence is detectable at extreme dilution of the fluorescing species, this method may measure with high precision dissociation constants as small as 10^{-8} to $10^{-9}M$. A similarly powerful method utilizes the change in the polarization of fluorescence when an antigen labeled with a fluorescent substituent combines with its antibody (Dandliker et al., 1964; Kirszenbaum et al., 1960). Using this technique for the study of ovalbuminantiovalbumin association, Dandliker et al. obtained dissociation constants of the order of $10^{-8}M$, while Singer and Campbell had found for the same system $10^{-4}M$. Dandliker et al. believe that this large discrepancy is caused by the heterogeneity of the antibody and the difference in the response of the two methods to the wide range of affinities between the components of the system.

Additional insight into the nature of the antigen-antibody reaction may be obtainable from experiments with synthetic polypeptides. Copolypeptides of L-lysine and L-glutamic acid have been found to be antigens, although poly(L-lysine) and poly(α-L-glutamic acid) do not elicit antibody formation (Gill and Doty, 1961). Good antigens are also obtained from copolypeptides of either L-lysine or L-glutamic acid with an amino acid carrying a hydrophobic residue such as alanine or tyrosine (Gill and Matthews, 1963). The antibodies produced in response to these materials cross-react with polypeptides of a range of compositions, but precipitation is clearly maximized with the homologous antigen. Typical data illustrating the extent of the specificity of this interaction are plotted in Fig. VIII.13, which shows the amount of two anti-polypeptide antibodies precipitated as a function of the concentration of various polypeptide antigens.

The use of synthetic polypeptide antigens allowed Sela (1969) to demonstrate the distinction between antigenic determinants depending on the sequence of amino acid residues and determinants depending on conformational properties. Thus a branched polyalanine carrying L-tyrosyl-L-alanyl-L-glutamic acid side chains is an example of an antigen with a "se-

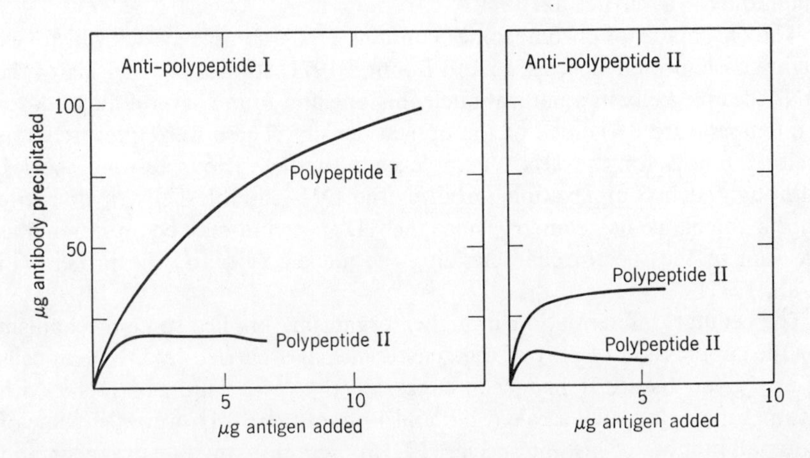

Fig. VIII.13. Specificity of interaction of synthetic polypeptides with their antibodies. Polypeptide I contained 42 mole % glutamic acid, 28 mole % lysine, and 30 mole % alanine. Polypeptide II contained 62 mole % glutamic acid, 33 mole % lysine, and 5 mole % phenylalanine (Gill and Matthews, 1963).

quential determinant," while the α-helical polypeptide with a repeating Tyr-Ala-Glu sequence is an antigen with a conformational determinant. The antibodies to the two substances exhibit practically no cross reaction.

4. Nucleoproteins and Their Analogs

Association of nucleic acids with proteins is involved in the formation of two essential structures of the living cell, the ribosomes and the chromosomes. The high complexity of these particles renders their study very difficult and this has provided an incentive for the investigation of association complexes formed in model systems containing nucleic acids or synthetic polynucleotides and synthetic polypeptides. Using this approach, it was found that addition of polypeptides of basic amino acids to DNA leads to biphasic melting curves, where the two T_m values are characteristic of free and complexed DNA, independently of the ratio in which they are mixed (Olins et al. 1967, 1968). Polylysine is more effective than polyarginine in stabilizing the double-helical structure of DNA. The two polypeptides exhibit also different specificities in their association with DNA samples of varying base compositions, polylysine interacting preferentially with A-T-rich DNA, while polyarginine binds more strongly to G-C-rich samples (Leng and Felsenfeld, 1966b). The association complexes contain amino acid and nucleotide residues in stoichiometric equivalence, whatever the proportion in which the DNA and the polypeptide are mixed. They form swollen spherical particles with a radius of 1700 Å and have CD spectra quite different from that of DNA alone, with an optical activity increased by two orders of magnitude (Shapiro et al., 1969).

The chromatin of chromosomes contains DNA complexed with histones, highly basic protein species. Li and Bonner (1971) studied the melting of the DNA double helix in a natural nucleohistone and found several maxima for the temperature derivative of the optical density. These were interpreted as melting points for the DNA complexes with the various histone species. Although such complexation stabilizes the DNA double helix, it does not appear to change its geometry, since the CD spectrum may be interpreted as the sum of the spectra characterizing the nucleic acid and the protein (Li et al., 1971).

The cellular differentiation in higher organisms implies some mechanism for the suppression of part of the genetic message, carried by DNA, in cells of any given tissue. It had been suggested that association of DNA with histones may play such a role (Shih and Bonner, 1970). However, in view of the small number of histone species (10–15) present in any one organism and the relatively slight difference in their affinities to DNA regions with different base compositions, von Hippel and McGhee (1972) doubt this interpretation and suggest that the formation of nucleohistone serves to control the general structure of chromosomes. On the other hand, some proteins acting as repressors of specific genes have been isolated; the most complete investigation has been carried out on the repressor for the *lac* gene directing the biosynthesis of the enzyme β-galactosidase, which catalyzes the hydrolysis of lactose in the bacterium *Escherichia coli* (Müller-Hill, 1971; von Hippel and McGhee, 1972). It has been estimated that this repressor has an association constant with the section of DNA coding for this enzyme (the *lac* operon) in the range of 10^{11}–$10^{12}M^{-1}$, while its affinity for other sections of DNA is lower by about five or six orders of magnitude. Since the geometry of the outside of the DNA double helix is rather insensitive to the nucleotide sequence, this high discrimination seems to suggest local melting of the DNA and association of the repressor with a single-stranded portion of the chain. Biological control over the synthesis of the enzyme is maintained by an ingenious mechanism in which a substance, derived from the substrate of the enzyme, complexes with the repressor and reduces its affinity for the *lac* operon. In this way, the repressor action is eliminated and the biosynthesis of the enzymes is induced. Inducers may exhibit very large affinities for the repressor; for example, isopropyl-1-thio-β-D-galactopyranoside forms with the *lac* repressor a complex with a dissociation constant of $1.8 \times 10^{-6}M$.

Our current views of the structure and function of bacterial ribosomes have been reviewed by Garrett and Wittmann (1973). Ribosomes are large nucleoprotein particles which provide the site at which the nucleotide sequence in "messenger" RNA is "translated" into an amino acid sequence during the synthesis of a protein. The ribosome contains two subunits with sedimentation constants $\mathbf{s}=30$ and $\mathbf{s}=50$, but each of these subunits has an exceedingly complex structure. The $\mathbf{s}=30$ subunit, with a molecular weight of about 0.9×10^6, contains one RNA chain (M.W. = 0.55×10^6) and 21 distinct protein species. The $\mathbf{s}=50$ subunit, with a molecular weight of 1.55×10^6, contains two different RNA chains (M.W. = 1.1×10^6 and 4×10^4) and between 30 and 35 distinct protein species. It is astonishing that efforts to reconstitute ribosomes, with all their functional properties, from the RNA and the many protein species have been crowned with success (Nomura, 1973). In this reconstitution it is essential to add the proteins in a specific order to the nucleic acid. The ribosome is by far the most complex biological structure which has been observed to assemble spontaneously from its macromolecular components.

Finally, we come to a consideration of viruses, nucleoprotein particles whose RNA or DNA component is endowed with the capacity to enter the genetic apparatus of a specific host cell of a living organism and modify it so that the cells propagate the virus. The degree of complexity of the virus structure varies within wide limits. Some viruses are composed of only two kinds of molecular species, a single chain of RNA and a large number of identical protein subunits. Caspar and Klug (1962) pointed out that the two most efficient designs for a biological container for the nucleic acid composed of a large number of identical protein subunits are helical tubes and icosahedral shells. Other viruses are much more complex, containing many different protein species and exhibiting a complex morphology. Examples of these modes of virus assembly (for a review see Eiserling and Dickson, 1972) are represented diagramatically in Fig. VIII.14.(*a*) Tobacco mosaic virus (TMV) has the shape of an elongated hollow cylinder with a diameter of 150 Å and a length of 3000 Å, containing 2130 identical protein subunits with a molecular weight of 17,500 and a single chain of RNA with a molecular weight of about 2×10^6. Crystallographic analysis shows that the protein is arranged in a helical array and that the RNA, which also forms a spiral, is embedded between the protein units (Franklin et al., 1958). (*b*) Particles with 180 protein subunits arranged with icosahedral symmetry are characteristic of a number of small spherical plant RNA viruses such as brome mosaic virus (BMV), cucumber mosaic virus (CMV), and turnip yellow mosaic virus (TYMV).

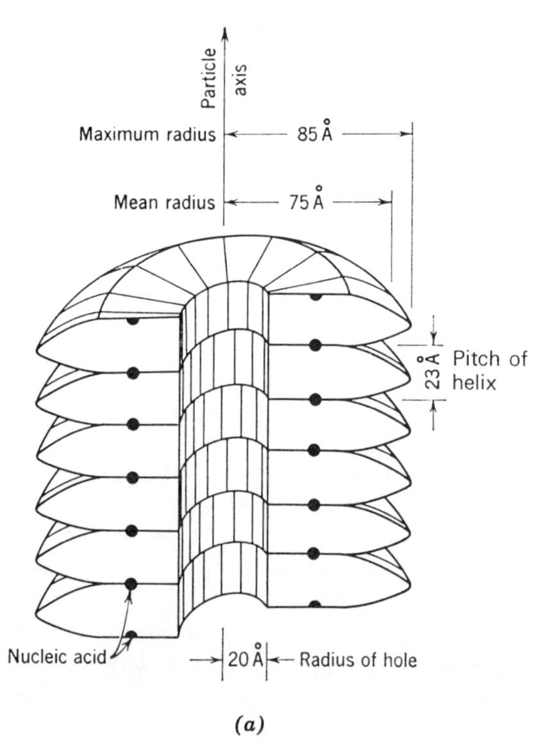

(*a*)

Fig. VIII.14. Structures of some representative viruses. (*a*) A section of the tobacco mosaic virus particle.

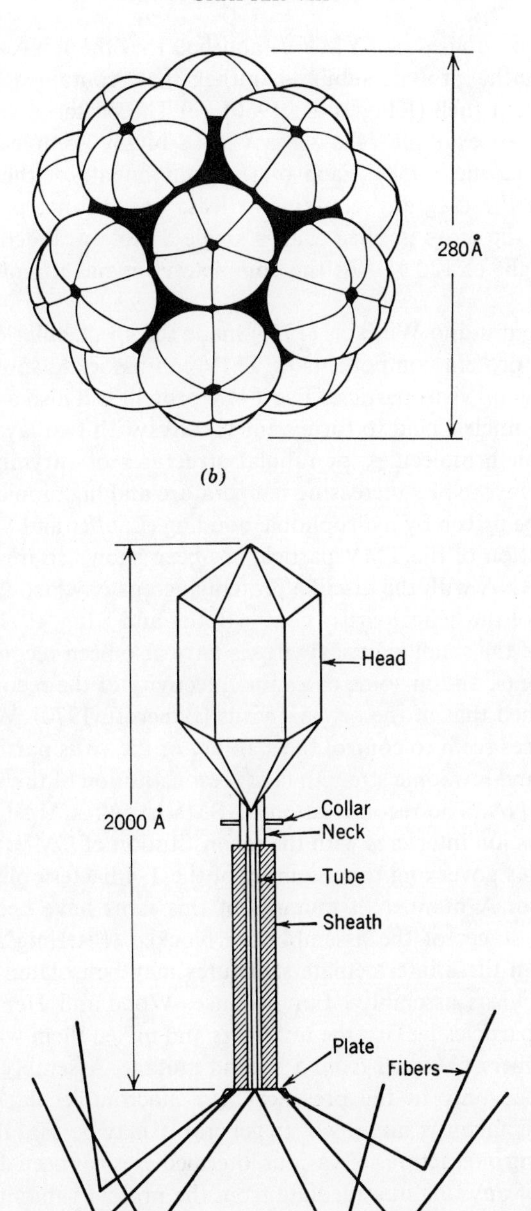

Fig. VIII. 14 (continued). (*b*) A small spherical plant virus, such as turnip yellow mosaic virus. (*c*) The bacterial virus T-4. Note the difference in the scale to which the three structures are drawn.

Crystallographic studies of TYMV established that the RNA chain is interlaced between the protein subunits, rather than contained in the cavity inside the protein shell (Klug et al., 1966). (*c*) The bacterial virus (bacteriophage) T-4 is an example of a virus with a highly complex morphology (Brenner and Horne, 1959). Each of the components of the structure, as represented in the diagram, contains at least one distinct protein species; the "head" of the virus is filled with a single double-stranded DNA with a molecular weight of 120×10^6; the complete virus has a molecular weight of 215×10^6.

Fraenkel-Conrat and Williams (1955) made the spectacular discovery that the RNA and protein components of TMV can associate spontaneously to reform the original virus particle. The TMV protein will also associate in the absence of the nucleic acid to form trimers, discs with two layers, each containing 17 protein molecules, or tubular structures of varying length. This aggregation is favored by increasing temperature and high ionic strength and is believed to be driven by hydrophobic bonding (Lauffer and Stevens, 1968). The reconstitution of the TMV particle has been shown to involve an interaction of the RNA with the disclike protein aggregate, which is transformed into a section of the helical virus "coat" (Butler and Klug, 1971).

A number of the small spherical viruses have also been reconstituted from their components, and in some cases the infectivity of the reconstituted particles approached that of the original virus (Bancroft, 1970). With these viruses, ionic forces seem to control the stability of the virus particle, since low temperatures and low ionic strength favor reconstitution of the virus from the protein and RNA. The reconstitution of BMV requires Mg^{2+}, whereas the presence of this ion interferes with the reconstitution of CMV.

The principles governing the assembly of the T-4 bacteriophage are much less understood. A number of mutants of this virus have been isolated in which various stages of the assembly are blocked (Eiserling and Dickson, 1970), and from these intermediate structures may be isolated and used for studies of the virus assembly . For instance, Wood and Henninger (1969) isolated virus particles lacking the tail fibers and mixed them with a preparation of these fibers obtained from a second mutant. Assembly of infectious virus took place only in the presence of a macromolecular extract. The function of this factor is unknown. In general it may be said that, while assembly involving interactions of various intermediates has been demonstrated, the assembly of any one intermediate from the protein subunits has not yet been accomplished.

Cahpter IX

CHEMICAL KINETICS IN MACROMOLECULAR SOLUTIONS

The bulk of the literature dealing with the physical chemistry of macromolecular solutions emphasizes the physical properties of these systems and entirely disregards their chemical reactivity. This seems unfortunate since studies of the reaction kinetics in macromolecular solutions yield information pertinent to a number of the problems which we considered in preceding chapters of this book. As we shall see, the reactivity of a chain molecule may depend on both its configuration and its conformation. In the case of polyelectrolytes, kinetic studies may yield new insight into the magnitude of electrostatic interactions and may, in fact, constitute a valuable method for characterizing the distribution of the electrostatic potential. In this case and others, the study of chemical reactivity may provide methods for the exploration of physical properties which are not easily investigated by other means.

In discussing reaction rates involving polymers, the point of departure is the generalization that the chemical characteristics of a functional group should not be altered when it is attached to a molecule of high molecular weight, provided that the mode of attachment introduces no steric hindrance effects. This principle was shown early to hold for the hydroxyl and carboxyl groups in polyesterification (Flory, 1939) and for the glycosidic linkages in cellulose degradation (Freudenberg et al., 1930; Freudenberg and Blomquist, 1935). The special effects with which we shall be concerned may then arise from two causes. First, the collision frequency between interacting functional groups may be altered for a variety of reasons which will be described on the following pages. Second, the effective solvent medium in the immediate vicinity of the polymer chain may be significantly different from the pure solvent.†

† Spectroscopic properties of groups adsorbed or covalently bound to macromolecules may be used to characterize the "local polarity". Thus, the quantum yield and the wavelength maximum of the fluorescence of 1-anilinonaphthalene-7-sulfonate bound to the active site of various enzymes dissolved in water have shown that the properties of the local medium are comparable to those of water-dioxane mixtures with a composition ranging

In reviewing studies on reaction kinetics in solutions of macromolecules, we shall exclude from our discussion polymerization, polymer degradation, and polymer crosslinking processes which are beyond the scope of this book. We shall also make only cursory references to the most interesting example of macromolecular reactivity, the catalytic activity of enzymes.

A. RATES OF CONFORMATIONAL TRANSITIONS IN DISSOLVED MACROMOLECULES

Since conformational transitions may be regarded as unimolecular chemical processes, it is convenient to consider them before treating reactions involving several reagents. As we shall see, such transitions encompass a wide variety of phenomena. We shall present here only some characteristic examples, referring the reader for a fuller treatment to a review of this field (Morawetz, 1972).

1. Rates of Conformational Transitions in Randomly Coiled Chains

If we consider the dynamics of conformational interconversion in the backbone of a randomly coiled chain, we encounter a conceptual difficulty. As shown schematically in Fig. IX.1*a*, a rotation around a single bond would require a large part of the chain to swing through the viscous solvent with a prohibitive expenditure of energy. It was suggested (Schatzki, 1965; Helfand, 1971), therefore, that two conformational transitions take place simultaneously in a "crankshaft-like motion" involving only a short segment of the chain (Fig. IX.1*b*). Such a mechanism would, however, imply a substantially higher activation energy, so that rotation around any given bond in the backbone of a long-chain molecule would have to be much less frequent than in a small molecule, where rotation around one bond only is possible.

This dilemma can be subjected in some special cases to direct spectroscopic examination. In a polyamide of piperazine:

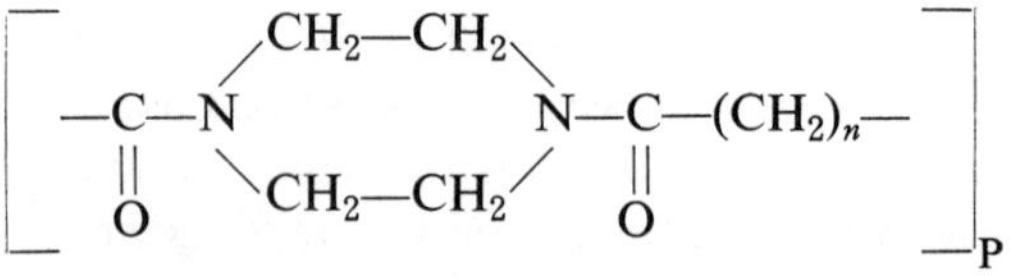

$$\left[\begin{array}{c} -\underset{\displaystyle \overset{\|}{O}}{C}-N \overset{\diagup CH_2-CH_2 \diagdown}{\underset{\diagdown CH_2-CH_2 \diagup}{}} N-\underset{\displaystyle \overset{\|}{O}}{C}-(CH_2)_n- \end{array} \right]_P$$

the methylene groups lying *cis* or *trans* to the carbonyl oxygen are in different magnetic environments and will be characterized by separate NMR peaks, provided the constant for rotation around the amide bond is small com-

from 40 to 80% dioxane (Turner and Brand, 1968). More recently, Štrop *et al* (1974) have used the location of the absorption maximum of a charge transfer complex attached to flexible chain molecules to characterize the polarity of the microenvironment of poly(2-vinylpyridine) and poly(4-vinyl-pyridine) in a variety of solvent media. It would seem most desirable to combine for the same polymer such spectroscopic techniques with studies of the kinetics of reactions sensitive to solvent effects.

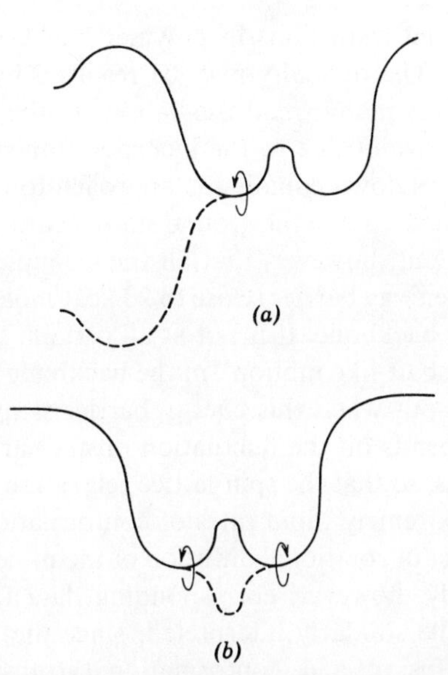

Figure. IX.1. Schematic representation of conformational transition in a flexible chain molecule: (*a*) rotation around a single bond; (*b*) correlated rotation around two bonds.

pared to the difference between their chemical shifts. On the other hand, if this rotation is very rapid, the magnetic environment is effectively averaged and the two peaks coalesce to a single peak with an intermediate location. It is possible, therefore, to derive the activation parameters for the hindered rotation from the temperature dependence of the NMR spectrum. Use of this method yielded very similar results for piperazine polyamides and their analog, diacetylpiperazine (Miron et al., 1969). In another study, employing polyamides with azobenzene residues in the backbone of the chain, the rate of the *cis-trans* isomerization was followed by ultraviolet spectroscopy. Again, there was no significant difference between the behavior of these residues in the chain molecules and that of their analogs (Tabak and Morawetz, 1970). A result bearing on the same problem was also obtained by comparing the rate of racemization of atropisomeric groups in the polymer and its analog, depicted below (Schulz, 1972):
The optical activity of the polymer was observed to decay only 40% as fast as that of the analog, but even this difference is trivial compared with that ex-

pected for conformational transitions in polymer backbones requiring two simultaneous rotations. The difficulty may be resolved by assuming that a hindered rotation involves many small oscillations in the internal angles of rotation and that the restraint due to the incorporation of the bond into a long-chain molecule slows down equally the approach to the transition state and the return of a strained state to the ground state (Morawetz, 1972).

It should be pointed out, however, that all the examples given above involve a very substantial energy barrier (close to 20 kcal/mole) for the hindered rotation in the polymer backbone. It is not at all certain, therefore, whether the concept of a "crankshaft-like motion" in the backbone of dissolved polymers can also be ruled out when this energy barrier is quite low. Nuclear magnetic relaxation depends on the fluctuation of nuclear magnetic dipoles over very short distances, so that the spin-lattice relaxation time may be used as a measure of these extremely rapid rates of conformational transitions in polymers where the effect of rotational diffusion of the molecule as a whole is negligible. Unfortunately, however, corresponding data for low molecular weight analogs cannot be similarly interpreted, since the relative contributions of rotational diffusion and conformational transitions cannot be separated. Nevertheless, the NMR relaxation method has yielded most interesting data (Anderson et al., 1970). For similar solvents, the effect of solvent viscosity on the rate of hindered rotation is clearly evident. Changes in the solvent medium may also have a pronounced influence on conformational mobility if they alter the equilibrium distribution of conformations. This effect was strikingly demonstrated on poly(ethylene oxide) in the water-methanol system, where the rate of hindered rotation goes through a maximum as the composition of the solvent mixture is being varied (Liu and Anderson, 1970).

Hindered rotation in side chains attached to polymers generally takes place at rates similar to those in small molecules if obvious steric hindrance effects are avoided. However, an exception to this generalization may be observed if the transition is sensitive to the polarity of the medium. In this case, the "local polarity" may depend on the nature of the chain backbone, and the rate of the transition may be used to characterize the properties of this local medium (Mikeš et al., 1974).‡

2. Kinetics of Transformations Involving Helical Conformations

A theory of the kinetics of helix-coil transitions was formulated by Schwarz (1965) on the basis of the treatments of helix-coil equilibria discussed in Section III.C.5. Schwarz found that helix nucleation is not significant under

‡The photochromism of spyropyran residues attached to polymers can also be used to characterize the local polarity (P. H. Vandewijer and G. Smets, *J. Polym. Sci.*, **C22**, 231 (1968)).

conditions employed in relaxation studies, so that the helical content changes at a rate depending on the rate constants for helix propagation and depropagation, k_1 and k_{-1}, where k_1/k_{-1} is the parameter j of (III.35), and the number of helical sequences. Since these sequences vary in length, the kinetics of the relaxation process will be complex, but the mean relaxation time τ^* (the reciprocal of the initial first-order rate constant) may be expressed as

$$\frac{1}{\tau^*} = k_1 [(j - 1)^2 + 4\zeta] \tag{IX.1}$$

where ζ is the nucleation parameter of (III.35). At the midpoint of the helix coil transition, $j = 1$ and τ^* assumes its maximum value, $\tau^*_{\max} = 1/4k_1\zeta$.

The helix-coil transition rates of polypeptides have been studied by temperature-jump relaxation (Lumry et al. 1964), by the frequency dependence of sound absorption (Hammes and Roberts, 1969); Barksdale and Steuhr, 1972. and by dielectric dispersion (Schwarz and Seelig, 1968; Wada et al., 1972). The ultrasonic results are interpreted by

$$\frac{\alpha}{f^2} = P + \frac{Q}{1 + (f/f_c)^2} \tag{IX.2}$$

where α is the absorption coefficient for the sound of frequency f, P and Q are constants, and $f_c = (k_1 + k_{-1})/2\pi$. The dielectric dispersion is affected by the rate of helix-coil transition, since a strong electrical field will shift the helix-coil equilibrium toward the helical form with its larger dipole moment. All these techniques yield $\tau^*_{\max}$ values between 10^{-8} and 10^{-6} sec, and since ζ is typically 10^{-4}, this would lead to k_1 in the range of $10^{10}/4 - 10^{12}/4$ sec^{-1}.

However, NMR spectra of polypeptide solutions yield two peaks of α-CH and NH protons in the helix-coil transition range. This seems to imply that the rate constant for this transition is small compared to the difference of the chemical shifts in the two conformations (i.e., $k_1 < 10^2$ sec^{-1}), much smaller than estimated by the methods discussed above. Three suggestions have been made to resolve the discrepancy:

(a) Since the nucleation parameter ζ is known to be of the order of 10^{-4}, the "cooperative length" (see p. 168) is of the order of 100 monomer residues. Unless the polypeptide is very long compared to this cooperative length, a significant fraction of the chains will contain no helical segments. Since nucleation of the helix is relatively slow, the NMR spectrum will contain for the α-H (or the NH) two peaks, one characteristic of the random coil chain, the other representing a weighted average of the helical and random coil sequences in the partially helical chains. This interpretation is supported by the observation that the location of one of the peaks remains fixed, while that of the other moves gradually from the location characterizing the helical form to that observed in the random coil as the polypeptide passes through the helix-coil transition (Ferretti and Jernigan, 1973).

(b) According to Ullman (1970), the double peaks arise from the polydispersity of the polypeptide samples, as well as from the fact that the stability of the helical structure decreases as the end of the molecular chain is approached. This interpretation is supported by the observation that a narrow fraction of poly(γ-benzyl-L-glutamate), prepared by gel permeation chromatography, does not exhibit double peaks for the α-CH and NH protons in the helix-coil transition range (Nagayama and Wada, 1973).

(c) Bradbury et al. (1969) suggested that the doubling of the NMR peaks results from a slow protonation of peptide residues by the helix-breaking acid, and this interpretation was seemingly supported by the similarity of NMR spectra of poly(L-alanine) and poly(DL-alanine) in the same solvent medium (Tam and Klotz, 1971). However, Bradbury et al. (1973) observed double peaks also during the helix-coil transition in a system containing no acid, so that this interpretation is unlikely to be correct.

We pointed out in Section III. C.1 that poly(L-proline) can exist as a right-handed or left-handed helix, but that junctions between helical segments wound in opposite directions require an energy estimated as 7 kcal/mole. When the solvent medium is altered so as to favor helix-helix transition, the rate of this process is found to be extremely slow, requiring typically times of the order of hours at ordinary temperatures. There are two reasons for this small velocity. First, the helix-helix transition involves rotation around the amide bond characterized by an energy barrier of about 20 kcal/mole. Second, since initiation of helix reversal in the middle of the chain is energetically unfavorable, most reversals must be initiated at chain ends; if the chains are long, the concentration of such initiation sites is, therefore, quite low.

Winklmair et al. (1971) showed that, for chains much shorter than the cooperative length (estimated as 320 proline residues) and for small perturbations of the solvent medium, the helix reversal initiated at the chain ends is rate limiting. For monodisperse poly(L-proline) samples, the process then follows first-order kinetics. The point of helix reversal may be considered to move in a random walk along the polypeptide chain; it will disappear either when the entire chain has undergone helix-helix transition, or when the junction returns to the chain end where it was initiated, leaving the polypeptide conformation unchanged. Theory predicts that, for a chain of length Z in which helix-helix transition has been initiated, the probability that it will propagate along the whole chain is proportional to $1/Z^2$. Thus, the relaxation time for the helix-helix transition should be proportional to the square of the chain length, and this prediction is borne out by experimental data (Fig. IX. 2). On the other hand, if the polymer chain is very long and the perturbations are large, the kinetics also reflect the time required for helix reversal in any one chain; the process then exhibits an initial accelerating phase, arising from

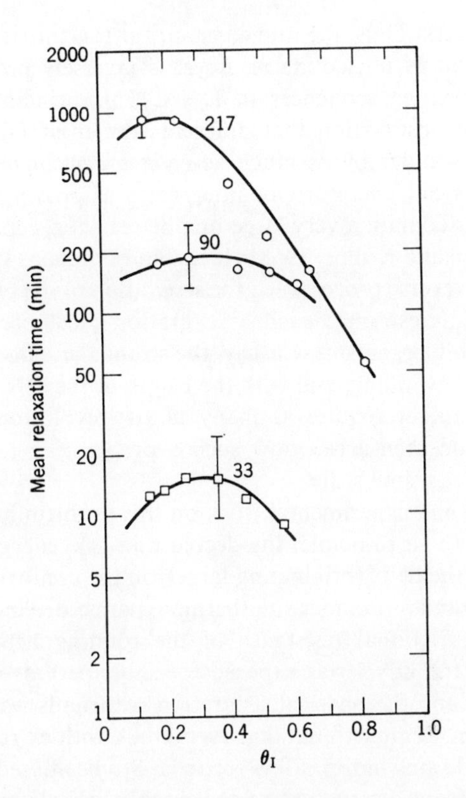

Fig. IX.2. Rate of helix-helix transition as a function of θ_I, the fraction initially present as poly(L-proline)I at 70°C. The degree of polymerization is specified above each curve.

a growing number of chains in which helix reversal has been initiated.

The formation of double-stranded helices from polyriboadenylic and polyribouridylic acid follows second-order kinetics, showing that the nucleation of the double helix is rate determining, whereas the subsequent "zippering up" of the double helix is a very fast process (Blake and Fresco, 1966). The rate of formation of the double helical structure *decreases with increasing temperature,* as expected for a process in which nucleation is rate-limiting, since the minimum length of helical nuclei which can survive increases as the melting point is approached. According to Flory and Weaver (1960), the rate constant for such a process will, therefore, have a temperature dependence given by $k = \text{const} \exp [- A/\boldsymbol{R}T(T_m - T)]$.

When the double helix is to be reconstituted from the separated strands of a DNA species which has a nonrepeating sequence of nucleotide residues, a given nucleotide in one of the strands has to encounter the unique corresponding nucleotide residue in the complementary strand to initiate the renaturation process. Thus, the rate constant for this initiation, based on the total concentration of nucleotide residues, is inversely proportional to the length of nonrepeating sequences in DNA. This principle has been well documented. The observation that different species of DNA, with similar chain length, may differ by as much as *eight orders of magnitude* in their renaturation rates can, therefore, be interpreted as proving that the rapidly renaturing species contain a very large number of short repeating nucleotide sequences (Britten and Kohne, 1968; Hearst and Botchan, 1973).

The rate of the reverse process (i.e., the separation of the two DNA strands) is of particular interest since such a separation must occur during DNA duplication in living organisms. Clearly, the strands of a double helix cannot separate without unwinding, and with the length of the DNA found in living cells this process might require as many as 10^5 revolutions of the double helix. The question then arises how such a process can take place on the biologically required time scale.

The theoretical and experimental work on this problem has been reviewed by Crothers (1969). In principle, the decrease in free energy accompanying the unwinding of the helix (originating largely in the conformational entropy of the isolated strands) provides the thermodynamic driving force, which is balanced by the frictional resistance of the rotating helix. Experimental studies, in which the DNA was exposed for short periods of time to denaturing conditions and the physical separation of strands was determined by ultracentrifugation, demonstrated that even DNA with extremely high molecular weights could unwind in a few seconds. Studies based on the spectral change resulting from destruction of the double-helical structure are more difficult to interpret, since they may also reflect local unstacking of nucleotide residues which does not lead to a separation of the macromolecular chains. The relaxation times observed by this technique after a temperature jump have been found to decrease by two orders of magnitude as the perturbation is being increased. This has been interpreted as indicating that the unstacking of the nucleotides can take place, on large perturbations, with little untwisting of the double helix. Under these conditions, the frictional resistance of the rotating double helix controls the rate of unwinding, and the time for chain separation becomes proportional to the square of the chain length, as required by theory (Crothers and Spatz, 1971).‡

The reconstitution of the collagen triple helix from the separated strands usually leads to extensive chain aggregation. Nevertheless, collagen renaturation can be accomplished in dilute solutions and in close proximity to the equilibrium helix-coil transition temperature (Beier and Engel, 1966). The

‡These studies may not be relevant to biological DNA replication, which requires the presence of a "DNA unwinding protein" (cf., e.g., R. C. Reuben and M. L. Gefter, *J. Biol. Chem.*, **249**, 3843 (1974)).

formation of the triple helix requires several hours under these conditions. Some mechanism is necessary, therefore, to accelerate the formation of collagen after the synthesis of the three chains in the living cell. It appears that nature has solved this problem by having the cell synthesize "procollagen" chains with highly cohesive ends, at which the nucleation of the triple helix takes place. These ends are then eliminated by enzymatic breakdown after the collagen particle has been produced (Bellamy and Bornstein, 1971).

3. Rates of Conformational Transitions in Globular Proteins

In considering conformational transitions in globular proteins, it is convenient to discuss separately the complete breakdown of the native structure to the denatured state and the much more limited alterations in the molecular structure, which may be of great importance in the biological function of these macromolecules.

When denaturation and its reversal involve only two states with no stable intermediates, the reaction rate should follow first-order kinetics in both directions. This is the behavior found by Tanford et al. (1966) for the unfolding and refolding of ribonuclease in guanidine hydrochloride solution. Here the rate of denaturation was proportional to $(\text{Gua})^5$, while its reversal depended on $(\text{Gua})^{-11}$. It should be pointed out, however, that the existence of a reaction intermediate can be missed if it resembles the spectral properties (by which the reaction is being followed) of either the native or the fully denatured state. For instance, the thermal denaturation of ribonuclease was long considered as a single-step process, but a careful study showed that it involves a very fast step followed by a second process slower by 3–4 orders of magnitude (Tsong et al., 1971).

It seems highly improbable that the newly synthesized polypeptide chain of a protein arrives at the native conformation by random conformational transitions, since such a process would be too slow for the requirements of a living cell. Lewis et al., (1971) have suggested that the native conformation may be "nucleated" by amino acid sequences, which have a strong tendency to form α-helical or other ordered structures. This step would then be followed by interactions between such nuclei, facilitated by amino acid sequences which tend to form "β-folds" in the polypeptide chain. Several experimental results are in accord with a scheme of this type. Epstein et al., (1971) studied the renaturation of acid-denatured nuclease and found that the process takes place in two steps with half-lives of 55 and 350 msec at 25°C. The fast step is temperature independent and may be described as an entropy-driven nucleation. In another study, Shen and Hermans (1972) found that the rate of renaturation of myoglobin depends on the time allowed after acid denaturation has seemingly run to completion. This suggests that some elements of the native structure, which are not observed spectroscopically, are only very slowly obliterated under denaturation conditions.

We pointed out (p. 409–410) that association of low molecular weight species with globular proteins is accompanied, in a number of cases, by a conformational transition of the macromolecule, and the rate of such a process is of great interest. A pioneering study of such an effect involved the photochemical decomposition of the CO complex of hemoglobin (Gibson, 1959). This decomposition leaves the hemoglobin in the conformation characteristic of the CO complex, and the transformation of this conformation to the molecular structure assumed by unliganded hemoglobin can be followed spectroscopically. The rate constant for this process is 220 sec^{-1} at pH 9 and 1 °C. Since the primary product of the carboxyhemoglobin decomposition is a weaker base than the unliganded hemoglobin in its equilibrium conformation, the uptake of hydrogen ions can also be utilized to monitor the transition (Gray, 1970).

The association of an enzyme with its substrate (or an allosteric effector) is frequently followed by conformational changes which can be studied kinetically. The data could be interpreted as indicating either that the bound molecule *induces* the conformational transition or that it *stabilizes* one of two (or more) conformations which exist in equilibrium with each other, so that formation of the complex leads to a shift in this equilibrium. In the case of the enzyme chymotrypsin, there are numerous indications that at least two conformations exist in the absence of substrate (Hess et al., 1970). Fersht and Requenna (1971) found that the attack of this enzyme on a fluorescent inhibitor bound at the catalytic site led to a biphasic change of fluorescence. The fast process was taken as reflecting association with the enzyme which was originally present in the catalytically active conformation, while the slow process was interpreted as due to the shift in the equilibrium between the active and inactive forms produced by association of the active form with the inhibitor. The relative magnitudes of the spectral changes in the two processes could then be used to estimate the equilibrium between the two conformational states.

Most of these minor conformational transitions in proteins take place over times of the order of milliseconds. Occasionally, however, much slower processes are also observed. With the enzyme ribonuclease, which contains 124 amino acid residues, the splitting of the peptide bond connecting residues 20 and 21 leads to two inactive fragments which regain the enzymatic activity on mixing. When changes occurring after such mixing are monitored spectroscopically, a first-order process with a half-life of 90 sec is observed. Obviously, this reflects a slow conformational change following the association of the two enzyme fragments (Richards and Logue, 1962).

B. REACTIONS OF POLYMERS WITH SMALL MOLECULES AND IONS

1. Isotope Exchange

The hydrogen atoms in protein molecules fall into three categories with respect to the ease with which they undergo exchange with the hydrogen atoms of water. At one extreme are the hydrogen atoms of —COOH, —NH_2, —OH, and —SH groups, which can exchange at rates that are essentially diffusion controlled. At the other extreme are hydrogens attached to carbon atoms, which exchange at negligible rates unless the solution is very basic. An intermediate position is occupied by the hydrogens of the peptide groups, which may exchange at rates accessible to observation by conventional techniques. It would be expected that this process would be drastically slowed down if a peptide group participates in an intramolecular hydrogen bond or if it is buried in the interior of a globular protein structure. This consideration provided the motivation for the study of hydrogen exchange, pioneered at the Carlsberg Laboratory in Copenhagen, particularly by Linderstrøm-Lang, who developed a delicate method for following the exchange process. In this procedure the protein is first treated at elevated temperature with D_2O. The deuterated protein is then dissolved in H_2O, and aliquots are removed at specified time intervals, frozen at $-60\,^\circ C$ to stop the exchange reaction, and subjected to high vacuum sublimation. The density of the water in the sublimate is then determined by a density gradient technique. It is also possible to follow specifically the deuteration of the peptide groups by infrared spectroscopy. In a third technique, hydrogen-tritium exchange is followed by scintillation counting after separation of the protein from the water in which it is dissolved by rapid gel filtration. An excellent review of theoretical aspects and experimental results of hydrogen exchange in proteins has been written by Hvidt and Nielsen (1966).

Poly(DL-alanine) may be used as a model for the behavior of peptide residues in randomly coiled chains in aqueous solution (Bryan and Nielsen, 1960). Although this polymer exhibits some hypochromicity which is indicative of α-helical sequences in the chain (Gratzer and Doty, 1963), these helices must be too unstable to impede isotopic exchange, since urea has no effect on the exchange rate, which is also similar to that observed in di- and tripeptides. The reaction is catalyzed by H^+ and OH^- ions with catalytic coefficients of $k_{H^+} = 0.8M^{-1}\ sec^{-1}$ and $k_{OH^-} = 5.4 \times 10^7M^{-1}\ sec^{-1}$ at 20°C. While k_{OH^-} appears to be temperature independent, $k_{H^+} = 0.059M^{-1}\ sec^{-1}$ at 0°C (Englander and Poulsen, 1969). On the other hand, poly(α-L-glutamic acid), which undergoes a coil-helix transition when its ionization is suppressed by lowering the pH, may be used to assess the extent to which isotopic exchange is impeded in the helical conformation (Nakanishi et al., 1972). In the pH range of helix-coil transition, the exchange deviates sharply from first-order kinetics and depends strongly on chain length. This behavior reflects the decreasing stability of the helical form as the end of the chain is approached, and Nakanishi et al., used their kinetic data to estimate the rate constant of the exchange as a function of the position of a peptide group in the chain (Fig. IX.3). The extrapolated value for the terminal residue is identical to the rate observed in the randomly coiled poly(α-DL-glutamic acid)

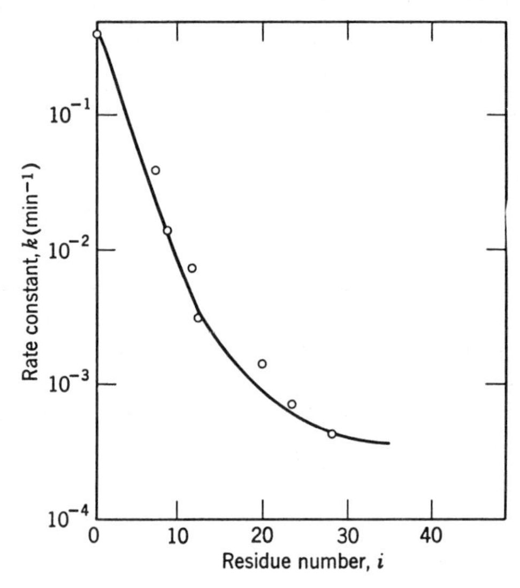

Fig. IX.3. Rate constants for deuterium exchange in poly(α-L-glutamic acid) as a function of the number i of monomer residues separating a peptide residue from the chain end (temp. 25°C, pD, 4.0).

and is three orders of magnitude faster than in residues very distant from the chain ends, where the helix has maximum stability. However, for reasons which are not clear, estimates of helix content based on isotopic exchange rates yield higher values than rotatory dispersion data.

The interpretation of isotopic exchange rates of globular proteins is generally based on a model (Hvidt and Nielsen, 1966) in which a given hydrogen exists either in conformations in which no exchange can take place (designated by N) or in conformations (designated by I) in which it is exposed to water and is as free to exchange as a random coil polypeptide. We have then, for any given hydrogen,

$$\mathrm{N} \underset{k_2}{\overset{k_1}{\rightleftarrows}} \mathrm{I} \xrightarrow{k_3} \text{exchange} \qquad \text{(IX.3)}$$

where k_3 represents the poly(DL-alanine) exchange rate constant at a particular pH.† Two limiting cases may be considered.

(a) If $k_3 \gg k_2$, the N→I transition is rate determining. Thus, the exchange rate should be independent of pH in the range in which the native form of the protein is stable.

(b) If $k_2 \gg k_3$, the exchange rate should reflect the N⇄I equilibrium and should exhibit the same pH dependence as k_3.

In principle, each exchangeable hydrogen may be characterized by a discrete rate constant for isotopic exchange. However, these individual rate constants are accessible only in special cases. Bradbury and Chapman (1972) followed by NMR spectroscopy the exchange with D_2O of the C-2 protons in the imidazole groups of histidine-105, histidine-12, histidine-119, and histidine-48‡ of ribonuclease and obtained rate constants of 0.48, 0.48, 0.19, and 0.25 sec^{-1}, respectively, at 37°C and pH 8.5. However, generally applicable methods for following isotopic exchange allow only a small number of parameters to be extracted from the kinetic data. The hydrogens are, therefore, treated as if they could be placed into a small number of groups with characteristic reactivities, and the dependence of the fraction f_u of unexchanged hydrogens at time t is represented by

$$f_u = \frac{\sum_i n_i \exp(-k_i t)}{\sum_i n_i} \qquad \text{(IX.4)}$$

where n_i is the number of hydrogens exchanging with a rate constant k_i. In an early study on insulin, Linderstrøm-Lang (1955) found 60 hydrogens per insulin monomer to exchange too fast to be followed and fitted his data by assigning to the next groups of 6, 15, and 8 hydrogens rate constants in the ratio of 320:20:1 at 38.6°C and pH 3. For many proteins the pH dependence of the hydrogen exchange was found to be similar to that of k_3, so that it reflects the equilibrium concentration of I conformations. In that case, plots of f_u as a function of log k_3t (or, more precisely, log $k_3't$) obtained at different pH values should be superimposable; and if data are obtained over a broad range of pH, such a plot will represent a "relaxation spectrum" of the exchangeable hydrogens (Hvidt and Wallevik, 1972).

However, the assumption that $k_2 \gg k_3$ is not universally applicable. In

†Strictly speaking, k_3 should be replaced by $k_3' = k_3 \exp(-\Delta G_{el}^{\ddagger}/\boldsymbol{RT})$, where $\Delta G_{el}^{\ddagger}$ represents the electrostatic free energy of activation due to the interaction of the pH-dependent net protein charge with the catalyzing H^+ or OH^- ions. Most workers seem to have neglected this factor in interpreting their data.

‡The number in this conventional designation specifies the location of the amino acid residue, counted from the amino terminal of the chain.

the case of myosin, Segal and Harrington (1967) found that the exchange rate attains a limiting value above pH 7, so that the N → I transition is rate limiting. Estimated half-lives for this process are extremely long, in the range of 20–2000 min. With trypsin, only the fastest exchanging hydrogens show rates which are strongly pH dependent (Lenz and Bryan, 1969). It has also been demonstrated that the binding of ligands may protect a substantial number of hydrogen atoms in globular proteins against isotopic exchange, reflecting, presumably, an increase in conformational rigidity (Schechter et al., 1969).

The use of poly(DL-alanine) as a model for the behavior of solvent-exposed peptide residues in proteins is strongly supported by the observation that differences between the isotopic exchange rates in poly(DL-alanine) and poly-(DL-lysine) may be almost entirely accounted for by the polylysine charge (Englander and Poulsen, 1969). Thus, changes in the local water structure due to the hydrophobic side chains do not appear to be important in this case. On the other hand, comparisons of the behavior of polymers and that of their low molecular weight analogs seem to be dangerous. Thus, isotopic exchange is 100 times slower in poly(*N*-isopropylacrylamide) than in *N*-isopropylpropionamide (Scarpa et al., 1967), and the exchange in poly-(*N*-vinylacetamide) is 20 times slower than that in *N*-methylacetamide (Hvidt and Corett, 1970).

Measurements of the isotopic exchange rate have recently led to a definitive resolution of an old controversy concerning the structure of collagen. This protein has, as discussed in Section III.C.2, a coiled-coil structure which is stabilized by the interaction of glycine residues placed in every third position of the polypeptide chains. However, it was not clear whether one or two interchain hydrogen bonds are formed for every three amino acid residues. Hydrogen-exchange data have now proved that the model with two hydrogen bonds is correct (Yee et al., 1974).

Isotopic exchange studies in nucleic acids and synthetic polynucleotides have also led to some interesting insights. That the hydrogen atoms engaged in the hydrogen bonding of the two polynucleotide strands of the double helix would exchange slowly would have been expected, but it came as a surprise that all of the exocyclic amino and endocyclic imino hydrogens of the purine and pyrimidine bases exchange at unusually slow rates. Englander and von Hippel (1972) studied DNA at 0°C and pH 7.6 and found that the hydrogens which are not involved in hydrogen bonds (2 for a G-C and 1 for and A-T pair) exchange with a half-life of 25 sec, while those participating in the hydrogen bonds (3 for a G-C and 2 for an A-T pair) have a half-life of 330 sec. It is even stranger to find that in the complex of polydeoxyriboguanidylic and polydeoxyribocytidylic acid isotopic exchange follows first-order kinetics, so that all of the amino and imino hydrogens react at the same rate (Englander et al., 1972). There is no correlation between helix stability as characterized by its melting point and the rate of isotopic exchange. It appears, therefore, that the process which has to precede isotopic exchange is distinct from the base unstacking required for the disruption of the helical structure.

2. Effects of Conformation of Globular Proteins on Their Reactivity

The unique conformation characterizing proteins in their native state may profoundly modify the reactivity of the functional groups attached to the polypeptide backbone. Such modifications may be accounted for in two ways. First, a reactive group may find itself in the interior of the globular structure and may be inaccessible to a given chemical reagent as long as the conformation of the native state is maintained. Second, the reactivity of a given functional group may be either enhanced or reduced by its juxtaposition to another group. Again, the effect is a result of the precisely defined macromolecular conformation and disappears when the protein is denatured. A large number of examples could be cited to illustrate these effects; we shall limit ourselves to a few characteristic cases.

Many proteins contain thiol groups of cysteine residues. Such groups, when attached to small molecules, are readily oxidized by a variety of reagents. Yet, in native globular proteins, such thiol groups are frequently "masked" so that their reactivity is observed only when the protein is denatured. The voluminous literature dealing with this effect has been reviewed by Barron (1951). However, in these early studies no means existed for identifying the amino acid residue which failed to react, nor was it possible to correlate the kinetic data with the location of the unreactive group in the molecular structure of the protein. When methods became available for placing the amino acid residue which had been chemically modified in the amino acid sequence of the protein and when molecular structures of proteins were solved by x-ray crystallography, the information obtained from the relative ease with which various groups engaged in chemical reactions became much more valuable.

Crestfield et al., (1963) subjected ribonuclease to the action of bromoacetate under mild conditions and observed that either histidine-12 or histidine-119 was carboxymethylated, but that reaction of one of these groups blocked the reactivity of the other. Both modified proteins were catalytically inactive. It was concluded that histidine-12 and histidine-119 must lie very close to each other in the catalytically active site and that histidine-105, which remained unreacted, must be buried in the interior of the enzyme structure. These predictions were fully confirmed when the ribonuclease structure was determined by Kartha et al., (1967).

Studies of protein reactivity led also to important conclusions concerning the relation between the molecular structure in the crystalline state and that in solution. When myoglobin was carboxymethylated, 5 of the 12 histidines failed to react. (Hugli and Gurd, 1970). Four of the 5 were identical when crystalline and dissolved myoglobin were used, but the fifth was different. While one may argue that a residue which is reactive in solution is inaccessible in the crystalline state because of intermolecular contacts, the reverse observation (a loss of reactivity of a residue when a protein dissolves) seems to be a clear indication of a significant conformational transition.

The reactivity of amino acid residues in protein solutions appears to be extraordinarily sensitive to small changes in molecular structure. All three methionine residues in human myoglobin are unreactive toward bromoacetate, being buried in the hydrophobic core of the molecule. Yet removal of the heme group exposes them to carboxymethylation, although optical activity data indicate that the globin retains the myoglobin structure (Harris and Hill, 1969). The thiol in cysteine-93 of the β chain of hemoglobin reacts 80 times more rapidly with *p*-mercuribenzoate if the hemoglobin has oxygen bound to the heme groups (Antonini and Brunori, 1969), and methionine-80 of cytochrome is similarly exposed to carboxymethylation in the oxidized form of the protein (Ando et al., 1966). The conversion of chymotrypsinogen to chymotrypsin, which involves a very small conformational change, removes tyrosine-94 from access to the nitration reagent tetranitromethane (Shlyapnikov et al., 1968). Finally, it was found that in staphylococcal nuclease, which contains seven tyrosines, only tyrosine-85 is nitrated in the free enzyme, but that in the presence of a competitive inhibitor the previously unreactive tyrosine-115 is nitrated instead. This striking effect must be ascribed to a conformational change induced by the binding of the inhibitor (Cuatrecasas et al., 1968).

The ability of globular proteins to bind selectively small molecules with closely defined geometry provided an opportunity for the development of a most valuable technique designated as "affinity labeling" (Singer, 1967). The reagent is here designed so that a reactive group is placed at a predetermined distance from a moiety which fulfills the steric requirements of a ligand for the binding site of the protein. It is then expected that an amino acid residue at a specified distance from the binding site will be modified, and identification of this residue provides information about the regions of the protein chain which participate in the formation of this site. For example, affinity labeling of chymotrypsin was accomplished using reagents I and II:

C_6H_5—CH_2—CH—C(=O)—CH_2Cl
with NH—SO_2—C_6H_4—CH_3 on CH

I

$C(=O)$—$C(CH_3)_2$—NH—C(=O)—CH_2Br
with O—C_6H_4—NO_2 on the first C

II

Reagent I contains the phenylalanyl residue which associates specifically with the binding site of the enzyme and the chloromethyl function was found to alkylate histidine-57 (Ong et al., 1965). The ester of reagent II is attacked by chymotrypsin with acylation of serine-195 in the catalytic site of the enzyme (cf. p. 456). The bromomethyl function was then found to alkylate methionine-192 (Lawson and Schramm, 1965). Histidine-57 and methionine-192 are both part of the active site, and the results demonstrate the precision with which the reaction may be directed toward a specific amino acid.

The technique of affinity labeling has also been most valuable in the study of antibodies. Singer and Doolittle (1966) worked with antibodies against the dinitrophenyl (DNP) group and found that affinity labeling modified both the light and the heavy chains of the antibody. This result was surprising at the time it was obtained, but the participation of both chains in the antibody binding site has since been substantiated by other means (Hochman et al., 1973). Detailed interpretation of the results of affinity labeling of antibodies has been greatly facilitated by the availability of molecularly homogeneous immunoglobulin preparations, some of which were found to bind specifically common antigenic determinants. Givol et al., (1971) and Haimovich et al., (1972) used a myeloma IgG protein, which was found to bind the DNP hapten, with reagents III, IV and V:

DNP—N(H)—CH_2—CH_2—BAA

III

DNP—N(H)—CH_2—CH_2—CH_2—CH_2—CH(COOH)—BAA

IV

DNP—N(H)—CH_2—CH_2—CH(BAA)—C(=O)—N(H)—BAA

V

BAA= —N(H)—C(=O)—CH_2—Br

containing one or two bromoacetamido (BAA) groups. Reagent III was found to label tyrosine-34 of the light chain, while reagent IV reacted with lysine-54 of the heavy chain of the antibody. Both these residues are located, as would be expected, within the variable region of the amino acid sequence by which the specificity of the antibody is determined. Moreover, the bifunctional reagent V reacted with both amino acid residues, producing a cross-link between the light and the heavy chain. Thus, these two residues must have been located within about 5 Å of each other in the IgG structure.

The labeling reagents described above will attack only nucleophilic substituents of amino acid residues, and it is desirable, therefore, to develop reagents which can interact with the hydrocarbon side chains of alanine, leucine, and so forth. This can be accomplished by the use of reagents which generate, on photolysis, carbenes or nitrenes (Knowles, 1972). One such reagent has been shown to label an alanyl group near the active site of a rabbit antibody (Fleet et al., 1972).

The most interesting aspect of chemical interaction of globular proteins and small molecules is, of course, enzymatic catalysis. Although a general discussion of this subject is beyond the scope of this book, it is appropriate to mention briefly some aspects of the action of the enzyme chymotrypsin, since considerable effort has been expended on attempts to produce polymers which will mimic the catalytic activity of chymotrypsin to some extent.

Chymotrypsin catalyzes the hydrolysis of certain *N*-acylated amino acid amides or esters, the derivatives of L-tyrosine and L-phenylalanine being particularly good substrates. Much of the investigation of chymotrypsin action also employed the "unnatural" substrate *p*-nitrophenyl acetate. The enzyme is formed in nature form its catalytically inactive precursor, chymotrypsinogen, by the scission of four peptide bonds (cf. Fig. III.32*b*). The mechanism of chymotrypsin action is understood in considerable detail on the basis of chemical evidence (Hess, 1971) and crystallographic data concerning the conformational change accompanying the activation of chymotrypsinogen (Kraut, 1971; Wright, 1973).

The hydrolysis of a substrate catalyzed by chymotrypsin takes place in two steps. In the first step, the acyl group of the substrate acylates the hydroxyl of the highly reactive serine-195. This reaction is aided by the hydrogen bonding of this hydroxyl to one of the imidazole nitrogens of histidine-57, and this imidazole group must, therefore, be present in its basic form.‡ The second step, in which the acyl-enzyme is hydrolyzed to its original form, employs the imidazole of histidine-57 as a general base catalyst. Using the chy-

‡The importance of the hydrogen bond has recently been confirmed by an NMR study of the ionization behavior of histidine-57 (G. Robbilard and R. G. Shulman, *J. Mol. Biol.*, **86**, 519 (1974)).

motrypsin attack on *p*-nitrophenyl acetate as an example, we can express the two steps of the process as follows:

$$\mathrm{ChT{-}OH + O_2N{-}C_6H_4{-}O{-}\overset{\overset{\displaystyle O}{\|}}{C}{-}CH_3 \rightarrow ChT{-}O\overset{\overset{\displaystyle O}{\|}}{C}CH_3 + O_2N{-}C_6H_4{-}OH} \quad \text{(IX.5a)}$$

$$\mathrm{ChT{-}O{-}\overset{\overset{\displaystyle O}{\|}}{C}CH_3 \xrightarrow{OH^-} ChT{-}OH + CH_3COO^-} \quad \text{(IX.5b)}$$

It should be emphasized that both steps depend on the precise conformation of the native enzyme. In the denatured state, the serine-195 residue exhibits no unusual tendency to be acylated and deacylation is slowed down a millionfold (Bender et al., 1962). During the activation of chymotrypsinogen, the scission of the peptide bond between residues 16 and 17 produces a cationic charge at the α-amino group of isoleucine-16. This forms an ion pair with aspartate-194, and it is this process which produces the slight conformational change required for catalytic activity. In the pH range 9–10 the catalytic activity is lost at a rate which parallels the conversion of the isoleucine-16 amino group to its basic form.

One point deserves special attention. Since imidazole and its derivatives are also catalysts for the hydrolysis of *p*-nitrophenyl acetate, the action of chymotrypsin is often compared, by those engaged in a search for enzyme models, to that of imidazole, as if the enzyme merely provided a binding site for the substrate and a favorable environment for this nucleophile. This approach is quite misleading. In the reaction catalyzed by imidazole, *N*-acetylimidazole is a reaction intermediate (Jencks, 1969), but crystallographic data show (Steitz et al., 1969) that histidine-57 is too far from the carbonyl group of the substrate to be acylated. Imidazole is also incapable of catalyzing the solvolysis of alkyl esters and amides, which are readily attacked by chymotrypsin.

3. Reactions of Flexible Chain Polymers with Low Molecular Weight Species

We saw in our discussion of the titration behavior of chain molecules (Section VII.C.2) that the electrostatic free energy required to remove a hydrogen ion from a highly charged polyion may produce a large shift in the ionization equilibrium. This shift will have a pronounced effect on the rate of a process in which the conjugate base of an ionizable function is the reactive species, if we compare the reaction rates of an ionizable chain molecule and its monofunctional analog at the same pH. The principle may be illustrated by comparing the quaternization rates in aqueous solutions of poly(4-vinylpyridine) and 4-picoline, its monofunctional analog (Ladenheim et al., 1959). The 4-picoline is characterized by $pK = 5.2$, so that at pH 3 only one picoline molecule in 150 is in the basic form which participates in the reaction. With poly(4-vinylpyridine), however, the electrostatic repulsion between the polycation and hydrogen ions tends to counteract the buildup of a high density of ionized groups on the chain molecule, so that at pH 3 and an ionic strength of 0.003 there are still 45% of pyridine residues in their basic form. The rate of quaternization with an uncharged alkyl halide will, therefore, drop off with decreasing pH much more slowly for the polymer than for its analog. The reaction of the polymer would also be expected to exhibit a much larger salt effect, since addition of neutral salts will reduce the electrostatic free energy of ionization, reduce the dissociation of the cationic pyridinium residues, and lead, as a result, to a decrease in the reaction rate. For instance, an increase in the ionic strength from 0.003 to 0.15 reduces by a factor of three the fraction of pyridine residues in poly(4-vinylpyridine) which remain at pH 3 in their basic form and can function in a chemical reaction as nucleophiles.

However, even when the reaction of a polyion with a second uncharged reagent is compared with the rate of the corresponding reaction of a monofunctional analog *at the same degree of ionization*, some striking differences may be observed. To avoid complications resulting from the change in the nature of the polymeric species with the progress of the reaction, it is best to concentrate on an analysis of initial reaction rates. A study of this type, carried out on the bromine displacement in α-bromoacetamide by partially ionized poly(methacrylic acid):

$$\begin{array}{l} \quad\;\; CH_3 \qquad\quad CH_3 \qquad\quad CH_3 \\ -CH_2-\overset{|}{\underset{|}{C}}-CH_2-\overset{|}{\underset{|}{C}}-CH_2-\overset{|}{\underset{|}{C}}- + BrCH_2CONH_2 \rightarrow \\ \quad\;\; COOH \qquad COO^- \qquad COO^- \end{array}$$

$$\begin{array}{l} \quad\;\; CH_3 \qquad\quad CH_3 \qquad\quad CH_3 \\ -CH_2-\overset{|}{\underset{|}{C}}-CH_2-\overset{|}{\underset{|}{C}}-CH_2-\overset{|}{\underset{|}{C}}- + Br^- \\ \quad\;\; COOH \qquad CO \qquad\quad COO^- \\ \qquad\qquad\quad\;\; | \\ \qquad\qquad\quad OCH_2CONH_2 \end{array} \tag{IX.6}$$

revealed that the reactivity of the carboxylate groups decreased sharply with an increasing degree of ionization of the polymeric acid (Ladenheim and Morawetz, 1959). On the other hand, when partially ionized poly(4-vinylpyridine) acted as a nucleophilic catalyst in the hydrolysis of 2,4-dinitrophenyl acetate, the apparent second-order rate constant, based on the concentration of un-ionized pyridine residues, was found to increase with an increase in the fraction of these residues present in the form of their conjugate base (Letsinger and Savereide, 1962). These two results have the common feature that the basic groups attached to the polymer tend to react with uncharged low molecular weight reagents at a rate which decreases with an increasing charge density (positive or negative) along the macromolecular chain (cf. Fig. IX.4). This seems reasonable if we consider the α-bromoaceta-

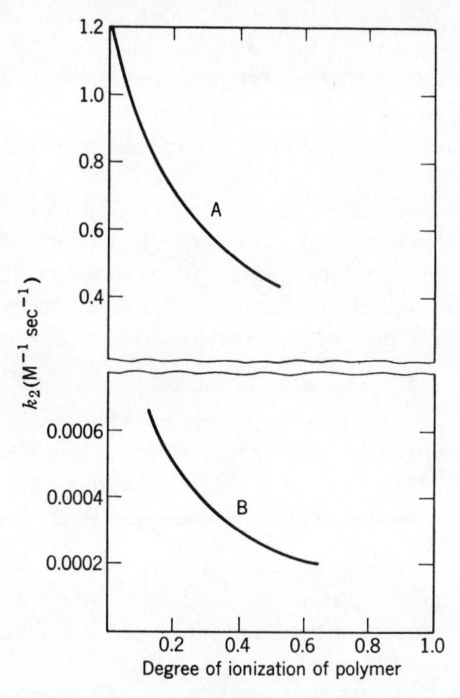

Fig. IX.4. Reactivity of polyions with small molecules. (A)Poly(4-vinylpyridine) with 2,4-dinitrophenyl acetate (50% ethanol, ionic strength 0.04, temp., 36.8°C, rate constant based on concentration of un-ionized pyridine residues). (B) Poly(methacrylic acid) with α-bromoacetamide (water solution, ionic strength 0.148, temp. 50°C, rate constant based on concentration of ionized carboxyls).

mide as a relatively nonpolar cosolvent added to the aqueous medium. We would then expect that the region surrounding the macromolecule would have an increasing tendency to exclude this reagent as the polarity of the polymer is being increased with an increasing degree of ionization. In nonaqueous media, where the polarity of the solvent is more similar to that of the reagent, the drift in the reaction rate constant may be diminished or even eliminated entirely. This is found to be the case in the quaternization of poly-(p-chloromethylstyrene) with triethylamine (Alger, 1964) and the quaternization of poly(4-vinylpyridine)with butyl bromide (Arcus and Hall, 1964).

Effects of a different type are encountered when polyions react with charged low molecular weight species. Qualitatively, it would be expected that electorstatic interactions would hinder reactions between species which carry charges of the same sign, while accelerating processes between positively and negatively charged reagents. Additions of electrolyte tend to reduce long-range forces between charged reagents, so that reactions involving species with charges of the same sign are accelerated, whereas those between oppositely charged species are slowed down. With reactions of simple ions the phenomenon is referred to as the "primary salt effect," which may be accounted for quantitatively (Bell, 1941), using the theory of absolute reaction rates and assigning to the reagents and the transition state complex activity coefficients which depend on the ionic strength as predicted by the Debye—Hückel theory.

The modification of reaction rates by electrostatic forces will be particularly large if one of the reagents is a polyion with a high charge density. A typical example is the hydroxyl-ion-catalyzed hydrolysis of pectins, the partially esterified derivatives of polygalacturonic acid, which are widely distributed in plants. The polymer may be represented schematically as follows:

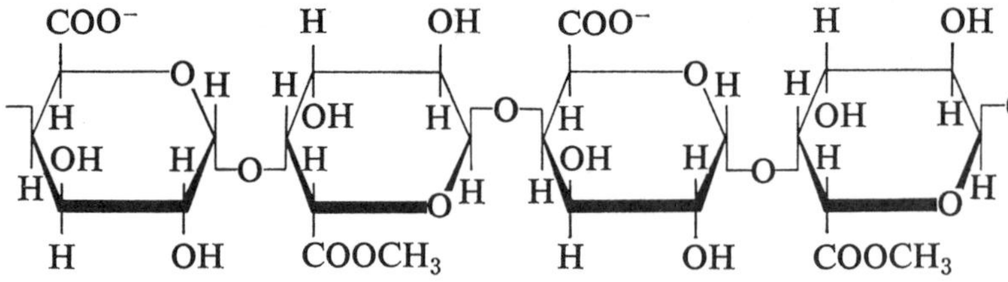

COO^- H OH COO^- H OH
O H H O H H
H OH H —O— H OH H —O—
OH H —O— H OH H —O— H
H O H H O H
H OH $COOCH_3$ H OH $COOCH_3$

and it was found that the apparent second-order rate constant for its basic hydrolysis falls off rapidly with the progress of the reaction (Deuel et al., 1953). This is what would be expected as a result of the mutual repulsion of the catalyzing hydroxyl ion and the polymer, which acquires an increasing negative charge as the ester groups are being hydrolyzed. Addition of neutral salts reduces this electrostatic interaction and leads to an acceleration of the process, particularly in its later stages (Deuel et al., 1953). A similar inhibition in the reaction of a polyion with a small anionic species was observed in the bromine displacement by the carboxylate group of partially ionized poly(methacrylic acid). This reaction proceeds at an appreciable rate with the uncharged α-bromoacetamide; but if the anionic α-bromoacetate is used as the second reagent, the bromine displacement by the carboxylate groups of the polyanion is too slow to be detected (Ladenheim and Morawetz, 1959). An analogous phenomenon was reported by Lovrien and Waddington (1964) in their study of copolymers of *N*-azobenzeneacrylamide:

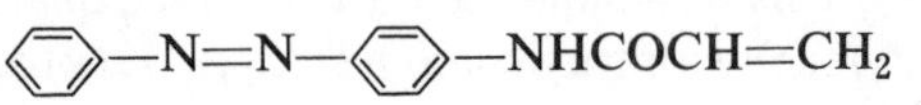

⬡—N=N—⬡—NHCOCH=CH_2

with acrylic or methacrylic acid. On irradiation with UV light, the azobenzene is isomerized from the *trans* to the *cis* form, and the reverse process, which occurs in the dark, is subject to both hydrogen ion and hydroxyl ion catalysis. Attachment of the azobenzene function to a polymeric acid tends to repress the catalytic effect of the hydroxyl ions which are repelled from the polyanion, and the minimum reaction rate is, therefore, shifted to much higher pH values.

Studies have also been reported in which a polyion reacts with a reagent of opposite charge. This is the situation in the reaction of partially ionized poly(4-vinylpyridine) with α-bromoacetate (Ladenheim et al., 1959), where the mutual electrostatic attraction of the two reagents leads to an acceleration of the reaction rate. Therefore, the addition of neutral salt has an inhibiting effect.

A quantiative theory of electrostatic effects on the reaction rate of a polyion with a small ion was proposed by Katchalsky and Feitelson (1954) for the interpretation of the hydroxyl-ion-catalyzed hydrolysis of pectin. Their treatment may be expressed by

$$k = k_0 \exp\left[\frac{-\Delta G_{\mathrm{el}}^{\ddagger}(\alpha_i)}{RT}\right] \tag{IX.7}$$

where k is the rate constant observed with the polyion at a degree of ionization α_i, while k_0 is the rate constant which would be observed in the hydrolysis of an uncharged ester. The electrical free energy of activation $\Delta G_{\mathrm{el}}^{\ddagger}(\alpha_i)$, required to bring the catalyzing hydroxyl ion to the ester group against the repulsion of the polyanion, is assumed to be equal to the electrical free energy required to remove a hydrogen ion from a carboxyl group against the field of the polyion, that is, the electrical free energy of ionization

$$\Delta G_{\mathrm{el}}^{\ddagger}(\alpha_i) = \Delta G_{\mathrm{el}}^{i}(\alpha_i) \tag{IX.8}$$

From a comparison of (IX.7) with (VII.36) we conclude that the apparent second-order rate constant for the basic hydrolysis of pectin should change with the progress of the reaction by the same factor as the apparent ionization constant, that is,

$$\frac{k}{K_{\mathrm{app}}} = \mathrm{const} \tag{IX.9}$$

This treatment, which should apply generally to reactions of polyions with small species carrying a single negative charge, is an extension of an analogous suggestion made by Ingold(1930) for thc basic hydrolysis of monoesters of dicarboxylic acids. It implies that the electrostatic potential at the point to which the hydroxyl ion has to be brought to attack the ester function is the same as the potential at the location of the acid group whose ionization constant is being compared to the hydrolytic rate constant. For the hydrolysis of esters attached to polyions this is a reasonable assumption. However, in other cases of reactions of polyions with small ionized species, $\Delta G^{\ddagger}_{el}$ need not be equal to ΔG^{i}_{el}. For instance, in the reaction of poly(4-vinylpyridine) with α-bromoacetate, the anionic charge of the α-bromoacetate need not be brought nearly as close to the backbone of the polymer to form the transition state as in the association of a hydrogen ion with a pyridine residue. It would then be expected that $\Delta G^{\ddagger}_{el}$ should be appreciably smaller than ΔG^{i}_{el}. This expectation is in accord with experimental data (Ladenheim et al., 1959) which show that the polyion charge has a distinctly greater effect on the ionization equilibrium of poly(4-vinylpyridine) than on its reaction rate with α-bromoacetate.

Interesting effects are encountered when a polymer carries groups which exert a catalytic action on a transformation of a small molecule and when this "substrate" is also attracted to the polymer domain. A system of this kind was first investigated by Letsinger and Savereide (1962), who studied the solvolysis of the 3-nitro-4-acetoxybenzenesulfonic acid (NABS) anion in the presence of poly(4-vinylpyridine). The polymer was found to be a much more powerful catalyst when it was partially ionized than when all the pyridine groups were in the basic form, and this was interpreted as a result of the attraction of the NABS to the polycation, where the basic pyridine residues could then function as nucleophilic catalysts for ester solvolysis.

The cooperative effect involving the cationic centers and the nucleophilic functions attached to the polymer backbone would be expected to depend on the spacing of these groups. This factor was investigated in a study in which 4(5)-vinylimidazole polymers:

$$(\text{—}CH_2\text{—}CH\text{—})_n$$

(imidazole ring attached to CH: HN N)

of varying chain length were used as the catalytic species (Overberger and Okamoto, 1972). As illustrated Fig. IX.5, the cooperative effect between the basic nucleophile residues and the cationic residues attracting the NABS substrate develops only gradually as the degree of polymerization of the polymer is increased.

The forces attracting a low molecular weight reagent to a polymer carrying catalytic groups need not, of course, be electrostatic. Hydrophobic bonding has been found to be most effective for this purpose, as first demonstrated by Kirsh et al. (1967, 1968), who found that the efficiency with which pyridine residues of poly(4-vinylpyridine) attack *p*-nitrophenyl acetate is sharply

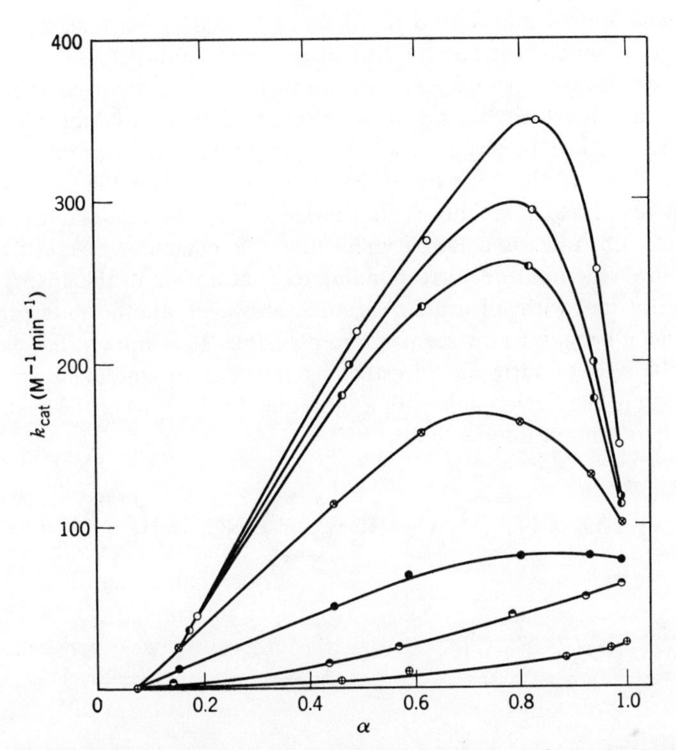

Fig. IX.5. Dependence of the catalytic coefficient for the hydrolysis of NABS on the fraction α of the monomer residues of poly(4(5)-vinylimidazole) present in their basic form and on the degree of polymerization $\bar{P}$ of the catalyst. (⊕) $\bar{P} = 2$; (◒)$\bar{P} = 4.3$; (●)$\bar{P} = 8.2$; (⊗) $\bar{P} = 14$; (◐)$\bar{P} = 27$; (◐)$\bar{P} = 37$; (○) high polymer.

increased over the catalytic efficiency of the low molecular weight analog, 4-methylpyridine, when the polymer is partially quaternized with benzyl chloride. As would be expected, this effect is rapidly reduced on addition of alcohol to the aqueous medium. Similar effects result when the ester whose solvolysis is being catalyzed is made more hydrophobic (Overberger and Morimoto, 1971; Overberger et al., 1971; Kunitake and Shinkai, 1971). A particularly high catalytic efficiency of a synthetic polymer was reported by Kiefer et al. (1972), who used poly(ethylene imine) carrying hydrophobic and imidazole substituents to accelerate the hydrolysis of 2-hydroxy-5-nitrophenyl sulfate. The results of this study cannot be accounted for by the concentration of the substrate in the polymer domain with its high local concentration of catalyzing nucleophiles, but must involve another principle, either a change in the reaction mechanism or the effect of a particularly favorable local solvent medium. A striking analogous medium effect was discovered by O'Connor et al. (1973), who found that the hydrolysis of dinitrophenyl sulfate may be up to 70 times as fast in benzene containing alkylammonium carboxylate micelles as in water solution.

All the research groups working in this field have reported that the catalyzed reaction velocity **v** tends to a maximum at high substrate concentration (S). The kinetic data may generally be represented by a linear relation between 1/**v** and 1/(S), a behavior which is formally analogous to the kinetics of many enzyme-catalyzed reactions. This has led some investigators to postulate a polymer-substrate association preceding the chemical reaction, with the maximum reaction rate corresponding to a saturation of the adsorption sites of the polymer with substrate molecules. However, another interpretation of the kinetic behavior now seems more plausible. It is known that catalysis of the solvolysis of nitrophenyl esters by pyridine or imidazole involves the acylation of the nucleophile with a subsequent solvolysis of the acylimidazole or acylpyridinium intermediate:

$$RC(=O)O-C_6H_4-NO_2 + \text{imidazole (N, NH)} \longrightarrow HO-C_6H_4-NO_2 + \text{imidazole (N, NC(=O)R)} \longrightarrow \quad \text{(IX. 10a)}$$

$$RC(=O)O-C_6H_4-NO_2 + \text{pyridine (N)} \longrightarrow {}^{-}O-C_6H_4-NO_2 + \text{pyridinium } (\overset{+}{N}C(=O)R) \longrightarrow \quad \text{(IX. 10b)}$$

Most investigators have followed the reaction by the release of nitrophenol, and the rate of this process may be limited by the velocity with which the acylated nucleophiles are deacylated. A detailed study of the acylation and deacylation of poly(4(5)-vinylimidazole) has revealed a number of interesting effects (Overberger et al., 1973; Overberger and Glowaky, 1973). If the ester of a long-chain fatty acid was used as the substrate with an excess of polyvinylimidazole, the nitrophenol release accelerated with the progress of the reaction, since the hydrophobic bonding of the substrate to the polymer increased with an increasing degree of acylation of the imidazole residues. The rate of polymer deacylation was found to be governed by intramolecular interaction of acylated and free imidazole residues, and this rate decreased sharply with an increasing chain length of the acyl groups. Presumably, the long hydrophobic residues led to a pronounced shrinkage of the polymer chain in the aqueous medium, and the decreasing polarity of the polymer domain was unfavorable for the deacylation process.

The catalytic action of polyvinylimidazole on nitrophenyl esters has occasionally been compared to that of the enzyme chymotrypsin. The two processes are quite different, however, in that the imidazole residue of the enzyme has only a catalytic function and does not play the role of an acyl acceptor (see p.456–457). In a recent study, Kunitake et al., (1974) have come a step closer toward mimicking the mechanism of the enzyme. Their synthetic polymer catalyst carries both imidazole and phenyl hydroxamate substituents; the attack on nitrophenyl acetate leads first to acetylation of the hydroxamate and then to deacetylation due to intramolecular imidazole catalysis:

$$\text{CO-N(OH)-C}_6\text{H}_5 \cdots \text{(imidazole: N, NH)} + O_2N\text{-C}_6\text{H}_4\text{-OCCH}_3(=O) \longrightarrow \text{CO-N(OCCH}_3(=O))\text{-C}_6\text{H}_5 \cdots \text{(imidazole: N, NH)} \quad \text{(IX. 11a)}$$

$$\text{CO-N(OCCH}_3(=O))\text{-C}_6\text{H}_5 \cdots \text{(imidazole: :N, NH)} + H_2O \longrightarrow \text{CO-N(OH)-C}_6\text{H}_5 \cdots \text{(imidazole: N, NH)} + CH_3COOH \quad \text{(IX. 11b)}$$

This polymer is also superior to polyvinylimidazole in that the hydroxamate group should be able to attack alkyl esters which are stable in the presence of imidazole. Nevertheless, synthetic polymers are still very far removed in catalytic power from enzymes, which owe their characteristics to a high degree of conformational rigidity by which cooperative effects of the substituents attached to the macromolecular chain may be maximized. It remains to be seen how far the synthetic polymer chemist can go in his attempt to duplicate the catalytic efficiency of biological macromolecules.

All the reactions discussed above involve substantial activation energies, so that a very large number of collisions of the interacting species are required before the reaction occurs. In cases where the activation energy is negligible, so that a large fraction of collisions lead to a chemical transformation, the process becomes diffusion controlled, resulting in some characteristic phenomena. A typical example of a process of this type is the interaction of small fluorescing molecules with florescence-quenching groups attached to a randomly coiled macromolecule. Since the lifetime of the excited state is extremely short, the probability that an excited molecule will lose its excitation energy before fluorescence occurs depends on the concentration of quenching groups in its immediate vicinity. The fluorescence intensity is, therefore, a function of the spatial distribution of polymer chain segment densities in the system. Data obtained by Duportail et al. (1971) in an experiment of this type were consistent with a Gaussian distribution of segment densities within randomly coiled polymer chains and indicated that extensive interpenetration of the molecular coils begins when the volume available to a coil is a sphere with a radius of $1.3\langle s^2\rangle^{1/2}$. If the polymer carries the fluorescent group and the quenching agent is a small molecule, the quenching reflects the relative solvation of the polymer by the solvent and the quenching agent (Moldovan and Weill, 1971). Diffusion-controlled reactions between small free radicals (such as $OH^{\cdot}$) and randomly coiled polymers have also been investigated (Behzadi and Schnabel, 1973). Since the free radical is likely to react with a polymer segment in the outer reaches of a macromolecular coil, before diffusing toward the center of the macromolecule, the chain segments in the interior of the coil contribute relatively less to the reaction rate. Thus, for a given number of macromolecules, the reaction rate increases less rapidly than in proportion to the chain length, and the reaction is accelerated when the polymer chain expands in better solvent media.

C. INTRAMOLECULAR INTERACTION EFFECTS ON POLYMER REACTIVITY

1. Neighboring Group Effects on the Reactivity of Chain Molecules

If several kinds of functional groups which may react with one another are attached to the backbone of the same polymer molecule, the probability of collisions between these groups will be increased, leading to correspondingly high reaction rates. This effect, which may be particularly pronounced if the interacting groups are part of neighboring monomer residues, was first recognized in a study of the hydrolysis of acrylic or methacrylic acid copolymers containing a small proportion of *p*-nitrophenyl methacrylate (Morawetz and Zimmering, 1954; Zimmering et al., 1957). This hydrolysis was expected to be very slow, in view of the repulsion of hydroxyl ions from the negatively charged polymer chain, but was found to be much more rapid than the hydrolysis of *p*-nitrophenyl esters of monocarboxylic acids. The pH dependence of the reaction rate showed that the velocity of the hydrolysis is proportional to the fraction of ionized carboxyls in the macromolecule. This indicates that the reaction does not involve hydroxyl ions and that the process is governed by the attack of an ionized carboxyl on the ester function. Bender and Neveu (1958) later demonstrated that the first product in this type of reaction is an acid anhydride, as represented by

$$-CH_2\overset{\overset{\displaystyle CH_3}{|}}{C}CH_2CHCH_2- \quad \text{with } O{=}C(OR) \text{ on the quaternary C and } C{=}O(O^-) \text{ on the CH} \longrightarrow -CH_2\overset{\overset{\displaystyle CH_3}{|}}{C}CH_2CHCH_2 \text{ (cyclic anhydride } O{=}C{-}O{-}C{=}O\text{)} + RO^- \longrightarrow$$

$$\xrightarrow{H_2O} -CH_2\overset{\overset{\displaystyle CH_3}{|}}{C}CH_2CHCH_2- \text{ (} O{=}C{-}OH \text{ and } C{=}O{-}O^-\text{)} + ROH \qquad \text{(IX.12)}$$

Monoesters of dicarboxylic acids, in which the ester and carboxyl groups have a spacing similar to that in the copolymer, behave in an analogous fashion. Figure IX.6. compares the pH dependence of the hydrolysis rate of the

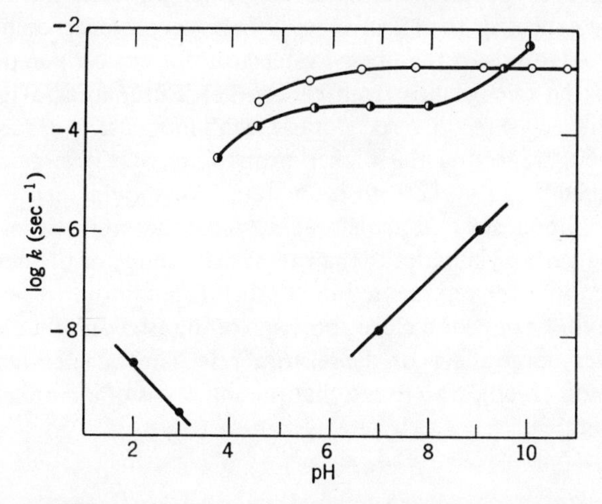

Fig. IX.6. pH dependence of the rate constants for the hydrolysis of *p*-nitrophenyl esters at 0°C. (●) *p*-Nitrophenyl pivalate; (◑) mono-*p*-nitrophenyl glutarate; (○) copolymer of acrylic acid with *p*-nitrophenyl methacrylate.

p-nitrophenyl methacrylate-acrylic acid copolymer with the behavior of the pivalic acid ester (analogous to the methacrylate residue in the polymer chain) and the glutaric acid monoester (analogous to a chain segment containing a methacrylic ester and an acrylic acid unit):

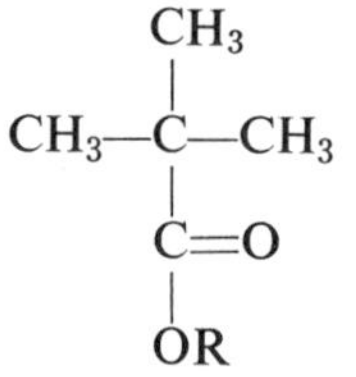

$$\begin{array}{c} CH_3 \\ | \\ CH_3{-}C{-}CH_3 \\ | \\ C{=}O \\ | \\ OR \end{array} \qquad\qquad \begin{array}{ccc} CH_2 & {-}CH_2{-} & CH_2 \\ | & & | \\ C{=}O & & C{=}O \\ | & & | \\ OR & & OH \end{array}$$

Pivalic acid ester 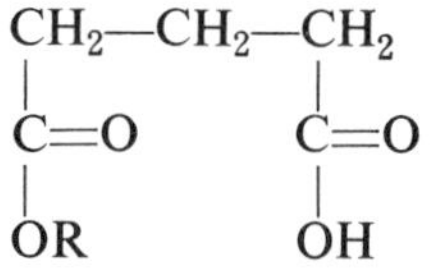 Glutaric acid monoester

The hydrolysis rate of the pivalate is proportional to (H^+) below pH 4 and to (OH^-) above pH 6, while the hydrolysis rate of the glutaric monoester rises with increasing pH, levels off to a constant rate between pH 6 and 8, and increases again in more basic solutions. This means that the carboxylate attack on the ester function is rate determining up to pH 8 and the hydroxyl ion attack makes a significant contribution to the observed reaction velocity only in more basic media. In the case of the copolymer, the plateau in the reaction rate extends beyond pH 10, indicating that the direct attack of hydroxyl ion on groups attached to the anionic polymer is effectively inhibited by the Coulombic repulsion discussed in the preceding section.

As would be expected, the effectiveness of the neighboring carboxylate in catalyzing the solvolysis of the phenyl ester group is critically dependent on the spacing of the two groups from each other. Succinic acid monoesters are 120–200 times as reactive as glutaric acid monoesters (Gaetjens and Morawetz, 1960), reflecting the higher probability of a five-membered as against a six-membered cyclic transition state. We should then expect that copolymers of monoesters of maleic acid would be labilized much more by neighboring carboxylate attack than the ester groups in the acrylic acid-methacrylate copolymer represented in (IX.12). In addition, steric restraints which either favor or impede a close approach of the ester and the carboxylate may produce very large effects on the reaction rate. This was demonstrated by Bruice and Pandit (1960), who found that the solvolysis of a monoester anion of the following type:

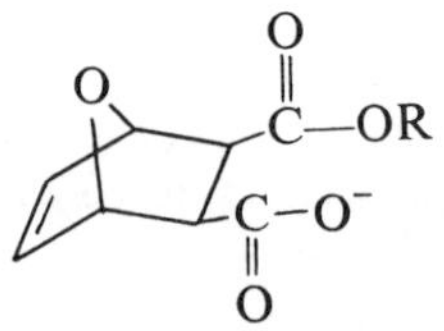

$$\text{[bicyclic structure: O-bridged ring bearing } {-}\overset{O}{\overset{\|}{C}}{-}OR \text{ and } {-}\underset{O}{\underset{\|}{C}}{-}O^- \text{]}$$

was 230 times as rapid as that of the anion of the corresponding succinic acid monoester. The same principle may lead to a pronounced dependence of the hydrolysis rate of ester-acid copolymers, depending on the relative steric configurations of the carbon atoms to which the ester and carboxylate groups are attached. Thus, the hydrolysis of copolymers of methacrylic acid containing about 1% of phenyl methacrylate residues deviates sharply from first-order kinetics and behaves, in fact, as if about 20% of the ester groups were 10 times as susceptible to neighboring carboxylate attack as the remaining ester groups (Morawetz and Gaetjens, 1958; Gaetjens and Morawetz, 1961). It seems that the more reactive structure corresponds to the isotactic configuration of the chain backbone. It is interesting that the deviation of the hydrolytic rate from first-order kinetics is observed in the case of methacrylic acid copolymers with both acrylic and methacrylic esters, but *not* for acrylic acid copolymers with either type of monomeric ester. This is a rather remarkable result, since one might have expected that the attack of a neighboring acrylate on a methacrylic ester and the attack of a neighboring methacrylate on an acrylic ester unit:

$$\sim CH_2-CH(C(=O)O^-)-CH_2-C(CH_3)(C(=O)OR)-CH_2\sim \qquad \sim CH_2-C(CH_3)(C(=O)O^-)-CH_2-CH(C(=O)OR)-CH_2\sim$$

would be subject to similar steric restraints. The fact that these two types of copolymers differ so strikingly in the sensitivity of their reaction rates to stereoisomerism suggests that the cumulative effect of a number of consecutive monomer units in a methacrylic chain leads to steric restraints which cannot be simulated in simple molecular analogs.

In the hydrolysis of acrylic acid copolymers with a small proportion of *N*-acrylyl-*p*-nitroaniline (Westhead and Morawetz, 1958) the pH profile of the reaction rate is the mirror image of that shown in Fig. IX 6, that is, the reaction rate increases with decreasing pH, reaches a plateau in the range in which the carboxyls are virtually un-ionized, and increases again only in strongly acid solutions where hydrogen ion catalysis becomes significant. The mono-*p*-nitroanilide of glutaric acid did not exhibit any neighboring group effect on the reactivity of the nitroanilide group; apparently in this case the steric restraints in the polymeric chain are highly favorable for the neighboring group interaction.

In the examples cited above, care was taken to study copolymers with a very small proportion of the reactive monomer residues, so that stereoisomerism constituted the only cause of variation in their chemical environment. When the reactive groups are more densely spaced along the polymer backbone, analysis of the kinetic pattern becomes much more difficult. For instance, during the solvolysis of poly(methyl methacrylate) an ester (E) group flanked by ester or acid (A) residues with the various relative steric configurations may occur in the ten triads depicted in Fig. IX.7. If each structure

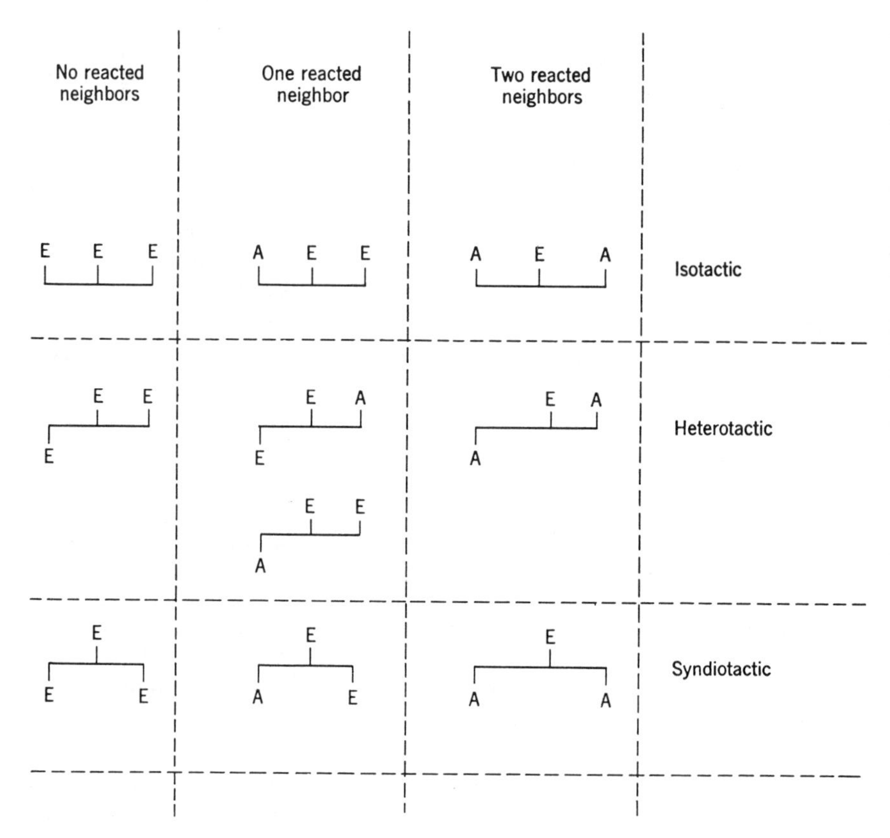

Fig. IX.7. Schematic representation of the different chemical and steric environments of ester (E) residues of poly(methyl methacrylate) during hydrolysis to carboxylic acid (A).

is characterized by a discrete rate constant, experimental precision is clearly insufficient for a rigorous analysis of the data.

An ingenious suggestion for surmounting this difficulty has been advanced by Robertson and Harwood (1971). In their method a polymer with the ester group tagged with ^{14}C is employed, and the solvolysis is allowed to proceed only to low conversion before the product is remethylated with unlabeled diazomethane. The procedure may be repeated a number of times, and the loss of radioactivity of the polymer then reveals the effect of stereoisomerism on the reactivity of the ester residues, while possible effects of neighboring carboxyls, formed as a result of the solvolysis, are minimized. With this technique, Robertson and Harwood arrived at relative rate constants of 43:2.7:1 for the acid-catalyzed solvolysis of the central ester of isotactic, heterotactic, and syndiotactic triads.

Alternatively, the experimentalist may use for his study a polymer of high stereoregularity and obtain data on the effect which one or two converted nearest-neighbor monomer residues exert on the reactivity of a chain substituent. In an outstanding study of this type, Klesper et al. (1970b) studied the solvolysis of syndiotactic poly(methyl methacrylate). The sequence distribution of hydrolyzed and unreacted monomer residues was then determined by NMR spectroscopy, and it was demonstrated that this distribution can be varied, for a given degree of conversion, by the choice of reaction conditions. Hydrolysis of methyl methacrylate-methacrylic acid copolymers in relatively concentrated KOH led to a copolymer in which alternation of ester and acid units was favored, while in solutions in which the base was only sufficient to neutralize the carboxyl groups, the resulting polymer tended to contain blocks of ester or acid units. This result indicates that, under conditions where the reaction is governed by the attack of hydroxide ion on the ester, neighboring charged carboxylates are inhibitory. When, however, only enough base is added to neutralize the carboxyl groups, the attack of a neighboring carboxylate on the ester is rate determining, leading to solvolysis of consecutive ester groups. A tendency toward blocks of hydrolyzed ester residues must also characterize the acid solvolysis of isotactic poly(methyl methacrylate), since the reaction has a pronounced accelerating phase indicative of an activation of neighboring esters by carboxyl groups (Robertson and Harwood, 1971).

An impressive attempt was made by Goldstein et al. (1971) to take into account both the effects of stereoisomerism and the effects of chemical differences of neighboring groups in the analysis of the kinetics of the aminolysis of poly(vinylene carbonate). This reaction is much faster with the polymer than with its low molecular weight analog, ethylene carbonate—an observation previously interpreted as due to the effect of neighboring unreacted monomer residues. Goldstein et al. synthesized the three stereoisomeric trimers of vinylene carbonate and extracted from the data of their aminolysis nine rate constants, corresponding to the reactivity of vinylene carbonate residues with none, one, or two reacted nearest neighbors in the three stereoisomeric environments.

Whereas the aminolysis of a monomer residue in poly(vinylene carbonate) reduces sharply the reactivity of neighboring residues, the inverse is the case in the hydrolysis of poly(vinyl acetate), where the hydroxyl groups formed during the reaction have an accelerating effect on the reactivity of neighboring ester functions. As a result, partially saponified poly(vinyl acetate) can be shown by infrared spectroscopy to have a much higher fraction of ester groups flanked by other unchanged ester residues then would be expected from a statistical distribution of hydrolyzed groups (Nagai and Sagane, 1955). The accelerating effect of neighboring hydroxyls on ester hydrolysis may also be demonstrated by the acceleration of the reaction rate in the initial stages of the process (Minsk et al., 1941; Fujii et al., 1963). A similar observation was reported with the hydrolysis of poly(vinyl acetal) in dilute acid solutions (Smets and Petit, 1959). With both poly(vinyl acetate) and poly(vinyl acetal) the syndiotactic polymer seems to be the more reactive species (Fujii et al., 1963).

A neighboring group effect of a different type seems to result when the primary alcohol group in an anhydroglucose residue of cellulose is oxidized to carboxyl. According to Rånby and Marchessault (1959), this modification stabilizes the glucosidic linkage between residues B and C, while the linkage between residues A and B becomes more labile in an acid medium:

COOH
A B C
CH_2OH CH_2OH

A small number of carboxyl groups have, therefore, been suggested to be the cause of "weak links" in the cellulose chain. A study of the pH-dependence of the hydrolysis rate of alginic acid shows also that *un-ionized* carboxyl activates the glucosidic linkage (Smidsrød et al., 1966).

Activation of a polymer chain substituent by a concerted attack involving two neighboring groups has been clearly demonstrated in only one case: the condensation of a copolymer of maleic acid and *N*-acrylyl-*p*-nitroaniline to the corresponding imide (Westhead and Morawetz, 1958). The rate of this reaction was found to be maximized at half-ionization of the carboxyls, suggesting that the mechanism requires the anilide to be attacked simultaneously by one ionized and one un-ionized carboxyl:

CII_3, CH, C, CH, C=O, O=C, C=O, NH, OH, $^-$O, R → CH_3, CH, C, CH, C=O, O=C, N, C=O, R, OH + OH^- (ix. 13)

The hydrolysis of partially neutralized acrylic acid-ethyl acrylate copolymers seems also to be fastest at half-neutralization of the carboxyl groups (Smets and Van Humbeeck, 1963), so that cooperative bifunctional catalysis may also be rate determining in this case.

2. Statistical Analysis of the Kinetic Pattern

In the preceding section we discussed a number of cases in which a chain molecule carries a large number of substituents whose reactivity depends on whether their nearest neighbors have been converted to another group. Provided that effects of stereoisomerism can be neglected, we have then three rate constant to consider: k_0, characteristic of groups with two unreacted nearest neighbors; k_1 for groups lying next to one reacted and one unreacted residue; and k_2, describing the reactivity of groups lying between two reacted residues of the chain molecule.

Consider first the case where a reactive group is strongly activated by the functional groups formed by the conversion of its nearest neighbor, so that $k_2 \approx k_1 \gg k_0$. We may then regard the slow reaction of a residue in the middle of an unreacted sequence as a "nucleation" process, which is followed by a more rapid "chain reaction" of a series of residues, each activated by the reaction of the preceding unit. Let us denote by (A) the total number of unreacted groups and by (B) the number of sequences consisting of one or more consecutive reacted groups. In the initial stages of the reaction, when (A) deviates only slightly from its original value $(A)_0$ and $(A) \gg (B)$, we may approximate $(B) = k_0(A)_0t$, and we have then for the observed reaction rate

$$-\left[\frac{d(A)}{dt}\right]_{(A)\to(A)_0} \approx k_0(A)_0 + 2k_1k_0(A)_0t \qquad \text{(IX.14)}$$

The conversion will, therefore, be a quadratic function of time in the initial phases of the reaction. The acceleration observed in the hydrolysis of poly(vinyl acetate) and poly(vinyl acetal), as discussed in the preceding section, conforms to this kinetic pattern. If k_1/k_0 could be made much larger than the degree of polymerization, reaction of any chain substituent would induce rapid conversion of all the reactive groups attached to the polymer and the reaction rate should then be proportional to the length of the molecular chain. Although such a phenomenon does not seem to have been described to date, there can be little doubt that polymers could be prepared which would behave in this a manner.

A number of investigators have treated the theory of the kinetic pattern for an infinite chain with substituents which have their reactivities altered by a reaction of their nearest neighbors. The complexity of the problem may be indicated by a simple example. If we denote by O a reacted monomer residue, the processes

$$\text{OAAAAAO} \longrightarrow \text{OAAOAAO}$$
$$\text{OAAAAAO} \longrightarrow \text{OAOAAAO}$$

are equally probable, but the first leads to a product in which all the A residues have one reacted neighbor, whereas the second results in a sequence in which one A has two reacted neighbors, two A groups have one reacted neighbor, and one A has no reacted neighbors. These two sequences will be quite different, therefore, in their subsequent kinetic behavior. The most frequently used expressions for the course of the reaction are those of Keller (1962, 1963) and of McQuarrie et al. (1965). According to Keller, the fraction of unreacted residues N_0 and N_1, which have no reacted nearest neighbors and one reacted nearest neighbor, respectively, may be expressed by

$$N_0(t) = \exp\left[-(2k_1 - k_0)t - \frac{2}{k_0}(k_1 - k_0)e^{-k_0t} - 1\right] \qquad \text{(IX.15a)}$$

$$N_1(t) = 2[\exp(k_0t) - 1]\exp\left[-(2k_1 + k_0)t - \frac{2}{k_0}(k_1 - k_0)(e^{-k_0t} - 1)\right] \qquad \text{(IX.15b)}$$

The general expression for N_2, the fraction of unreacted residues with two reacted neighbors, is much more complex. It assumes a simpler form, however, for two special cases of interest.

For $k_1 = k_2 = 0$:

$$N_2(t) = \tfrac{1}{2} + [\exp(-k_0t) - \tfrac{3}{2}]\exp[2(e^{-k_0t} - 1)] \qquad \text{(IX.15c)}$$

For $k_0 = k_1$:

$$N_2(t) = \frac{2k_0\exp(-2k_0t)}{k_2 - 2k_0} - \frac{2k_0\exp(-3k_0t)}{k_2 - 3k_0} + \frac{2k_0^2\exp(-k_2t)}{(k_2 - 2k_0)(k_2 - 3k_0)} \qquad \text{(IX.15d)}$$

Relation (IX.15d) predicts that one-third of the functional groups will remain unreacted at infinite time if the reactivity is unaffected by one neighboring reacted group but is completely eliminated when both nearest neighbors have reacted. This is the model which has been used to interpret the hydrolysis of polymethacrylamide in strong base solutions (Arcus, 1949). If even one reacted neighboring residue leads to complete inhibition, (IX.15b) and (IX.15c) predict that for $t \to \infty$ a fraction $N_1 + N_2 = (1 + e^{-2})/2$ (i.e., 56.8%) of the potentially reactive groups will remain unreacted. For some much more complicated cases, the kinetic course of the reaction has been obtained by computer simulation (Litmanovich et al., 1969; Goldstein et al., 1971), and Klesper et al. (1971, 1972) have developed procedures for the evaluation of sequence distributions during the reaction of polymers subject to neighboring group effects.

Comparison of theoretical predictions with experimental data is rendered difficult by several factors. In many cases, the product obtained when a polymeric reaction has gone to completion is insoluble in the solvent media for the original polymeric reagent, so that the reaction may be observed over only a limited range of conversion. This is so, for instance, in the hydrolysis of poly(vinyl acetate). In other cases, the reaction can be followed over its entire course in a single solvent medium, but neighboring group effects are not sufficiently pronounced (or experimental data sufficiently precise) to allow the evaluation of three distinct rate constants. For instance, in the quaternization of poly(4-vinylpyridine) with *n*-butyl bromide (see p. 460), the data may be interpreted by $k_1/k_0 = 1$ and $k_2/k_0 = 0.32$ (Arends, 1963), but no claim is made that other ratios of the three rate constants would not yield an equally acceptable fit of the observed kinetic curve. Finally, ample evidence demonstrates the sensitivity of reaction rates to the stereoisomerism of polymeric reagents. This complicates the situation to such an extent that any interpretation of data collected over a broad range of conversion usually becomes highly uncertain when both neighboring group effects and effects resulting from stereoisomerism are known to be pronounced.

3. Effects of Non-neighboring Groups

The neighboring group activations considered in Section IX.B.1 involved situations where a neighboring group attack on a reactive function leads to a five-membered or six-membered cyclic transition state. Five and six-membered rings form very easily, and such neighboring group effects will, therefore, be highly efficient. The effect of a favorable spacing of the activating group and the reactive function may be illustrated by comparing the hydrolysis rate in monophenyl succinate and monophenyl glutarate with the catalytic power of acetate ion in the hydrolysis of phenyl acetate (Gaetjens and Morawetz, 1960). Such a comparison shows that the neighboring carboxylate group in the succinate and the glutarate is 4000 and 30 times, respectively, as effective as 1*M* acetate. However, the effectiveness of neighboring groups in accelerating reaction rates would be expected to decline abruptly if the transition state involves rings of larger size. The relative ease with which rings of various sizes may be formed was studied by Stoll and Rouvé (1935) and by Spanagel and Carothers (1935), who found that rings containing 9–12 atoms form with great difficulty and that the probability of ring formation passes through a second flat maximum (much lower than that characterizing five-and six-membered rings) for rings containing about 18 atoms. These complicated effects are caused by bond angle restrictions and steric hindrance. For long chains, the first-order rate constant k_1 for the closure of a ring by reaction of two functional groups attached to the chain ends may be formally expressed by $k_1 = k_2^\circ c_{\text{eff}}$, where k_2° is the rate constant for the bimolecular process of low molecular weight species analogous to the chain ends, and c_{eff} is the "effective local concentration" of one chain end in the neighborhood of the other. The parameter c_{eff} is related to the distribution function $\mathscr{W}(h)$ of chain end displacements by

$$c_{\text{eff}} = \lim_{h \to 0} \frac{1000}{\boldsymbol{N}} \left(\frac{\mathscr{W}(h)\, dh}{4\pi h^2 dh} \right) \tag{IX.16}$$

where c_{eff} is expressed in moles per liter.† In the absence of an excluded volume effect (i.e., in Θ-solvents) we obtain, by using for $\mathscr{W}(h)$ relations (III.7) and (III.9),

$$c_{\text{eff}} = \frac{1000}{\boldsymbol{N}} \left(\frac{3}{2\pi \langle h^2 \rangle} \right)^{3/2} \tag{IX.17}$$

This value of c_{eff} corresponds to a particle enclosed in a sphere with a radius of $1.1\langle h^2 \rangle^{1/2}$. It leads then to a first-order rate constant for ring closure, given by

$$k_1 = \frac{1000 k_2^\circ}{\boldsymbol{N}} \left(\frac{3}{2\pi \langle h^2 \rangle} \right)^{3/2} \tag{IX.18}$$

For long, randomly coiled chain molecules in Θ-solvents, $\langle h^2 \rangle = C_\infty Z b^2$ (III.16), so that the rate constant for ring closure should be inversely proportional to the 3/2 power of the chain length (Kuhn, 1934; Jacobsen and Stockmayer, 1950). In good solvent media, where the excluded volume effect is operative, ring closure would be expected to decay faster with increasing chain length, but no statistical treatment of this problem exists. Computer simulation (Wall et al. 1955; Hiley and Sykes, 1961; Domb et al. 1965; Fischer, 1966) leads to the prediction that, in strong solvent media and for very long chains, ring closure probabilities are proportional to Z^{-a}, with a estimated by different workers as lying in the range $1.8 \leqslant a \leqslant 2.0$. Since Flory's theory of the excluded volume effect predicts $\langle h^2 \rangle$ to be proportional to $Z^{1.2}$ for long chains in good solvent media, the functional relationship between k_1/k_2° and $\langle h^2 \rangle$ appears to be similar in good and poor solvents. We may then regard the k_1/k_2° ratio as providing essentially similar information on chain flexibility and the excluded volume effect as measurements of the extension of the macromolecular coil derived, for instance, from the angular dependence of light scattering intensities or intrinsic viscosity.

Experimental studies of the rates of cyclization involving the interaction of groups attached to the ends of long-chain molecules are, in general, not possible. This is so because the intramolecular reaction becomes dominant and the intermolecular process can be neglected only at concentrations so low that analytical methods are unavailable for following the process. An outstanding exception is a study of the cyclization of a DNA species in which two short strands, complementary to each other, extend from the two ends

†Here it is assumed that the ring closure reaction has a high activation energy, so that the process is not kinetically controlled by the rate of conformational transitions of the polymer chain. For a theory of reactions in which conformational transitions of the polymer are rate limiting, see G. Wilenski and M. Fixman, *J. Chem. Phys.*, **60**, 866, 878 (1974).

of the double-helical structure. In this case the rate at which the cyclic DNA forms can be followed at very high dilution, and the corresponding bimolecular process can be studied on the DNA broken by shearing forces into halves. The experimental results have been reported to be in good agreement with (IX.18) (Wang and Davidson, 1966), although it is hard to believe that the excluded volume effect is negligible in the case of these highly charged chain molecules.

Although we cannot follow cyclization rates in synthetic polymers carrying reactive groups at the chain ends, similar information can be obtained in a less direct manner. For this purpose, reaction kinetics can be studied in highly dilute solutions of random copolymers carrying a small number of reactive and catalytic chain substituents distributed at random along the chain backbone. We may treat the interaction of a reactive group on the nth monomer residue with catalytic groups on the jth residue as equivalent to the probability of interaction of groups attached to the ends of a chain with $j - n$ segments. In the absence of an excluded volume effect, the rate constant characterizing a given reactive group attached to a chain of P monomer residues becomes, in analogy to (IX.18),

$$k_1 = \frac{1000k_2^\circ}{N}\left(\frac{3\mathrm{P}}{2\pi\langle h^2\rangle}\right)^{3/2}\left(\sum_{j=1}^{j=n-x'} x^{-3/2} + \sum_{j=n+x'}^{j=\mathrm{P}} x^{-3/2}\right) + k_1' \quad \text{(IX.19)}$$

where $x = |j - n|$, x' is the smallest value for the separation of interacting groups for which the statistical theory of cyclization is applicable, and k_1' is the contribution to the rate constant from catalytic groups separated by fewer than x' monomer residues from the reactive group.

Since the catalytic groups are distributed at random along the chain, the spacing from these groups will be different for each reactive group and (IX. 19) will yield rate constants characterized by a distribution function $W(k_1)$. Thus, the fraction y of groups left unreacted at time t will be given by

$$\mathrm{y} = \int W(k_1) \exp(-k_1 t)\, dk_1 \quad \text{(IX.20)}$$

exhibiting a characteristic curvature on a first-order plot. For good solvent media, where cyclization is being impeded by a large excluded volume effect, (IX.19) must be replaced by an expression in which a "ring closure exponent" a larger than 3/2 is used; this has the effect of broadening the distribution function $W(k_1)$ and increasing the curvature of the first-order plot of the course of the reaction.

The approach outlined above has been tested experimentally (Goodman and Morawetz, 1970, 1971), using acrylamide or methacrylamide copolymers containing about 1 mole % of residues carrying nitrophenyl ester groups or pyridine residues which catalyze the hydrolysis of these esters. Computer analyses by Goodman and Morawetz and by Sisido (1972) showed that the curvatures of first-order plots were consistent with a ring closure exponent $a = 2$ in a good solvent medium. In a medium close to the Θ point, a value of $a = 1.6$ was estimated, slightly larger than predicted by theory. Since the initial apparent first-order rate constant must be proportional to the fraction of residues which carry catalytic substituents, one can also estimate the "maximum effective local concentration," $c_{\mathrm{eff}}^{\max}$, of catalytic groups in the neighborhood of a reactive group attached to a polymer in which every other residue carries the catalytic substituent. This value of $c_{\mathrm{eff}}^{\max}$ was found to lie in the range 2.2–2.8M in good solvents and in the range of 3.0–4.5M close to the Θ point. Since each monomer residue corresponds to two bonds in the chain backbone, $\langle h^2 \rangle / \mathrm{P} = 2C_\infty b^2$ (see p. 125), and we obtain for very long chains

$$c_{\mathrm{eff}}^{\max} = 2 \cdot \left(\frac{1000}{N}\right)\left(\frac{3}{4\pi C_\infty b^2}\right)^{3/2} \int_{x=x'}^{x=\infty} x^{-3/2}\, dx$$
$$= \frac{4000}{Nb^3}\left(\frac{3}{4\pi C_\infty}\right)^{-3/2} (x')^{-1/2} \qquad \text{(IX.21)}$$

where the contribution to $c_{\mathrm{eff}}^{\max}$ of residues spaced more closely than x' has been neglected. When the reasonable values of $C_\infty = 6$, $x' = 10$ are used, this leads to $c_{\mathrm{eff}}^{\max} = 4.6M$, in satisfactory agreement with experimental results.

It is also possible for two nonneighboring groups attached to a polymer chain to attack cooperatively a small molecular species. The pH dependence of the solvolysis of nitrophenyl acetate in poly(4(5)-vinylimidazole) solutions appears to involve cooperative attack by a neutral and an anionic imidazole moiety on the ester function (Overberger et al., 1965, 1966). A similar cooperativity was described for the esterolytic catalysis in solutions of copolymers of vinyl imidazole with 4-vinylphenol or with vinyl alcohol (Overberger et al., 1967).

D. REACTIONS OF TWO LOW MOLECULAR WEIGHT SPECIES IN THE PRESENCE OF POLYIONS

The presence of macromolecules in dilute solution has, in general, little effect on the reactions of small molecules with each other. Although the presence of the polymer may increase appreciably the macroscopic viscosity of the system, the mobility of small molecular species usually remains almost unchanged (cf. Section VI. A.5). Moreover, even if the mobility of the reactive species were reduced, there might be no appreciable effect on the reaction rate, since most chemical processes have an activation energy which requires a very large number of molecular collisions before the reaction will occur. Although a lowered mobility of the reactive species will reduce the rate at which they approach one another, it will also diminish the velocity with which they diffuse apart, so that the collision frequency will not be altered (Flory, 1953d).

A polymer may, however, produce a large effect on the reaction rate of two small species if it exerts forces on both of them, so that their spatial distribution in the system is modified and their collision frequency is altered. These forces may be of various types, but investigations have centered on the consequences of hydrophobic bonding and Coulombic interactions of charged species. If both reagents are attracted to the polymer, reaction rates will be increased. If one of the reagents is attracted, while the other is being repelled from the polymer, the reaction will be inhibited. Kinetic effects will generally be small if both reagents are repelled from the polymer domain. This pattern of behavior is quite similar to that observed in micellar solutions (Cordes and Dunlap, 1969; Morawetz, 1969; Fendler and Fendler, 1970).

The importance of hydrophobic bonding in enhancing the catalytic effect of polyions was demonstrated by Sakurada et al. (1966) and by Yoshikawa and Kim (1966) in their studies of acid-catalyzed ester hydrolysis in the presence of polysulfonic acids. As would be expected, catalysis was most effective in systems in which either the substrate or the polymer carried hydrophobic substituents, since the ester was then drawn into the polymer domain and thus into a region of high local hydrogen ion concentration. An extreme case is the catalysis of the alkaline fading of triphenylmethane dyes in solutions of polycations with hydrophobic substituents (Okubo and Ise, 1973). In that case, the hydrophobic bonding leads to a binding of the cationic dye by the polycation in spite of mutual electrostatic repulsion, and the fading reaction is accelerated because of the high local hydroxide ion concentration. The identical effect has been demonstrated in solutions containing cationic micelles (Menger and Portnoy, 1968).

Kinetic effects in polyelectrolyte solutions which depend on electrostatic interaction of charged reagents with the polyions have been reported for a number of systems (Morawetz, 1969). Some years ago, it was suggested (Morawetz and Westhead, 1955) that such an effect should be tested for the benzidine rearrangement, which is known to follow the mechanism

$$C_6H_5\text{—NH—NH—}C_6H_5 \underset{-H^+}{\overset{+H^+}{\rightleftharpoons}} C_6H_5\text{—}\overset{+}{N}H_2\text{—NH—}C_6H_5$$

$$\xrightarrow{H^+} H_2N\text{—}C_6H_4\text{—}C_6H_4\text{—}\overset{+}{N}H_3 \qquad (IX.22)$$

and Arcus et al. (1964) found, in fact, that the reaction velocity is more than 100 times higher in poly(styrenesulfonic acid) than in benzenesulfonic acid solutions of the same normality. Similarly, the Canizzaro reaction with the mechanism

$$\underset{\underset{O}{\|}}{HC}-\underset{\underset{O}{\|}}{CH} \underset{-OH^-}{\overset{+OH^-}{\rightleftharpoons}} \underset{\underset{O}{\|}}{HC}-\overset{\overset{H}{|}}{\underset{\underset{OH}{|}}{C}}-O^- \xrightarrow{OH^-} H_2O + \underset{\underset{O}{\|}}{HC}-\overset{\overset{H}{|}}{\underset{\underset{O^-}{|}}{C}}-O^-$$

$$\longrightarrow {}^-OCH_2\underset{\underset{O}{\|}}{C}O^- \xrightarrow{H_2O} HOCH_2\underset{\underset{O}{\|}}{C}O^- + OH^- \qquad \text{(IX.23)}$$

was found to be much faster in a solution of poly(vinylbenzyltrimethylammonium hydroxide) than in a solution of its low molecular weight analog (Arcus and Jackson, 1964). An example of a system in which polyions exert an inhibitory effect is provided by the solvolysis of a cationic ester in a polycarboxylic acid solution (Morawetz and Shafer, 1963). Here the inhibition is clearly the result of the concentration of the ester in the neighborhood of the polyanion from which the catalytic hydroxide ions are being repelled. A similar principle seems to be involved in the observation that cationic dyes are protected from attack by solvated electrons if polyanions are added to the system (Balasz et al., 1968).

In the examples cited above, hydrophobic bonding is superimposed on long-range Coulombic interactions, and it is impossible to estimate the contribution of each of these forces to the effects observed. Hence it is of particular interest to consider the acceleration or inhibition produced by polyelectrolytes on reactions of compact inorganic species, where the data might be interpretable on the basis of electrostatic interactions only. The reactions of Co(III) complexes are particularly suitable for this purpose, since these reactions involve ions with many different combinations of charges and since the high extinction coefficients of these reagents make it possible for studies to be carried out at very high dilution. With the reaction

$$Co(NH_3)_5Cl^{2+} + Hg^{2+} + H_2O \rightarrow Co(NH_3)_5H_2O^{3+} + HgCl^+ \qquad \text{(IX.24)}$$

in the presence of poly(vinyl sulfonate) (PVS) and poly(methacryloxyethyl sulfonate) (PMES), Morawetz and Vogel (1969) found that the process could be accelerated by factors of up to 176,000 and 24,000 in solutions of PVS and PMES, respectively. As would be expected, the catalytic effect of the polyion is reduced on addition of other electrolytes which compete with the reagent ions for inclusion into the polyion domain. Figure IX.8, which shows how the acceleration factor depends on the polyion concentration, is particularly illuminating. At low polyion concentrations, where only a portion of the reagent ions is concentrated in the polyion domains, the reaction rate is approximately proportional to the concentration of polyelectrolyte. However, when the polyelectrolyte is sufficiently concentrated to capture essential-

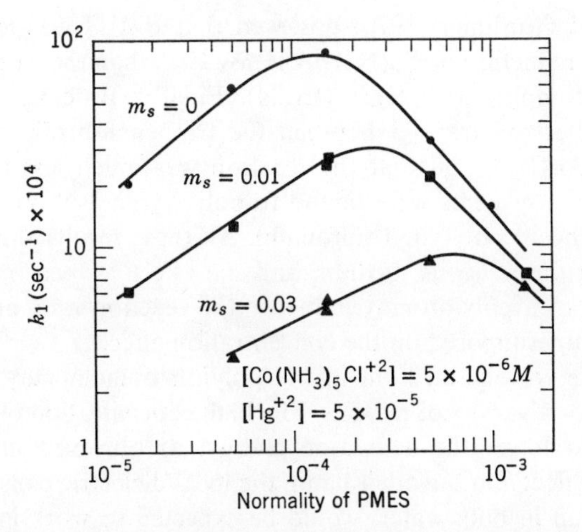

Fig. IX.8. Effect of PMES on the Hg^{2+}-induced aquation of $Co(NH_3)_5Cl^{2+}$ in solutions containing various concentrations of $NaClO_4$ (m_s) at 5°C.

ly all of the reagent ions, further polyelectrolyte addition will require these reagents to be shared among an increasing number of polymer domains; the local reagent concentration in these domains will then be inversely proportional to the polyelectrolyte concentration, and this will be reflected by a corresponding decrease in the reaction rate. Morawetz and Vogel showed that the catalytic effect of the polyion may be interpreted by two parameters, one characterizing the mean association constant of the reagent ions with the polyion, and the other representing the "effective local concentration" of counterions in the polyion domain. Although this is clearly a rather crude model, the reagent ion-polyion association constant estimated from the kinetic data was found to be in gratifying agreement with the results of dialysis equilibrium measurements. At high polyelectrolyte concentration, where all the reagent ions are in the polymer domain, the catalytic effect should be independent of added electrolyte, reflecting merely the distribution of the fixed charges in the isolated polyion. The data plotted in Fig. IX.8 are consistent with this concept.

Several studies of the effect of polyelectrolytes on electron transfer processes have also been reported. The data show that different reactions involving reagents with the same ionic charges have somewhat different susceptibilities to polyion catalysis, even when the polyelectrolyte is in large excess, so that virtually all the reagent ions must be situated in the polyion domain. For instance, Gould (1970) found that the reduction of different $Co(NH_3)X^{3+}$ complexes by Cr^{2+} was accelerated in the presence of poly(vinyl sulfonate), under his experimental conditions, by factors ranging from 1000 to 4500.

Morawetz and Gordimer (1970) observed that PMES was less efficient in catalyzing the reduction of $Co(NH_3)_5Cl^{2+}$ by Fe^{2+} than the interaction of the same cobalt complex with Hg^{2+} (IX.24). Finally, Brückner et al. (1970), who studied electron transfer between the triphenanthroline complexes of Co(III) and Co(II), found that the reaction was much less powerfully catalyzed when the reagents were bound to poly(styrene sulfonate) than when they were bound to poly(vinyl sulfonate). All these results show that differences between the aqueous medium and the effective local medium in the neighborhood of a polyion may influence the reaction rate, and such influences will be superimposed on the concentration effect.

The effective solvent medium in the polyion domain may influence the reaction rate for a variety of reasons. The high concentration of ionic charges should tend to lower the activation energy, as observed in the classical Brønsted salt effect. On the other hand, the local dielectric constant, substantially lower than in bulk water, would be expected to work in the opposite direction. Changes in ion hydration in the formation of the transition-state complex should also be sensitive to medium effects and might make substantial contributions to changes in the activation parameters. In view of the complexity of the problem, attempts to draw mechanistic conclusions from the effect of polyions on $\Delta H^\ddagger$ and $\Delta S^\ddagger$ of ionic reactions (Ise, 1971) are quite unwarranted. With the reagent ions in excess, the activation parameters reflect partially the thermodynamics of reagent ion-polyion association. With the polyion in excess, they characterize the change produced in $\Delta H^\ddagger$ and $\Delta S^\ddagger$ when the reagent ion interaction is made to proceed entirely in the polymer domain. These effects are quite specific. The reaction of $Co(NH_3)_5Cl^{2+}$ with Hg^{2+} has $\Delta H^\ddagger$ reduced by 1.5 kcal/mole and by 3.9 kcal/mole in the presence of a large excess of PMES and PVS, respectively (Morawetz and Vogel, 1969), while the reaction of $Co(NH_3)_5Cl^{2+}$ with Fe^{2+} has $\Delta H^\ddagger$ *increased* by 3.8 kcal/mole in the presence of an excess of PVS (Morawetz and Gordimer, 1970).

We may, of course, express the effect of polyions on ionic reactions in terms of the activity coefficients of the reagents, γ_A and γ_B, and the transition-state complex, $\gamma^\ddagger_{AB}$ (Ise, 1971). When the activity coefficients are taken as equal to unity in the absence of polyelectrolytes, the acceleration factor, k_2/k_2°, produced by polyions is, in terms of the ionic activity coefficients equal to $\gamma_A\gamma_B/\gamma^\ddagger_{AB}$. In the case of the reaction of ammonium and cyanate ion to form urea, the transition-state complex has no net charge, so that $\gamma^\ddagger_{AB}$ may be assumed to remain at unity. The deceleration of the reaction in the presence of polyions is then predictable from the activity coefficients of ammonium cyanate (Okubo and Ise, 1972). In other cases, such as reaction (IX.24), no direct test of this approach is possible, since we cannot estimate the activity coefficient of the highly charged transition-state complex. However, simple consideration shows that this formulation is in qualitative accord with the simple model described above, where the two counterion reagents are either "outside" or "inside" the polyion domain. As long as the reagent ions are in excess over the polyion, its presence will affect mostly the activity coefficient, $\gamma^{\ddagger}_{AB}$, of the more highly charged transition-state complex. Since $\gamma^{\ddagger}_{AB}$ would be expected to be inversely proportional to the polyion concentration, k_2/k_2° will be proportional to the concentration of the polyelectrolyte, c_p. However, with the polyions in excess, all the ionic activity coefficients, γ_A, γ_B, $\gamma^{\ddagger}_{AB}$, should be proportional to $1/c_p$ and k_2/k_2° should then also vary as $1/c_p$.

A very important discovery was reported by Patel et al. (1973), who studied the effect of a poly(sulfonic acid) on the interaction of Co(III) complexes with various reducing agents. With $Ru(NH_3)_6^{2+}$, known to react by the "outer sphere mechanism" in which the bimolecular interaction of oxidant and reductant is rate determining, the polyanion led to a sharp acceleration of the reaction. However, in reactions of $Cr(H_2O)_6^{2+}$ with Co(III)-chloropentammine complexes, where the rate-limiting step is a slow unimolecular decomposition of a reversibly formed binuclear complex of the two reagents, no polyelectrolyte catalysis was observed. It appears, therefore, that the susceptibility to polyion catalysis may serve as a valuable diagnostic tool of the reaction mechanism.

A different type of information can be obtained from experiments in which one of the counterions is a short-lived fluorescing species, while another counterion is able to quench the fluorescence. As long as the polyion is in excess over the fluorescing species, both the fluorescing ion and the quencher will be concentrated in the polyion domain and the quenching efficiency will be greatly enhanced. However, when the concentration of the quenching ion is sufficiently increased, it displaces the fluorescent ion from the polymer domain into a space where the concentration of the quencher is low. This leads to a dramatic increase in fluorescence intensity (Taha and Morawetz, 1971). The result, illustrated in Fig. IX.9, proves that the exchange of counterions inside and outside the polymer domain is slow compared to the excited lifetime of the fluorescing species.

Specific complex formation with ligands attached to a polymer chain may produce high local concentrations of catalytically active complex ions which react rapidly with a second species drawn by some other forces into the polymer domain. Such a phenomenon has been described by Pecht et al. (1967) for the oxidation of ascorbate and other anionic species by polyhistidine complexed with Cu(II). The Cu(II) complex of poly(L-lysine) has been similarly shown to be an efficient catalyst for the oxidation of 3,4-dihydroxyphenylalanine (Nozawa and Hatano, 1971) and the hydrolysis of phenylalanine methyl ester (Nozawa et al., 1972). This system is of special interest,

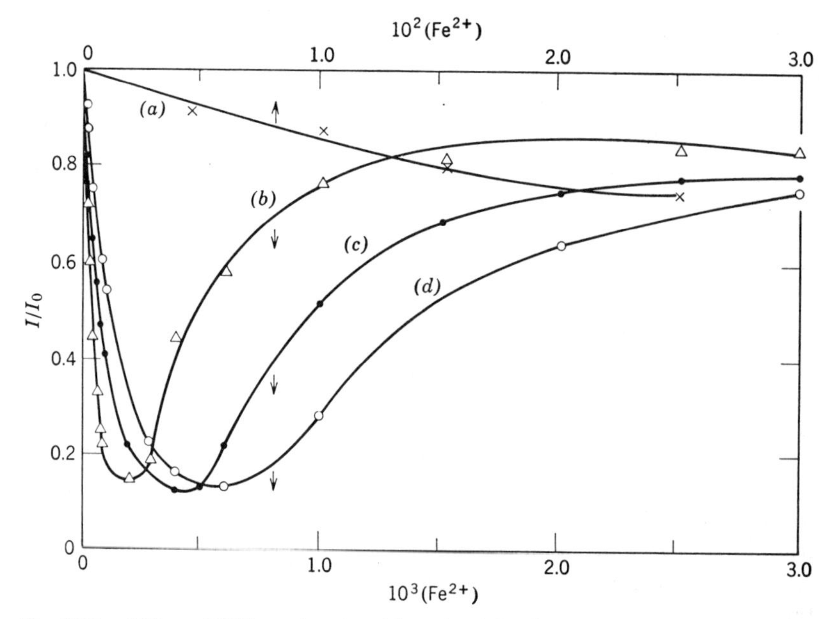

Fig. IX.9. Effect of PVS on the quenching of quinine fluorescence by Fe^{2+} in the presence of $10^{-3}M$ $HClO_4$. (*a*) No PVS; (*b*) $10^{-3}N$ PVS; (*c*) $2 \times 10^{-3}N$ PVS; (*d*) $3 \times 10^{-3}N$ PVS. Under the conditions of the experiment, quinine carries a double positive charge.

since the catalytic effect exhibits some stereospecificity in both cases. The ratio of the reaction rates for the D and L isomers of the substrates reaches maximum values of 1.5 and 2.3, respectively—modest compared to the sharp discrimination between stereoisomers exhibited by enzymes. Nevertheless, the effect is of great potential significance.†

Principles completely different from those discussed above are involved in the effect of polymers on the photochemical behavior of small molecules. For instance, ascorbic acid can photoreduce the cationic triphenylmethane dyes when they are bound to poly (methacrylic acid), although no reaction

†Recently, E. Tsuchida, H. Nishide and H. Nishikawa (*Macromolecules*, in press) have studied the catalysis of xylenol oxidation by atmospheric oxygen in dimethylsulfoxide solutions containing Cu(II) and partially quaternized poly(4-vinylpyridine). The catalytic power of the Cu(II)polymer complex increased sharply with the degree of quaternization, although the affinity of the oxidizable substrate for the polymer would not be expected to increase. Also, spectroscopic studies were interpreted as showing that the rate-determining step in this catalytic process is an *intramolecular* electron transfer within the Cu(II) complex containing a xylenoxy group and two pyridine residues of the polymer as ligands. The authors suggest that the electrostatic driving force towards polyion expansion reduces the stability of the Cu(II) complex and thus facilitates the reaction. Alternatively, it can be argued that the change in reactivity reflects a change in the solvent properties of the polymer domain with a change in the degree of quaternization.

takes place in the absence of the polymer (Oster and Bellin, 1957). Several factors seem to play a role in making the dye associated with a polymeric chain more susceptible to photoreduction. The photoreduction involves a long-lived excited state whose formation is favored by a rigid medium; attachment to a polymer backbone then acts to impede internal rotations. Flash spectroscopy experiments have also demonstrated that the concentration of the metastable species is unaffected by atmospheric oxygen when the dye is bound to a polymer, whereas it would be effectively quenched by minute traces of oxygen in solutions containing the free dye molecule (Oster and Oster, 1962).

E. CHEMICAL INTERACTION BETWEEN GROUPS ATTACHED TO DIFFERENT FLEXIBLE CHAIN MOLECULES

In considering reactions involving groups attached to different polymer molecules in dilute solutions, we have to draw a sharp distinction between processes characterized by high and by low activation energies. If the activation energy is high, the interacting groups will approach each other and separate a very large number of times before a chemical event takes place, so that the reaction rate is a measure of the statistical probability that these groups will be placed in juxtaposition to each other. Studies of the reaction rates of polymeric reagents should then furnish information concerning the extent to which the probability of such contacts between functional groups is altered because of their attachment to two different chain molecules. On the other hand, if the activation energy is low, a chemical reaction is likely to occur during the first approach of the interacting groups, and the rate of such a reaction will characterize the velocity of diffusive processes governing the encounter of these groups. This is the situation which will be typical of radical-radical interactions such as those involved in the termination step of the free-radical polymerization of vinyl compounds.

The very large second virial coefficients characterizing flexible chain polymers in good solvent media indicate that the macromolecular coils strongly resist mutual interpenetration in dilute solutions (see p. 190). One should expect, therefore, that functional groups attached to a polymer which find themselves, at any given moment, in the interior of the randomly coiled chain will be effectively shielded from interaction with functional groups attached to another chain molecule. As a result, the rate constants for such interactions should be smaller for bimolecular processes involving macromolecules than for analogous reactions of low molecular weight species and should decrease with the chain length of the macromolecular reagents. Attempts at theoretical treatments of this "kinetic excluded volume effect" have been published by

Wetmur (1971) and by Morawetz et al. (1973). Moreover, since the central segments of a chain molecule are less likely to find themselves at the periphery of the molecular coil than is a segment close to the end of the chain, reactive chain substituents attached at random along the length of the polymer chain should be characterized by different reaction rate constants.

For a proper study of the effects described above, it is important that no permanent bond results from the reaction of groups attached to two polymer chains, since such a linkage would modify profoundly the probability that other groups, attached to the same macromolecules, will react with each other. To avoid such a complication, Morawetz and Song (1966) and Cho and Morawetz (1973) studied reaction rates in solutions containing two acrylamide copolymers, one containing a small number of reactive nitrophenyl ester groups, and the other carrying a small number of imidazole or pyridine substituents that act as nucleophilic catalysts for ester solvolysis. The results of these studies were in sharp contrast to expectations based on the consideration outlined above: (a) no deviations from first-order kinetics were observed in good solvents; (b) the reaction rate of the polymeric reagents in water solution, a good solvent medium, was very similar to rates observed with the low molecular weight analogs; (c) contrary to expectation, the reduction in the reaction rate of the polymers, as compared to that of the analogs, was more pronounced and more dependent on the chain length of the polymer in ethylene glycol, a poorer solvent than water. These results seem at variance with our theoretical concepts, and no convincing explanation can be offered at this time. The conclusion that the probability of an encounter between substituents of similar long-chain polymers seems to be unaffected by the excluded volume effect has been confirmed by a study of solutions containing two modified polystyrenes, carrying a small number of dinitrophenol and dimethylbenzylamine substituents, respectively (Worsfold, 1974). Using the shift in the spectrum of the phenolic residue to determine the extent of its association with the amine groups, it was found that this association equilibrium was essentially the same for the groups attached to the polymer chains as for analogous small molecules.

In the examples cited above, the interacting groups were attached to similar polymer chains. When the chains carrying the reactive functions are different, the energetic interaction between them would be expected to have a large effect on the rate of their chemical interaction. We have noted (p. 90) that the mixing of different polymers is generally endothermic and that frequently such mixing is, in fact, so difficult that it results, even at fairly high dilution, in the formation of two liquid phases in systems containing a solvent and two polymers. We should then expect that chemical interaction between polymeric reagents would be very slow in systems of this type. The reverse effect would be expected for the reaction of groups attached to polymers which mix with a large evolution of heat. Such systems were studied by Letsinger and Klaus (1964) and by Letsinger and Wagner (1966), who used solutions containing anionic polymers with reactive ester substituents and partially ionized poly(*N*-vinylimidazole). In this case the attractive forces between the acidic and the basic polymer led to association of the two chain molecules, resulting in very efficient catalysis of the ester solvolysis. The reaction rate reached a maximum when complex formation was complete, and further increase in the concentration of the catalytic polymer was without effect. As expected, poly(acrylic acid) competed with the anionic chains carrying ester groups for association with the polycations and functioned, therefore, as a competitive inhibitor.

Reactions with very low activation energy involving two macromolecular reagents in dilute solution are of special interest because the termination step in the free-radical-initiated polymerization of vinyl compounds falls into this category. Benson and North (1959, 1962) showed that the termination rate constant in the polymerization of methyl methacrylate is inversely proportional to the viscosity of the solvent, so that the process must be diffusion controlled. They concluded also that the limiting factor is not the diffusion of the entire macromolecules toward each other, but a segmental diffusion within the coiled chains which brings the reactive chain ends to the surface of the coil, where they are exposed to mutual interaction. This point was confirmed by the observation that the termination rate constant in the polymerization of alkyl methacrylates decreases rapidly on extending the alkyl residue (North and Reed, 1963), although the diffusion rates of the polymer molecules should be very insensitive to the length of the alkyl group.

Experimental evidence showing how the rates of diffusion-controlled reactions depend on the length of polymeric reagents was provided by Fischer (1966) and by Borgwardt et al. (1969). Fischer studied the decay of radicals in polymerizing acrylic acid by ESR spectroscopy and found that the termination rate constant for propagating polymer radicals was three times smaller than the rate constant for the combination of $CH_3CH_2\dot{C}HCOOH$ radicals. No estimate for the average length of the chain radicals is given in this report, but the chains were undoubtedly relatively short. In the study of Borgwardt et al., an aqueous solution of poly(ethylene oxide) samples of different chain lengths (up to a degree of polymerization of 3000) was subjected to pulse radiolysis. The $OH^{\cdot}$ radicals produced in this way abstracted hydrogen atoms from the polymer, and the decay of the $—CH_2\dot{C}HO—$ radicals was followed by UV spectroscopy. The rate constant for the radical-radical combination was found to be inversely proportional to the 0.53 power of the molecular weight of the polymer, although it appeared to become less sensitive to the size of very long chain molecules.

A number of investigators have tried to formulate a quantitative theory for the diffusion-controlled reaction of polymeric reagents. The various factors to be considered are discussed by Horie et al. (1973), who also provide a valuable bibliography of this field. According to their treatment, the rate constant may be considered as a product of three factors: the first a function of the excluded volume effect (as characterized by the expansion coefficient), the second a function of the chain length, and the third a function of the frictional coefficient of a statistical chain segment and the temperature. However, Horie et al., refer to the internal viscosity of the polymer chain (see p. 314) as an "obscure concept" and disregard completely its role in determining the rate of segmental diffusion within the individual polymer coils. On the other hand, in a treatment of diffusion-controlled reactions in which the internal viscosity is taken into account (but the excluded volume effect is neglected), Burkhart (1965) arrives at the implausible conclusion that the rate constant should pass through a maximum with an extension of the chain length. In all these treatments, the segmental diffusion is treated as taking place independently within the two interacting polymers. Since chain entanglements must, in fact, play an important role, it is questionable whether models lending themselves to mathematical analysis can be considered realistic.

In discussing the renaturation of DNA from the two separated polynucleotide strands (p. 445), we pointed out that the rate-determining step is the nucleation of the double helix from two corresponding sequences of complementary nucleotide residues. In a study of this process with DNA samples of different chain length, Wetmur and Davidson (1968) noted that the rate constant for this nucleation was inversely proportional to the square root of the molecular weights of the reagents. Wetmur and Davidson (1968) and Wetmur (1971) attempted to interpret their results as a consequence of the excluded volume effect. However, the nucleation of the double helix is known to be essentially a diffusion-controlled process, and the rate of segmental diffusion in the polynucleotide strands is, therefore, probably rate limiting. In this context, the similarity in the molecular weight dependence of the nucleation of the DNA double helix and the radical-radical interaction rates studied by Borgwardt et al., is highly suggestive.

In concluding this discussion, we should emphasize that we have considered only rates of processes taking place in solutions in which the volume of the system is large compared to the total volume occupied by the swollen macromolecular coils. If the molecular chains are very long, this condition may imply very high dilution. At higher concentrations, where the chain molecules are compelled to interpenetrate, diffusion-controlled reactions may be subject to constraints which are even more difficult to analyze.

REFERENCES

Abe, A. (1968), *J. Am. Chem. Soc.*, **90,** 2205.

Abe, A. (1970), *Polym. J.*, **1,** 232.

Abe, A., and P. J. Flory (1965), *J. Am. Chem. Soc.*, **87,** 1838.

Abe, A., R. L. Jernigan, and P. J. Flory (1966), *J. Am. Chem. Soc.*, **88,** 631.

Ackermann, T., and M. Ruterjans (1964), *Z. Phys. Chem. Frankfurt*, **41,** 116.

Ackers, G. K. (1970), *Adv. Protein Chem.*, **24,** 343.

Adair, G. S. (1925), *J. Biol. Chem.*, **63,** 529.

Adams, E. T., Jr. (1965), *Biochemistry*, **4,** 1646, 1655.

Adams, E. T. Jr., and H. Fujita (1963), in *Ultracentrifugal Analysis in Theory and Experiment*, J. W. Williams, ed., Academic Press, New York, p. 119.

Adams, M. J., T. L. Blundell, E. J. Dodson, G. G. Dodson, M. Vijayan, E. N. Baker, M. M. Harding, D. C. Hodgkin, B. Rimmer, and S. Sheats (1969), *Nature*, **224,** 491.

Adams, M. J., G. C. Ford, R. Koekoek, P. J. Lentz, Jr., A. McPherson, Jr., M. G. Rossmann, I. E. Smiley, R. W. Schewitz, and A. J. Wonacott (1970), *Nature*, **227,** 1098.

Adams, M. J., M. Buehner, K. Chandrasekhar, G. C. Ford, M. L. Hackett, A. Liljas, M. G. Rossmann, I. E. Smiley, W. S. Allison, J. Everse, N. O. Kaplan, and S. S. Taylor (1973), *Proc. Natl. Acad. Sci. U.S.*, **70,** 1968.

Adamson, A. W. (1960), *Physical Chemistry of Surfaces*, Interscience, New York, pp. 76–81.

Akkerman, F., D., T. F. Pals, and J. J. Hermans (1952), *Rec. Trav. Chim.*, **71,** 56.

Albright, D. A., and J. W. Williams (1967), *J. Phys. Chem.*, **71,** 2780.

Albertsson, P. Å. (1958), *Biochim. Biophys. Acta.*, **27,** 378.

Albertsson, P. Å. (1970), *Adv. Protein Chem.*, **24,** 309.

Albertsson, P. Å. (1971), *Partition of Cell Particles and Macromolecules*, 2nd ed., Almquist and Wiksel, Stockholm.

Alexander, P., and S. F. Hitch (1952), *Biochim. Biophys. Acta*, **9,** 229.

Alexandrowicz, Z. (1959), *J. Polym. Sci.*, **40,** 91.

Alexandrowicz, Z. (1960), *J. Polym. Sci.*, **43,** 337.

Alexandrowicz, Z. (1962), *J. Polym. Sci.*, **56,** 115.

Alexandrowicz, Z. (1967), *J. Chem. Phys.*, **47,** 4377.

Alexandrowicz, Z., and A. Katchalsky (1963), *J. Polym. Sci.*, **A1,** 3231.

Alfrey, T., Jr., and G. Goldfinger (1944), *J. Chem. Phys.*, **12,** 205.

Alfrey, T., Jr., and H. Morawetz (1952), *J. Am. Chem. Soc.*, **74,** 436.

Alfrey, T., Jr., A. I. Goldberg, and J. A. Price (1950), *J. Colloid Sci.*, **5,** 251.

Alfrey, T., Jr., P. W. Berg, and H. Morawetz (1951), *J. Polym. Sci.*, **7,** 543.

Alger, M. S. M. (1964), Ph. D. Thesis, University of London.

Allegra, G., P. Ganis, and P. Corradini (1963), *Makromol. Chem.*, **61,** 225.

Allegra, G., P. Corradini, and P. Ganis (1966), *Makromol. Chem.*, **90,** 60.

Allegra, G. E. Benedetti, and C. Pedone (1970), *Macromolecules*, **3,** 727.

Allegra, G., S. Brückner, and V. Crescenzi (1972), *Eur. Polym. J.*, **8,** 1255.

Allegra, G., M. Calligaris, L. Randaccio, and G. Moraglio (1973), *Macromolecules*, **6,** 397.

Allinger, N. L., M. A. Wilson, F. A. Van Catledge, and J. A. Hirsch (1967), *J. Am. Chem. Soc.*, **89**, 4345.

Allinger, N. L., J. A. Hirsch, and M. A. Miller (1968), *J. Am. Chem. Soc.*, **90**, 1199.

Allis, J. W., and J. D. Ferry (1965), *J. Am. Chem. Soc.*, **87**, 4681.

Ambrose, E. J., and A. Elliott (1951), *Proc. Roy Soc. (London)*, **A205**, 47.

Ambrose, E. J., C. H. Bamford, A. Elliott, and W. E. Hanby (1951), *Nature*, **167**, 264.

Ambrose, J. F., G. B. Kistiakowski, and A. G. Kridl (1951), *J. Am. Chem. Soc.*, **71**, 1232.

Amelunxen, R. E. (1967), *Biochim. Biophys. Acta*, **139**, 24.

Anderson, J. E., and K. J. Liu (1970), *Macromolecules*, **3**, 163.

Anderson, J. E., K. J. Liu, and R. Ullman (1970), *Discuss, Faraday Soc.*, **49**, 257.

Anderson, N. S., J. W. Campbell, M. M. Harding, D. A. Rees, and J. W. B. Samuel (1969), *J. Mol. Biol.*, **45**, 85.

Anderson, W. B., A. B. Schneider, M. Emmer, R. L. Perlman, and I. Paston (1971), *J. Biol. Chem.*, **246**, 5929.

Ando, K., M. Matsuhara, and K. Okunuki (1966), *Biochim. Biophys. Acta*, **118**, 256.

Andress, K. R. (1928), *Z. Phys. Chem. (Leipzig)*, **A136**, 279.

Andress, K., and L. Reinhardt (1930), *Z. Phys. Chem. (Leipzig)*, **A151**, 425.

Andrews, P. (1965), *Biochem. J.*, **96**, 595.

Anfinsen, C. B. (1967), *Harvey Lectures*, **61**, 95.

Antonini, E., and M. Brunori (1969), *J. Biol. Chem.*, **244**, 3909.

Antonini, E., and M. Brunori (1970), *Ann. Rev. Biochem.*, **39**, 977.

Antonini, E., and M. Brunori (1971), *Hemoglobin and Myoglobin in Their Interaction with Ligands*, North Holland, Amsterdam.

Anufrieva, E. V., M. V. Volkenstein, M. G. Krakovyak, and T.V. Sheveleva (1969), *Dokl. Akad. Nauk SSSR*, **186**, 854.

Anufrieva, E. V., M. V. Volkenstein, Yu. Ya. Gotlib, M. G. Krakovyak, S. S. Skorokhodov, and T. V. Sheveleva (1970), *Dokl. Akad. Nauk SSSR*, **194**, 1108.

Applequist, J., and P. Doty (1962), in *Polyamino Acids, Polypeptides and Proteins*, M. A. Stahman, ed., University of Wisconsin Press, Madison, p. 161.

Applequist, J., and T. G. Mahr (1966), *J. Am. Chem. Soc.*, **88**, 5419.

Archibald, W. J. (1947), *J. Phys. Chem.*, **51**, 1204.

Arcus, C. L. (1949), *J. Chem. Soc.*, 2732.

Arcus, C. L., and W. A. Hall (1964), *J. Chem. Soc.*, 5995.

Arcus, C. L., and B. A. Jackson (1964), *Chem. Ind. (London)*, 2022.

Arcus, C. L., T. L. Howard, and D. S. South (1964), *Chem. Ind (London)*, 1756.

Arends, C. B. (1963), *J. Chem. Phys.*, **38**, 322.

Armstrong, R. W., T. Kuruscev, and U. P. Strauss (1970), *J. Am. Chem. Soc.*, **92**, 3174.

Arnon, R., and M. Sela (1969), *Proc. Natl. Acad. Sci., U.S.*, **62**, 163.

Arnone, A., C. J. Bier, F. A. Cotton, E. E. Hazen, Jr., D. C. Richardson, and J. S. Richardson (1969), *Proc. Natl. Acad. Sci., U.S.*, **64**, 420.

Asakura, S. (1961), *Biochim. Biophys. Acta*, **52**, 65.

Ash, K. O. (1968), *Science*, **162**, 452.

Assarson, P. G., P. S. Leung, and G. J. Safford (1969), *Polym. Prepr., Am. Chem. Soc., Div. Polym. Chem.,* **10,** (2) 1241.

Auer, H. E., and P. Doty (1966), *Biochemistry,* **5,** 1716.

Auer, R. L. and C. S. Gardner (1955), *J. Chem. Phys.,* **23,** 1546.

Avery, O. T., C. M. MacLeod, and M. MeCarthy (1944), *J. Exp. Med.,* **79,** 137.

Bacskai, R., L. P. Lindeman, and D. L. Rabenstein (1972), *J. Polym. Sci.,* **10,** 1297.

Badger, R. M., and S. H. Bauer (1937), *J. Chem. Phys.,* **5,** 839.

Bailey, F. E., Jr., R. D. Lundberg, and R. W. Callard (1964), *J. Polym. Sci.,* **A2,** 845.

Balasz, F. A., J. V. Davies, G. D. Phillips, and D. S. Schaufele (1968), *J. Chem. Soc.,* **C1424,** 1429.

Ballard, D. G. H., G. D. Wignall, and J. Schelten (1973), *Eur. Polym. J.,* **9,** 965.

Bamford, C. H., W. E. Hanby, and F. Happey (1951), *Proc. Roy. Soc. (London),* **A205,** 30.

Banaszak, L. J., and F. R. N. Gurd (1964), *J. Biol. Chem.,* **239,** 1836.

Banaszak, L. J., P. A. Andrews, J. W. Burgner, E. H. Eyler, and F. R. N Gurd (1963), *J. Biol. Chem.,* **258,** 3307.

Bancroft, J. B. (1970), *Adv. Virus Res.,* **16,** 99.

Banks, W., C. T. Greenwood, and J. Sloss (1970), *Makromol. Chem.,* **140,** 109, 119.

Barker, J. A. (1952), *J. Chem. Phys.,* **20,** 1526.

Barker, J. A., and W. Fock (1953), *Discussi. Faraday Soc.,* **15,** 188.

Barker, J. A., and F. Smith (1954), *J. Chem. Phys.,* **22,** 375.

Barker, J. A., I. Brown, and F. Smith (1953), *Discussi. Faraday Soc.,* **15,** 142.

Barkin, S. M. (1957), Ph. D. Thesis, Polytechnic Institute of Brooklyn.

Barkin, S. M., H. P. Frank, and F. R. Eirich (1955), *Ric. Sci. Suppl.,* **25,** 1, "Symposio internazionale di chimica macromoleculare."

Barksdale, A. D., and J. Steuhr (1972), *J. Am. Chem. Soc.,* **94,** 3334.

Barone, G., and E. Rizzo (1973), *Gaz. Chim. Ital.,* **103,** 401.

Barone, G., V. Crescenzi, and F. Quadrifoglio (1965), *Ric. Scient.,* **35,** (II-A) 1069.

Barrales-Rienda, J. M., and D. C. Pepper (1967), *Polymer,* **8,** 351.

Barron, E. S. G. (1951), *Adv. Enzymol.* **11,** 201.

Bartell, L. S., and D. A. Kohl (1963), *J. Chem. Phys.,* **39,** 3097.

Bates, F. L., D. French, and R. E. Rundle (1943), *J. Am. Chem. Soc.,* **65,** 142.

Bates, T. W., and K. J. Ivin (1967), *Polymer,* **8,** 263.

Bawn, C. E. H., and A. Ledwith (1962), *Quart. Rev. (London),* **16,** 427–428.

Bawn, C. E. H., and M. A. Wajid (1956), *Trans. Faraday Soc.,* **52,** 1658.

Becker, R. R., and M. A. Stahmann (1952), *J. Am. Chem. Soc.,* **74,** 38.

Behrens, O. K., and E. L. Grinnan (1969), *Ann. Rev. Biochem.,* **38,** 83.

Behzadi, A., and W. Schnabel (1973), *Macromolecules,* **6,** 824.

Beier, G., and J. Engel (1966), *Biochemistry,* **5,** 2744.

Beijerinck, M. W. (1910), *Kolloid-Z.,* **7,** 16.

Bell, R. P. (1941), *Acid and Base Catalysis,* Clarendon Press, Oxford, pp. 21–35.

Bellamy, G., and P. Bornstein (1971), *Proc. Natl. Acad. Sci. U.S.,* **68,** 1138.

Bello, J., D. Harker, and E. de Jarnette (1961), *J. Biol. Chem.*, **236,** 1358.

Bender, M. L., and M. C. Neveu (1958), *J. Am. Chem. Soc.*, **80,** 5388.

Bender, M. L., G. R. Schonbaum, and B. Zerner (1962), *J. Am. Chem. Soc.*, **84,** 2540.

Benedetti, E., C. Pedone, and G. Allegra (1970), *Macromolecules,* **3,** 16.

Benesch, R., and R. E. Benesch (1963), *J. Mol. Biol.*, **6,** 498.

Benesch, R., and R. E. Benesch (1969), *Nature,* **221,** 618.

Benesch, R., R. E. Benesch, and Y. Enoki (1968), *Proc. Natl. Acad. Sci. U.S.*, **61,** 1102.

Benoit, H. (1950), *J. Chim. Phys.*, **47,** 719.

Benoit, H. (1951*a*), *Ann. Phys. (Paris),* **6, 591.**

Benoit, H. (1951*b*), *J. Chim. Phys.*, **48,** 612.

Benoit, H. (1952), *J. Chim. Phys.*, **49,** 517.

Benoit, H. (1953), *J. Polym. Sci.*, **11,** 507.

Benoit, H., and P. Doty (1953), *J. Phys. Chem.*, **57,** 958.

Benoit, H., and G. Weill (1957), *Collect. Czech. Chem. Commun.*, **22,** 35 (special issue).

Benoit, H., and C. Wippler (1960), *J. Chim. Phys.*, **57,** 524.

Benoit, H., Z. Grubisic, P. Rempp, D. Decker, and J. G. Zilliox (1966), *J. Chim. Phys.*, **63,** 1507.

Benoit, H., L. Freund, and G. Spach (1967), in *Poly-α-Amino Acids,* G. D. Fasman, ed., Marcel Dekker, New York, pp. 105–155.

Benoit, H., D. Decker, J. S. Higgins, C. Picot, J. P. Cotton, B. Farnoux, G. Jannink, and R. Ober (1973), *Nature-Phys. (London),* **245,** 13.

Benson, S. W., and A. M. North (1959), *J. Am. Chem. Soc.*, **81,** 1339.

Benson, S. W., and A. M. North (1962), *J. Am. Chem. Soc.*, **84,** 935.

Beredjick, N. (1963), in *Newer Methods of Polymer Characterization,* B. Ke, ed., Interscience, New York, Chap. 16.

Beredjick, N., and H. E. Ries, Jr. (1962), *J. Polym. Sci.*, **62,** 564.

Beredjick, N., and C. Schuerch (1968), *J. Am. Chem. Soc.*, **80,** 1933.

Berek, D., I. Novak, Z. Grubisic-Gallot, and H. Benoit (1970), *J. Chromatogr.*, **53,** 55.

Bernal, J. D., and I. Fankuchen (1941), *J. Gen. Physiol.*, **25,** 111.

Bernal, J. D., and R. H. Fowler (1933), *J. Chem. Phys.*, **1,** 515.

Berry, G. C. (1966), *J. Chem. Phys.*, **44,** 4550.

Berry, G. C., and T. G. Fox (1968), *Adv. Polym. Sci.*, **5,** 261.

Berry, K. L., and J. H. Peterson (1951), *J. Am. Chem. Soc.*, **73,** 5195.

Bersohn, R., and I. Isenberg (1961), *J. Chem. Phys.*, **35,** 1640.

Berson, S. A., and R. S. Yalow (1961), *Nature,* **191,** 1392.

Bertland, A. U., and H. M. Kalckar (1968), *Proc. Natl. Acad. Sci. U.S.*, **61,** 629.

Bevington, J. C., H. W. Melville, and R. P. Taylor (1954), *J. Polym. Sci.*, **12,** 449.

Beychok, S. (1965), *Proc. Natl. Acad. Sci. U.S.*, **53,** 999.

Beychok, S. (1967), in *Poly-α-Amino Acids,* G. D. Fasman, ed., Marcel Dekker, New York, 1967, Chap. 7.

Beychok, S., and E. R. Blout (1961), *J. Mol. Biol.*, **3,** 769.

Beychok, S., J. Mc D. Armstrong, C. Lindblow, and J. T. Edsall (1966), *J. Biol. Chem.*, **241,** 5150.

Bhattacharya, D. N., C. L. Lee, J. Smith, and M. Szwarc (1964), *Polymer,* **5,** 54.

Bhattacharya, D. N., C. L. Lee, J. Smith, and M. Szwarc (1965), *J. Phys. Chem.,* **69,** 612.

Bianchi, U., and A. Peterlin (1968), *J. Polym. Sci.,* **A-2, 6,** 1759.

Biddle, D., and T. Nordström (1970), *Ark. Kemi,* **32,** 359.

Bier, M. (1959), *Electrophoresis: Theory, Method and Applications,* Academic Press, New York.

Binder, H. (1960), German Patent 974, 244.

Bird, G. R., and E. R. Blout (1959), *J. Am. Chem. Soc.,* **81,** 2499.

Birks, J. B. (1970), *Prog. React. Kinet.,* **5,** 181.

Birktoft, J. J., and D. M. Blow (1972), *J. Mol. Biol.,* **68,** 187.

Birktoft, J. J., B. W. Matthews, and D. M. Blow (1969), *Biochim. Biophys. Res. Commun.,* **36,** 131.

Birshtein, T. M., and P. L. Luisi (1964), *Vysokomol. Soedin.,* **6,** 1238.

Birshtein, T. M., and O. B. Ptitsyn (1966), *Conformations of Macromolecules,* Interscience, New York.

Bisschops, J. (1954), *J. Polym. Sci.,* **12,** 583.

Bisschops, J. (1955), *J. Polym. Sci.,* **17,** 89.

Bjerrum, N. (1923), *Z. Phys. Chem.,* **106,** 219.

Bjerrum, N. (1926), *Proc. Danish Royal Soc., Math.-Phys. Commun.,* **7,** No. 9.

Björnesjö, K. B., and T. Teorell (1945), *Ark. Kemi,* **A19,** No. 34.

Blais, P., and R. St. John Manley (1966), *J. Polym. Sci.,* **A-2, 4,** 1022.

Blake, A., and A. R. Peacocke (1966), *Biopolymers,* **4,** 1091.

Blake, A., and A. R. Peacocke (1968), *Biopolymers,* **6,** 1225.

Blake, C. C. F., G. A. Mair, A. C. T. North, D. C. Phillips, and V. R. Sarma (1967a), *Proc. Roy. Soc. (London),* **B167,** 365.

Blake, C. C. F., L. N. Johnson, G. A. Mair, A. C. T. North, D. C. Phillips, and V. R. Sarma (1967b), *Proc. Roy. Sec. (London),* **B167,** 378.

Blake, R. D., and J. R. Fresco (1966), *J. Mol. Biol.,* **19,** 145.

Bloomfield, V., and B. H. Zimm (1966), *J. Chem. Phys.,* **44,** 315.

Blout, E. R. (1960), in *Rotary Optical Dispersion,* C. Djerassi, ed., McGraw-Hill, New York, p. 238.

Blout, E. R. (1964), *Biopolym. Symp.,* **1,** 397.

Blout, E. R., C. de Lozé, S. M. Bloom, and G. D. Fasman (1960), *J. Am. Chem. Soc.,* **82,** 3787.

Blout, E. R., J. P. Carver, and J. Gross (1963), *J. Am. Chem. Soc.,* **85,** 644.

Blow, D. M. and T. A. Steitz (1970), *Ann. Rev. Biochem.,* **39,** 63.

Boeder, P. (1932), *Z. Phys.,* **75,** 258.

Bohak, Z., and E. Katchalski (1963), *Biochemistry,* **2,** 228.

Bohn, L. (1966), *Kolloid-Z.,* **213,** 95.

Bohon, R. L., and W. F. Claussen (1951), *J. Am. Chem. Soc.,* **73,** 1571.

Bonner, R. U., M. Dimbat, and G. H. Stross (1958), *Number Average Molecular Weights,* Interscience, New York, pp. 191–259.

Booth, C., G. Gee, and G. R. Williamson (1957), *J. Polym. Sci.,* **23,** 3.

Booth, F. (1950), *Proc. Roy. Soc. (London)*, **A203**, 533.

Booth, F. (1951), *J. Chem. Phys.*, **19**, 391, 1327, 1615.

Borgwardt, U., W. Schnabel, and A. Henglein (1969), *Makromol. Chem.*, **127**, 176.

Bornstein, P. (1967), *Biochemistry*, **6**, 3082.

Bovey, F. A. (1962), *J. Polym. Sci.*, **62**, 197.

Bovey, F. A. (1966), *Pure Appl. Chem.*, **16**, 417.

Bovey, F. A. (1972), *High Resolution NMR of Macromolecules*, Academic Press, New York.

Bovey, F. A., and G. V. D. Tiers (1960), *J. Polym. Sci.*, **44**, 173.

Bovey, F. A., E. W. Anderson, D. C. Douglass, and J. A. Manson (1963), *J. Chem. Phys.*, **39**, 1199.

Bovey, F. A., F. P. Hood, E. W. Anderson, and L. C. Snyder (1965), *J. Chem. Phys.*, **42**, 3900.

Bovey, F. A., J. J. Ryan, G. Spach, and F. Heitz (1971), *Macromolecules*, **4**, 433.

Boyer, R. F., and R. S. Spencer (1948), *J. Polym. Sci.*, **3**, 97.

Boyd, G.E. (1974), *Polyelectrolytes*, E. Selegny, ed., D. Reidel, Dordrecht-Holland, p. 135.

Bradbury, E. M., A. R. Downie, A. Elliott, and W. E. Hanby (1961), *Proc. Roy. Soc. (London)*, **A259**, 110.

Bradbury, E. M., B. G. Carpenter, and H. Goldman (1968a), *Biopolymers*, **6**, 837.

Bradbury, E. M., B. G. Carpenter, and R. M. Stephens (1968b), *Biopolymers*, **6**, 905.

Bradbury, E. M., C. Crane-Robinson, L. Paolillo, and P. Temussi (1973), *Polymer*, **14**, 303.

Bradbury, J. H., and B. E. Chapman (1972), *Biochem. Biophys. Res. Commun.*, **49**, 891.

Bradbury, J. H., and N. L. R. King (1969), *Nature (London)*, **223**, 5211.

Bradbury, J. H., and N. L. R. King (1971), *Aust. J. Chem.*, **24**, 1703.

Bradbury, J. H., M. D. Fenn, and A. G. Moritz (1969), *Aust. J. Chem.*, **22**, 2443.

Bradley, D. F., and M. K. Wolf (1959), *Proc. Natl. Acad. Sci. U.S.*, **45**, 944.

Brady, G. W., and R. Salovey (1964), *J. Am. Chem. Soc.*, **86**, 3499.

Brady, G. W., and R. Salovey (1967), *Biopolymers*, **5**, 331.

Brady, G. W., E. Wasserman, and J. Wallensdorf (1967), *J. Chem. Phys.*, **47**, 855.

Brahms, J. (1968), *J. Chim. Phys.*, **65**, 107.

Brahms, J., J. C. Maurizot, and A. M. Michelson (1967), *J. Mol. Biol.*, **25**, 481.

Brändén, C. I., H. Eklund, B. Nordström, T. Boive, G. Söderlund, E. Zappezauer, I. Ohlsson, and Å. Åkeson (1973), *Proc. Natl. Acad. Sci. U.S.*, **70**, 2439.

Brandts, J. F. (1964), *J. Am. Chem. Soc.*, **86**, 4291.

Brandts, J. F., and L. Hunt (1967), *J. Am. Chem. Soc.*, **89**, 4826.

Brant, D. A., and W. L. Dimpfl (1970), *Macromolecules*, **3**, 655.

Brant, D. A., W. G. Miller, and P. J. Flory (1967), *J. Mol. Biol.*, **23**, 47.

Braun, D. (1959), *Makromol. Chem.*, **30**, 85.

Brenner, S., and R. W. Horne (1959), *Biochim. Biophys. Acta*, **34**, 103.

Bresler, S. E. (1958), *Discussi. Faraday Soc.*, **25**, 158.

Bresler, S. E., L. M. Pyrkov, and S. Ya. Frenkel (1960), *Vysokomol. Soedin.*, **2**, 216.

Breslow, E., S. Beychok, K. Hardman, and F. R. N. Gurd (1965), *J. Biol. Chem.*, **204**, 304.

Brewster, J. H. (1959), *J. Am. Chem. Soc.*, **81**, 5475.

Brillouin, L. (1922), *Ann. Phys.*, **17,** 88.

Britten, R. J., and D. E. Kohne (1968), *Science,* **161,** 529.

Brown, C. P., A. R. Mathieson, and J. C. J. Thynne (1955), *J. Chem. Soc.*, 4141.

Brown, F. R., III, J. P. Carver, and E. R. Blout (1969), *J. Mol. Biol.*, **39,** 307.

Brown, J. F., Jr. (1963), *J. Polym. Sci.*, **C1,** 83.

Brown, L., and I. F. Trotter (1956), *Trans. Faraday Soc.*, **52,** 537.

Brownlee, G. G., F. Sanger, and B. G. Farrell (1968), *J. Mol. Biol.*, **34,** 379.

Brückner, S., V. Crescenzi, and F. Quadrifoglio (1970), *J. Chem., Soc.*, A 1168.

Bruice, T. C., and U. K. Pandit (1960), *J. Am. Chem. Soc.*, **82,** 5858.

Bryan, W. P., and S. O. Nielsen (1960), *Biochim. Biophys. Acta,* **42,** 552.

Bucci, E., C. Fronticelli, E. Chiancone, J. Wyman, E. Antonini, and A. Rossi-Fanelli (1965), *J. Mol. Biol.*, **12,** 183.

Buchdahl, R., H. A. Ende, and L. H. Peebles (1961), *J. Phys. Chem.*, **65,** 1468.

Buchdahl, R., H. A. Ende, and L. H. Peebles (1963a), *J. Polym. Sci.*, **C1,** 143.

Buchdahl, R., H. A. Ende, and L. H. Peebles (1963b), *J. Polym. Sci.*, **C1,** 153.

Bueche, F. (1962), *Physical Properties of Polymers,* Interscience, New York: (a) pp. 66–69. (b) pp. 223–224.

Bull, H. B. (1947), *Adv. Protein Chem.*, **3,** 95.

Bungenberg de Jong, H. G. (1937), *Kolloid-Z.*, **79,** 223; **80,** 221, 350.

Bungenberg de Jong, H. G. (1949), in *Colloid Science,* H. R. Kruyt, ed., Elsevier, Amsterdam, Chap. 10.

Bungenberg de Jong, H. G., and H. R. Kruyt (1930), *Kolloid-Z.*, **50,** 39.

Burn, C. W. (1939), *Trans. Faraday Soc.*, **35,** 482.

Burn, C. W. (1955), *J. Polym. Sci.*, **16,** 323.

Bunn, C. W., and D. R. Holmes (1958), *Discuss. Faraday Soc.*, **25,** 95.

Bunn, C. W., and E. R. Howells (1955), *J. Polym. Sci.*, **18,** 307.

Burkhart, R. D. (1965), *J. Polym. Sci.*, **A3,** 883.

Burow, S. P., A. Peterlin, and D. T. Turner (1965), *Polymer,* **6,** 35.

Bushuk, W., and H. Benoit (1958), *Can. J. Chem.*, **36,** 1616.

Butchard, C. G., R. A. Dwek, S. J. Ferguson, P. W. Kent, R. J. P. Williams, and A. V. Xavier (1972), *FEBS Lett.*, **25,** 91.

Butler, J. A. V. (1940), *J. Gen. Physiol.*, **24,** 189.

Bulter, P. J. G., and A. Klug (1971), *Nature- New Biol.*, **229,** 47.

Butler, W. T. (1970), *Biochemistry,* **9,** 44.

Cairns, J. (1961), *J. Mol. Biol.*, **3,** 756.

Cairns, J. (1962), *J. Mol. Biol.*, **4,** 407.

Campbell, I. D., C. M. Dobson, R. J. P. Williams, and A. V. Xavier (1973), *J. Magn. Resonance,* **11,** 172.

Campbell, T. W. (1958), U.S. Patent 2,831,825, assigned to E. I. du Pont de Nemours and Co.

Cantor, C. R., and I. Tinoco, Jr. (1967), *Biopolymers,* **5,** 821.

Cantow, H. J. (1959), *Makromol. Chem.*, **30,** 169.

Carpenter, D. K., and J. H. Hsieh (1972), *Polym. Prepr., Am. Chem. Soc., Div. Polym. Chem.,* **13,** 981.

Carpenter, D. K., and W. R. Krigbaum (1958), *J. Chem. Phys.,* **28,** 513.

Carr, H. Y., and E. M. Purcell (1954), *Phys. Rev.,* **94,** 630.

Carver, J. P., and E. R. Blout (1967), *Treatise on Collagen,* G. M. Ramachandran, ed., Academic Press, New York, Chap. 9.

Casassa, E. F. (1955), *J. Chem. Phys.,* **23,** 596.

Casassa, E. F. (1956), *J. Am. Chem. Soc.,* **78,** 3980.

Casassa, E. F. (1962), *Polymer,* **3,** 625.

Casassa, E. F. (1966), *J. Chem. Phys.,* **45,** 2811.

Casassa, E. F., and H. Eisenberg (1960), *J. Phys. Chem.,* **64,** 753.

Casassa, E. F., and Y. Tagami (1969), *Macromolecules,* **2,** 14.

Caspar, D. L. D., and A. Klug (1962), *Cold Spring Harbor Symp. Quant. Biol.,* **27,** 1.

Cerf, R. (1957), *J. Polym. Sci.,* **23,** 125.

Cerf, R. (1959), *Adv. Polym. Sci.,* **1,** 382.

Chamberlin, M. J. (1965), *Fed. Proc.,* **24,** 1446.

Chandrasekhar, S. (1943), *Rev. Mod. Phys.,* **15,** 20.

Chang, S. Y., and H. Morawetz (1956), *J. Phys. Chem.,* **60,** 782.

Changeux, J. P., J. C. Gerhart, and H. K. Schachman (1968), *Biochemistry,* **7,** 531.

Chapman, R. E., Jr., L. C. Klotz, D. S. Thompson, and B. H. Zimm (1969), *Macromolecules,* **2,** 637.

Chargaff, E. (1950), *Experientia,* **6,** 201.

Chargaff, E. (1955), in *The Nucleic Acids,* Vol. 1, E. Chargaff and J. N. Davidson, eds., Academic Press, New York, pp. 348–360.

Charney, E., J. B. Milstein, and K. Yamaoka (1970), *J. Am. Chem. Soc.,* **92,** 2657.

Cheesman, D. F., and J. T. Davies (1954), *Adv. Protein Chem.,* **9,** 440.

Chen, R. F. (1967), in *Fluorescence: Theory, Instrumentation and Practice,* G. G. Guilbault, ed., Marcel Dekker, New York, Chap 11.

Chiancone, E., J. B. Wittenberg, B. A. Wittenberg, E. Antonini, and J. Wyman (1966), *Biochim. Biophys. Acta,* **117,** 379.

Chipman, D. M., and N. Sharon (1969), *Science,* **165,** 454.

Cho, J. R., and H. Morawetz (1973), *Macromolecules,* **6,** 628.

Chou, P. Y., and G. D. Fasman (1974), *Biochemistry,* **13,** 211, 222.

Christensen, J. J., R. M. Izatt, and L. D. Hansen (1967), *J. Am. Chem. Soc.,* **89,** 213.

Chu, P., and J. A. Marinsky (1967), *J. Phys. Chem.,* **71,** 4352.

Chuang, J. C., and H. Morawetz (1973), *Macromolecules,* **6,** 43.

Chujo, R., S. Satoh, T. Ozeki, and E. Nagai (1962), *J. Polym. Sci.,* **61,** S12.

Ciardelli, F. Salvadori, C. Carlini, and E. Chiellini (1972), *J. Am. Chem. Soc.,* **94,** 6536

Citri, N. (1973), *Adv. Enzymol.,* **37,** 397.

Clark, H. G. (1965), *Makromol. Chem.,* **86,** 107.

Cohen, C, (1955), *Nature,* **175,** 129.

Cohen, C., and K. C. Holmes (1963), *J. Mol. Biol.,* **6,** 423.

Cohen, C., and A. G. Szent-Gyorgyi (1957), *J. Am. Chem. Soc.*, **79**, 248.

Cohen, S. G., and R. M. Schultz (1968), *J. Biol. Chem.*, **243**, 2607.

Cohn, E. J., and J. T. Edsall (1943), *Proteins, Amino Acids and Peptides*, Reinhold, New York.

Cohn, E. J., and J. D. Ferry (1943), in *Proteins, Amino Acids and Peptides*, E. J. Cohn and J. T. Edsall, eds, Reinhold, New York, Chap. 24.

Cohn, E. J., F. R. N. Gurd, D. M. Surgenor, B. A. Barnes, R. K. Brown, G. Derouaux, J. M. Gillespie, F. W. Kahnt, W. F. Lever, C. H. Liu, D. Mittelman, R. F. Mouton, K. Schmid, and E. Uroma (1950), *J. Am. Chem. Soc.*, **72**, 465.

Cohn, M. (1963), *Biochemistry*, **2**, 623.

Cohn, M., and *J.* S. Leigh (1962), *Nature*, **193**, 1037.

Coleman, B. D. (1958), *J. Polym. Sci.*, **31**, 155.

Coleman, B. D., and T. G. Fox (1963), *J. Chem. Phys.*, **38**, 1065.

Coleman, D. L., and E. R. Blout (1968), *J. Am. Chem. Soc.*, **90**, 2405.

Colin and H. Gaultier de Claubry (1814), *Ann. Chim. (Paris)*, **90**, 87.

Coll, H. (1967), *Makromol. Chem.*, **109**, 38.

Coover, H. W. Jr., and J. B. Dickey (1956), U.S. Patent 2,742,444, assigned to Eastman Kodak Co.

Copp, J. L., and D. H. Everett (1953), *Discuss. Faraday Soc.*, **15**,174.

Corbett, J. W., and J. H. Wang (1956), *J. Chem. Phys.*, **25**, 422.

Cordes, E. H., and R. B. Dunlap (1969), *Acc. Chem. Res.*, **2**, 329.

Corey, R. B., and L. Pauling (1955), *Rend. ist. Lombardo sci.*, **89**, 10.

Corradini, P., P. Ganis, C. Pedone, and G. Diana (1963), *Makromol. Chem.*, **61**, 242.

Corradini, P., G. Natta, P. Ganis, and P. A. Temussi (1967), *J. Polym. Sci.*, **C16**, 2477.

Cottrell, F. R., E. W. Merrill, and K. A. Smith (1969), *J. Polym. Sci.*, **A-2, 7**, 1415.

Coulombeau, C., and A. Rassat (1966), *Bull. Soc. Chim. France*, 3752.

Cowan, P. M., and S. McGavin (1955), *Nature*, **176**, 501.

Cowie, J. M. G., and S. Bywater (1970), *J. Polym. Sci.*, **C30**, 85.

Cox, H. L., and L. H. Cretcher (1926), *J. Am. Chem. Soc.*, **48**, 451.

Cox, R. A., and A. R. Peacocke (1957), *J. Polym. Sci.*, **23**, 765.

Crabbé, P. (1964), *Tetrahedron*, **20**, 1211.

Crawford, I. P., and J. Ito (1964), *Proc. Natl. Acad. Sci. U.S.*, **51**, 390.

Crawford, I. P., and C. Yanofsky (1958), *Proc. Natl. Acad. Sci. U.S.*, **44**, 1161.

Crescenzi, V., and P. J. Flory (1964), *J. Am. Chem., Soc.* **86**, 141.

Crescenzi, V., F. Quadrifoglio, and F. Delben (1972), *J. Polym. Sci.*, **A-2, 10**, 357.

Crestfield, A. M., W. H. Stein, and S. Moore (1963), *J. Biol. Chem.*, **238**, 2421.

Creswell, C. J., and A. L. Allred (1962), *J. Phys. Chem.*, **66**, 1469.

Crichton, R. R. (1973), *Angew. Chem., Int. Ed.*, **12**, 57.

Crick, F. H. C. (1963), *Science*, **139**, 461.

Crisp, D. J. (1958), in *Surface Phenonema in Biology and Chemistry*, J. E. Danielli, K. G. A. Pankhurst, and A. C. Riddiford, eds,. Pergamon Press, New York, pp. 23–54.

Crothers, D. M. (1969), *Acc. Chem. Res.* **2**, 225.

Crothers, D. M., and H. C. Spatz (1971), *Biopolymers,* **10,** 1949.
Crowley, J. D., G. S. Teague, Jr., and J. W. Lowe, Jr. (1966), *J. Paint Technol.,* **38,** 269.
Crugnola, A., and F. Danusso (1968), *J. Polym. Sci.,* **B6,** 535.
Cuatrecasas, P. (1972), *Proc. Natl. Acad. Sci. U.S.,* **69,** 318, 1277.
Cuatrecasas, P., S. Fuchs, and C. B. Anfinsen (1968), *J. Biol. Chem.,* **243,** 4787.
Cummings, D. J., and L. M. Kozloff (1960), *Biochim. Biophys. Acta,* **44,** 445.
Dandliker, W. B., H. C. Shapiro, J. W. Meduski, R. Alonso, G. A. Feigen, and J. R. Hamrick, Jr. (1964), *Immunochemistry,* **1,** 165.
Daniel, E. (1969), *Biopolymers,* **7,** 359.
Danon, J. and R. Derai (1969), *Eur. Polym. J.,* **5,** 659.
Dastoli, F. R., D. V. Lopiekes, and S. Price (1968), *Biochemistry,* **7,** 1160.
Davern, C. I., and M. Meselson (1960), *J. Mol. Biol.,* **2,** 153.
David, C., M. Lempereur, and G. Geuskens (1972), *Eur. Polym. J.,* **8,** 417.
Davidson, B., and G. D. Fasman (1967), *Biochemistry,* **6,** 1616.
Davies, J. T., and J. Llopis (1954), *Proc. Roy. Soc (London),* **A227,** 537.
Davies, R. J. H., and N. Davidson, (1971), *Biopolymers,* **10,** 1455.
Davis, R. W., and N. Davidson (1968), *Proc. Natl. Acad. Sci. U.S.,* **60,** 243.
Davison, P. E., and D. Freifelder (1967), *J. Mol. Biol.,* **5,** 635.
Dea, I. C. M., S. Arnott, J. M. Guss, and E. A. Balasz (1973), *Science,* **179,** 560.
Deb, P. C., and S. R. Palit (1973), *Makromol. Chem.,* **166,** 227.
DeBolt, L. C. and P. J. Flory (1975), in preparation
Debye, P. (1915), *Ann. Phys.* [4], **46,** 809.
Debye, P. (1944), *J. Appl. Phys.,* **15,** 338.
Debye, P. (1947), *J. Phys. Colloid Chem.,* **51,** 18.
Debye, P., and A. M. Bueche (1948), *J. Chem. Phys.,* **16,** 573.
Debye, P., and F. Bueche (1952), *J. Chem. Phys.,* **20,** 1337.
Debye, P., and E. Hückel (1923), *Phys. Z.,* **24,** 185.
Delmas, G., D. Patterson, and D. Böhme (1962a), *Trans. Faraday Soc.,* **58,** 2116.
Delmas, G., D. Patterson, and T. Somcynsky (1962b), *J. Polym. Sci.,* **57,** 79.
Derksen, J. C., and J. R. Katz (1932), *Rec. Trav. Chim.,* **51,** 523.
De Santis, P., E. Giglio, A. M. Liquori, and A. Ripamonti (1963), *J. Polym. Sci.,* **A1,** 1383.
De Santis, P., E. Giglio, A. M. Liquori, and A. Ripamonti (1965), *Nature,* **206,** 456.
Desmyter, A., and J. H. van der Waals (1958), *Rec. Trav. Chem.,* **77,** 53.
Deuel, H., K. Hutschnecker, and J. Solms (1953), *Z. Elektrochem.,* **57,** 172.
Deutsche, C. W., D. A. Lightner, R. W. Woody, and A. Moscowitz (1969), *Ann. Rev. Phys. Chem.,* **20,** 407.
De Voe, H., and I. Tinoco, Jr. (1962), *J. Mol. Biol.,* **4,** 500.
Dickerson, R. E., and I. Geis (1969), *The Structure and Action of Proteins,* Harper and Row, New York.
Dickerson, R. E., T. Takano, D. Eisenberg, O. B. Kallai, L. Samson, A. Cooper, and E. Margoliash (1971), *J. Biol. Chem.,* **246,** 1511.

Dickey, J. B., T. E. Stanin, and H. W. Coover, Jr. (1949), U.S. Patent 2,487,859, assigned to Eastman Kodak Co.

Dietrich, H. V., and F. Cramer (1954), *Chem., Ber.*, **87**, 806.

Djerassi, C., L. E. Geller, and E. J. Eisenbraun (1960), *J. Org. Chem.*, **25**, 1.

Dobry, A., and F. Boyer-Kawenoki (1947), *J. Polym. Sci.*, **2**, 90.

Doi, R. H., and S. Spiegelman (1962), *Science*, **138**, 1270.

Domb, C., J. Gillis, and G. Wilmer (1965), *Proc. Phys. Soc. (London)*, **85**, 625.

Dondos, A. (1969), *Eur. Polym. J.*, **5**, 767.

Dondos, A., and H. Benoit (1971), *Macromolecules*, **4**, 279.

Dondos, A., D. Froelich, P. Rempp, and H. Benoit (1967), *J. Chim. Phys.*, **64**, 1012.

Donnan, F. G. (1934), *Z. Phys. Chem. (Leipzig)*, **A168**, 369.

Donovan, J. W., M. Laskowski, Jr., and H. A. Scheraga (1961), *J. Am. Chem. Soc.*, **83**, 2686.

Doolittle, A. K. (1944), *Ind. Eng. Chem.*, **36**, 239.

Doolittle, A. K. (1946), *Ind. Eng. Chem.*, **38**, 535.

Doolittle, R. F. (1973), *Adv. Protein Chem.*, **27**, 2.

Dorman, D. E., D. A. Torchia, and F. A. Bovey (1973), *Macromolecules*, **6**, 80.

Dorrington, K. J., and C. Tanford (1970), *Adv. Immunol.*, **12**, 333.

Doscher, M. S., and F. M. Richards (1963), *J. Biol. Chem.*, **238**, 2399.

Doskočilová, D., J. Stokr, B. Schneider, H. Pivcová, M. Kolinský, J. Petránek, and D. Lím (1967), *J. Polym. Sci.*, **C16**, 215.

Doskočilová, D., S. Sykora, H. Pivcová, B. Obereigner, and D. Lím (1968), *J. Polym. Sci.*, **C23**, 365.

Doty, P., and R. F. Steiner (1950), *J. Chem., Phys.*, **18**, 1211.

Doty, P., and R. F. Steiner (1952), *J. Chem., Phys.*, **20**, 85.

Doty, P., and J. T. Yang (1956), *J. Am. Chem. Soc.*, **78**, 498.

Doty, P., H. Wagner and S. J, Singer (1947), *J. Phys. Colloid Chem.*, **51**, 32.

Doty, P., A. M. Holtzer, J. H. Bradbury, and E. R. Blout (1954), *J. Am. Chem. Soc.*, **76**, 4493.

Doty, P., J. H. Bradbury, and A. M. Holtzer (1956), *J. Am. Chem. Soc.*, **78**, 947.

Doty, P., S. Wada, J. T. Yang, and E. R. Blout (1957), *J. Polym. Sci.*, **23**. 851.

Doty, P., H. Boedtker, J. R. Fresco, R. Haselkorn, and M. Litt (1959), *Proc. Natl. Acad. Sci. U.S.*, **45**, 488.

Doty, P., J. Marmur, J. Eisner, and C. Schildkraut (1960), *Proc, Natl. Acad. Sci. U.S.*, **46**, 461.

Dove, W. F., and N. Davidson (1962a), *J. Mol. Biol.*, **5**, 467.

Dove, W. F., and N. Davidson (1962b), *J. Mol. Biol.*, **5**, 479.

Downie, A. R., and A. A. Randall (1959), *Trans. Faraday Soc.*, **55**, 2132.

Downie, A. R., A. Elliott, W. E. Hanby, and B. R. Malcolm (1957), *Proc. Roy. Soc. (London)*, **A242**, 325.

Doyle, B. B., W. Traub, G. P. Lorenzi, and E. R. Blout (1971), *Biochemistry*, **10**, 3052.

Drenth, J., J. N. Jansonius, R. Koekoek, H. M. Swen, and B. G. Wolthers (1968), *Nature,* **218,** 929.

Dubin, P. L., and U. P. Strauss (1970), *J. Phys. Chem.,* **74,** 2842.

Dubin, S. B., J. H. Lunacek, and G. B. Benedek (1967), *Proc. Natl. Acad. Sci. U.S.,* **57,** 1164.

Dubos, R. J. (1950), *Louis Pasteur, Free Lance of Science,* Little, Brown, Boston. Mass., Chap. 6.

Dunkel, W. (1928), *J. Phys. Chem.,* **A138,** 42.

Duportail, G., D. Froelich, and G. Weill (1971), *Eur. Polym. J.,* **7,** 977.

Ebashi, S. and M. Endo (1968), *Prog. Biophys. Mol. Biol.,* **18,** 123.

Eberson, L. (1959), *Acta Chem. Scand.,* **13,** 211.

Edelman, G. M. (1972), *Science,* **180,** 830.

Edelman, G. M., and J. A. Gally (1962), *J. Exp. Med.,* **116,** 207.

Edelman, G. M., and W. O. McClure (1968), *Acc. Chem. Res.,* **1,** 65.

Edelman, G. M., B. A. Cunningham, G. N. Reeke, Jr., J. W. Beeker, M. J. Waxdal, and J. L. Wang (1972), *Proc. Natl. Acad. Sci. U.S.,* **69,** 2580.

Edsall, J. T. (1962) Arch. Biochem. Biophys., Suppl.**1,** 12.

Edsall, J. T. (1972), *J. Hist. Biol.,* **5,** 205.

Edsall, J. T., and J. Foster (1948), *J. Am. Chem., Soc.,* **70,** 1860.

Ehresmann, C., P. Fellner, and J. P. Ebel (1970), *Nature,* **227,** 1321.

Ehrlich, P., and J. J. Kurpen (1963), *J. Polym. Sci.,* **A1,** 3217.

Eigen, M., and G. Schwarz (1957), *J. Colloid Sci.,* **12,** 181.

Einstein, A. (1905), *Ann. Phys.* [4], **17,** 549.

Einstein, A. (1906a), *Ann. Phys.* [4], **19,** 289.

Einstein, A. (1906b), *Ann Phys.* [4], **19,** 371.

Einstein, A. (1910), *Ann. Phys.* [4], **33,** 1275.

Einstein, A. (1911), *Ann. Phys.* [4], **34,** 591.

Eisenberg, D., and W. Kauzmann (1969), *The Structure and Properties of Water,* Oxford University. Press.

Eisenberg, H. (1957), *J. Polym. Sci.,* **23,** 579.

Eisenberg, H. (1962), *J. Chem. Phys.,* **36,** 1837.

Eisenberg, H. (1971), *Acc. Chem. Res.,* **4,** 379.

Eisenberg, H., and G. Felsenfeld (1967), *J. Mol. Biol.,* **30,** 17.

Eisenberg, H., and G. R. Mohan (1959), *J. Phys. Chem.,* **63,** 671.

Eisenberg, H., and J. Pouyet (1954), *J. Polym. Sci.,* **13,** 85.

Eiserling, F. A., and R. C. Dickson (1972), *Ann. Rev. Biochem.,* **41,** 467.

Eldridge, J. E., and J. D. Ferry (1954), *J. Phys. Chem.,* **58,** 992.

Eldridge, R. J., and F. E. Treloar (1970), *J. Phys. Chem.,* **74,** 1446.

Elias, H. G., and J. Gerber (1968), *Makromol. Chem.,* **112,** 122.

Elias, H. G., and E. Männer (1960), *Makromol. Chem.,* **40,** 207.

Engel, J., J. Kurtz, E. Katchalski, and A. Berger (1966), *J. Mol. Biol.,* **17,** 255.

Englander, J. J., and P. H. von Hippel (1972), *J. Mol. Biol.,* **63,** 171.

Englander, J. J., N. R. Kallenbach, and S. W. Englander (1972), *J. Mol. Biol.,* **63,** 153.

Englander, S. W., and A. Poulsen (1969), *Biopolymers,* **7,** 379.

Epand, R. F., and H. A. Scheraga (1968), *Biopolymers,* **6,** 1551.

Epstein, C. J., R. F. Goldberger, D. M. Young, and C. B. Anfinsen (1962), *Arch. Biochem. Biophys. Suppl.,* No. **1,** 223.

Epstein, H. F., A. N. Schechter, R. F. Chen, and C. B. Anfinsen (1971), *J. Mol. Biol.,* **60,** 499.

Erlanger, B. F. (1967), *Proc. Natl. Acad. Sci. U.S.,* **58,** 703.

Ermolenko, J. N., and M. S. Katibnikov (1962), *Vysokomol. Soedin.,* **4,** 1249.

Evans, D. R., S. G. Warren, B. F. P. Edwards, C. H. McMurray, P. H. Bethge, D. C. Wiley, and W. N. Lipscomb (1973), *Science,* **179,** 683.

Evans, M. T. A., J. Mitchell, P. R. Musselwhite, and L. Irons (1970), in *Surface Chemistry of Biological Systems,* M. Blank, ed., Plenum Press, New York, p. 1.

Ewart, R. H., C. P. Roe, P. Debye, and J. M. McCartncy (1946), *J. Chem. Phys.,* **14,** 687.

Eyring, E. J., H. Kraus, and T. J. Yang (1968), *Biopolymers,* **6,** 703.

Fabrizzi, L., P. Paoletti, M. C. Zobrist, and G. Schwarzenbach (1973), *Helv. Chim. Acta,* **56,** 670.

Falk, M. and T. A. Ford (1966), *Can. J. Chem.,* **44,** 1699.

Fasman, G. D. (1962a), *Nature,* **193,** 681.

Fasman, G. D. (1962b), in *Polyamino Acids, Polypeptides and Proteins,* M. A. Stahmann, ed., University of Wisconsin Press, p. 221.

Fasman, G. D., and E. R. Blout (1960), *J. Am. Chem., Soc.,* **82,** 2262.

Fasman, G. D., C. Lindblow, and E. Bodemheimer (1962), *J. Am. Chem. Soc.,* **84,** 4977.

Fasman, G. D., R. J. Foster, and S. Beychok (1966a), *J. Mol. Biol.,* **19,** 240.

Fasman, G. D., E. Bodenheimer, and A. Pesce (1966b), *J. Biol. Chem.,* **241,** 916.

Feder, J., L. R. Garrett, and B. S. Wildi (1971), *Biochemistry,* **10,** 4552.

Felsenfeld, G. and G. L. Cantoni (1964), *Proc. Natl. Acad, Sci. U.S.,* **51,** 818.

Felsenfeld, G., and H. T. Miles (1967), *Ann. Rev. Biochem.,* **36,** Pt. 2, 407.

Felsenfeld, G., and A. Rich (1957), *Biochim. Biophys. Acta,* **26,** 457.

Felsenfeld, G., and G. Sandeen (1962), *J. Mol. Biol.,* **5,** 587.

Fendler, E. J., and J. H. Fendler (1970), *Adv. Phys. Org. Chem.,* **8,** 271.

Ferguson, R. C. (1971), *Macromolecules,* **4,** 324.

Fernandez-Moran, H., E. F. J. van Bruggen, and M. Ohtsuki (1966), *J. Mol. Biol.,* **16,** 191.

Ferretti, J. A., and R. L. Jernigan (1973), *Macromolecules,* **6,** 687.

Ferry, J. D. (1970), *Viscoelastic Properties of Polymers,* 2nd Ed., Wiley, New York: (a) Chap. 9; (b) pp. 252–257.

Ferry, J. D., D. C. Udy, F. C. Wu, G. E. Heckler, and D. F. Fordyce (1951), *J. Colloid Sci.,* **6,** 429.

Ferry, J. D., S. Shulman, K. Gutfreund, and S. Katz (1952), *J. Am. Chem., Soc.,* **74,** 5709.

Fersht, A. R., and Y. Requenna (1971), *J. Mol. Biol.,* **60,** 279.

Fetters, L. J. (1972), *J. Polym. Sci.,* **B10,** 577.

Fetters, L. J., and H. Yu (1966), *Polym. Prepr., Am. Chem. Soc. Div. Polym. Chem.,* **7,** 443.

Fetters, L. J., and H. Yu (1971), *Macromolecules,* **4,** 385.

Fiers, W., and R. L. Sinsheimer (1962), *J. Mol. Biol.,* **5,** 424.

Figini, R. V. (1967), *J. Polym. Sci.,* **C16,** 2049.

Figini, R. V. and G. V. Schulz (1960), *Mokromol. Chem.*, **41,** 1; *Z. phys. Chem.* (*Frankfurt*) **23,** 233.

Fischer, E. (1894), *Ber.*, **27,** 2985.

Fischer, H. (1966), *Makromol. Chem.*, **98,** 179.

Fish, W. W., K. G. Mann, and C. Tanford (1969), *J. Biol. Chem.*, **244,** 4989.

Fisher, M. E. (1961), *J. Chem. Phys.*, **45,** 1469.

Fitzgerald, E. B., and R. M. Fuoss (1954), *J. Polym. Sci.*, **14,** 329.

Fixman, M. (1955), *J. Chem. Phys.*, **23,** 1656.

Fixman, M. (1960), *J. Polym. Sci.*, **47,** 91.

Fixman, M. (1964), *J. Chem. Phys.*, **41,** 3772.

Flaschner, O. (1908), *Z. Phys. Chem.* (*Leipzig*), **62,** 493.

Fleischer, D., and R. C. Schulz (1972), *Makromol. Chem.*, **152,** 311.

Fleet, G. W., J. R. Knowles, and R. R. Porter (1972), *Biochem. J.*, **128,** 499.

Flory, P. J. (1939), *J. Am. Chem. Soc.*, **61,** 3334.

Flory, P. J. (1940a), *J. Am. Chem. Soc.*, **62,** 1561.

Flory, P. J. (1940b), *J. Am. Chem. Soc.*, **62,** 2261.

Flory, P. J. (1942), *J. Chem. Phys.*, **10,** 51.

Flory, P. J. (1949). *J. Chem. Phys.*, **17.** 303.

Flory, P. J. (1953), *Principles of Polymer Chemistry*, Cornell University Press, Ithaca, N. Y.: (a) p. 56; (b) pp. 596–602; (c) Chap. XIV; (d) p. 76.

Flory, P. J. (1953e), *J. Chem. Phys.*, **21,** 162.

Flory, P. J. (1956), *Proc. Roy. Soc. (London)*, **A234,** 73.

Flory, P. J. (1969), *Statistical Mechanics of Chain Molecules*, Interscience, New York.

Flory, P. J. (1970), *Macromolecules*, **3,** 613.

Flory, P. J. (1971), *Pure Appl. Chem.*, **26,** 309.

Flory, P. J., and S. Fisk (1966), *J. Chem. Phys.*, **44,** 2243.

Flory, P. J., and W. R. Krigbaum (1950), *J. Chem. Phys.*, **18,** 1086.

Flory, P. J., and J. E. Osterheld (1954), *J. Phys. Chem.*, **58,** 653.

Flory, P. J., and A. Vrij (1963), *J. Am. Chem. Soc.*, **85,** 3548.

Flory, P. J., and E. S. Weaver (1960), *J. Am. Chem. Soc.*, **82,** 4518.

Flory, P. J., O. K. Spurr, Jr., and D. K. Carpenter (1958), *J. Polym. Sci.*, **27,** 231.

Flory, P. J., R. A. Orwoll, and A. Vrij (1964), *J. Am. Chem. Soc.*, **86,** 3515.

Flory, P. J., J. E. Mark, and A. Abe (1966), *J. Am. Chem. Soc.*, **88,** 639.

Fordham, J. W. L., P. H. Burleigh, and C. L. Sturm (1959), *J. Polym. Sci.*, **41,** 73.

Forget, B. G., and S. M. Weissman (1967), *Science*, **158,** 1695.

Fowler, R. H., and E. A. Guggenheim (1939), *Statistical Thermodynamics*, Cambridge University Press, New York, pp. 576–581.

Fox, T. G. (1962), *Polymer*, **3,** 111.

Fox, T. G., and V. R. Allen (1964), *J. Chem. Phys.*, **41,** 344.

Fox, T. G, and P. J. Flory (1951), *J. Am. Chem. Soc.*, **73,** 1909.

Fox, T. G. B. S. Garrett, W. E. Goode, S. Gratch, J. F. Kincaid, A. Spell, and J. D. Stroupe (1958), *J. Am. Chem. Soc.*, **80,** 1768.

Fox, T. G, J. B. Kinsinger, H. F. Mann, and E. M. Schuele (1962), *Polymer,* **3,** 71.

Fraenkel-Conrat, H., and R. C. Williams (1955), *Proc. Natl. Acad. Sci., U.S.,* **41,** 690.

Frank, H. P., S. Barkin, and F. R. Eirich (1957), *J. Phys. Chem.,* **61,** 1375.

Frank, H. S. (1970), *Science,* **169,** 635.

Frank, H. S., and M. J. Evans (1945), *J. Chem. Phys.,* **13,** 507.

Frank, H. S., and W. Y. Wen (1957), *Discuss. Faraday Soc.,* **24,** 133.

Franklin, R. E., A. Klug, J. T. Finch, and K. C. Holmes (1958), *Discuss. Faraday Soc.,* **25,** 197.

Franks, F., ed. (1972), *Water, a Comprehensive Treatise,* Vol. I: *The Physics and Chemistry of Water,* Plenum Press, New York.

Franta, E. (1966), *J. Chim. Phys.,* **63,** 595.

Fraser, R. D. B., B. S. Harrap. T. P. Macrae, F. H. C. Stewart, and E. Suzuki (1967), *Biopolymers,* **5,** 251.

Frederick, J. E., T. F. Reed, and O. Kramer (1971), *Macromolecules,* **4,** 242.

Fredericq. E. (1956), *Arch. Biochem. Biophys.,* **65,** 218.

Frenkel, S. Ya., V. G. Baranov, and T. I. Volkov (1967), *J. Polym. Sci.,* **C16,** 1655.

Fresco, J. R., B. M. Alberts, and P. Doty (1960), *Nature,* **188,** 4745.

Freudenberg, K., and G. Blomquist (1935), *Ber.,* **B68,** 2070.

Freudenberg, K., W. Kuhn, W. Durr, and G. Steinbrunn (1930), *Ber.,* **63,** 1510.

Freudenberg, K., E. Schauf, G. Dumpert, and T. Ploetz (1939), *Naturwissenschaften,* **27,** 850.

Frey, M., P. Wahl, and H. Benoit (1964), *J. Chim. Phys.,* **61,** 1005.

Frieden, C. (1971), *Ann. Rev. Biochem.,* **40,** 653.

Friend, J. A., J. A. Larkin, A. Maroudas, and M. L. McGlashan (1963), *Nature,* **198,** 683.

Frisch, H. L., and R. Simha (1956), in *Rheology; Theory and Applications,* Vol. 1, F. R. Eirich, ed., Academic Press, New York, Chap. 14.

Fruton, *J. S. (1974), Acc. Chem. Res.,* **7,** 241.

Frye, C. L., and J. M. Klosowski (1971), *J. Am. Chem. Soc.,* **93,** 4599.

Fujii, K., T. Mochizuki, S. Imoto, J. Ukida, and M. Matsumoto (1962), *Makromol. Chem.,* **51,** 225.

Fujii, K., J. Ukida, and M. Matsumoto (1963), *Makromol. Chem.,* **65,** 86.

Fujimoto, T., and M. Nagasawa (1967), *J. Phys. Chem.,* **71,** 4274.

Fujita, H. (1962), *The Mathematical Theory of Sedimentation Analysis,* Academic Press, New York.

Fujita, H. (1969), *J. Phys. Chem.,* **73,** 1759.

Fujiwara, Y., and P. J. Flory (1970), *Macromolecules,* **3,** 43.

Fuoss, R. M. (1948), *J. Polym. Sci.,* **3,** 603.

Fuoss, R. M., and H. Sadek (1949), *Science,* **110,** 552.

Fuoss, R. M., A. Katchalsky, and S. Lifson (1951), *Proc. Natl. Acad. Sci. U.S.,* **37,** 579

Gabor, G. (1968), *Biopolymers,* **6,** 809.

Gadd, G. E. (1971), in *Encyclopedia of Polymer Science and Technology,* Vol. 15, Interscience, New York, pp. 224–253.

Gaetjens, E., and H. Morawetz (1960), *J. Am. Chem. Soc.*, **82,** 5328.

Gaetjens, E., and H. Morawetz (1961), *J. Am. Chem. Soc.*, **83,** 1738.

Gaines, G. L., Jr. (1969), *J. Polym. Sci.*, **A-2, 7,** 1379.

Gall, W. E., and G. M. Edelman (1970), *Biochemistry,* **9,** 3188.

Gallot, B., and C. Sadron (1971), *Macromolecules,* **4,** 515.

Gallot, Y., M. Leng, H. Benoit, and P. Rempp (1962), *J. Chim. Phys.*, **59,** 1093.

Gally, J. A., and G. M. Edelman (1964), *Biopolym. Symp.* No. **1,** 367.

Gally, J. A., and G. M. Edelman (1965), *Biochim. Biophys. Acta,* **94,** 175.

Gans, R. (1928), *Ann. Physik.* [4], **86,** 628.

Ganser, V., J. Engel, D. Winklmair, and G. Krause (1970), *Biopolymers,* **9,** 329.

Gardon, J. L., and S. G. Mason (1957), *J. Polym. Sci.*, **26,** 255.

Garrett, B. S., W. E. Goode, S. Gratch, J. F. Kincaid, C. L. Levesque, A. Spell, J. D. Stroupe, and W. H. Watanabe (1959), *J. Am. Chem. Soc.*, **81,** 1007.

Garrett, R. A., and H. G. Wittman (1973), *Adv. Protein Chem.*, **27,** 278.

Gaylord, N., and H. Antropiusova (1969), *Macromolecules,* **2,** 443.

Gaylord, N. G., and H. F. Mark (1959), *Linear Stereoregular Addition Polymers* (Polymer Reviews, Vol. 1), Interscience, New York.

Gee, G. (1942), *Trans. Faraday Soc.*, **38,** 418.

Gee, G. (1944), *Trans. Faraday Soc.*, **40,** 468.

Gee, G., and L. R. G. Treloar (1942), *Trans. Faraday Soc.*, **48,** 147.

Gee, G., W. C. E. Higginson, and G. T. Merrall (1959), *J. Chem. Soc.*, 1345.

Geiduschek, E. P. (1961), *Proc. Natl. Acad. Sci. U.S.*, **47,** 950.

Geiduschek, E. P. (1962), *J. Mol. Biol.*, **4,** 467.

Gellert, M. (1967) *Proc. Natl. Acad. Sci. U.S.*, **57,** 148.

Gerber, J., and H. G. Elias (1968), *Makromol. Chem.*, **112,** 142.

Gerhart, J. C., and A. B. Pardee (1962), *J. Biol. Chem.*, **237,** 891.

Gerhart, J. C., and H. K. Schachman (1968), *Biochemistry,* **7,** 538.

Germer, H., K. H. Hellwege, and U. Johnsen (1963), *Makromol. Chem.*, **60,** 106.

Gerngross, O., K. Herrmann, and R. Lindemann (1932), *Kolloid-Z.*, **60,** 276.

Giacometti, G., A. Turolla, and R. Boni (1968), *Biopolymers,* **6,** 441.

Gibbs, J. H., and E. A. DiMarzio (1959), *J. Chem. Phys.*, **30,** 271.

Gibson, Q. H. (1959), *Biochem. J.*, **71,** 293.

Gilbert, G. A., and J. V. R. Marriott (1948), *Trans. Faraday Soc.*, **44,** 84.

Gill, S. J., and D. S. Thompson (1967), *Proc. Natl. Acad. Sci. U.S.*, **57,** 562.

Gill, T. J., III, and P. Doty (1964), *J. Biol. Chem.*, **236,** 2677.

Gill, T. J., III, and Z. S. Matthews (1963), *J. Biol. Chem.*, **238,** 1373.

Gill, T. J., III, and G. S. Omenn (1965), *J. Am. Chem. Soc.*, **87,** 4188.

Girolamo, M., and J. R. Urwin (1971), *Eur. Polym. J.*, **7,** 693.

Givol, D., F. De Lorenzo, F. Goldberger, and C. B. Anfinsen (1965), *Proc. Natl. Acad. Sci. U.S.*, **53,** 676.

Givol, D., P. H. Strausbauch, E. Hurwitz, M. Wilchek, J. Haimovich, and H. N. Eisen (1971), *Biochemistry,* **10,** 3461.

Glover, C. A., and R. A. Stanley (1961), *Anal. Chem.*, **33,** 447.

Goates, J. R., R. J. Sullivan, and J. B. Ott (1959), *J. Phys. Chem.*, **63,** 589.

Goebel, K. D., and D. A. Brant (1970), *Macromolecules,* **3,** 634.

Goebel, K. D., and W. G. Miller (1970), *Macromolecules,* **3,** 64.

Goldstein, B. N., A. N. Goryunov, Yu. Ya. Gotlib, A. M. Elyashevich, T. P. Zubova, and A. I. Koltzov (1971), *J. Polym. Sci.*, **A-2, 9,** 769.

Goldstein, M. (1953), *J. Chem. Phys.*, **21,** 1255.

Golub, E. I. (1964), *Biopolymers,* **2,** 113.

Gomatos, P. J., and I. Tamm (1963), *Proc. Natl. Acad. Sci. U.S.*, **50,** 878.

Goodman, M., and S.-C. Chen (1970), *Macromolecules,* **3,** 398.

Goodman, M., and M. Fried (1967), *J. Am. Chem. Soc.*, **89,** 264.

Goodman, M., and I. Listowsky (1962), *J. Am. Chem. Soc.*, **84,** 3770.

Goodman, M., and E. E. Schmitt (1959), *J. Am. Chem. Soc.*, **81,** 5507.

Goodman, M., E. E. Schmitt, and D. A. Yphantis (1962), *J. Am. Chem. Soc.*, **84,** 1288.

Goodman, M., A. M. Felix, C. M. Deber, A. R. Brause, and G. Schwartz (1963), *Biopolymers,* **1,** 371.

Goodman, M., A. Abe, and Y.-L. Fan (1967), *Macromolecular Reviews,* Vol. 1, Interscience, New York, pp. 1–33.

Goodman, M., A. S. Verdini, C. Toniolo, W. D. Phillips, and F. A. Bovey (1969), *Proc. Natl. Acad. Sci. U.S.*, **64,** 444.

Goodman, M., F. Naider and R. Rupp (1971), *Bioorg. Chem.*, **1,** 310.

Goodman, N., and H. Morawetz (1970), *J. Polym. Sci.*, **C31,** 177.

Goodman, N., and H. Morawetz (1971), *J. Polym. Sci.*, **A-2, 9,** 1657.

Gordy, W. (1941), *J. Chem. Phys.*, **9,** 215.

Gottlieb, Yu. Ya., and P. Wahl (1963), *J. Chim. Phys.*, **60,** 849.

Gould, E. S. (1970), *J. Am. Chem. Soc.*, **92,** 6797.

Graham, T. (1862), *Ann.*, **121,** 1.

Gratzer, W. B. (1967), *Proc. Roy. Soc. (London),* **A297,** 163.

Gratzer, W. B., and P. Doty (1963), *J. Am. Chem. Soc.*, **85,** 1193.

Gray, H. B. and J. Vinograd (1971), *J. Mol. Biol.*, **62,** 1.

Gray, R. D. (1970), *J. Biol. Chem.*, **245,** 2914.

Greenfield, N., B. Davidson, and G. D. Fasman (1967), *Biochemistry,* **6,** 1630.

Gregor, H. P., and M. Frederick (1957), *J. Polym. Sci.*, **23,** 451.

Griffith, J. H., and B. G. Rånby (1959), *J. Polym. Sci.*, **38,** 107.

Grimley, T. B. (1961), *Trans. Faraday. Soc.*, **57,** 1974.

Grosius, P., Y. Gallot, and A. Skoulios (1969), *Makromol. Chem.*, **127,** 94.

Grosius, P., Y. Gallot, and A. Skoulios (1970a), *Makromol. Chem.*, **132,** 35.

Grosius, P., Y. Gallot, and A. Skoulios (1970b), *Makromol. Chem.*, **136,** 191.

Gross, L. M., and U. P. Strauss (1966), in *Chemical Physics of Ionic Solutions,* B. E. Conway and R. G. Barradas, eds., Wiley, New York, pp. 361–389.

Gruber, E., R. Gruber, and J. Schurz (1972), *Makromol. Chem.*, **158,** 81.

Grubisic, L., and H. Benoit (1968), *Compt. Rend., Ser. C,* **266,** 1275.

Guggenheim, E. A. (1935), *Proc. Roy. Soc. (London)*, **A148**, 304.

Guinier, A. (1939), *Ann. Phys. (Paris)*, [11], **12**, 161.

Gurd, F. R. N., and D. S. Goodman (1952), *J. Am. Chem. Soc.*, **74**, 670.

Gurd, F. R. N., T. E. Hugli, and R. A. Bradshaw (1968), *Polym. Prepr., Am. Chem. Soc., Div. Polym. Chem.*, **9**, (1), 123.

Haas, H. C., and R. L. MacDonald (1970), *J. Polym Sci.*, **B8**, 425.

Haas, H. C., R. D. Moreau, and N. W. Schuber (1967), *J. Polym. Sci.*, **A-2, 5**, 915.

Haas, H. C., C. K. Chiklis, and R. D. Moreau (1970a), *J. Polym. Sci.*, **A-1, 8**, 1131.

Haas, H. C., M. J. Manning, and M. H. Mach (1970b), *J. Polym. Sci.*, **A-1, 8**, 1725.

Haas, H. C., R. L. MacDonald, and A. N. Schuler (1971), *J. Polym. Sci.*, **A-1, 9**, 959.

Haimovich, J., H. N. Eisen, E. Hurwitz, and D. Givol (1972), *Biochemistry*, **11**, 2389.

Hall, B. D., and S. Spiegelman (1961), *Proc. Natl. Acad. Sci. U.S.*, **47**, 137.

Hall, C. E. (1960), *J. Biophys. Biochem. Cytol.*, **7**, 613.

Hall, C. E., and P. Doty (1958), *J. Am. Chem. Soc.*, **80**, 1269.

Hall, C. E., and H. S. Slayter (1959), *J. Biochem. Biophys. Cytol.*, **5**, 11.

Hammes, G. G., and P. B. Roberts (1969), *J. Am. Chem. Soc.*, **91**, 1812.

Hanes, C. S. (1937), *New Phytol.*, **36**, 101, 189.

Hanson, J., and J. Lowy (1963), *J. Mol. Biol.*, **6**, 46.

Harned, H. S., and B. B. Owen (1958), *The Physical Chemistry of Electrolytic Solutions*, 3rd ed., Reinhold, New York, p. 415.

Harpst, J. A., A. I. Krasna, and B. H. Zimm (1968), *Biopolymers*, **6**, 595.

Harrington, R. E., and B. H. Zimm (1965), *J. Phys. Chem.*, **69**, 161.

Harrington, W. F., and N. V. Rao (1970), *Biochemistry*, **9**, 3714.

Harrington, W. F., and M. Sela (1958), *Biochim. Biophys. Acta*, **27**, 24.

Harrington, W. F., and P. H. von Hippel (1961a), *Arch. Biochem. Biophys.*, **92**, 100.

Harrington, W. F., and P. H. von Hippel (1961b), *Adv. Protein Chem.*, **16**, 1.

Harris, C. M., and R. L. Hill (1969), *J. Biol. Chem.*, **244**, 2195.

Harris, F. E., and S. A. Rice (1954), *J. Phys. Chem.*, **58**, 725.

Harrison, M. A., P. H. Morgan, and G. S. Park (1972), *Eur. Polym. J.*, **8**, 1361.

Harrison, S. C., and E. R. Blout (1965), *J. Biol. Chem.*, **240**, 299.

Hartmann, H., and R. Jaenicke (1956), *Z. Phys. Chem. (Frankfurt)*, **6**, 220.

Harwood, H. J. (1971), in *NMR Basic Principles and Progress*, Vol. 4: *Natural and Synthetic High Polymers*, P. Diehl, E. Fluck and R. Kosfeld, eds., Springer, Berlin, p. 71.

Hashimoto, M., and J. Aritoni (1966), *Bull. Chem. Soc. Jap.*, **39**, 2707.

Haugland, R. P., J. Yguerabide, and L. Stryer (1969), *Proc. Natl. Acad. Sci. U.S.*, **63**, 23.

Haurowitz, F., and F. Breinl (1933), *Hoppe-Seylers Z. Physiol. Chem.*, **214**, 111.

Hayes, R. A. (1961), *J. Appl. Polym. Sci.*, **5**, 318.

Hayes, T. L., J. C. Murchio, F. T. Lindgren, and A. V. Nichols (1959), *J. Mol. Biol.*, **1**, 297.

Hearst, J. E. (1963), *J. Chem. Phys.*, **38**, 1062.

Hearst, J. E., and M. Botchan (1973), *Acc. Chem. Res.*, **6**, 293.

Hearst, J. E., and W. H. Stockmayer (1962), *J. Chem. Phys.*, **37**, 1425.

Hearst, J. E., and J. Vinograd (1961), *Proc. Natl. Acad. Sci. U.S.*, **47,** 825, 1005.

Hearst, J. E., J. B. Ifft, and J. Vinograd (1961), *Proc. Natl. Acad. Sci. U.S.*, **47,** 1015.

Hearst, J. E., C. W. Schmid, and F. P. Rinehart (1968), *Macromolecules,* **1,** 491.

Heatley, F., and F. A. Bovey (1968), *Macromolecules,* **1,** 301.

Heatley, F., R. Salovey, and F. A. Bovey (1969), *Macromolecules,* **2,** 619.

Heine, S., D. Kratky, G. Porod, and P. J. Schmitz (1961), *Makromol. Chem.,* **44/46,** 682.

Heisel, F. and G. Laustriat (1969), *J. Chim. Phys.,* **66,** 1881.

Helfand, E. (1971), *J. Chem. Phys.,* **54,** 4651.

Helinski, D. R., and D. B. Clewell (1971), *Ann. Rev. Biochem.,* **40,** 899.

Hellfritz, H. (1951), *Makromol. Chem.,* **7,** 191.

Hellwege, K. H., U. Johnsen, and K. Kolbe (1966), *Kolloid-Z.,* **214,** 45.

Helms, J. B. and G. Challa (1972), *J. Polym. Sci.,* **A-2, 10,** 1447.

Hengstenberg. J., and E. Schuch (1964), *Makromol. Chem.,* **74,** 55.

Henry, D. C. (1931), *Proc. Roy. Soc. (London),* **A133,** 106.

Herma, H., V. Gröbe, and R. Schmolke (1966), *Faserforsch. Textiltech.,* **17,** 56.

Hermans, J., Jr. (1962), *J. Colloid Sci.,* **17,** 638.

Hermans, J. J. (1955), *J. Polym. Sci.,* **18,** 527.

Hermans, J. J. (1963), *J. Colloid Sci.,* **18,** 433.

Hermans, J. J., and H. A. Ende (1963a), in *Newer Methods of Polymer Characterization,* B. Ke. ed., Interscience, New York, Chap. 13.

Hermans, J. J., and H. A. Ende (1963b), *J. Polym. Sci.,* **C1,** 161.

Hermans, J. J., and H. Fujita (1955), *Proc. Roy. Netherland Acad. Sci.,* **B58,** 182.

Hermans, J. J., and J. T. G. Overbeek (1948), *Rec. Trav. Chim.,* **67,** 761.

Herrmann, K., O. Gerngross, and W. Abitz (1930), *Z. Phys. Chem. (Leipzig),* **B10,** 371.

Hershey, A. D., and E. Burgi (1960), *J. Mol. Biol.,* **2,** 143.

Hershey, A. D., and E. Burgi (1965), *Proc. Natl. Acad. Sci. U.S.,* **53,** 325.

Herskovits, T. T. (1962), *Arch. Biochem. Biophys.,* **97,** 474.

Hess, G. P. (1971), in *The Enzymes,* 3rd ed., Vol. III, P. D. Boyer, ed., Academic Press, New York p. 213.

Hess, G. P., J. McConn, E. Ku, and G. McConkey (1970), *Phil. Trans. Roy. Soc. (London),* **B257,** 89.

Heymann, E. (1935), *Trans. Faraday Soc.,* **31,** 846.

Higuchi, W. I., M. A. Schwartz, E. G. Rippie, and T. Higuchi (1959), *J. Phys. Chem.,* **63,** 996.

Hildebrand, J. H. (1929), *J. Am. Chem. Soc.,* **51,** 66.

Hildebrand, J. H. (1953), *Discuss. Faraday Soc.,* **15,** 9.

Hildebrand, J. H., and R. L. Scott (1950), *The Solubility of Nonelectrolytes,* American Chemical Society Monograph No. 17, Reinhold, New York, Chaps. 7, 8.

Hildebrand, J. H., and R. L. Scott (1962), *Regular Solutions,* Prentice-Hall, Englewood Cliffs, N. J.

Hiley, B. J., and M. F. Sykes (1961), *J. Chem. Phys.,* **34,** 1531.

Hilschmann, N., and L. C. Craig (1965), *Proc. Natl. Acad. Sci. U.S.,* **53,** 1403.

Hirano, T., P. H. Khanh, and T. Tsuruta (1972), *Makromol. Chem.,* **153,** 331.

Hirayama, F. (1965), *J. Chem. Phys.*, **42**, 3163.

Hirooka, M., H. Yabuuchi, J. Iseki, and Y. Nakai (1968), *J. Polym. Sci.*, **A-1, 6,** 1381.

Hirschman, S. Z., M. Gellert, S. Falkow, and G. Felsenfeld (1967), *J. Mol. Biol.*, **28**, 469.

Hitchcock, S. E. (1973), *Biochemistry*, **12**, 2509.

Ho, F. F. L. (1971), *J. Polym Sci.*, **B9**, 491.

Hochman, J., D. Inbar, and D. Givol (1973), *Biochemistry*, **12**, 1130.

Hoffman, S. J., and R. Ullman (1970), *J. Polym. Sci.*, **C31**, 205.

Holcomb, D. N., and S. N. Timasheff (1968), *Biopolymers*, **6**, 513.

Holftyzer, P. J., and D. W., van Krevelen, (1970), *IUPAC International Symposium on Macromolecules, Leyden, Book of Abstracts*, p. 563.

Holleman, T. (1960), *Rec. Trav. Chim.*, **79**, 1301.

Hollo, J., and J. Szejtli (1958), *Die Stärke*, **10**, 49 (cf. *C. A.*, **53**, 12437).

Holtzer, A., and M. F. Emerson (1969), *J. Phys. Chem.*, **73**, 26.

Hoover, M. F. (1970), *J. Macromol. Sci. Chem.*, **A4**, 1324.

Horie, K., I. Mita, and H. Kambo (1973), *Polym. J.*, **4**, 341.

Horn, P. (1955), *Ann. Phys. (Paris)*, [12], **10.** 386.

Horn, P., H. Benoit, and G. Oster (1951), *J. Chim. Phys.*, **48**, 1.

Hotta, H. (1954), *Bull. Chem. Soc. Jap.*, **27**, 80.

Howard, F. B., J. Frazier, M. N. Lipsett, and H. T. Miles (1964), *Biochem. Biophys. Res. Commun.*, **17**, 93.

Howard, F. B., J. Frazier, M. F. Singer, and H. T. Miles (1966), *J. Mol. Biol.*, **16**, 415.

Howard, G. J., and D. O. Jordan (1954), *J. Polym. Sci.*, **12**, 209.

Huang, W. M. and P. O. P. Ts'o (1966), *J. Mol. Biol.*, **16**, 523.

Hückel, E. (1924), *Phys. Z.*, **25**, 204.

Huestis, W. H., and M. A. Raftery (1973), *Biochemistry*, **12**, 2531.

Huggins, M. L. (1942a), *J. Phys. Chem.*, **46**, 151.

Huggins, M. L. (1942b), *Ann. N. Y. Acad. Sci.*, **43**, 1

Huggins, M. L. (1942c), *J. Am. Chem. Soc.*, **64**, 1712.

Huggins, M. L. (1942d), *J. Am. Chem. Soc.*, **64**, 2716.

Huggins, M. L. (1948), *J. Phys. Chem.*, **51**, 248.

Huggins, M. L., G. Natta, V. Desreux, and H. Mark (1962), *J. Polym. Sci.*, **56**, 153.

Hughes, L. J., R. H. Anderatta, and H. A. Scheraga (1972), *Macromolecules*, **5**, 187.

Hugli, T. E., and F. R. N. Gurd (1970), *J. Biol. Chem.*, **245**, 1930, 1939.

Huglin, M. B., ed. (1972), *Light Scattering from Polymer Solutions*, Academic Press, New York.

Huizenga, J. R., P. F. Grieger, and F. T. Wall (1950), *J. Am. Chem. Soc.*, **72**, 2636.

Husemann, E., B. Fritz, R. Lippert, and B. Pfannemüller (1958), *Makromol. Chem.*, **26**, 199.

Huxley, H. E. (1972), in *The Structure and Function of Muscle*, G. H. Bourne, ed., Academic Press, New York, Chap. 7.

Hvidt, A., and R. Corett (1970), *J. Am. Chem. Soc.*, **92**, 5546.

Hvidt, A., and S. O. Nielsen (1966), *Adv. Protein Chem.*, **21**, 287.

Hvidt, A., and K. Wallevik (1972), *J. Biol. Chem.*, **247**, 1530.

Hyde, A. J., J. H. Ryan, and F. T. Wall (1958), *J. Polym. Sci.*, **33**, 129.

Ifft, J. B., D. H. Voet, and J. Vinograd (1961), *J. Phys. Chem.*, **65**, 1138.

Imai, K., and M. Matsumoto (1961), *J. Polym. Sci.*, **55**, 335.

Imai, N., and T. Onishi (1959), *J. Chem. Phys.*, **30**, 1115.

Immergut, E. H., B. G. Rånby, and H. F. Mark (1953), *Ind. Eng. Chem.*, **45**, 2483.

Inagaki, H., T. Miyamoto, and S. Ohta (1966), *J. Phys. Chem.*, **70**, 3420.

Inagaki, H., H. Matsuda, and F. Kamiyama (1968), *Macromolecules*, **1**, 520.

Ingold, C. K. (1930), *J. Chem. Soc.*, 1375.

Ingram, V. M. (1959), *Biochim. Biophys. Acta*, **36**, 402.

Ingwall, R. T., H. A. Scheraga, N. Lotan, A. Berger, and E. Katchalski (1968), *Biopolymers*, **6**, 331.

Inman, R. B. (1967), *J. Mol. Biol.*, **28**, 103.

Inoue, Y., and A. Nishioka (1972), *Polym. J.*, **3**, 149.

Inoue, Y., I. Ando, and A. Nishioka (1972), *Polym. J.*, **3**, 246.

Ise, N. (1971), *Adv. Polym. Sci.*, **7**, 536.

Ise, N., and T. Okubo (1965), *J. Phys. Chem.*, **69**, 4102.

Ise, N., and T. Okubo (1967), *J. Phys. Chem.*, **71**, 1287, 1886.

Ise, N., and T. Okubo (1968), *J. Phys. Chem.*, **72**, 1361.

Isemura, T., and A. Imanishi (1958), *J. Polym. Sci.*, **33**, 337.

Isihara, A. (1950a), *J. Chem. Phys.*, **18**, 1446.

Isihara, A. (1950b), *J. Phys. Soc. Jap.*, **5**, 201.

Isihara, A., and R. Koyama (1956), *J. Chem. Phys.*, **25**, 712.

Ivin, K. J., and M. Navrátil (1970), *J. Polym. Sci.*, **A-1**, **8**, 3373.

Iwamoto, R., Y. Saito, H. Ishihara, and H. Tadokoro (1968), *J. Polym. Sci.*, **A-2**, **6**, 1509.

Jacobsen. H., and W. H. Stockmayer (1950), *J. Chem. Phys.*, **18**, 1600.

Jacobson, B. (1953), *Rev. Sci. Instr.*, **24**, 949.

Janeschitz-Kriegl, H. (1969), *Adv. Polym. Sci.*, **6**, 170.

Jardetzky, O., and N. G. Wade-Jardetzky (1971), *Ann. Rev. Biochem.*, **40**, 605.

Jayme, G., and F. Lang (1955), *Kolloid-Z.*, **144**, 75.

Jayme, G., and K. Neuschäffer (1957), *Makromol. Chem.*, **23**, 71.

Jeffery, P. D., and J. H. Coates (1966), *Biochemistry*, **5**, 489.

Jencks, W. P. (1969), *Catalysis in Chemistry and Enzymology*, McGraw-Hill, New York, pp. 67–71.

Jennings, B. R., G. Spach, and T. M. Schuster (1968), *Biopolymers*, **6**, 635.

Jensen, R. H., and N. Davidson (1966), *Biopolymers*, **4**, 17.

Jerrard, H. G. (1959), *Chem. Rev.*, **59**, 345.

Jessup, R. S. (1958), *J. Res. Natl. Bur. stand.*, **60**, 47.

Jirgensons, B. (1965), *J. Biol. Chem.*, **240**, 1064.

Jirgensons, B. (1966), *J. Biol. Chem.*, **241**, 147.

Joesten, M. D., and R. S. Drago (1962), *J. Am. Chem. Soc.*, **84**, 3817.

Johnson, L. F., F. Heatley, and F. A. Bovey (1970), *Macromolecules*, **3**, 175.

Johnson, R. M., J. L. Schrag, and J. D. Ferry (1970), *Polym. J.*, **1**, 742.

Johnson, W. R., and C. C. Price (1960), *J. Polym. Sci.*, **45**, 217.

Johnston, J. P., and A. G. Ogston (1946), *Trans. Faraday Soc.*, **42**, 789.

Jones, G. (1962), *J. Appl. Polym. Sci.*, **6**, 15.

Jordan, C. F., L. S. Lehrman, and J. H. Venable, Jr. (1972), *Nature—New Biol.*, **226**, 67.

Jost, K. (1958), *Rheol. Acta*, **1**, 303.

Jou, W. M., G. Haegeman, M. Ysebaert, and W. Fiers (1972), *Nature*, **237**, 82.

Joynson, M. A., A. C. T. North, V. R. Sarma, E. R. Dickerson, and L. K. Steinrauf (1970), *J. Mol. Biol.*, **50**, 137.

Kallos, J. (1964), *Biochim. Biophys. Acta*, **89**, 364.

Kang, A. H., and J. Gross (1970), *Biochemistry*, **9**, 796.

Karlson, R. H., K. S. Norland, G. D. Fasman, and E. R. Blout (1960), *J. Am. Chem. Soc.*, **82**, 2268.

Kartha, G., J. Bello, and D. Harker (1967), *Nature*, **213**, 862.

Katchalski, E. (1951), *Adv. Protein Chem.*, **6**, 123.

Katchalski, E., M. Sela, H. I. Silman, and A. Berger (1964), in *The Proteins*, Vol. 2, H. Neurath, ed., 2nd ed., Academic Press, New York, Chap. 10.

Katchalsky, A. (1971), *Pure Appl. Chem.*, **26**, 327.

Katchalsky, A., and H. Eisenberg (1951), *J. Polym. Sci.*, **6**, 145.

Katchalsky, A., and J. Feitelson (1954), *J. Polym. Sci.*, **13**, 385.

Katchalsky, A., and S. Lifson (1956), *J. Polym. Sci.*, **11**, 409.

Katchalsky, A., and I. R. Miller (1954), *J. Polym. Sci.*, **13**, 57.

Katchalsky, A., O. Künzle, and W. Kuhn (1950), *J. Polym. Sci.*, **5**, 283.

Katchalsky, A., Z. Alexandrowicz, and O. Kedem (1966), in *Chemical Physics of Ionic Solutions*, B. E. Conway and R. G. Barradas, eds., Wiley, New York, p. 295.

Kato, T., K. Miyaso, and M. Nagasawa (1968), *J. Phys. Chem.*, **72**, 2161.

Katz, J. R., and J. C. Derksen (1931), *Rec. Trav. Chim.*, **50**, 149.

Katz, J. R., and J. C. Derksen (1932), *Rec. Trav. Chim.*, **51**, 513.

Katz, J. R., and A. Weidinger (1932), *Rec. Trav. Chim.*, **51**, 847.

Katz, J. R., J. C. Derksen, and W. F. Bon (1931), *Rec. Trav. Chim.*, **50**, 725, 1138.

Kauzmann, W. (1959), *Adv. Protein Chem.*, **14**, 1.

Kauzmann, W. J., J. E. Walter, and H. Eyring (1940), *Chem. Revs.*, **26**, 339.

Kawaguchi, Y., and M. Nagasawa (1969), *J. Phys. Chem.*, **73**, 4382.

Kaye, W., A. J. Havlik, and J. B. McDaniel (1971), *J. Polym. Sci.*, **B9**, 965.

Kedem, O., and A. Katchalsky (1955), *J. Polym. Sci.*, **15**, 321.

Keith, H. D., R. G. Vadimsky, and F. J. Padden, Jr. (1970), *J. Polym. Sci.* **A-2, 8**, 1687.

Keller, J. B. (1962), *J. Chem. Phys.*, **37**, 2584.

Keller, J. B. (1963), *J. Chem. Phys.*, **38**, 325.

Kendrew, J. C. (1963), *Science*, **139**, 1259.

Kendrew, J. C., R. E. Dickerson, B. E. Strandberg, R. G. Hart, D. R. Davies, D. C. Phillips and V. C. Shore (1960), *Nature*, **185**, 422.

Kendrew, J. C., H. C. Watson, B. E. Strandberg, R. E. Dickerson, D. C. Phillips, and V. C. Shore (1961), *Nature*, **190**, 666.

Kennedy, J. P., L. S. Minckler, Jr., G. Wenless, and R. M. Thomas (1964), *J. Polym. Sci.*, **A2,** 1441, 2093.

Kenner, R. A., and A. A. Aboderin (1971), *Biochemistry,* **10,** 4433.

Kerker, M. (1969), *The Scattering of Light and Other Electromagnetic Radiation,* Academic Press, New York: (a) Chap. 8; (b) pp. 487–504, (c) pp. 583–587.

Kern, R. J. (1956), *J. Polym. Sci.,* **21,** 19.

Kern, R. J. (1958), *J. Polym. Sci.,* **33,** 524.

Kern, R. J., and R. J. Slocombe (1955), *J. Polym. Sci.,* **15,** 183.

Kern, W., W. Gruber, and H. O. Wirth (1960), *Makromol. Chem.,* **31,** 198.

Kerr, W. R. (1949), *Nature,* **164,** 757.

Kershaw, R. W., and G. N. Malcolm (1968), *Trans. Faraday Soc.,* **64,** 323.

Kiefer, H. C., W. I. Congdon, I. S. Scarpa, and I. M. Klotz (1972), *Proc. Natl. Acad. Sci. U.S.,* **69,** 2155.

Kim, O. K. (1973), Private communication.

Kim, S. H., F. L. Suddath, G. J. Quigley, J. L. Sussman, A. H. J. Wang, N. C. Seeman and A. Rich (1974), *Science,* **185,** 435.

Kimball, G. E., M. Cutler, and H. Samelson (1952), *J. Phys. Chem.,* **56,** 47.

King, N. L. R. and J. H. Bradbury (1971), *Nature (London),* **229,** 5284.

Kinsinger, J. B., and R. E. Hughes (1963), *J. Phyc. Chem.,* **67,** 1922.

Kirkwood, J. G. (1934), *J. Chem. Phys.,* **2,** 351.

Kirkwood, J. G., and J. Riseman (1948), *J. Chem., Phys.,* **16,** 565.

Kirkwood, J. G., and V. Shumaker (1952), *Proc. Natl. Acad. Sci. U.S.,* **38,** 863.

Kirsh, Y. E., V. A. Kabanov, and V. A. Kargin (1967), *Proc. Acad. Sci. USSR, Chem. Sec.* (Engl. transl.), **177,** 976.

Kirsh, T. E., V. A. Kabanov, and V. A. Kargin (1968), *Vysokomol. Soedin.,* **A10,** 349.

Kirsh, Yu. E., O. P, Komarova, and G. M. Lukovkin (1973) *Eur. Polym. J.,* **9,** 1405.

Kirste, R., and O. Kratky (1962) *Z. phys. Chem. (Frankfurt),* **31,** 363.

Kirste, R., and W. Wunderlich (1964), *Makromol. Chem.,* **73,** 240.

Kirste, R. G., W. A. Kruse, and J. Schelten (1972), *Makromol. Chem.,* **162,** 299.

Kirszenbaum, F., J. Dandliker, and W. B. Dandliker (1969), *Immunochemistry,* **6,** 125.

Kisselev, N. A., L. P. Gavrilova, and A. Spirin (1961), *J. Mol. Biol.,* **3,** 778.

Kleine, J., and H. H. Kleine (1959), *Makromol. Chem.,* **30,** 23.

Kleinschmidt, A., D. Lang, and R. K. Zahn (1961), *Z. Naturforsch.,* **16b,** 730.

Kleinschmidt, A. K., D. Lang, D. Tacherts, and R. K. Zahn (1962), *Biochim. Biophys. Acta,* **61,** 857.

Klesper, E., A. Johnsen, and W. Gronski (1970a), *J. Polym. Sci.,* **B 8,** 369.

Klesper, E., W. Gronski, and V. Barth (1970b), *Makromol. Chem.,* **139,** 1.

Klesper, E., W. Gronski, and V. Barth (1971), *Makromol. Chem.,* **150,** 223.

Klesper, E., A. Johnsen, and W. Gronski (1972), *Makromol. Chem.,* **160,** 167.

Klotz, I. M., and H. A. Fiess (1951), *J. Phys. Colloid Chem.,* **55,** 101.

Klotz, I. M., N. R. Langerman, and D. W. Darnall (1970), *Ann. Rev. Biochem.,* **39,** 25.

Klotz, L. C. (1969), *Biopolymers,* **7,** 265.

Klotz, L. C., and B. H. Zimm (1972), *Macromolecules,* **5,** 471.

Klug, A., W. Longley, and R. Leberman (1966), *J. Mol. Biol.,* **15,** 315.

Knowles, J. R. (1972), *Acc. Chem. Res.,* **5,** 155.

Kobayashi, M., K. Tsumura, and H. Tadokoro (1968), *J. Polym. Sci.,* **A-2, 6,** 1493.

Koenig, J. L., A. C. Angood, J. Semen, and J. B. Lando (1969), *J. Am. Chem. Soc.,* **91,** 7250.

Kohlrausch, K. W. F. (1932), *Z. Phys. Chem.* (*Leipzig*), **B18,** 61.

Kohn, M. C. (1973), *J. Polym. Sci., Polym. Phys. Ed.,* **11,** 2339.

Konings, W. N., R. J. Siezen, and M. Gruber (1969), *Biochim. Biophys. Acta,* **194,** 376.

Kontos, E. (1959), Ph. D. Thesis, Columbia Univerity, New York.

Kornberg, A. (1959), *Rev. Mod. Phys.,* **31,** 200.

Kornberg, A. (1961), *Enzymatic Synthesis of DNA,* Wiley, New York.

Koshland, D. E., Jr., G. Nemethy, and D. Filmer (1966), *Biochemistry,* **5,** 365.

Kotaka, T., H. Ohnuma, and H. Inagaki (1969), *Polymer,* **10,** 517.

Kotaka, T., T. Tanaka, H. Ohnuma, Y. Murakami, and H. Inagaki (1970), *Polym. J.,* **1,** 245.

Kotaka, T., H. Ohnuma, and H. Inagaki (1971), *Colloidal and Morphological Behavior of Block and Graft Copolymers,* G.E. Molau, ed., Plenum Press, New York, p. 259.

Kotaka, T., T. Tanaka, and H. Inagaki (1971), *Polym. J.,* **3,** 327.

Kotin, L., and M. Nagasawa (1962), *J. Chem. Phys.,* **36,** 873.

Kotliar, A. M., and H. Morawetz (1955), *J. Am. Chem., Soc.,* **77,** 3692.

Kozak, D., J. Kristan, and D. Dolar (1971), *Z. phys. Chem.* (*Frankfurt*), **76,** 85.

Kraemer, E. O., and J. R. Fanselow (1928), *J. Phys. Chem.,* **32,** 894.

Krakauer, H., and J. M. Sturtevant (1968), *Biopolymers,* **6,** 491.

Krasna, A. I. (1970), *Biopolymers,* **9,** 1029.

Kratky, O. (1948), *J. Polym. Sci.,* **3,** 195.

Kratky, O. (1960), *Angew. Chem.,* **72,** 467.

Kratky, O. (1962a), *Kolloid-Z.,* **182,** 7.

Kratky, O. (1962b), *Z. Elektrochem.,* **64,** 880.

Kratky, O., and W. Kreutz (1960), *Z. Elektrochem.,* **64,** 880.

Kratky, O., and G. Porod (1949), *Rec. Trav. Chim.,* **68,** 1106.

Kratky, O., and H. Sand (1960), *Kolloid-Z.,* **172,** 18.

Kratky, O., and W. Worthmann (1947), *Monatsh. Chem.,* **76,** 263.

Kratky, O., G. Porod, A. Sekora, and B. Paletta (1955), *J. Polym. Sci.,* **16,** 163.

Kratochvil, P., V. Petrus, P. Munk, M. Bohdanecký, and K. Šolc (1967), *J. Polym. Sci.,* **C16,** 1157.

Kratochvil, P., M. Bohdanecký, K. Šolc, M. Kolinský, M. Ryska, and D. Lim (1968), *J. Polym. Sci.,* **C23,** 9.

Krause, S. (1961), *J. Phys. Chem.,* **65,** 1618.

Krause, S. (1972), *J. Macromol. Sci., Revs. Macromol. Chem.,* **C7,** 251.

Krause, S., and C. T. O'Konski (1959), *J. Am. Chem. Soc.,* **81,** 5082.

Krause, S., and C. T. O'Konski (1963), *Biopolymers,* **1,** 503.

Kraut, J. (1971), in *The Enzymes,* 3rd ed., Vol. III, D. P. Boyer, ed., Academic Press, New York, p. 165.

Krigbaum, W. R. (1954), *J. Am. Chem. Soc.*, **76,** 3758.

Krigbaum, W. R. (1955), *J. Chem. Phys.*, **23,** 2113.

Krigbaum, W. R., and D. K. Carpenter (1955), *J. Phys. Chem.*, **59,** 1166.

Krigbaum, W. R., and D. O. Geymer (1959), *J. Am. Chem. Soc.*, **81,** 1859.

Krigbaum, W. R., D. K. Carpenter, and S. Newman (1958), *J. Phys. Chem.*, **62,** 1586.

Krimm, S., and C. M. Venkatachalam (1971), *Proc. Natl. Acad. Sci. U.S.*, **68,** 2468.

Kuge, T., and S. One (1960), *Bull. Chem. Soc. Jap.*, **33,** 1269, 1273.

Kuhn, L. P., and R. E. Bowman (1961), *Spectrochim. Acta,* **17,** 650.

Kuhn, W. (1934), *Kolloid-Z.*, **68,** 2.

Kuhn, W. (1958), *Ann. Rev. Phys. Chem.*, **9,** 417.

Kuhn, W., and F. Grün (1942), *Kolloid-Z.*, **101,** 248.

Kuhn, W., and H. Kuhn (1943), *Helv. Chim. Acta,* **26,** 1394.

Kuhn, W., and H. Kuhn (1945a), *Helv. Chim. Acta,* **28,** 97.

Kuhn, W., and H. Kuhn (1945b), *Helv. Chim. Acta.* **28,** 1533.

Kuhn, W., and H. Kuhn (1946), *Helv. Chim. Acta,* **29,** 71, 609, 830.

Kuhn, W., O. Künzle, and A. Katchalsky (1948), *Helv. Chim. Acta,* **31,** 1994.

Kulkarni, R. K., and H. Morawetz (1961), *J. Polym. Sci.*, **54,** 491.

Kunitake, T., and S. Shinkai (1971), *J. Am. Chem. Soc.*, **93,** 4247, 4256.

Kunitake, T., Y. Okahata, and R. Ando (1974), *Macromolecules,* **7,** 140.

Kuntz, I., and N. F. Chamberlain (1974), *Polym. Prepr., Am. Chem. Soc., Div. Polym. Chem.*, **15,** 474.

Kurata, M., and W. H. Stockmayer (1963), *Adv. Polym. Sci.*, **3,** 196.

Kurata, M., and H. Yamakawa (1958), *J. Chem. Phys.*, **29,** 311.

Kurata, M., H. Yamakawa, and H. Utiyama (1959), *Makromol. Chem.*, **34,** 139.

Kurata, M., W. H. Stockmayer, and A. Roig (1960), *J. Chem. Phys.*, **33,** 151.

Kurtz, J., A. Berger, and E. Katchalski (1956), *Nature.* **178,** 1066.

Küster, F. W. (1894), *Ann. Chem.*, **283,** 360.

Łabudzińska, A., and A. Ziabicki (1971), *Kolloid-Z., Z. Polym.*, **243,** 21.

Łabudzińska, A., A. Wasiak, and A. Ziabicki (1967), *J. Polym. Sci.*, **C16,** 2835.

Ladenheim, H., and H. Morawetz (1959), *J. Am. Chem. Soc.*, **81,** 4860.

Ladenheim, H., E. M. Loebl, and H. Morawetz (1959), *J. Am. Chem. Soc.*, **81,** 20.

Lamb, Sir H. (1945), *Hydrodynamics,* 1st American ed., Dover, New York, p. 597.

Landau, L. D., and E. M. Lifshitz (1958), *Statistical Physics,* Pergamon Press, London, 1958, pp. 478–482.

Lando, J. B., J. Semen, and B. Farmer (1970), *Macromolecules,* **3,** 524.

Lando, J. B., J. L. Koenig, and J. Semen (1973), *J. Macromol. Sci.—Phys.*, **B7,** 319.

Landsteiner, K. (1946), *The Specificity of Serological Reactions,* 2nd ed., Harvard University Press, Cambridge, Mass.

Langridge, R., H. R. Wilson, C. W. Hooper, M. H. F. Wilkins, and L. D. Hamilton (1960a), *J. Mol. Biol.*, **2,** 19.

Langridge, R., D. A. Marvin, W. E. Seeds, H. R. Wilson, C. W. Hooper, M. H. F. Wilkins, and L. D. Hamilton (1960b), *J. Mol. Biol.*, **2,** 38.

Lapanje, S. (1962), *Vestn. Slov. Kem. Drus.*, **9,** 5.

Lapanje, S., and C. Tanford (1967), *J. Am. Chem. Soc.*, **89**, 5030.

Laskowski, M., Jr., S. J. Leach and H. A. Scheraga (1960), *J. Am. Chem. Soc.*, **82**, 571.

Latt, S. A., D. S. Auld, and B. L. Vallee (1972), *Biochemistry*, **11**, 1305.

Lauffer, M. A., and C. L. Stevens (1968), *Adv. Virus Res.*, **13**, 1.

Lawson, W. B., and H. J. Schramm (1965), *Biochemistry*, **4**, 377.

Leach, S. J., and H. A. Scheraga (1960), *J. Am. Chem. Soc.*, **82**, 4790.

Lehrman, L. S. (1971), *Proc. Natl. Acad. Sci. U.S.*, **68**, 1886.

Lemberg, R., and J. W. Legge (1949), *Hematin Compounds and Bile Pigments*, Interscience, New York.

Leng, M., and H. Benoit (1962a), *J. Polym. Sci.*, **57**, 263.

Leng, M., and H. Benoit (1962b), *J. Chim. Phys.*, **59**, 929.

Leng, M., and G. Felsenfeld (1966a), *J. Mol. Biol.*, **15**, 455.

Leng, M., and G. Felsenfeld (1966b), *Proc. Natl. Acad. Sci. U.S.*, **56**, 1325.

Leng, M., C. Strazielle, and H. Benoit (1963), *J. Chim. Phys.*, **60**, 501.

Lenz, D. E., and W. D. Bryan (1969), *Biochemistry*, **8**, 1123.

Lenz, R. W. (1967), *Organic Chemistry of Synthetic High Polymers*, Interscience, New York: (a) pp. 710–719; (b) pp. 719–723.

Leray, J. (1957), *J. Polym, Sci.*, **23**, 167.

Letsinger, R. L., and I. Klaus (1964), *J. Am. Chem. Soc.*, **86**, 3884.

Letsinger, R. L., and T. J. Savereide (1962), *J. Am. Chem. Soc.*, **84**, 3122.

Letsinger, R. L., and T. E. Wagner (1966), *J. Am. Chem. Soc.*, **88**, 2062.

Levinthal, C., and P. F. Davison (1961), *J. Mol. Biol.*, **3**, 674.

Levitzki, A., and D. E. Koshland, Jr. (1969), *Proc. Natl. Acad. Sci. U.S.*, **62**, 1121.

Lewis, P. N., F. A. Momany, and H. A. Scheraga (1971), *Proc. Natl. Acad. Sci. U.S.*, **68**, 2293.

Lewis, W. H. P. (1971), *Nature* (*London*), **230**, 215.

Leyte, J. C., L. U. Zuiderberg, and M. van Reisen (1968), *J. Phys. Chem.*, **72**, 1127.

Li, H. J., and J. Bonner (1971), *Biochemistry*, **10**, 1461.

Li, H. J., and D. M. Crothers (1969), *Biopolymers*, **8**, 217.

Li, H. J., I. Isenberg, and W. C. Johnson, Jr. (1971), *Biochemistry*, **10**, 2587.

Li, H.-M., B. Post, and H. Morawetz (1972), *Makromol. Chem.*, **154**, 89.

Lifson, S. (1957), *J. Chem. Phys.*, **27**, 700.

Lifson, S. (1958), *J. Chem. Phys.*, **29**, 89.

Lifson, S. (1959), *J. Chem. Phys.*, **30**, 964.

Lifson, S. (1963), *Biopolymers*, **1**, 25.

Lifson, S., and I. Oppenheim (1960), *J. Chem. Phys.*, **33**, 109.

Lifson, S., and B. H. Zimm (1963), *Biopolymers*, **1**, 15.

Liljas, A., K. K. Kennan, P. C. Bergston, K. Fridborg, B. Stranberg U. Carlblom, L. Järup, S. Lövgren, and M. Pelef (1972), *Nature—New Biol.*, **235**, 131.

Lin, O. C. C. (1970), *Macromolecules*, **3**, 80.

Linderstrøm-Lang, K. (1955), *Chem. Soc.* (*London*) *Spec. Publ.*, **2**, 1.

Linderstrøm-Lang, K., and J. A. Schellman (1959), *The Enzymes*, P. D. Boyer, H. Lardy, and K. Myrbäck, eds., 2nd ed., Vol. 1, Academic Press, New York, p. 443.

Lindskog, S., and B. G. Malmström (1962), *J. Biol. Chem.*, **237,** 1129.

Lipmann, F (1973), *Acc. Chem. Res.*, **6,** 361.

Lipscomb, W. N. (1970), *Acc. Chem. Res.*, **3,** 81.

Lipscomb, W. N., J. A. Hartsuck, F. A. Quiocho, and G. N. Reeke, Jr. (1969), *Proc. Natl. Acad. Sci. U.S.*, **64,** 28.

Lipscomb, W. N., G. N. Reeke, Jr., J. A. Hartsuck, F. A. Quiocho, and P. H. Bethge (1970), *Phil. Trans. Roy .Soc. (London)*, **B257,** 177.

Liquori, A. M. (1966), *J. Polym. Sci.*, **C12,** 209.

Liquori, A. M., G. Anzuino, V. M. Coiro, M. d'Alagni, P. De Santis, and M. Savino (1965), *Nature*, **206,** 358.

Liquori, A. M., P. De Santis, A. L. Kovacs, and L. Mazzarella (1966), *Nature*, **211,** 1059.

Litmanovich, A. D., N. A. Platé, O. V. Noah, and V. I. Golyakov (1969), *Eur. Polym. J. Suppl.*, p. 517.

Litt, M. H., and J. W. Summers (1973), *J. Polym. Sci., Polym. Chem. Ed.*, **11,** 1359.

Liu, H. Z., and K. J. Liu (1968), *Makromolecules*, **1,** 157.

Liu, K. J. (1969), *Makromol. Chem.*, **126,** 189.

Liu, K. J., and J. E. Anderson (1970), *Macromolecules*, **3,** 163.

Liu, K. J., and J. L. Parsons (1969), *Macromolecules*, **2,** 529.

Lloyd, P. H., R. N. Prutton, and A. R. Peacocke (1968), *Biochem. J.*, **107,** 353.

Löhr, G., and G. V. Schulz (1964), *Makromol. Chem.*, **77,** 240.

Long, F. A., and W. F. McDevit (1952), *Chem. Rev.*, **51,** 119.

Longworth, J. W. (1966), *Biopolymers*, **4,** 1131.

Longsworth, L. G. (1941), *Ann. N. Y. Acad. Sci.*, **41,** 167.

Lotan, N., F. A. Momany, J. F. Yan, G. Vanderkooi, and H. A. Scheraga (1969), *Biopolymers*, **8,** 26.

Lovrien, R., and J. C. Waddington (1964), *J. Am. Chem. Soc.*, **86,** 2315.

Lowe, W. G. (1943), U.S. Patents 2,311,058; 2,311,059.

Lowey, S., J. Kucera, and A. Holtzer (1963), *J. Mol. Biol.*, **7,** 234.

Lowey, S., H. S. Slayter, A. G. Weeds, and H. Baker (1969), *J. Mol. Biol.*, **42,** 1.

Luisi, P. L. (1972), *Polymer*, **13,** 232.

Lumley, J. L. (1973), *Macromol. Rev.*, **7,** 263.

Lumry, R., and H. Eyring (1954), *J. Phys. Chem.*, **58,** 110.

Lumry, R., R. R. Legare, and W. G. Miller (1964), *Biopolymers*, **2,** 489.

Lundberg, R. D., F. E. Bailey, and R. W. Callard (1966), *J. Polym. Sci.*, **A-1, 4,** 1563.

Lütje, H., and G. Meyerhoff (1963), *Makromol. Chem.*, **68,** 180.

Luzzati, V. (1961), *Acta Cryst.*, **13,** 939.

Luzzati, V., M. Cesari, G. Spach, F. Masson, and J. M. Vincent (1961a), *J. Mol. Biol.*, **3,** 566.

Luzzati, V., A. Nicolaieff, and F. Masson (1961b), *J. Mol. Biol.*, **3,** 185.

Luzzati, V., J. Witz, and A. Nicolaieff (1961c), *J. Mol. Biol.*, **3,** 367.

Luzzati, V., J. Witz, and A. Nicolaieff (1961d), *J. Mol. Biol.*, **3,** 373.

Lyons, J. W., and L. Kotin (1965), *J. Am. Chem. Soc.*, **87,** 1670.

Madison, V., and J. Schellman (1972), *Biopolymers*, **11,** 1041.

Magasanik, B. (1955), in *The Nucleic Acids,* Vol. 1, E. Chargaff and J. Davidson, eds., Academic Press, New York, Chap. 11.

Malcolm, G. N., and J. S. Rowlinson (1957), *Trans. Faraday Soc.,* **53,** 921.

Malcolm, G. N., C. E. Baird, G. R. Bruce, K. G. Cheyne, R. W. Kershaw, and M. C. Pratt (1969), *J. Polym. Sci.,* **A-2, 7,** 1495.

Malmström, B. G. (1955), *Arch. Biochem.,* **58,** 381.

Mandel, M. (1967), *J. Polym. Sci.,* **C16,** 2955.

Mandel, M. (1970), *Eur. Polym. J.,* **6,** 807.

Mandel, M., and A. Jenard (1963), *Trans. Faraday Soc.,* **59,** 2158.

Mandel, M., and J. C. Leyte (1964), *J. Polym. Sci.,* **A1,** 2883, 3771.

Mandelkern, L., and P. J. Flory (1952), *J. Chem. Phys.,* **20,** 212.

Mandelkern, L., and T. G. Fox (1953), *J. Chem. Phys.,* **21,** 187.

Mandelkern, L., W. R. Krigbaum, H. A. Scheraga, and P. J. Flory (1952), *J. Chem. Phys.,* **20,** 1392.

Mandelkern, L., L. C. Williams, and S. G. Weissberg (1957), *J. Phys. Chem.,* **61,** 271.

Manning, G. (1969), *J. Chem. Phys.,* **51,** 924.

Marcker, K. (1960), *Acta Chem. Scand.,* **14,** 194.

Marcus, R. A. (1954), *J. Phys. Chem.,* **58,** 621.

Marcus, R. A. (1955), *J. Chem. Phys.,* **23,** 1057.

Mark, H., and A. V. Tobolsky (1950), *Physical Chemistry of High Polymeric Systems* (High Polymers, Vol. 2), Interscience, New York, p. 289.

Mark, J. E., and P. J. Flory (1965), *J. Am. Chem. Soc.,* **87,** 1415.

Markert, C. L. (1969), *Ann. N. Y. Acad. Sci.,* **151,** 14.

Markley, J. L., and O. Jardetzky (1970), *J. Mol. Biol.,* **50,** 223.

Markley, J. L., M. N. Williams, and O. Jardetzky (1970), *Proc. Natl. Acad. Sci. U.S.,* **65,** 645.

Marmur, J., and P. Doty (1962), *J. Mol. Biol.,* **5,** 109.

Marmur, J., and D. Lane (1960), *Proc. Natl. Acad. Sci. U.S.,* **46,** 453.

Marmur, J., R. Rownd, and C. L. Schildkraut (1963), in *Progress in Nuclei Acid Research,* Vol. 1, J. N. Davidson and W. E. Cohn, eds., Academic Press, New York, p. 232.

Maron, E., C. Shiozawa, R. Arnon, and M. Sela (1971), *Biochemistry,* **10,** 763.

Marvel, C. S., and C. G. Overberger (1946), *J. Am. Chem. Soc.,* **68,** 2106.

Marx-Figini, M., and G. V. Schulz (1966), *Biochim. Biophys. Acta,* **112,** 81; *Naturwiss.,* **53,** 466.

Massa, D. J., J. L. Schrag, and J. D. Ferry (1971), *Macromolecules,* **4,** 210.

Massoulié, J. (1968), *Eur. J. Biochem.,* **3,** 428.

Masuda, Y., T. Miyazawa, and M. Goodman (1969), *Biopolymers,* **8,** 515.

Mathieson, A. R., and J. C. J. Thynne (1956), *J. Chem. Soc.,* 3708.

Mathot, V., and D. Desmyter (1953), *J. Chem. Phys.,* **21,** 782.

Matsuzaki, K., T. Uryu, M. Okada, and H. Shiroki (1968), *J. Polym. Sci.,* **A-1, 6,** 1475.

Matthews, B. W., P. B. Sigler, R. Henderson, and D. M. Blow (1967), *Nature,* **214,** 652.

Matthews, B. W., J. N. Jansonius, P. M. Colman, B. P. Schoenborn, and D. Dupourque (1972a), *Nature—New Biol.,* **238,** 37.

Matthews, F. S., P. Argos, and M. Levine (1972b), *Cold. Spring Harbor Symp. Quant. Biol.*, **36**, 387.

Mayo, F. R., and F. M. Lewis (1944), *J. Am. Chem. Soc.*, **66**, 1594.

McClure, W. O., and G. M. Edelman (1967), *Biochemistry*, **6**, 559, 567.

McCormick, H. W. (1959a), *J. Polym. Sci.*, **36**, 341.

McCormick, H. W. (1959b), *J. Polym. Sci.*, **41**, 327.

McCormick, H. W., F. M. Brower, and L. Kin (1959), *J. Polym. Sci.*, **39**, 87.

McDonald, C. C., and W. D. Phillips (1967), *J. Am. Chem. Soc.*, **89**, 6332.

McDonald, C. C., W. D. Phillips, and J. Lazar (1967), *J. Am. Chem. Soc.*, **89**, 4166.

McDonald, C. C., W. D. Phillips, and J. D. Glickson (1971), *J. Am. Chem. Soc.*, **93**, 235.

McDowell, W. H., and W. O. Kenyon (1941), U.S. Patents, 2,234,186; 2,249,536; 2,249,-537; 2,249,538.

McHattie, L. A., and C. A. Thomas, Jr. (1964), *Science*, **144**, 142.

McIntyre, D., and F. Gornick, eds. (1964), *Light Scattering from Dilute Polymer Solutions*, Gordon and Breach, New York.

McQuarrie, D. A., J. P. Mc Tague, and H. Reiss (1965), *Biopolymers*, **3**, 657.

Means, G. E., D. S. Ryan, and R. E. Feeney (1914), *Acc. Chem. Res.*, **7**, 315.

Medalia, A. I., H. H. Freedman, and S. Sinha (1959), *J. Polym. Sci.*, **40**, 15.

Menger, F. M., and C. E. Portnoy (1968), *J. Am. Chem. Soc.*, **90**, 1875.

Merrett, F. M. (1954), *Trans. Faraday Soc.*, **50**, 759.

Merrett, F. M. (1957), *J. Polym. Sci.*, **24**, 467.

Meselson, M., F. W. Stahl, and J. Vinograd (1957), *Proc. Natl. Acad. Sci., U.S.*, **43**, 581.

Meyer, K. H., and O. Klemm (1940), *Helv. Chim. Acta.* **25**, 23.

Meyer, V. E., and G. G. Lowry (1965), *J. Polym. Sci.*, **A3**, 2843.

Meyerson, K. (1966), in *Polymer Handbook*, J. Brandrup and E. H. Immergut, eds., Interscience, New York, pp. 192, 205, 227.

Michelson, A. M., and C. Monny (1967), *Biochim. Biophys. Acta*, **149**, 107.

Middleton, W. J., and R. V. Lindsey, Jr. (1964), *J. Am. Chem. Soc.*, **86**, 4948.

Mikeš, F., P. Štrop, and J. Kálal (1974), *Makromol, Chem.*, **175**, 2375.

Mikus, R. F., R. M. Hixon, and R. E. Rundle (1946), *J. Am. Chem. Soc.*, **68**, 1115.

Mildwan, A. S., and M. Cohn (1970), *Adv. Enzymol.*, **33**, 1.

Miller, G. A., F. I. San Filippo, and D. K. Carpenter (1970), *Macromolecules*, **3**, 125.

Miller, I. R., and D. Bach (1973), in *Surface and Colloid Science*, Vol. 6, E. Matijevic, ed., Wiley, New York, p. 185.

Miller, R. L. (1962), *J. Polym. Sci.*, **56**, 375.

Miller, R. L. and L. E. Nielsen (1960), *J. Polym. Sci.*, **46**, 303.

Miller, R. L., and L. E. Nielsen (1961), *J. Polym. Sci.*, **55**, 643.

Miller, W. G., and C. V. Goebel, (1968), *Biochemistry*, **7**, 3925.

Miller, W. G., D. A. Brant, and P. J. Flory (1967), *J. Mol. Biol.*, **23**. 67.

Millet, F., and M. A. Raftery (1972), *Biochemistry*, **11**, 1639.

Millionova, M. I. (1964), *Biofizika*, **9**, 145 (Engl. transl.: *Biophysics*, **9**, 149).

Millionova, M. I., N. J. Andreeva, and L. A. Lebedev (1963), *Biofizika*, **8**, 430 (Engl. transl.: *Biophysics*, **8**, 478).

Milstien, J. B., and E. Charney (1969), *Macromolecules*, **2**, 678.

Minsk, L. M. ,W. J. Priest, and W. O. Kenyon (1941), *J. Am. Chem. Soc.*, **63**, 2715.

Miron, Y., B. R. McGarvey, and H. Morawetz (1969), *Macromolecules*, **2**, 154.

Mitchell, A. G., and W. F. K. Wynne-Jones (1953), *Discuss. Faraday Soc.*, **15**, 161.

Mitsui, Y, Y. Iitaka, and M. Tsuboi (1967), *J. Mol. Biol.*, **24**, 15.

Miyamoto, T., K. Kodama, and K. Shibayama (1970), *J. Polym. Sci.*, **A-2, 8**, 2095.

Miyazawa, T., and E.R. Blout (1961), *J. Am. Chem. Soc.*, **83**, 712.

Mizushima, S. (1954), *Structure of Molecules and Internal Rotation*, Academic Press, New York, pp. 43–44.

Mizushima, S., Y. Morino, I. Watanabe, T. Simanouti, and S. Yamaguchi (1949), *J. Chem. Phys.*, **17**, 591.

Mock, R. A., and C. A. Marshall (1954), *J. Polym. Sci.*, **13**, 263.

Mock, R. A., C. A. Marshall, and T. E. Slykhouse (1954), *J. Phys. Chem.*, **58**, 498.

Moffitt, W. (1956), *J. Chem. Phys.*, **25**, 467; *Proc. Natl. Acad. Sci. U.S.*, **42**, 736.

Moffitt, W., and J. T. Yang (1956), *Proc. Natl. Acad. Sci. U.S.*, **42**, 596.

Moha. P., G. Weill, and H. Benoit (1964), *J. Chim. Phys.*, **61**, 1239.

Molau, G. E. (1965), *J. Polym. Sci.*, **A3**, 1267, 4235.

Molau, G. E. (1970), in *Block Polymers*, G. E. Molau, ed., Plenum Press, New York, p. 79.

Molau, G. E., and W. M. Wittbrodt (1968), *Macromolecules*, **1**, 260.

Moldovan, L., and G. Weill (1971), *Eur. Polym. J.*, **7**, 1023.

Möller, W. J. H. M., G. A. J. Van Os, and J. T. G. Overbeek (1961), *Trans. Faraday Soc.*, **57**, 312, 325.

Molyneux, P., and H. P. Frank (1961), *J. Am. Chem. Soc.*, **83**, 3169, 3175.

Mommaerts, W. F. H. M. (1952), *J. Biol. Chem.*, **198**, 467.

Monod, J., J.-P. Changeux, and F. Jacob (1963), *J. Mol. Biol.*, **6**, 306.

Monod, J., J. Wyman, and J.-P. Changeux (1965), *J. Mol. Biol.*, **12**, 88.

Moore, D. E., and A. G. Parts (1968), *Polymer*, **9**, 52.

Moore, W. R., and R. Shuttleworth (1963), *J. Polym. Sci.*, **A1**, 733.

Moraglio, G., and G. Gianotti (1969), *Eur. Polym. J.*, **5**, 781.

Morawetz, H. (1954), *Ind. Chim. Belge*, **19**, 607.

Morawetz, H. (1969), *Adv. Catal.*, **20**, 341.

Morawetz, H. (1972), *Adv. Protein Chem.*, **26**, 243.

Morawetz, H. (1973), *Israel J. Sci.*, **11**, 173.

Morawetz, H., and E. Gaetjens (1958), *J. Polym. Sci.*, **32**, 526.

Morawetz, H., and R. H. Gobran (1954), *J. Polym. Sci.*, **12**, 133.

Morawetz, H., and R. H. Gobran (1955), *J. Polym. Sci.*, **18**, 455.

Morawetz, H., and G. Gordimer (1970), *J. Am Chem. Soc.*, **92**, 7532.

Morawatz, H., and W.L. Hughes, Jr (1952), *J. Phys. Chem.*, **56**, 64.

Morawetz, H., and E. Sammak (1957), *J. Phys. Chem.*, **61**, 1357.

Morawetz, H., and J. A. Shafer (1963), *J. Phys. Chem.*, **67**, 1293.

Morawetz, H., and W. R. Song (1966), *J. Am. Chem. Soc.*, **88**, 5714

Morawetz, H., and B. Vogel (1969), *J. Am. Chem. Soc.*, **91**, 563.

Morawetz H., and E. W. Westhead, Jr. (1955). *J. Polym. Sci.*, **16**, 273.

Morawetz, H., and P. E. Zimmering (1954), *J. Phys. Chem.*, **58**, 753.

Morawetz, H., A. M. Kotliar, and H. Mark (1954), *J. Phys. Chem.*, **58**, 619.

Morawetz, H., J. R. Cho, and P. J. Gans (1973), *Macromolecules*, **6**, 624.

Morton, M., T. E. Helminiak, S. D. Gadkary, and F. Bueche (1962), *J. Polym. Sci.*, **57**, 482.

Moscowitz, A. (1962), *Adv. Chem. Phys.*, **4**, 67.

Muirhead, H., and J. Greer (1970), *Nature*, **228**, 516.

Müller-Hill, B. (1971), *Angew. Chem., Int. Ed. Engl.*, **10**, 160.

Murano, M., and R. Yamadera (1968), *J. Polym. Sci.*, **A-1**, **6**, 843.

Myers, C. S. (1954), *J. Polym. Sci.*, **13**, 549.

Myrbäck, K. (1926), *Z. Physiol. Chem.*, **159**, 1.

Nagai, E., and N. Sagane (1955), *Kobunshi Kagaku*, **12**, 195.

Nagasawa, M., and Y. Eguchi (1967), *J. Phys. Chem.*, **71**, 880.

Nagasawa, M., and H. Fujita (1964), *J. Am. Chem. Soc.*, **86**, 3005.

Nagasawa, M., and A. Holtzer (1964a), *J. Am. Chem. Soc.*, **86**, 531.

Nagasawa, M., and A. Holtzer (1964b), *J. Am. Chem. Soc.*, **86**, 538.

Nagasawa, M., and A. Takahashi (1972), in *Light Scattering from Polymer Solutions*, M. G. Huglin, ed., Academic Press, New York, Chap. 16.

Nagasawa, M., A. Soda, and I. Kagawa (1958), *J. Polym. Sci.*, **31**, 439.

Nagasawa, M., M. Izumi and I. Kagawa (1959a), *J. Polym. Sci.*, **37**, 375.

Nagasawa, M., A. Takahashi, M. Izumi and I. Kagawa (1959b), *J. Polym. Sci.*, **38**, 213.

Nagasawa, M., T. Murase, and K. Kondo (1965), *J. Phys. Chem.*, **69**, 4005.

Nagasawa, M., I. Noda, T. Takahashi, and N. Shimamoto (1972), *J. Phys. Chem.*, **76**, 2286.

Nagayama, K., and A. Wada (1973), *Biopolymers*, **12**, 2443.

Nagel, R. L., J. B. Wittenberg, and H. M. Ranney (1965), *Biochim. Biophys. Acta*, **100**, 286.

Nakajima, A., and A. Saijyo (1968), *J. Polym. Sci.*, **A-2**, **6**, 735.

Nakanishi, M., M. Tsuboi, A. Ikegami, and M. Kanehisa (1972), *J. Mol. Biol.*, **64**, 363.

Nakazawa, A., and J. J. Hermans (1971), *J. Polym. Sci.*, **A-2**, **9**, 1871.

Nandi, U.S., J. C. Wang, and N. Davidson (1965), *Biochemistry*, **4**, 1687.

Natta, G. (1957), *Experientia Suppl.*, **7**, 21.

Natta, G. (1960), *Makromol. Chem.*, **35**, 94.

Natta, G., and P. Corradini (1955), *Makromol. Chem.*, **16**, 77.

Natta, G., and P. Corradini (1956), *J. Polym. Sci.*, **20**, 251.

Natta, G., and F. Danusso (1959), *J. Polym. Sci.*, **34**, 3.

Natta, G., P. Pino, P. Corradini, F. Danusso, E. Mantica, G. Mazzanti, and G. Moraglio (1955), *J. Am. Chem. Soc.*, **77**, 1708.

Natta, G., P. Corradini, and I. W. Bassi (1959), *Gazz. Chim. Ital.*, **89**, 784.

Natta, G., P. Corradini, and P. Ganis (1962), *J. Polym. Sci.*, **58**, 1191.

Natta, G., M. Peraldo, and G. Allegra (1964), *Makromol. Chem.*, **75**, 215.

Néel, J., and B. Sebille (1961), *J. Chim. Phys.*, **58**, 738.

Neely, W. B. (1963), *J. Polym. Sci.,* **A1,** 311.

Nelson, C. A., and J. P. Hummel (1962), *J. Biol. Chem.,* **237,** 1567.

Nemethy, G., and H. A. Scheraga (1962), *J. Chem. Phys.,* **36,** 3401.

Nemethy, G., S. J. Leach, and H. A. Scheraga (1966), *J. Phys. Chem.,* **70,** 998.

Neugebauer, T. (1942), *Ann. Phys.* [5], **42,** 509.

Neurath, H., J. P. Greenstein, F. W. Putnam, and J. O. Erickson (1944), *Chem. Rev,* **34,** 157.

Newitt, E. J., and V. Kokle (1966), *J. Polym. Sci.,* **A-2, 4,** 705.

Newman, M. S. (1956), *Steric Effects in Organic Chemistry,* Wiley, New York, pp. 4–9.

Newman, S., W. R. Krigbaum, and D. K. Carpenter (1956), *J. Phys. Chem.,* **60,** 648.

Nichol, L. W., J. L. Bethune, G. Kegeles, and E. L. Hess (1964), in *The Proteins,* Vol. II, H. Neurath, ed., 2nd ed., Academic Press, New York, p. 305.

Nishihara, K., and N. Sakota (1974), *J. Polym. Sci., Polym. Chem. Ed.,* **12,** 57.

Nishijima, Y., and G. Oster (1956), *J. Polym. Sci.,* **19,** 337.

Nishijima, Y., A. Teramoto, M. Yamamoto, and S. Hiratsuka (1967), *J. Polym. Sci.,* **A-2, 5,** 23.

Nishijima, Y., K. Mitani, S. Katayama, and M. Yamamoto (1970), *Rep. Prog. Polym. Phys. Jap.:* (a) **13,** 421; (b) **13,** 425.

Nishijima, Y., Y. Sasaki, M. Tsujisaki, and M. Yamamoto (1972), *Rep. Prog. Polym. Phys. Jap.,* **15,** 453.

Noble, R. W., M. Reichlin, and R. D. Schreiber (1972), *Biochemistry,* **11,** 3326.

Noda, I., and M. Nagasawa (1970), *Polym. J.,* **1,** 304.

Noda, I., M. Nagasawa, and M. Ota (1964), *J. Am. Chem. Soc.,* **86,** 5075.

Noda, I., Y. Yamada, and M. Nagasawa (1968), *J. Phys. Chem.,* **72,** 2890.

Noda, I., T. Tsuge, and M. Nagasawa (1970), *J. Phys. Chem.,* **74,** 710.

Nomura, M. (1973), *Science,* **179,** 864.

Nord, F. F., M. Bier, and N. Timasheff (1951), *J. Am. Chem. Soc.,* **73,** 289.

North, A. C. T., and D. C. Phillips (1969), *Prog. Biophys.,* **19,** 1.

North, A. M., and G. A. Reed (1963), *J. Polym. Sci.,* **A1,** 1311.

Nozaki, Y., and C. Tanford (1963), *J. Biol. Chem.,* **238,** 4074.

Nozakura, S., S. Ishihara, Y. Inaba, K. Matsumura, and S. Murahashi (1973), *J. Polym. Sci., Polym. Chem. Ed.,* **11,** 1053.

Nozawa, T., and M. Hatano (1971), *Makromol. Chem.,* **141,** 31.

Nozawa, T., Y. Akimoto, and M. Hatano (1972), *Makromol. Chem.,* **158,** 21.

O'Connor, C. J., E. J. Fendler, and J. H. Fendler (1973), *J. Org. Chem.,* **38,** 3371.

Ogata, R. T., and H. M. McConnell (1972), *Proc. Natl. Acad. Sci. U.S.,* **69,** 335.

Ohama, M., and T. Ozawa (1966), *J. Polym. Sci.,* **A-2, 4,** 817.

Ohta, Y., Y. Ogura, and A. Wada (1966), *J. Biol. Chem.,* **241,** 5915.

Oi, N., and J. F. Coetzee (1969), *J. Am. Chem. Soc.,* **91,** 2478.

Oka, S. (1942), *Proc. Phys-Math. Soc. Jap.,* **24,** 657.

Okada, R., Y. Toyoshima, and H. Fujita (1963), *Makromol. Chem.,* **59,** 137.

Okubo, T., and N. Ise (1972), *Proc. Roy. Soc. (London),* **A327,** 413.

Okubo, T., and N. Ise (1973), *J. Am. Chem. Soc.,* **95,** 2293.

Okuyama, K., N. Tanaka, T. Ashida, M. Kakudo, S. Sakakibara, and Y. Kishida (1972), *J. Mol. Biol.,* **72,** 571.

Olander, J. (1971), *Biochemistry,* **10,** 601.

Olins, D. E., A. L. Olins, and P. H. von Hippel (1967), *J. Mol. Biol.,* **24,** 151.

Olins, D. E., A. L. Olins, and P. H. von Hippel (1968), *J. Mol. Biol.,* **33,** 265.

Oncley, J. L., D. Gitlin, E. Ellenbogen, and F. R. N. Gurd (1952), *J. Phys. Chem.* **56,** 85.

Ong, E., E. Shaw, and G. Schoellmann (1965), *J. Biol. Chem.,* **240,** 694.

Ono, S., S. Tsuchihashi, and T. Kuge (1953), *J. Am. Chem. Soc.,* **75,** 3601.

Onsager, L. (1949), *Ann. N. Y. Acad. Sci.,* **51,** 627.

Ooi, T., R. A. Scott, G. Vanderkooi, and H. A. Scheraga (1967), *J. Chem. Phys.,* **46,** 4410.

Oosawa, F. (1971), *Polyelectrolytes,* Marcel Dekker, New York.

Oosawa, F., and M. Kasai (1962), *J. Mol. Biol.,* **4,** 10.

Oosawa, F., S. Asakura, K. Hotta, N. Imai, and T. Ooi (1959), *J. Polym. Sci.,* **37,** 323.

Oparin, A. T. (1957), *The Origin of Life on Earth,* Oliver and Boyd, London, Chap. 7.

Oriel, P. J., and E. R. Blout (1966), *J. Am. Chem. Soc.,* **88,** 2041.

Orofino, T. A., and P. J. Flory (1957), *J. Chem. Phys.,* **26,** 1067.

Orwoll, R. A., and P. J. Flory (1967), *J. Am. Chem. Soc.,* **89,** 6814, 6822.

Osaki, K., and J. L. Schrag (1971), *Polym. J.,* **2,** 54.

Osaki, K., Y. Mitsuda, R. M. Johnson, J. L. Schrag, and J. D. Ferry (1972), *Macromolecules,* **5,** 17.

Oster, G. (1950), *J. Gen. Physiol.,* **33,** 445.

Oster, G. (1957), *Trans. Faraday Soc.,* **47,** 660.

Oster, G., and J. S. Bellin (1957), *J. Am. Chem. Soc.,* **79,** 294.

Oster, G., and Y. Nishijima (1964), *Adv. Polym. Sci.,* **3,** 313.

Oster, G., and G. K. Oster (1962), in *Luminescence of Organic and Inorganic Materials,* H. Kallman and G. Spruch, eds., Wiley, New York, pp. 186–195.

Ostroy, S. E., N. Lotan, R. T. Ingwall, and H. A. Scheraga (1970), *Biopolymers,* **9,** 749.

Overberger, C. G., and R. C. Glowaky (1973), *J. Am. Chem. Soc.,* **95,** 6014.

Overberger, C. G., and H. Jabloner (1963), *J. Am. Chem. Soc.,* **85,** 3431.

Overberger, C. G., and J. A. Moore (1970), *Adv. Polym. Sci.,* **7,** 113.

Overberger, C. G., and M. Morimoto (1971), *J. Am. Chem. Soc.,* **93,** 3222.

Overberger, C. G., and Y. Okamoto (1971), *Macromolecules,* **5,** 363.

Overberger, C. G., and L. C. Palmer (1956), *J. Am. Chem. Soc.,* **78,** 666.

Overberger, C. G., T. St. Pierre, N. Vorchheimer, J. Lee, and S. Yaroslavsky (1965), *J. Am. Chem. Soc.,* **87,** 296.

Overberger, C. G., T. St. Pierre, C. Yaroslavsky, and S. Yaroslavsky (1966), *J. Am. Chem. Soc.,* **88,** 1184.

Overberger, C. G., J. C. Salamone, and S. Yaroslavsky (1967), *J. Am. Chem. Soc.,* **89,** 6231.

Overberger, C. G., M. Morimoto, I. Cho, and J. C. Salamone (1971), *J. Am. Chem. Soc.,* **93,** 3228.

Overberger, C. G., R. C. Glowaky, and P. H. Vandewyer (1973), *J. Am. Chem. Soc.,* **95,** 6008.

Owston, P. G. (1958), *Adv. Phys.,* **7,** 171.

Padlan, E. A., D. M. Segal, T. F. Spender, and D. R. Davies (1973), *Nature—New Biol.*, **245**, 165.

Paetkau, V. H. (1967), *Biochemistry*, **6**, 2767.

Pal, M. K., and M. Schubert (1963), *J. Phys. Chem.*, **67**, 1821.

Pals, D. T. F., and J. J. Hermans (1952), *Rec. Trav. Chim.*, **71**, 469.

Panzik, H. L., and J. E. Mulvaney (1972), *J. Polym. Sci., Polym. Chem. Ed.*, **10**, 3469.

Paolillo, L., P. Temussi, E. Trivellone, E. M. Bradbury, and C. Crane-Robinson (1973), *Macromolecules*, **6**, 831.

Pardee, A. B. (1968), *Science*, **162**, 632.

Parry, D. A. D., and A. Elliott (1965), *Nature*, **206**, 616.

Patel, R. C., G. Atkinson, and E. Baumgartner (1973), *Bioinorg. Chem.*, **3**, 1.

Patterson, D. (1967), *Rubber Chem. Technol.*, **40**, 1.

Patterson, D. (1969), *Macromolecules*, **2**, 672.

Patterson, R. W., and F. H. Abernathy (1970), *J. Fluid Mech.*, **43**, 689.

Pauling, L., and R. B. Corey (1951a), *Proc. Natl. Acad. Sci. U.S.*, **37**, 235.

Pauling, L., and R. B. Corey (1951b), *Proc. Natl. Acad. Sci. U.S.*, **37**, 241.

Pauling, L., and R. B. Corey (1951c), *Proc. Natl. Acad. Sci. U.S.*, **37**, 729.

Pauling, L. and R. B. Corey (1956), *Arch, Biochem. Biophys.*, **65**, 164.

Pauling, L., H. A. Itano, S. J. Singer, and L. C. Wells (1949), *Science*, **110**, 543.

Pchelin, V. A., V. N. Izmailova, and V. N. Merzlov (1963), *Dokl. Akad. Nauk SSSR*, **150**, 1307.

Pecht, I., A. Levitzki, and M. Anbar (1967), *J. Am. Chem. Soc.*, **89**, 1587.

Pederson, K. O. (1958), *J. Phys. Chem.*, **62**, 1282.

Perrin, F. (1908), *Compt. Rend.*, **146**, 967.

Perrin, F. (1929), *Ann. Phys. (Paris)*, [10], **12**, 169.

Perrin, F. (1934), *J. Phys. Radium*, [7], **5**, 497.

Perrin, F. (1936), *J. Phys. Radium* [7], **7**, 1.

Perry, S. V. (1967), *Prog. Biophys. Mol. Biol.*, **17**, 325.

Perutz, M. E. (1970), *Nature*, **228**, 726.

Perutz, M. F., M. G. Rossmann, A. F. Cullis, H. Muirhead, G. Will, and A. C. T. North (1960), *Nature*, **185**, 416.

Perutz, M. F., J. C. Kendrew, and H. C. Watson (1965), *J. Mol. Biol.*, **13**, 669.

Perutz, M. F., H. Muirhead, J. M. Cox, and L. C. G. Goaman (1968), *Nature*, **219**, 131.

Peterlin, A. (1938), *Z. Phys.*, **111**, 232.

Peterlin, A. (1953), *J. Polym. Sci.*, **10**, 425.

Peterlin, A. (1960), *J. Polym. Sci.*, **47**, 403.

Peterlin, A. (1961), *Ann. N. Y. Acad. Sci.*, **89**, 578.

Peterlin, A. (1963), *J. Chem. Phys.*, **39**, 224.

Peterlin, A. (1967), *J. Polym. Sci.*, **A-2, 5**, 179.

Peterlin, A. (1972), *J. Polym. Sci.*, **B10**, 101.

Peterlin, A. (1973), *J. Polym. Sci.*, **C43**, 187.

Peterlin, A., and H. A. Stuart (1939), *Z. Phys.*, **112**, 1.

Petersen, R. J., R. D. Corneliussen, and L. T. Rozelle (1969), *Polym. Preprs, Am. Chem. Soc., Div. Polym. Chem.*, **10**, (1), 385.

Peterson, J. M., and M. Fixman (1963), *J. Chem. Phys.*, **39**, 2516.

Peyser, P., D. J. Tutas, and R. R. Stromberg (1967), *J. Polym. Sci.*, **A-1**, **5**, 651.

Pfannemüller, B., and W. Burchard (1969), *Makromol. Chem.*, **121**, 1.

Pfannemüller, B., H. Mayerhöfer, and R. C. Schulz (1969), *Makromol. Chem.*, **121**, 147.

Pfannemüller, B., H. Mayerhöfer, and R. C. Schulz (1971), *Biopolymers*, **10**, 243.

Phibbs, M. K. (1955), *J. Phys. Chem.*, **59**, 346.

Piko, L., D. G. Blair, A. Tyler, and J. Vinograd (1968), *Proc. Natl. Acad. Sci. U.S.*, **59**, 838.

Pino, P. (1965), *Adv. Polym. Sci.*, **4**, 393.

Pino, P., F. Ciardelli, G. Montagnoli, and O. Pieroni (1967), *J. Polym. Sci.*, **B5**, 307.

Pino, P., C. Carlini, E. Chiellini, F. Ciardelli, and P. Salvadori (1968), *J. Am. Chem. Soc.*, **90**, 5025.

Pino, P., F. Ciardelli, and M. Zandomenghi (1970), *Ann. Rev. Phys. Chem.*, **21**, 561.

Pitzer, K. S. (1940), *Chem. Rev.*, **27**, 39; *J. Chem. Phys.*, **8**, 711.

Pivcová, H., M. Kolinský, D. Lím, and B. Schneider (1969), *J. Polym. Sci.*, **C22**, 1093.

Platzer, K. E. B., V. S. Ananthanarayanan, R. H. Andreatta, and H. A. Scheraga (1972), *Macromolecules*, **5**, 177.

Pople, J. A. (1951), *Proc. Roy. Soc. (London)*, **A205**, 163.

Porod, G. (1953), *J. Polym. Sci.*, **10**, 157.

Porter, R. R. (1972), *Science*, **180**, 713.

Porter, R. S., and J. F. Johnson (1966), *Chem. Rev.*, **66**, 1.

Pouchlý, J., and J. Biroš (1969), *J. Polym. Sci.*, **B7**, 463.

Price, C. C., and M. Osgan (1956), *J. Am. Chem. Soc.*, **78**, 4787.

Prigogine, I. (1957), *The Molecular Theory of Solution*, Interscience, New York.

Prigogine, I., A. Bellemans, and C. Naar-Colin (1957), *J. Chem. Phys.*, **26**, 751.

Privalov, P. L., and E. I. Tiktopulo (1970), *Biopolymers*, **9**, 127.

Ptitsyn, O. B. (1959), *Vysokomol. Soedin.*, **1**, 715 (English transl.: *Polym. Sci. USSR*, **1**, 259 (1961).

Ptitsyn, O. B., and Yu. E. Eizner (1959), *Zh. Tekhn. Fiz.*, **29**, 1117 [English transl.: *Soviet Phys. Tech. Phys.*, **4**, 1020 (1960)].

Pullman, B., and A. Pullman (1959), *Biochim. Biophys. Acta*, **36**, 343.

Putter, I., J. L. Markley, and O. Jardetzky (1970), *Proc. Natl. Acad. Sci. U.S.*, **65**, 395.

Quadrifoglio, F., V. Crescenzi, and F. Delben (1973), *Macromolecules*, **6**, 301.

Quiocho, F. A., and F. M. Richards (1964), *Proc. Natl. Acad. Sci. U.S.*, **52**, 833.

Quiocho, F. A., and F. M. Richards (1966), *Biochemistry*, **5**, 4062.

Ramachandran, G. N. (1967), *Treatise on Collagen*, G. N. Ramachandran, ed., Academic Press, New York, Chap. 3.

Ramachandran, G. N., and V. Sasisekharan (1961), *Nature*, **190**, 1004.

Ramachandran, G. N., and V. Sasisekharan (1968), *Adv. Protein Chem.*, **23**, 283.

Ramachandran, G. N., C. Ramakrishnan, and V. Sasisekharan (1963), *J. Mol. Biol.*, **7**, 95.

Ramachandran, G. N., C. M. Venketachalam, and S. Krimm (1966), *Biophys. J.*, **6,** 849.

Ramey, K. C., and J. Massick (1966), *J. Polym. Sci.*, **A-2, 4,** 155.

Rånby, B. G., and R. H. Marchessault (1959), *J. Polym. Sci.*, **36,** 561.

Randall, J. C. (1974), *J. Polym. Sci., Polym. Phys., Ed.*, **12,** 703.

Rao, V. S. R., N. Yathindra, and P. R. Sundarajan (1969), *Biopolymers*, **8,** 323.

Rayleigh, Lord (J. W. Strutt) (1871), *Phil. Mag.*, [4] **41,** 107, 224, 447.

Rayleigh, Lord (J. W. Strutt) (1881), *Phil. Mag.*, [5] **12,** 81.

Rayleigh, Lord (1918), *Phil. Mag.*, **35,** 373.

Reed, L. J., and D. J. Cox (1970), in *The Enzymes*, 3rd ed., Vol. I, P. D. Boyer, Ed., Chap. 4.

Reed, T. F., and J. E. Frederick (1971), *Macromolecules*, **4,** 72.

Reeves, R. E. (1949), *J. Am. Chem. Soc.*, **71,** 212.

Rein, H. (1938), U. S. Patent 2,140, 921.

Reiss, C., and H. Benoit (1968), *J. Polym. Sci.*, **C16,** 3079.

Rhodes, W. (1961), *J. Am. Chem. Soc.*, **83,** 3609.

Rice, S. A. (1955), *J. Polym. Sci.*, **16,** 94.

Rice, S. A., and F. E. Harris (1954), *J. Phys. Chem.*, **58,** 733.

Rice, S. A., and A. Wada (1958), *J. Chem. Phys.*, **29,** 233.

Rich, A., and F. H. C. Crick (1961), *J. Mol. Biol.*, **3,** 483.

Rich, A., D. R. Davies, E. H. C. Crick, and J. D. Watson (1961), *J. Mol. Biol.*, **3,** 71.

Richards, F. M., and A. D. Logue (1962), *J. Biol. Chem.*, **237,** 3693.

Rinaudo, M., and M. Milas (1969), *J. Chim. Phys.*, **66,** 1489.

Rinaudo, M. and M. Milas (1973), *Macromolecules*, **6,** 879.

Rinaudo, M., and C. Pierre (1969), *Compt. Rend.*, **C269,** 1280.

Riseman, J., and J. G. Kirkwood (1949), *J. Chem. Phys.*, **17,** 442.

Riseman, J., and J. G. Kirkwood (1950), *J. Chem. Phys.*, **18,** 512.

Robertson, A. B., and H. J. Harwood (1971), *Polym. Prepr., Am. Chem. Soc., Div. Polym. Chem.*, **12,** 620.

Robinson, C. (1956), *Trans. Faraday Soc.*, **52,** 571.

Robinson, C., J. C. Ward, and R. B. Beevers (1958), *Discuss. Faraday Soc.*, **25,** 29.

Robinson, D. R., and M. E. Grant (1966), *J. Biol. Chem.*, **241,** 4030.

Robinson, D. R., and W. P. Jencks, (1963), *J. Biol. Chem.*, **238,** PC1558.

Robinson, D. R., and W. P. Jencks, (1965), *J. Am. Chem. Soc.*, **87,** 2470.

Rolfe, R., and M. Meselson (1959), *Proc. Natl. Acad. Sci. U.S.*, **45,** 1039.

Rolfson, F. B., and H. Coll (1964), *Anal. Chem.*, **36,** 888.

Rosen, B., P. Kamath, and F. Eirich (1951), *Discuss. Faraday Soc.*, **11,** 135.

Rosenberg, A. (1966), *J. Biol. Chem.*, **241,** 5119, 5126.

Rosenberg, A., and R. Lumry (1964), *Biochemistry*, **3,** 1055.

Rosenbloom, J. and V. Schumaker (1967), *Biochemistry*, **6,** 276.

Rosenheck, K., and P. Doty (1961), *Proc. Natl. Acad. Sci. U.S.*, **47,** 1775.

Rosenheck, K., and G. Weber (1964), *Biopolym. Symp.*, **1,** 333.

Rouse, P. E., Jr. (1953), *J. Chem. Phys.*, **21,** 1272.

Rowland, F. W., and F. R. Eirich (1966), *J. Polym. Sci.*, **A-1, 4,** 2401.

Rudd, J. F. (1960), *J. Polym. Sci.*, **44,** 459.

Rudin, A., and H. L. W. Hoegy (1972), *J. Polym. Sci.*, **A-1, 10,** 217.

Rühlmann, A., D. Kukla, P. Schwager, K. Bartels, and R. Huber (1973), *J. Mol. Biol.*, **77,** 417.

Rundle, R. E., and R. R. Baldwin (1943), *J. Am. Chem. Soc.*, **65,** 554.

Rundle, R. E., and D. French (1943), *J. Am. Chem. Soc.*, **65,** 1707.

Rupley, J. A. (1968), *J. Mol. Biol.*, **35,** 455.

Rupley, J. A. (1969), *Structure and Stability of Biological Macromolecules,* S. M. Timasheff and G. D. Fasman, eds., Marcel Dekker, New York, pp. 291–352.

Sachs, S. B., A. Raziel, H. Eisenberg, and A. Katchalsky (1969), *Trans. Faraday Soc.*, **65,** 77.

Sadron, C. (1963), *Angew. Chem.*, **75,** 472.

Saeki, S., N. Kuwahara, S. Konno, and M. Kaneko (1973), *Macromolecules,* **6,** 246.

Sage, H. J., and S. J. Singer (1958), *Biochim. Biophys. Acta,* **29,** 663.

Saito, S. (1953), *Kolloid-Z.*, **133,** 12.

Saito, S. (1954), *Kolloid-Z.*, **137,** 98.

Saito, S. (1957), *Kolloid-Z.*, **154,** 19.

Sakai, T. (1968), *J. Polym. Sci.*, **A-2, 6,** 1535.

Sakato, K. and M. Kurata (1970), *Polym. J.*, **1,** 260.

Sakurada, I., Y. Sakaguchi, T. Ono, and T. Ueda (1966), *Makromol. Chem.*, **91,** 243.

Saludjian, P., and V. Luzzati (1966), *J. Mol. Biol.*, **16,** 681.

Sasakawa, S., and H. Walter (1972), *Biochemistry,* **11,** 2760.

Sarkar, P. K., and P. Doty (1966), *Proc. Natl. Acad. Sci. U.S.*, **55,** 981.

Sato, T., Y. Kyogoku, S. Higuchi, S. Mitsui, Y. Iitaka, and M. Tsuboi (1966), *J. Mol. Biol.*, **16,** 180.

Saunders, P. R. (1962), *J. Polym. Sci.*, **57,** 131.

Saunders, P. R. (1964), *J. Polym. Sci.*, **A2,** 3755.

Scarpa, J. S., D. D. Mueller, and I. M. Klotz (1967), *J. Am. Chem. Soc.*, **89,** 6024.

Scatchard, G. (1949), *Ann. N. Y. Acad. Sci.*, **51,** 660.

Scatchard, G., S. E. Wood, and J. M. Mochel (1939), *J. Phys. Chem.*, **43,** 119.

Scatchard, G., I. H. Scheinberg, and S. H. Armstrong (1949), *J. Am. Chem. Soc.*, **72,** 535, 540.

Schachman, H. K. (1959), *Ultracentrifugation in Biochemistry,* Academic Press, New York.

Schachman, H. K. (1963), *Biochemistry,* **2,** 887.

Schaefgen, J. R., and P. J. Flory (1948), *J. Am. Chem. Soc.*, **70,** 2709.

Schaefgen, J. R., and C. F. Trivisonno (1952), *J. Am. Chem. Soc.*, **74,** 2715.

Schatzki, T. (1965), *Polym. Prepr., Am. Chem. Soc., Div. Polym. Chem.*, **6,** 646.

Schechter, A. N., L. Moravek, and C. B. Anfinsen (1969), *J. Biol. Chem.*, **244,** 4981.

Schechter, I., and A. Berger (1968), *Biochem. Biophys. Res. Commun.*, **32,** 898.

Scheffler, I. E., E. L. Elson, and R. L. Baldwin (1968), *J. Mol. Biol.*, **36,** 291.

Schellman, J. A. (1955), *Compt. Rend. Trav. Lab. Carlsberg, Sér. Chim.*, **29,** No. 15.

Scheraga, H. A. (1955), *J. Chem. Phys.*, **23,** 1526.

Scheraga, H. A., and L. Mandelkern (1953), *J. Am. Chem. Soc.*, **75**, 179.
Scheraga, H. A., J. T. Edsall, and J. O. Gaddy, Jr. (1951), *J. Chem. Phys.*, **19**, 1101.
Schick, M. J. (1957), *J. Polym. Sci.*, **25**, 465.
Schildkraut, C. L., and S. Lifson (1965), *Biopolymers*, **3**, 195.
Schildkraut, C. L., J. Marmur, and P. Doty (1961), *J. Mol. Biol.*, **3**, 595.
Schildkraut, C. L., J. Marmur, and P. Doty (1962a), *J. Mol. Biol.*, **4**, 430.
Schildkraut, C. L., K. L. Wierzchowski, J. Marmur. D. M. Green, and P. Doty (1962b), *Virology*, **18**, 43.
Schimmel, P., and P. J. Flory (1968), *J. Mol. Biol.*, **34**, 105.
Schindewolf, U. (1954a), *Z. Phys. Chem. (Frankfurt)*, **1**, 129; *Z. Elektrochem.*, **58**, 697.
Schindewolf, U. (1954b), *Z. Phys. Chem. (Frankfurt)*, **1**, 134.
Schleich, T., and P. H. von Hippel (1969), *Biopolymers*, **7**, 861.
Schlessinger, J., and I. Z. Steinberg (1972), *Proc. Natl. Acad. Sci. U.S.*, **69**, 769.
Schlichting, H. (1960), *Boundary Layer Problems*, 4th ed., McGraw-Hill, New York, p. 16.
Schmid, C. W., and J. Hearst (1969), *J. Mol. Biol.*, **44**, 143.
Schmid, C. W., F. P. Rinehart, and J. E. Hearst (1971), *Biopolymers*, **10**, 883.
Schmitt, G. J., and C. Schuerch (1960), *J. Polym. Sci.*, **45**, 313.
Schmolka, I. R., and L. R. Bacon (1967), *J. Am. Oil Chem. Soc.*, **44**, 559.
Schmolke, R., W. Kimmer, P. Kuzay, and W. Hufenreuter (1971), *Plaste Kautsch.*, **18**, 95.
Scholtan, W. (1953), *Makromol. Chem.*, **11**, 131.
Scholtan, W. and H. Marzolph (1962), *Makromol. Chem.*, **57**, 52.
Schuerch, C. (1952), *J. Am. Chem. Soc.*, **64**, 5061.
Schultz, A. W., and U. P. Strauss (1972), *J. Phys. Chem.*, **76**, 1767.
Schulz, G. V. (1939), *Z. Phys. Chem. (Leipzig)*, **B43**, 25.
Schulz, G. V., and H. Baumann (1963), *Makromol. Chem.*, **60**, 120.
Schulz, G. V., and R. Kirste (1961), *Z. Phys. Chem. (Frankfurt)*, **30**, 171.
Schulz, G. V., and H. Marzolph (1954), *Z. Elektrochem.*, **58**, 211.
Schulz, G. V., K. V. Gunner, and H. Gerrens (1955), *Z. Phys. Chem. (Frankfurt)*, **4**, 192.
Schulz, R. C. (1972), *Pure Appl. Chem.*, **30**, 239.
Schulz, R. C., and R. H. Jung (1966), *Makromol. Chem.*, **96**, 295.
Schulz, R. C., and R. H. Jung (1968), *Makromol. Chem.*, **116**, 190.
Schulz, R. C., and E. Kaiser (1965), *Adv. Polym. Sci.*, **4**, 236.
Schumaker, V., and P. Adams (1968), *Biochemistry*, **7**, 3422.
Schumaker, V. N., and H. K. Schachman (1957), *Biochim. Biophys. Acta*, **23**, 628.
Schwarz, G. (1959), *Z. Phys. Chem. (Frankfurt)*, **19**, 286.
Schwarz, G. (1965), *J. Mol. Biol.*, **11**, 64.
Schwarz, G., and P. Seelig (1968), *Biopolymers*, **6**, 1263.
Schwarzenbach, G. (1970), *Pure Appl. Chem.*, **24**, 307.
Scott, R. L., and M. Magat (1949), *J. Polym. Sci.*, **4**, 555.
Segal, D. M., and W. F. Harrington (1967), *Biochemistry*, **6**, 768.
Sela, M. (1969), *Science*, **166**, 1365.

Sela, M., and A. Berger (1953), *J. Am. Chem. Soc.*, **75**, 6350.
Shapiro, J. T., M. Leng, and G. Felsenfeld (1969), *Biochemistry*, **8**, 3219.
Sharon, N., and H. Lis (1972), *Science*, **177**, 949.
Shashoua, V. E., and R. G. Beaman (1958), *J. Polym. Sci.*, **33**, 101.
Shearer, H. M. M., and V. Vand (1956), *Acta Cryst.*, **9**, 379.
Shechter, E., and E. R. Blout (1964), *Proc. Natl. Acad. Sci. U.S.*, **51**, 695, 794.
Shen, L. L., and J. Hermans, Jr. (1972), *Biochemistry*, **11**, 1836, 1842.
Shih, T. Y., and J. Bonner (1970), *J. Mol. Biol.*, **48**, 469.
Shlyapnikov, S. V., B. Beloun, B. Keil, and F. Šorm (1968), *Coll. Czech. Chem. Commun.*, **33**, 2292.
Shmueli, U., W. Traub, and K. Rosenheck (1969), *J. Polym. Sci.*, **A-2**, **7**, 515.
Shotton, D. M., and H. C. Watson (1970), *Nature*, **225**, 811.
Shultz, A. R., and P. J. Flory (1952), *J. Am. Chem. Soc.*, **74**, 4760.
Shultz, A. R., and P. J. Flory (1953), *J. Am. Chem. Soc.*, **75**, 3888.
Sigler, P. B., D. M. Blow, B. W. Matthews, and R. Henderson (1968), *J. Mol. Biol.*, **35**, 143.
Silberberg, A. (1968), *J. Chem. Phys.*, **48**, 2835.
Silberberg, A. (1971), *Pure Appl. Chem.*, **26**, 583.
Silberberg, A., J. Eliassaf, and A. Katchalsky (1957), *J. Polym. Sci.*, **23**, 259.
Simha, R. (1940), *J. Phys. Chem.*, **44**, 25.
Simmons, N. S., C. Cohen, A. G. Szent-Gyorgyi, D. B. Wetlaufer, and E. R. Blout (1961), *J. Am. Chem. Soc.*, **83**, 4706.
Simons, E. R., E. G. Schneider, and E. R. Blout (1969), *J. Biol. Chem.*, **244**, 4023.
Singer, S. J. (1962), *Adv. Protein Chem.*, **17**, 1.
Singer, S. J. (1967), *Adv. Protein Chem.*, **22**, 1.
Singer, S. J., and D. H. Campbell (1955), *J. Am. Chem. Soc.*, **77**, 3499, 4851.
Singer, S. J., and R. F. Doolittle (1966), *Science*, **153**, 13.
Siow, K. S., G. Delmas, and D. Patterson (1972), *Macromolecules*, **5**, 29.
Sisido, M. (1972), *Polym. J.*, **3**, 84.
Skeist, I. (1946), *J. Am. Chem. Soc.*, **68**, 1781.
Skoulios, A. E., G. Tsouledze, and E. Franta (1964), *J. Polym. Sci.*, **C4**, 507.
Slagowski, E., L. J. Fetters, and D. M. McIntyre (1971), *Polym. Prepr., Am. Chem. Soc., Div. Polym. Chem.*, **12**, No. 2, 753.
Slough, W. (1959), *Trans. Faraday Soc.*, **55**, 1030.
Slough, W. (1962), *Trans. Faraday Soc.*, **58**, 2360.
Small, T. A. (1953), *J. Appl. Chem.*, **3**, 71.
Smets, G., and B. Petit (1959), *Makromol. Chem.*, **33**, 41.
Smets, G., and W. Van Humbeeck (1963), *J. Polym. Sci.*, **A1**, 1227.
Smidsrød, O., and A. Haug (1971), *Biopolymers*, **10**, 1213.
Smidsrød, O., A. Haug, and B. Larsen (1966), *Acta Chem. Scand.*, **20**, 1026.
Smith, C. R. (1919), *J. Am. Chem. Soc.*, **41**, 135.
Smith, K. L., A. E. Winslow, and D. E. Peterson (1959), *Ind. Eng. Chem.*, **51**, 1361.

Smoluchowski, M. v. (1921), in *Handbuch der Elektrizitaet und des Magnetismus,* Vol. II. L. Graetz, ed., J. A. Barth, Leipzig. p. 366.

Snyder, R. G., and J. H. Schachtschneider (1965), *Spectrochim. Acta,* **21,** 169.

Sodek, J., R. S. Hodges, L. B. Smillie, and L. Jurasek (1972), *Proc. Natl. Acad. Sci. U.S.,* **69,** 3800.

Solc, K., and W. H. Stockmayer (1971), *J. Chem. Phys.,* **54,** 2756.

Spanagel, E. W., and W. H. Carothers (1935), *J. Am. Chem. Soc.,* **57,** 929.

Spencer, M., W. Fuller, M. H. F. Wilkins, and G. L. Brown (1962), *Nature,* **194,** 1014.

Spurlin, H. M. (1955), in *Cellulose and Cellulose Derivatives,* Part 3, E. Ott, H. M. Spurlin, and M. W. Graffin, eds. (High Polymers, Vol. 5) Interscience, New York, pp. 1077–1082.

Stadtman, E. R. (1966), *Adv. Enzymol.,* **28,** 43.

Stamm, A. J. (1952), in *Wood Chemistry,* L. E. Wise and E. C. Jahn, eds., American Chemical Society Monograph No. 97, 2nd ed., Reinhold, New York, pp. 226–277.

Stanton, G. W., W. Creek, and T. B. Lefferdink (1953), U. S. Patent 2,648,647.

Starkweather, H. W. (1959), *J. Appl. Polym. Sci.,* **2,** 129.

Staudinger, H. (1932), *Die Hochmolekularen Organischen Verbindungen,* Springer, Berlin, p. 39.

Staudinger, H., K. Frey, and W. Starck (1927), *Ber.,* **60,** 1787.

Staveley, L. A. K., W. I. Tupman, and K. R. Hart (1955), *Trans. Faraday Soc.,* **51,** 323.

Staverman, A. J. (1952), *Rec. Trav. Chim.,* **71,** 623.

Stein, R. S., and R. E. Rundle (1948), *J. Chem. Phys.,* **16,** 195.

Steinberg, I. Z. (1971), *Ann. Rev. Biochem.,* **40,** 83.

Steinberg, I. Z., and H. K. Schachman (1966), *Biochemistry,* **5,** 3728.

Steinberg, I. Z., W. F. Harrington, A. Berger, M. Sela, and E. Katchalski (1960), *J. Am. Chem. Soc.,* **82,** 5263.

Steiner, R. F., and R. F. Beers, Jr. (1961), *Polynucleotides,* Elsevier, Amsterdam.

Steiner, R. F., and H. Edelhoch (1962), *Chem. Rev.,* **62,** 457.

Steiner, R. F., and H. Edelhoch (1963), *Biochim. Biophys. Acta,* **66,** 341.

Steiner, R. F., R. E. Lipoldt, H. Edelhoch, and V. Frattali (1964), *Biopolym. Symp.,* No. **1,** 355.

Steitz, T. A., R. Henderson, and D. M. Blow (1969), *J. Mol. Biol.,* **46,** 337.

Stern, A., W. A. Gibbons and L. C. Craig (1968), *Proc. Natl. Acad. Sci. U.S.,* **61,** 738.

Stevens, C. L., and G. Felsenfeld (1964), *Biopolymers,* **2,** 293.

Stockmayer, W. H. (1945), *J. Chem. Phys.,* **13,** 199.

Stockmayer, W. H. (1949), *J. Chem. Phys.,* **17,** 588.

Stockmayer, W. H. (1950), *J. Chem. Phys.,* **18,** 58.

Stockmayer, W. H., and E. F. Casassa (1952), *J. Chem. Phys.,* **20,** 1560.

Stockmayer, W. H., and M. Fixman (1963), *J. Polym. Sci.,* **C1,** 137.

Stockmayer, W. H., L. D. Moore, Jr., M. Fixman, and B. N. Epstein (1955), *J. Polym. Sci.,* **16,** 517.

Stokes, Sir G. (1880), *Mathematical and Physical Papers,* Cambridge University Press, New York.

Stoll, M., and A. Rouvé (1935), *Helv. Chim. Acta,* **18,** 1087.

Stone-Masui, J., and A. Watillon (1968), *J. Colloid Interface Sci.*, **28,** 187.

Strauss, U. P., and P. Ander (1958), *J. Am. Chem. Soc.*, **80,** 6494.

Strauss, U. P., and S. Bluestone (1959), *J. Am. Chem. Soc.*, **81,** 5292.

Strauss, U. P., and Y. P. Leung (1965), *J. Am. Chem. Soc.*, **87,** 1476.

Strauss, U. P., and P. D. Ross (1959), *J. Am. Chem. Soc.*, **81,** 5295.

Strauss, U. P., N. L. Gershfeld, and H. Spira (1954), *J. Am. Chem. Soc.*, **76,** 5909.

Strauss, U. P., N. L. Gershfeld, and E. H. Crook (1956), *J. Phys. Chem.*, **60,** 577.

Strauss, U. P., D. Woodside, and P. Wineman (1957), *J. Phys. Chem.*, **61,** 1353.

Strazielle, C., and H. Benoit (1961), *J. Chim. Phys.*, **58,** 675, 678.

Stromberg, R. R., D. J. Tutas, and E. Passaglia (1965), *J. Phys. Chem.*, **69,** 3955.

Štrop, P., F. Mikeš, and J. Kálal (1974) Private Communication.

Stryer, L. (1965), *J. Mol. Biol.*, **13,** 482.

Stryer, L. (1968), *Science,* **162,** 526.

Stryer, L., and R. P. Haugland (1967), *Proc. Natl. Acad. Sci. U.S.*, **58,** 719.

Suelter, C. H. (1970), *Science,* **168,** 789.

Sueoka, N., J. Marmur, and P. Doty (1959), *Nature,* **183,** 1429.

Svedberg, T. (1926), *Z. Phys. Chem. (Leipzig)*, **121,** 65.

Svedberg, T. (1930), *Kolloid-Z.*, **51,** 10.

Svedberg, T., and R. Fåhraeus (1926), *J. Am. Chem. Soc.*, **48,** 430.

Svedberg, T., and J. B. Nichols (1926), *J. Am. Chem. Soc.*, **48,** 3081.

Svensson, H. (1939), *Kolloid-Z.*, **87,** 181.

Svensson, H. (1940), *Kolloid-Z.*, **90,** 141.

Sykora, S. (1968), *Collect. Czech. Chem. Commun.*, **33,** 3514.

Szasz, G. J., and N. Sheppard (1949), *J. Chem. Phys.*, **17,** 83.

Szer, W., and D. Shugar (1966), *J. Mol. Biol.*, **17,** 174.

Szwarc, M. (1966), *Pure Appl. Chem.*, **12,** 127.

Szwarc, M. (1970), *Science,* **170,** 23.

Tabak, D., and H. Morawetz (1970), *Macromolecules,* **3,** 403.

Tadokoro, H., T. Yasumoto, S. Murahashi, and I. Nitta (1960), *J. Polym. Sci.*, **44,** 266.

Tadokoro, H., Y. Takahashi, Y. Chatani, and H. Kakida (1967), *Makromol. Chem.*, **109,** 96.

Tadokoro, H., Y. Chatani, H. Kusanagi, and M. Yokoyama (1970), *Macromolecules,* **3,** 441.

Taft, R., and L. E. Malm (1939), *J. Phys. Chem.*, **43,** 499.

Taha, I. A. and H. Morawetz (1971), *J. Am. Chem. Soc.*, **93,** 829; *J. Polym. Sci.*, **A-2, 9,** 1669.

Takahashi, A., T. Kato, and M. Nagasawa (1967), *J. Phys. Chem.*, **71,** 2001.

Takahashi, A., N. Kato, and M. Nagasawa (1970), *J. Phys. Chem.*, **74,** 944.

Takahashi, A., T. Nakamura, and I. Kagawa (1972), *Polym. J.*, **3,** 207.

Takano, T., R. Swanson, O. B. Kallai, and R. E. Dickerson (1971), *Cold Spring Harbor Symp. Quant. Biol.*, **36,** 397.

Takenaka, H. (1957), *J. Polym. Sci.*, **24,** 321.

Talen, J. L., and A. J. Staverman (1965), *Trans. Faraday Soc.*, **61,** 2794.

Tam, J. W. O., and I. M. Klotz (1971), *J. Am. Chem. Soc.*, **93,** 1313.

Tanford, C. (1962), *J. Am. Chem. Soc.*, **84,** 4240.

Tanford, C. (1970), *Adv. Protein Chem.*, **24,** 1.

Tanford, C., and J. G. Buzzell (1956), *J. Phys. Chem.*, **60,** 225.

Tanford, C., and J. D. Hauenstein (1956), *J. Am. Chem. Soc.*, **78,** 5287.

Tanford, C., J. D. Hauenstein, and D. G. Rands (1955), *J. Am. Chem. Soc.*, **77,** 6409.

Tanford, C., P. K. De, and V. G. Taggart (1960), *J. Am. Chem. Soc.*, **82,** 6028.

Tanford, C., R. H. Pain, and N. S. Otchin (1966), *J. Mol. Biol.*, **15,** 489.

Tanford, C., K, Kawahara, S. Lapanje, T. M. Hooker, Jr., M. H. Farlengo, A. Salahuddin, K. C. Aune, and T. Tagaki (1967), *J. Am. Chem. Soc.*, **89,** 5023.

Taniuchi, H., and C. B. Anfinsen (1969), *J. Biol. Chem.*, **244,** 3864.

Taniuchi, H., and C. B. Anfinsen (1971), *J. Biol. Chem.*, **246,** 2291.

Tanner, J. E., K. J. Liu, and J. E. Anderson (1971), *Macromolecules*, **4,** 586.

Taylor, G. I. (1932), *Proc. Roy. Soc. (London)*, **A138,** 41.

Taylor, W. J. (1947), *J. Chem. Phys.*, **15,** 412.

Taylor, W. J. (1948), *J. Chem. Phys.*, **16,** 257.

Teale, F. W. J. (1960), *Biochem. J.*, **76,** 381.

Teale, F. W. J., and G. Weber (1957), *Biochem. J.*, **65,** 476.

Temin, H. M., and D. Baltimore (1972), *Adv. Virus Res.*, **17,** 129.

Terayama, H. (1952), *J. Polym. Sci.*, **8,** 243.

Theorell, H., and B. Chance (1951), *Acta Chem. Scand.*, **5,** 1127.

Thoma, J. A., and D. E. Koshland (1960), *J. Am. Chem. Soc.*, **82,** 3329.

Tiffany, M. L., and S. Krimm (1969), *Biopolymers*, **8,** 347.

Timasheff, S. N., and G. D. Fasman (1971), *Subunits in Biological Systems*, Part A, Marcel Dekker, New York.

Timasheff, S. N., J. Witz, and V. Luzzati (1961), *Biophys. J.*, **1,** 525.

Tinoco, I., Jr. (1959), *J. Am. Chem. Soc.*, **81,** 1540.

Tinoco, I., Jr. (1960), *J. Am. Chem. Soc.*, **82,** 4785.

Tinoco, I., Jr., A. Halpern, and W. I. Simpson (1962), in *Polyaminoacids, Polypeptides and Proteins*, M. A. Stahmann, ed., University of Wisconsin Press, p. 147.

Tiselius, A. (1932), *Kolloid-Z.*, **59,** 306.

Titani, K., E. Whitley, Jr., L. Avogadro, and F. W. Putnam (1965), *Science*, **149,** 1090.

Tomito, K., and A. Rich (1964), *Nature*, **201,** 1160.

Tompa, H. (1952), *J. Polym. Sci.*, **8,** 51.

Toms, B. A. (1949) *Proceedings of the International Congress on Rheology*, Vol. 2, North Holland, Amsterdam, p. 135.

Toniolo, C., M. L. Falxa, and M. Goodman (1968), *Biopolymers*, **6,** 1579.

Torchia, D. A., and F. A. Bovey (1971), *Polym. Prepr, Am. Chem. Soc., Div. Polym. Chem.*, **12** (1), 547

Tosi, C., P. Corradini, A. Valvassori, and F. Ciampelli (1969), *J. Polym. Sci.*, **C22,** 1085.

Traub, W., and K. A. Piez (1971), *Adv. Protein Chem.*, **25,** 243.

Traub, W., and M. Shmueli (1963), *Nature*, **198,** 1165.

Trautman, R., V. N. Schumaker, W. F. Harrington, and H. K. Schachman (1954), *J. Chem. Phys.*, **22,** 555.

Treloar, L. D. G. (1958), *The Physics of Rubber Elasticity,* 2nd ed, Clarendon Press, Oxford.

Tremblay, R., A. F. Sirianni, and I. E. Puddington (1958), *Can. J. Chem.*, **36,** 543, 725.

Trementozzi, Q. A., R. F. Steiner, and P. Doty (1952), *J. Am. Chem. Soc.*, **74,** 2070.

Trunkel, H. (1910), *Biochem. Z.*, **26,** 493.

Tschoegl, N. W. (1964), *J. Phys. Chem.*, **40,** 473.

T'so, P., S. A. Rapaport, and F. J. Bollum (1966), *Biochemistry,* **5,** 4153.

Tsong, T. Y., R, L. Baldwin, and E. L. Elson (1971), *Proc. Natl. Acad. Sci. U.S.*, **68,** 2712.

Tsuji, K. (1973), M. S. Thesis, University of Tokyo.

Tsvetkov, V. N. (1957), *J. Polym. Sci.*, **23,** 151.

Tsvetkov, V. N. (1962), *J. Polym. Sci.*, **51,** 727.

Tvetkov, V. N. (1964), in *Newer Methods of Polymer Characterization,* B. Ke, ed., Interscience New York, Chap. 14.

Tsvetkov, V. N. (1972), *Makromol. Chem.*, **160,** 1 (1972).

Tsvetkov, V. N., and V. P. Budtov (1964), *Vysokomol. Soedin.* **6,** 1209.

Tsvetkov, V. N., and S. I. Klenin (1958), *J. Polym. Sci.*, **30,** 187; *Zh. Tekhn. Fiz.*, **28,** 1019 [English transl.: *Soviet Phys. Tech. Phys.*, **3,** 949 (1959)].

Tsvetkov, V. N., and S. I. Klenin (1959), *Zh. Tekhn. Fiz.*, **29,** 1393 [(English transl.: *Soviet Phys. Tech. Phys.*, **4,** 1283 (1960)].

Tsvetkov, V. N., S. L. Magarik, N. N. Boitsova, and M. G. Okuneva (1961), *J. Polym. Sci.*, **54,** 635.

Tsvetkov, V. N., I. N. Shtennikova, E. I. Rjumtsev, and Yu. P. Gatmanchuk (1971), *Eur. Polym. J.*, **7.** 767.

Tsvetkov, V. N., E. I. Riumtsev, I. N. Shtennikova, E. V. Korneeva, B. A. Krentsel, and Yu. B. Amerik (1973), *Eur. Polym. J.*, **9,** 481.

Turner, D. C. and Brand (1968), *Biochemistry*, **7,** 3381.

Turska, E., and M. Łaczkowski (1957), *J. Polym. Sci.*, **23,** 285.

Uchida, T., Y. Kurita, and M. Kubo (1956), *J. Polym. Sci.*, **19,** 365.

Uda, K. and G. Meyerhoff (1961), *Makromol. Chem.*, **47,** 168.

Uhlenbeck, O. C., F. H. Martin, and P. Doty (1971), *J. Mol. Biol.*, **57,** 217.

Ukaji, T., and R. A. Bonham (1962), *J. Am. Chem. Soc.*, **84,** 3631.

Ullman, R. (1965), *J. Chem. Phys.*, **43,** 3161.

Ullman, R. (1969), *Macromolecules,* **2,** 27.

Ullman, R. (1970), *Biopolymers,* **9,** 471.

Urnes, P. J. and P. Doty (1961), *Advan. Protein Chem.*, **16,** 401.

Urnes, P. J., K. Imahori, and P. Doty (1961), *Proc. Natl. Acad. Sci. U.S.*, **47,** 1635.

Utiyama, H., N. Tagata, and M. Kurata (1969), *J. Phys. Chem.*, **73,** 1448.

Vala, M. T., and S. A. Rice (1963), *J. Chem. Phys.*, **39,** 2348.

Vala, M. T., J. Haebig, and S. A. Rice (1965), *J. Chem. Phys.*, **43,** 886.

Valeur, B., C. Noel, P. Monjol, and L. Monnerie (1971), *J. Chim. Phys.*, **68,** 97.

Vallee, B. L., and R. J. P. Williams (1968), *Proc. Natl. Acad. Sci. U.S.*, **59,** 498.

Van Amerongen, G. J. (1951), *J. Polym. Sci.,* **6,** 471.

Vand, V. (1948), *J. Phys. Colloid Chem.,* **52,** 300.

Vanderkooi, G., S. J. Leach, G. Nemethy, and H. A. Scheraga (1966), *Biochemistry,* **5,** 2991.

Van der Waals, J. H., and J. J. Hermans (1950), *Rec. Trav. Chim.,* **69,** 949, 971.

Van Holde, K. E., and R. L. Baldwin (1958), *J. Phys. Chem.,* **62,** 734.

Van Laar, J. J., and R. Lorenz (1925), *Z. Anorg, Allgem. Chem.,* **146,** 42.

Vanzo, E. (1966), *J. Polym. Sci.,* **A-1, 4,** 1727.

Varoqui, R., and U. P. Strauss (1968), *J. Phys. Chem.,* **72,** 2507.

Veis, A., and M. P. Drake (1963), *J. Biol. Chem.,* **238,** 2003.

Velick, S. F. (1953), *J. Biol. Chem.,* **203,** 563.

Velick, S. F. (1954), in *The Mechanism of Enzyme Action,* W. D. McElory and B. Glass, eds., Johns Hopkins University Press, Baltimore, p. 491.

Velick, S. F., C. W. Parker, and H. N. Eisen (1960), *Proc. Natl. Acad. Sci. U.S.,* **46,** 1470.

Vincent, J. P., and M. Lazdunski (1972), *Biochemistry,* **11,** 2967.

Vinograd, J., and R. Brunner (1966), *Biopolymers,* **4,** 131, 157.

Vinograd, J., and J. E. Hearst (1962), *Fortschr. Chem. Org. Naturst,* **20,** 372.

Vinograd, J., R. Brunner, R. Kent, and J. Weigle (1963), *Proc. Natl. Acad. Sci. U.S.,* **49,** 902.

Vinograd, J., J. Lebowitz, and R. Watson (1968), *J. Mol. Biol.,* **33,** 173.

Virk, P. S., E. W. Merrill, A. S. Mickley, K. A. Smith, and E. L. Mollo-Christensen (1967), *J. Fluid Mech.,* **30,** 305.

Volkenstein, M. V. (1963), *Configurational Statistics of Polymer Chains,* Interscience, New York; (a) p. 74; (b) p. 121; (c) pp. 301–390.

von Hippel. P. H. (1967), *Treatise on Collagen,* G. N. Ramachandran, ed., Academic Press, New York, Chap. 6.

von Hippel, P. H., and J. D. McGhee (1972), *Ann. Rev. Biochem.,* **41,** 231.

Waack, R., A. Rembaum, J. D. Coombes, and M. Szwarc (1957), *J. Am. Chem. Soc.,* **79,** 2026.

Wada, A. (1964), *Biopolymers,* **2,** 361.

Wada, A., I. Kawata, and K. I. Miura (1971), *Biopolymers,* **10,** 1153.

Wada, A., T. Tanaka, and H. Kihara (1972), *Biopolymers,* **11,** 587.

Wahl, P. (1966), *Compt. Rend.,* **263D,** 1525.

Wales, M., and K. Van Holde (1954), *J. Polym. Sci.,* **14,** 81.

Wales, M., F. T. Adler, and K. E. Van Holde (1951), *J. Phys. Colloid Chem.,* **55,** 145.

Walker, E. E. (1952), *J. Appl. Chem.,* **2,** 470.

Wall, F. T., and J. J. Erpenbeck (1959), *J. Chem. Phys.,* **30,** 634.

Wall, T. T., and D. F. Hornig (1965), *J. Chem. Phys.,* **43,** 2079.

Wall, F. T., L. A. Hiller, Jr., and D. J. Wheeler (1954), *J. Chem. Phys.,* **22,** 1036.

Wall, F. T., L. A. Hiller, Jr., and W. F. Atchison (1955), *J. Chem. Phys.,* **23,** 2314.

Wall, F. T., S. Windwer, and P. J. Gans (1962), *J. Chem. Phys.,* **37,** 1461.

Walrafen, G. E., (1968), *J. Chem. Phys.,* **48,** 244.

Wang, J. C. (1973), *Acc. Chem. Res.,* **6,** 252.

Wang, J. C., and N. Davidson (1966), *J. Mol. Biol.,* **14,** 469.

Wang, J. H. (1958), *J. Am. Chem. Soc.,* **80,** 3168.

Wang, Y.-C., and H. Morawetz (1975), *Makromol. Chem.,* in press.

Watanabe, W. H., C. F. Ryan, P. C. Fleischer, Jr., and B. S. Garrett (1961), *J. Phys. Chem.,* **65,** 896.

Watenpaugh, K. D., L. C. Siekers, J. R. Herriot, and L. H. Jensen (1972), *Cold Spring Harbor Symp. Quant. Biol.,* **36,** 359.

Watson, J. D. (1963), *Science,* **140,** 17.

Watson, J. D., and F. H. C. Crick (1953), *Nature,* **171,** 737.

Waugh, D. F. (1958), *Discuss. Faraday Soc.,* **25,** 186.

Waugh, D. F. (1961), *J. Phys. Chem.,* **65,** 1793.

Weber, F. N., Jr., D. W. Kupke, and J. W. Beams (1963), *Science,* **139,** 837.

Weber, G. (1952), *Biochem. J.,* **51,** 145, 155.

Weber, G. (1953), *Adv. Protein Chem.,* **8,** 416.

Weber, G. (1960), *Biochem. J.,* **75,** 335, 345.

Weber, G., and S. R. Anderson (1969), *Biochemistry,* **8,** 361, 371.

Weber, G., and D. J. R. Laurence (1954), *Biochem. J.,* **56,** XXXI.

Weber, R. E., and C. Tanford (1959), *J. Am. Chem. Soc.,* **81,** 3255.

Weibull, B., and B. Nycander, *Acta Chim. Scand.,* **8,** 847 (1954).

Weil, R., and J. Vinograd (1963), *Proc. Natl. Acad. Sci. U.S.,* **50,** 730.

Weimarn, P. von (1926), *Kolloid-Z,* **40,** 120; **41,** 148.

Weissberg. S. G., R. Simha, and S. Rothman (1951), *J. Res. Natl. Bur. Stand.,* **47,** 298.

Wenger, F. (1960a), *Makromol. Chem.,* **36,** 200.

Wenger, F. (1960b), *J. Am. Chem. Soc.,* **82,** 4281.

Wessling, R. A. (1970), *J. Appl. Polym. Sci.,* **14,** 1531.

Westheimer, F. H., and J. G. Kirkwood (1938), *J. Chem. Phys.,* **6,** 513.

Westheimer, F. H., and M. W. Shookhoff (1939), *J. Am. Chem. Soc.,* **61,** 555.

Westgren, A. (1914), *Z. Phys. Chem. (Leipzig),* **89,** 63.

Westhead, E. W., Jr., and H. Morawetz (1958), *J. Am. Chem. Soc.,* **80,** 237.

Westrum, E. F., Jr., and J. P. McCullough (1963), in *Physics and Chemistry of the Organic Solid State,* Vol. 1, D. Fox. M. M. Labes, and A. Weissberger eds., Interscience, New York, p. 106.

Wetlaufer, D. B., J. T. Edsall, and B. R. Hallingworth (1958), *J. Biol. Chem.,* **233,** 1421.

Wetmur, J. G. (1971), *Biopolymers,* **10,** 601.

Wetmur, J. G., and N. Davidson (1968), *J. Mol. Biol.,* **31,** 349.

White, F. H., Jr. (1960), *J. Biol. Chem.,* **235,** 383.

Wilkes, C. E., C. J. Carman, and R. A. Harrington (1973), *J. Polym. Sci.,* **C43,** 237.

Wilkins, M. H. F. (1963), *Science,* **140,** 941.

Williams, E. J., and J. F. Foster (1959), *J. Am. Chem. Soc.,* **81,** 865.

Williams, J. W., and W. M. Saunders (1954), *J. Phys. Chem.,* **58,** 854.

Williams, J. W., R. L. Baldwin, W. M. Saunders, and P. G. Squire (1952), *J. Am. Chem. Soc.,* **74,** 1542.

Williams, J. W., K. E. Van Holde, R. L. Baldwin, and H. Fujita (1958), *Chem. Rev.*, **58,** 715.

Williams, R. C., and R. L. Steere (1951), *J. Am. Chem. Soc.*, **73,** 2057.

Wilson, C. W., III (1963), *J. Polym. Sci.*, **A1,** 1305.

Winklmair, D., J. Engel, and V. Ganser (1971), *Biopolymers*, **10,** 721.

Wojtech, B. (1963), *Makromol. Chem.*, **66,** 181.

Wolff, C. (1969), *J. Chim. Phys.*, **65,** 1569.

Woller, P. B., and E. W. Garbisch, Jr. (1972), *J. Am. Chem. Soc.*, **94,** 5310.

Wood, W. B., and M. Henninger (1969), *J. Mol. Biol.*, **39,** 603.

Wood, W. W., W. Fickett, and J. G. Kirkwood (1962), *J. Chem. Phys.*, **20,** 561.

Worsfold, D. J. (1974), *J. Polym. Sci., Polym. Chem. Ed.* **12,** 337.

Wright, C. S., R. A. Alden, and J. Kraut (1969), *Nature*, **221,** 235.

Wright, H. T. (1973), *J. Mol. Biol.*, **79,** 1, 13.

Wyckoff, H. W., K. D. Hardman, N. M. Allewell, T. Inagami, L. N. Johnson, and F. M. Richards (1967), *J. Biol. Chem.*, **242,** 3984.

Yamakawa, H. (1965), *J. Chem. Phys.*, **42,** 1764.

Yamakawa, H. (1971), *Modern Theory of Polymer Solutions*, Harper and Row, New York.

Yamakawa, H., and M. Kurata (1960), *J. Chem. Phys.*, **32,** 1852.

Yamakawa, H., A. Aoki, and G. Tanaka (1966), *J. Chem. Phys.*, **45,** 1938.

Yamamoto, A., M. Fujii, G. Tanaka, and H. Yamakawa (1971), *Polym. J.*, **2,** 799.

Yan, J. F., G. Vanderkooi, and H. A. Scheraga (1968), *J. Chem. Phys.*, **49,** 2713.

Yan, J. F., F. A. Momany, and H. A. Scheraga (1970), *J. Am. Chem. Soc.*, **92,** 1109.

Yanari, S., and F. A. Bovey (1960), *J. Biol. Chem.*, **235,** 2818.

Yang, J. T. (1959), *J. Am. Chem. Soc.*, **81,** 3902.

Yang, J. T. (1967), in *Poly-α-Amino Acids*, G. D. Fasman, ed., Marcel Dekker, New York, Chap. 6.

Yathindra, N., and V. S. R. Rao (1970), *Biopolymers*, **9,** 783.

Yee, R. Y., S. W. Englander, and P. H. von Hippel (1974), *J. Mol. Biol.* **83,** 1.

Yguerabide, J., H. F. Epstein, and L. Stryer (1970), *J. Mol. Biol.*, **51,** 573.

Yokoyama, M., H. Ishihara, R. Iwamoto, and H. Tadokoro (1969), *Macromolecules*, **2,** 184.

Yoshida, T., S. Sakurai, T. Ukuda, and Y. Takagai (1962), *J. Am. Chem. Soc.*, **84,** 3590.

Yoshikawa, S., and O. K. Kim (1966), *Bull. Chem. Soc. Jap.*, **39,** 1515.

Yoshioka, K., and C. T. O'Konski (1968), *J. Polym. Sci.*, **A-2, 6,** 421.

Yphantis, D. A. (1960), *Ann. N.Y. Acad. Sci.*, **88,** 586.

Yu, N. T., and B. H. Jo (1973a), *Arch. Biochem. Biophys.*, **56,** 469.

Yu, N. T., and B. H. Jo (1973b), *J. Am. Chem. Soc.*, **95,** 5033.

Yu, N. T., B. H. Jo, and D. C. O'Shea (1972), *J. Mol. Biol.*, **70,** 117.

Yunis, J. J., and W. G. Yasmineh (1971), *Science*, **174,** 1200.

Zachau, H. G. (1969), *Angew. Chem., Int. Ed.*, **8,** 711.

Zambelli, A., A. L. Segre, M. Farina, and G. Natta (1967a), *Makromol. Chem.*, **110,** 1.

Zambelli, A., G. Natta, I. Pasquon, and R. Signorini (1967b), *J. Polym. Sci.*, **C16,** 2485.

Zana, R., C. Tondre, M. Rinaudo, and M. Milas (1971), *J. Chim. Phys.*, **68,** 1258.

Zimm, B. H. (1946), *J. Chem. Phys.*, **14,** 164.

Zimm, B. H. (1948), *J. Chem. Phys.*, **16,** 1099.

Zimm, B. H. (1956), *J. Chem. Phys.*, **24,** 269.

Zimm, B. H. (1960), *J. Chem. Phys.*, **33,** 1349.

Zimm, B. H., and J. K. Bragg (1959), *J. Chem. Phys.*, **31,** 526.

Zimm, B. H., and R. W. Kilb (1959), *J. Polym. Sci.*, **37,** 19.

Zimm, B. H., and S. A. Rice (1960), *Mol. Phys.*, **3,** 391.

Zimm, B. H., and W. H. Stockmayer (1959), *J. Chem. Phys.*, **17,** 1301.

Zimm, B. H., R. S. Stein, and P. Doty (1945), *Polym. Bull.*, **1,** 90.

Zimm, B. H., W. H. Stockmayer, and M. Fixman (1953), *J. Chem. Phys.*, **21,** 1716.

Zimmering, P. E., E. W. Westhead, Jr., and H. Morawetz (1957), *Biochim. Biophys. Acta.* **25,** 376.

Zwick, M. M. (1965), *J. Appl. Polym. Sci.*, **9,** 2393.

INDEX